THE NEW
STATISTICAL
ANALYSIS
OF DATA

Springer
New York
Berlin
Heidelberg
Barcelona
Budapest
Hong Kong
London
Milan
Paris
Santa Clara
Singapore
Tokyo

THE NEW STATISTICAL ANALYSIS OF DATA

T.W. Anderson

Stanford University

Jeremy D. Finn

State University of New York at Buffalo

Springer

T.W. Anderson
Stanford University
Department of Statistics
Stanford, CA 94305
USA

Jeremy D. Finn
State University of New York at Buffalo
Graduate School of Education
Buffalo, NY 14260-1000
USA

Library of Congress Cataloging-in-Publication Data
Anderson, T.W. (Theodore Wilbur), 1918–
    The new statistical analysis of data / T. W. Anderson, Jeremy D.
  Finn.
      p.   cm.
    Includes bibliographical references and index.
    ISBN 0-387-94619-5 (hardcover : alk. paper)
    1. Statistics.   I. Finn, Jeremy D.   II. Title.   III. Series.
    QA276.12.A46   1996
    519.5—dc20   95-44885

Printed on acid-free paper.

Production managed by Bill Imbornoni; manufacturing supervised by Joe Quatela.
Typeset by Integre Technical Publishing Co., Inc., Albuquerque, NM.
Printed and bound by R.R. Donnelley & Sons, Harrisonburg, VA.
Printed in the United States of America.

9 8 7 6 5 4 3

ISBN 0-387-94619-5 Springer-Verlag New York Berlin Heidelberg   SPIN 10650718

To Dorothy and Kristin

Preface

The Nature of the Book

This book is a text for a first course in statistical concepts and methods. It introduces the analysis of data and statistical inference and explains various methods in enough detail so that the student can apply them. Little mathematical background is required; only high school algebra is used. No mathematical proof is given in the body of the text, although algebraic demonstrations are given in appendices at the ends of some chapters. The exposition is based on logic, verbal explanations, figures, and numerical examples. The verbal and conceptual levels are higher than the mathematical level.

The concepts and methods of statistical analysis are illustrated by more than 100 interesting real-life data sets. Some examples are taken from daily life; many deal with the behavioral sciences, some with business, the health sciences, the physical sciences, and engineering. The exercises are of varying degrees of difficulty.

This book is suitable for undergraduates in many majors and for graduate students in the health sciences, behavioral sciences, and education. It has grown out of our experience over many years of teaching such courses.

An earlier text by T. W. Anderson and S. L. Sclove, *The Statistical Analysis of Data*, had similar objectives and was of a similar nature. The organization of *The New Statistical Analysis of Data* follows an outline like that of the former book. However, the explanations of statistical topics are more elementary, detailed, and comprehensive in the new book. Instead of one chapter on the organization of data, there is one chapter on the description of quantitative measurements and another chapter on qualitative variables. Probability distributions are treated more thoroughly with emphasis on the normal distribution. Many more examples appear throughout *the New Statistical Analysis of Data*.

Statistics and Computers

It is expected that in putting their statistical knowledge into practice students will use computers to carry out the procedures they have learned. Indeed, the computer has become an essential ingredient to researchers analyzing statistical data. The computer furnishes more accurate results than either hand computation or a calculator and is more efficient with respect to time and effort.

Most courses provide an opportunity for students to learn about statistical computing together with the statistical content itself. At the present time there are a number of computer packages available for performing basic statistical procedures, each with its advantages and disadvantages. Rather than basing the entire text on one of these packages, the authors are preparing a separate volume with instructions for a general purpose package, SPSS, which will be available from the publisher. The instructor who wishes to use SPSS as part of the course will find the self-teaching supplement invaluable. The instructor who has a preference for another statistical package (or none) will still find the textbook appropriate.

The SPSS package is user-friendly and will carry out virtually all of the procedures described in this text. The *SPSS Guide to the New Statistical Analysis of Data* provides step-by-step instructions on the use of the program for performing these analyses plus guidelines for locating and interpreting results in the output files. Many of the illustrations in the guide use specific data sets and analyses presented in the text. Alternative data sets are presented to give the reader additional experience with the methodology; a number of recommended "computer exercises" are furnished.

The Organization of the Book

Part One, "Introduction," shows how statistical methods are used in several substantive fields for answering important questions. Part Two, "Descriptive Statistics," considers the organization of data and summarization by means of descriptive statistics; the student learns how to approach information that comes in numerical form. Both univariate and multivariate data are considered. Part Three, "Probability," develops the ideas of probability to form the basis of statistical inference. Part Four, "Statistical Inference," begins by considering the use of sample characteristics to estimate population characteristics; the idea of variability of sample quantities leads directly to confidence intervals. Part Four also discusses tests of hypotheses and the allied concepts of significance level and power. This part treats some of the basic methods for means, proportions, and variances. Part Five, "Statistical Methods for Other Problems," includes techniques such as chi-square tests, analysis of variance, regression analysis, and sample surveys.

The Use of the Book

The 17 chapters in the book provide enough material for a course of two semesters or three quarters although it is anticipated that the most common use will be for a one-semester introductory course. In some chapters there are starred sections (*) that may be omitted without affecting the understanding of subsequent sections; some such material is put into appendices in order not to burden the main development.

Chapter 1 presents examples of the use of statistics in studies of general interest and some basic concepts of statistics.

Chapters 2 through 6 are largely *descriptive statistics*. Most beginning students will start with these chapters. Association is discussed at length in Chapters 5 and 6 that include many examples of both numerical and categorical data.

Chapters 7, 8, and 9 on probability and probability distributions provide the theoretical background for statistical inference. If the course is of an applied nature, some of the more formal parts of these chapters may be given less emphasis.

The ideas of estimation and testing hypotheses in Chapters 10 and 11 are the basic ingredients of *inferential statistics* and require in-depth attention. If time is at a premium, the sections on nonparametric procedures (Sections 10.5 and 11.6) can be omitted.

Chapter 12 dealing with statistical inference based on two samples presents useful methods. Again, the instructor may omit the material on nonparametric tests (Section 12.5) if time is short. Chapter 13 on variances in one and two populations presents chi-square and F-tests which are based on an assumption of normal populations. An instructor may choose to emphasize certain topics, for example, the F-test. Chapter 14 presents statistical inference for "contingency" tables. Chapter 15 discusses simple regression and correlation. Chapter 16 is on the analysis of variance; the nonparametric material is optional (Section 16.3).

Chapter 17 presents an introduction to the basic ideas of sample surveys. In a one quarter or one-semester course in certain disciplines (e.g., sociology or management), the instructor may prefer to include this material instead of the analysis of variance chapter.

The chapters are divided into sections (numbered 1.1 and 1.2 in Chapter 1, for example). Many sections are in turn presented in subsections. The effect of this hierarchy of chapters, sections, and subsections is to form an outline of the material. The summary at the end of each chapter reviews that chapter's contents and helps the student determine whether the important concepts have been learned.

The material of the book is organized so that the 17 chapters provide a basis for courses of varying lengths. Some guidelines for coverage are as follows.

A one-year course (two semesters or three quarters). All 17 chapters can be covered in a one-year course. Some instructors may wish to omit the starred sections and possibly de-emphasize other relatively advanced sections, such as Sections 6.3 and 6.4 on effects of a third variable on the relationship between two given variables.

A two-quarter course. One can go in the direction of either breadth or depth. A survey course covering all 17 Chapters would mean (i) the omission of starred sections, (ii) de-emphasis of Chapter 7, and (iii) omission of peripheral topics in Chapters 13–16. An instructor can design a course which goes in the direction of depth rather than breadth by omitting one or more of Chapters 13–17.

A one-semester course. The first twelve chapters comprise a text for a one-semester course. Some instructors may wish to include smatterings of later chapters, such as 13, 14, 15, or 16.

A one-quarter course. The first twelve chapters provide a basis for such a course. The instructor can omit starred sections and de-emphasize Chapter 7. Again, some teachers may wish to include parts of Chapters 13, 14, 15, or 16.

Exercises

In order to learn statistical concepts and methods, hands-on experience with many data sets, real and simulated, is essential. To provide practice, numerous exercises are provided at the end of each chapter. These apply the ideas presented in the chapter and demonstrate the utility of the statistical methods to a broad range of problems.

Answers to Selected Exercises appear at the end of the book. The instructor has many options in assigning exercises of varying difficulty, either with or without answers available. The instructor can obtain a complete Solutions Manual from the publisher.

Acknowledgments

We are indebted to Stanley L. Sclove and the many colleagues, teaching assistants, typists, and reviewers who helped in the preparation of the texts preceding this volume. We wish to thank the staff of Springer-Verlag, Bill Imbornoni, and several anonymous reviewers for their many helpful comments and suggestions in preparing the present text, and Marjorie Weinstock for assistance with the references. A special note of gratitude is extended to Ingram Olkin for encouraging us to begin this book and for useful suggestions during its preparation.

T.W.A.
J.D.F.
March 1996

Contents

Introduction

1

The Nature of Statistics

Introduction

Statistics enters into almost every phase of life in some way. A daily news broadcast may start with a weather forecast and end with an analysis of the stock market. In a newspaper at hand we see in the first five pages stories on an increase in the wholesale price index, an increase in the number of *habeas corpus* petitions filed, new findings on mothers who smoke, the urgent need for timber laws, a state board plan for evaluation of teachers, popularity of new commuting buses, a desegregation plan, and sex bias. Each article reports some information, proposal, or conclusion based on the organization and analysis of numerical data.

Statistics in systematic and penetrating ways provides bases for investigations in many fields of knowledge, such as social, physical, and biological sciences, engineering, education, business, medicine, and law. Information on a topic is acquired in the form of numbers; an analysis of these data is made in order to obtain a better understanding of the phenomenon of interest, and some conclusions may be drawn. Often generalizations are sought; their validity is assessed by further investigation. In the following section some examples of such studies are described.

A definition of statistics is *making sense out of figures.* Statistics is the methodology which scientists and mathematicians have developed for interpreting and drawing conclusions from data. This first chapter begins with a survey of some uses of statistics and concludes with a discussion of some of the general ideas underlying the subject of statistics.

1.1   Some Examples of the Use of Statistics

The large-scale trial of the Salk polio vaccine, the Surgeon General's report on smoking and health, the many political polls conducted by elected officials and by candidates for office, the use of educational screening devices such as the College Board Entrance Examinations, and the use of the census, consumer price indices, unemployment rates, and population projections in assessing human progress and future potentialities all attest to the importance of statistical methods in matters of everyday importance. We discuss several classic applications of statistics.

Political Polls

Political polls illustrate some uses of statistics in the field of political science and political sociology. During the period of January 7–8, 1972, each person in a national sample of 1380 adults, 18 years of age and older, was interviewed. Each respondent was asked a question which has been asked about every president since Franklin Delano Roosevelt held office: "Do you approve or disapprove of the way the president is handling his job?" The response in January, 1972, regarding Richard Nixon's performance, was 49% approving and 39% disapproving, the remaining 12% having no opinion. The pollster [Gallup (1972)] noted that Mr. Nixon was somewhat less popular among college students than he was among the whole adult population.

Some of the early political polls were made in order to forecast outcomes of elections. In 1920, 1924, 1928, and 1932 *The Literary Digest* (a magazine now defunct) successfully predicted the winner of the United States presidential election, forecasting the popular vote with only a small percentage of error—less than 1% in 1932.

Then came 1936. The contest was between the incumbent, Franklin D. Roosevelt, and Alfred E. Landon. An October 31 headline read

> # Landon   1,293,669
> # Roosevelt   972,897
>
> # Final Return in The Digest's
> # Poll of Ten Million Voters

The poll results were listed by state. The Landon vote, Roosevelt vote, and votes for minor party candidates were given; the total number of such "votes" was 2,376,523. This information is partially summarized in Table 1.1.

TABLE 1.1

Summary of Results of The Literary Digest *Poll*

	Landon	Roosevelt
Number of states "carried"	32	16
Number of electoral votes	370	161

source: *The Literary Digest* (1936).

The *Digest* reported the figures; they did not "weight, adjust, or interpret" them. However, it was hard for any reader not to take the poll results as a prediction of the actual vote.

The rest is history. Instead of being a Landon victory, the election was a landslide for Roosevelt. The *Digest* poll had 42.9% for Roosevelt; the actual vote was 62.5% for Roosevelt! Instead of 161 electoral votes, Roosevelt actually received 523; instead of carrying only 16 states, he carried 46.

What happened? Why was there such a great difference between the percentages in the poll and the percentages of actual vote? To answer this we examine *The Literary Digest*'s procedure. Ballots were mailed out to some ten million persons; each recipient was asked to fill out the ballot and mail it back. Every third registered voter in Chicago and every registered voter in Allentown, Pennsylvania, were polled; names were drawn from telephone books, from the rosters of clubs and associations, from city directories, lists of registered voters, and other lists.

A major fault with this procedure was that the lists used for mailing tended to have more people with higher incomes than voters as a whole and in this election—because of the New Deal—those with above-average incomes tended to vote Republican more than those with lower income. Another questionable feature was that only one-fourth of the recipients of the ballots returned them; maybe those who did not respond had political opinions somewhat different from those who responded.

Among the respondents to the poll, most had voted for Hoover in 1932. This was a sure sign that the sample was *biased*, that is, not representative of all voters, since Roosevelt had won in 1932. On November 14 the *Digest* said they "were willing to overlook this inconsistency." The final outcome was the demise of the *Literary Digest* in 1937.

Later Pre-election Polls.   Even before the *Literary Digest* debacle, improved methods of polling had been developed for use in election prediction, market research, and for other purposes. Mail questionnaires with their inherent uncertainty of response were replaced by direct interviews. Interviewers were sent out to question respondents met on the streets or in homes. Their selections were guided by quotas; for example, in a given day an interviewer might be asked to interview 25 men and 25 women, of whom 30 lived in the city and 20 in the suburbs. These polls were able to provide much useful information on behavior of consumers and opinions of the citizenry, but forecasting voting provided a more acid test.

In 1948 it was Dewey vs. Truman. Three leading pollsters were George Gallup (Institute of Public Opinion), Elmo Roper, and Archibald Crossley. As early as September a Dewey victory was predicted, and interest in the election waned.

It was a close election; so close, indeed, that an early *Chicago Tribune* headline read: "Dewey wins." Truman won, however. Table 1.2 shows the popular vote together with the reports of the various polls.[1] In calculating the percentages those who were undecided were omitted; and Gallup included only Thurmond and Wallace in "Other candidates." Again the polls were wrong. What happened?

Subsequent analysis showed that in spite of the quotas some parts of the electorate were over- or under-represented in the polls; for instance, too many of those interviewed were college educated, and there were too few with only grade school education. Another contributor to the error in forecasting was the time element. One-seventh of the voters made up their minds only during the last two weeks preceding the election, and three-quarters of these decided in favor of Truman. Roper,

---

[1]We should note that even if the popular vote is predicted well, the problem of predicting the outcome of a U.S. presidential election is complicated by the system of electoral votes. Dewey could have won by carrying Ohio, California, and Illinois, each of which he lost by less than 1%!

TABLE 1.2

1948 Poll Results and Actual National Vote

	Dewey	*Truman*	*Other candidates*
Actual national vote	45.1%	49.5%	5.4%
Crossley	49.9%	44.8%	5.3%
Gallup	49.5%	44.5%	5.5%
Roper	52.2%	37.1%	10.7%

source: McCarthy (1949).

however, closed down his operation September 9, two months before the election; and Gallup and Crossley did not gather new information during the last two weeks. In the latest poll 15% were undecided, and it was assumed that those would be split in the same proportions as those who had already decided; this contributed about 1.5 percentage points to the error of predictions.

In the case of quota sampling there may be biases due to the interviewer's selection of persons to interview; he or she may prefer well-dressed or pleasant-looking individuals; these are factors not controlled by the quotas. There is also a chance factor in whom the interviewer meets. The factor of chance, luck, or probability comes into almost every sampling method. It gives rise to what we call *sampling error*. "Sampling error" refers to the inherent variation between an estimate of some characteristic as computed from a sample and the actual value of that characteristic in the population from which the sample was drawn. (See Section 1.2.) When the sampling error is due to "chance or luck," the laws of probability can be used to assess the possible magnitude of the error. Thus one important function of statistical methods is to estimate characteristics of a population, based on a sample of data. A second function is to assess the likely accuracy of that estimate by evaluating the sampling error.

In these pre-election polls the sample sizes were large enough so that the sampling error was relatively small. The biases introduced by underrepresentation of some portions of the electorate and by the fact that the pollsters completed their interviewing early, caused the gross error. Interestingly enough, the public opinion polls were similarly unsuccessful in predicting the outcome of the 1970 general election in the United Kingdom, and for the same reasons [The Market Research Society (1972)].

Modern Refinements.   Social scientists, market researchers, and pollsters have developed better methods of selecting interviewees and more sophisticated modes of inquiry. Many political polls today do not rely simply on voters' answers to the question of whom they will vote for. The interviewer may ask the occupation of the respondent, the number of dependents (both children and senior citizens), and information concerning union affiliations.

The voter may be asked to state what issue is most important in the current campaign or be asked to rate the candidates as to how closely they agree with the voter's own position on this issue. The questioning may relate to several issues, and then by weighting each issue a score can be obtained which indicates the voter's relation to each candidate. Strengths and weaknesses of candidates are assessed by asking a voter about ways to convince a neighbor to vote for the favored candidate. These indirect methods, based on scientific studies of the motives and attitudes that underlie choice behavior, provide more power to the researcher when the time comes to calculate a prediction.

A Panel Study of Voting Behavior

Pre-election polls simply find the intentions of a sample of persons (that is, report opinion) and then predict the vote (that is, predict action). In addition to sample bias, there are two further sources of discrepancy that may contaminate this procedure. One is the discrepancy between present and future—a prediction is being made, that is, present data are being brought to bear on the future. The other is the possible discrepancy of people's stated opinions, their actual opinions, and their actions. Polls ascertain people's stated opinions or intentions, whereas what is being predicted is an action, actually voting.

Social scientists have undertaken to study voting on a more scientific basis in order to learn how voters behave and why. One such study was an investigation of the evolution of opinion regarding the 1940 election among persons in Erie County, Ohio. The purpose of the investigation was to study how opinions are formed and changed.

Every fourth house was sampled to obtain 3000 adults. From these, four groups of 600 persons each were selected by sampling in such a way that the groups were matched on important characteristics. The study was a *dynamic* study in the sense that it was carried out over time rather than at a single point in time. Each of the three groups was interviewed once, one group in July, one in August, and the third in October. The fourth group comprised a *panel*, or group of people interviewed successively over time, on seven occasions, about one month apart.[2] The last of these seven interviews was in November, after the election. Out of the original 600 persons in the panel 483 were avail-

---

[2]Study of the same observations over time is generally referred to as *longitudinal* research. In contrast, study of observations at one time point is termed *cross-sectional* research.

able for the interviews in October and November. some of the results of the next-to-last interview and the actual vote are reported in Table 1.3. Among the 65 persons who in October said that they would not vote ("No vote"), 6 voted Republican and 59 did not vote. The 6 who voted Republican said Republican party members persuaded them.

TABLE 1.3

Tabulation of Persons Interviewed After 1940 Election

	Vote intention in October	Actual vote in November
Republican	229	232
Democrat	167	160
Don't know	22	—
No vote	65	91
	483	483

source: Lazarsfeld, Berelson, and Gaudet (1968), p. xxiii. Copyright 1968 by Columbia University Press. Reprinted with permission of the publisher.

The data gathered in this study were much more complete than the information usually obtained by a political poll. They can be used for answering questions about voting behavior as well as suggesting problems to investigate. For example, the data indicated that Catholics tended to vote Democratic and upper-income people tended to vote Republican. Especially interesting were people subject to cross pressures. Would Democrats in the upper income bracket be more likely to change their intention than Democrats in the lower income bracket? With richer data such as these, it is possible to use more sophisticated techniques, such as the construction of indices, composite measures, relating the propensity to vote for one party or the other to various social, economic, and religious factors.

The Polio Vaccine Trial

The largest public health experiment to date took place in 1954. The amount of publicity given the evaluative report, produced by the Poliomyelitis Evaluation Center at the University of Michigan, was at that time unequalled for a scientific work. The cost of the study, involving nearly 2,000,000 children, has been estimated to be about $5,000,000.

The question at hand was how effective the Salk vaccine might be in preventing the occurrence of poliomyelitis. The original plan for this massive field trial was to administer the Salk polio vaccine to children in the second grade of school. First and third graders would not be inoculated, but would be kept under observation for the occurrence of poliomyelitis. The rate of occurrence among the second graders would then be compared with that among the first and third graders. The second graders constituted the *experimental group*; the first and third graders constituted the *control group.*

Use of a control group is efficacious only if the control group differs from the experimental group in no important way, except that it is not subjected to the procedure under investigation. In the original plan the control and experimental groups would be observed for the same time period and in the same geographical areas. However, the following questions were raised:

- Might the act of inoculation itself increase or decrease the chances of contracting polio?
- Might second graders (about 7 years old) differ from first and third graders (about 6 and 8 years old) in the incidence of polio?
- Might knowledge by the children (and their parents) of whether they had been inoculated affect the risk?
- Might the diagnosis of physicians be affected by knowledge of whether children had been inoculated?

Since these questions indicated possible defects in the plan, another plan was proposed in which children of the first, second, and third grades would be treated alike. One half of all the children in the study were to be inoculated with a placebo, an inert solution of appearance similar to the polio vaccine. The incidence of polio in the placebo group would be compared with that in the vaccine group.

In implementing this plan it was important that there also be no biasing factors that would influence the choice of which children received the vaccine. Possibly some physicians practiced in areas of higher susceptibility to polio than others. Children in different areas may differ in regard to overall health. There was the possibility that some physicians might give the placebo to the healthier children.

These problems were overcome by placing in each one of a number of boxes 50 vials, 25 filled with placebo and 25 with vaccine. The vials were labelled with code numbers; only certain persons at the Evaluation Center knew which vials were placebos. The vials were positioned randomly within the box. In effect, a nurse or doctor picking out a vial was doing it blindfolded. This procedure served to eliminate possible bias in the reporting of apparent cases of polio, since the diagnosing doctor would have had no way of knowing whether the child had received placebo or vaccine.

Results.   Some results of the improved experiment are summarized in Table 1.4. (In some areas the original plan was followed.) While the

TABLE 1.4

Incidence of Paralytic Polio

	Inoculated with vaccine	Inoculated with placebo
Number of children inoculated	200,745	201,229
Number of cases of paralytic polio	33	110

source: Francis, *et al.* (1957).

number of children inoculated with the vaccine was about the same as the number inoculated with the placebo, the number of cases of paralytic polio among those not vaccinated was about $3\frac{1}{3}$ the number of cases among those vaccinated. This was convincing evidence that the vaccine was effective.

On the basis of this statistical evaluation, public health authorities instituted a campaign to have every child vaccinated. In time, the original Salk vaccine (which had various defects) was replaced by the Sabin preparation. Now the fearsome polio is virtually unknown in the United States.

Smoking and Health

In the early part of this century statisticians concerned with vital statistics noted an increase in lung cancer. In the last several decades a great many studies have been concerned with the possible effects of smoking on health. The primary focus of much of this research has been the relationship, if any, between lung cancer and smoking. In 1962 the Surgeon General of the United States Public Health Service, Dr. Luther Terry, formed an advisory committee to review and evaluate the relevant data. *Smoking and Health*, the 1964 report of that committee, has received an amount of publicity and discussion surpassing even the attention given the polio vaccine trial.

Some of the conclusions of the report are stated in such a way as to identify cigarette smoking as a *cause* of certain diseases; for example (p. 37), "Cigarette smoking is causally related to lung cancer in men; the magnitude of the effect of cigarette smoking far outweighs all other factors. The data for women, though less extensive, point in the same direction. The risk of developing lung cancer increases with duration of smoking and the number of cigarettes smoked per day, and is diminished by discontinuing smoking." The report also points out that there is a definite relationship between other diseases and smoking.

What methods are used to reach such conclusions? First it is noted whether or not there is an *association* between smoking and the incidence of a given disease. Table 1.5 illustrates this; Hammond and Horn (1958) observed (via volunteers from the American Cancer Society) 187,783 men over a period of 44 months, determining at the outset the smoking habits of the men studied. The death rate from lung cancer was ten times as great for cigarette smokers as for nonsmokers. A next step corroborates this evidence by considering the death rates for those who smoke different amounts of cigarettes, as shown in Table 1.6.

TABLE 1.5

Death Rates from Lung Cancer. Rates per 100,000 Men per year, based on observation of 187,783 men 50–69 years old over 44 months

Never smoked	Cigars only	Pipes only	Cigarettes and pipes or cigars	Cigarettes only
13	13	39	98	127

source: Table 7 from Hammond and Horn, *Journal of the American Medical Association* 166: 1294–1308. Copyright 1958, American Medical Association.

TABLE 1.6

Death Rates from Lung Cancer by Current Daily Cigarette Smoking. Rates per 100,000 men per year, based on observation of 187,783 men 50–69 years old over 44 months

Less than $\frac{1}{2}$ pack	$\frac{1}{2}$ to 1 pack	1 to 2 packs	More than 2 packs
95	108	229	264

source: Table 8 from Hammond and Horn, *Journal of the American Medical Association* 166: 1294–1308. Copyright 1958, American Medical Association.

Many questions remain to be answered. Might smoking and lung cancer be related to something else, something which causes an individual both to smoke and to contract lung cancer? For instance, many persons who smoke a great deal also drink heavily; alcohol consumption might be a common factor associated with both smoking and lung cancer. However, reports have shown that smokers who drink little are more likely to contract lung cancer than are nonsmokers who drink little, and the same was observed among those who drink a lot.

Other possible causes of lung cancer, such as air pollution, have also been studied. Death rates from lung cancer are indeed higher in urban than in rural areas. However, in both urban and rural areas the incidence of lung cancer is higher for smokers than for nonsmokers. Differences both in levels of air pollution and in smoking habits of people in urban and rural areas account for some of the difference in death rates due to lung cancer.

Project "Head Start"

In the 1960s a nationwide program for disadvantaged youngsters of preschool age, administered by the Office of Economic Opportunity through local Head Start agencies, was begun. Most of the youngsters who participated in the program were four-year olds whose parents were earning less than $4000 per year. The programs provided were similar to those in nursery schools. The philosophy of the project was that, while with middle- and upper-class children teachers can build upon awareness developed in the home, with lower-class youngsters teachers must build that awareness itself; Head Start centers in some areas were working with children who did not know what an orange was, had never held a crayon or seen a museum.

A team of investigators undertook assessment of the program by comparing Head Start youngsters with their counterparts who had not been in Head Start. A sample of 104 Head Start Centers was selected for study from the total of 12,927 centers. A total of 1980 children who had gone on to the first, second, and third grades were compared with 1983 non-Head Start children in the same grades and schools. The children in both groups were given a series of six tests of ability and achievement in language and of self-concept and attitudes.

As an example of the results, consider the Metropolitan Readiness Test, given to 864 children in first grade; of these, 432 children had been in full-year Head Start programs in 27 centers and 432 children of similar backgrounds had not been in programs [Westinghouse Learning Corporation/Ohio University (1969), p. 153]. The average school readiness score for Head Start children was 51.74 and for the other children was 48.46. These scores indicated that the Head Start children were a little more ready than similar children not in such programs. Although the difference in average scores was not very large, analysis indicated it was significant in the sense that it was unlikely that the average scores of two sets of 432 children picked at random from two groups of equally prepared students would differ as much, that is, 3.28 points on average.

Questions raised but not answered completely in the Head Start evaluation concern the source of the effects: Did the early intervention *cause* the difference in readiness levels? What are the mechanisms through which the improved outcomes were obtained? Not all of the Head Start youngsters were assigned to the program or nonprogram group at random so that, strictly speaking, there was no control group; the nonprogram group might better be termed a *comparison group*. Differential outcomes may be attributable to the program intervention *or* to possible bias in the way youngsters were chosen for the Head Start program. Further, even if Head Start is responsible for superior levels of school readiness, further research is needed to assess whether the effects are due to the age of the children (i.e., *any* cognitive stimulation at four years of age may have the same impact), to the social interaction that is necessarily part of the preschool program, or to the specific academic components of the Head Start intervention.

"A Minority of One Versus a Unanimous Majority"

Does group pressure influence individual behavior? The experiments conducted by Solomon Asch at Swarthmore attempted to answer this question.

In the first experiment, subjects were asked to compare lengths of line segments. The experiment consisted of sequences of 18 trials. On each trial a test line was to be compared with 3 comparison lines by each of 8 persons in turn, the test line being exactly equal in length to one of the 3 comparison lines. In each sequence of trials, 7 persons had been instructed by the experimenter to give incorrect responses on a prescribed dozen of the 18 trials. The eighth person, who was the experimental subject, was not instructed and considered the activity *bona fide* comparisons of lengths. Imagine yourself as the eighth person in one of these tests, finding that your seven predecessors have unanimously made an "obviously" incorrect choice on a trial! On 12 of the 18 trials the subject turned out to be "a minority of one versus a unanimous majority." Fifty such sequences were carried out. In most of these, the subject under pressure conformed to some extent and made errors of comparison. There were other sequences, however, in which all 8 subjects made legitimate comparisons, that is, no such pressure to conform was exerted on the last person; this is called the "control" situation. Table 1.7 gives the frequency of the number of errors made by persons in the experimental and control situations.

There was a clear and strong effect of the "unanimous majority." The investigation showed that the phenomenon was worth studying and led to questions for further research. What variables affect this phenomenon? What will be the effect of variations in group composition, for example, different size groups, or the presence of a "partner"? In Table 1.7 great individual differences are apparent: even under pressure 13 of 50 subjects made no errors, whereas one subject made 11 errors; this suggested questions about what types of persons are most likely, or least likely, to conform.

TABLE 1.7
Numbers of Errors Made by Last Person in Each Set of Trials

	Frequencies	
No. of errors	*Groups with persons under group pressure*	*Groups with persons not under group pressure*
0	13	35
1	4	1
2	5	1
3	6	0
4	3	0
5	4	0
6	1	0
7	2	0
8	5	0
9	3	0
10	3	0
11	1	0
12	0	0
	50	37

source: Asch (1951).

Deciding Authorship

The "Federalist Papers" played an important role in the history of the United States. Written in 1787–1788 by Alexander Hamilton, John Jay, and James Madison, the purpose of these 77 newspaper essays was to persuade the citizens of the State of New York to ratify the newly written Constitution of our emerging nation. These essays were signed with the pen name Publius, and published as a book which also contains eight essays by Hamilton. The question of whether Hamilton or Madison wrote 12 of the "Publius" essays has been a matter of dispute. Standard methods of historical research have not settled the problem.

Mosteller and Wallace (1964) applied statistical methods towards its solution. The problem is a difficult one, for Hamilton and Madison used the same style, standard phrases, and sentence structure, which were characteristic of most educated Americans of their time; for example, their average sentence lengths in the undisputed papers were 34.5 and 34.6 words, respectively. However, there were some subtle differences of style. Scholars had noticed, for instance, that Hamilton tended to use "while," and Madison, "whilst." The pair of "marker" words, however, was not decisive because in some papers neither word appeared. After painstaking analysis Mosteller and Wallace found that 30 words differed substantially in the frequency of use by the two authors.

Table 1.8 gives the rate of use of nine words by the two authors: the rate 3.24 is obtained by taking the number of times "upon" was used by Hamilton divided by the number of 1000's of words in the Hamilton essays. Note that Madison used "upon" infrequently but used "on" more frequently than Hamilton.

The frequencies of occurrence of the 30 words were combined into an index in such a way that the index was large for papers known to have been written by Hamilton and small for Madison papers. In fact, the score ranged from 0.3120 to 1.3856 for Hamilton papers and from −0.8627 to 0.1462 for Madison papers. Except for one paper, the scores of the disputed papers went from −0.7557 to −0.0145. Thus, these disputed papers were assigned to Madison. The paper with a score of 0.3161 was also assigned to Madison on the basis of further investigation and with less assurance.

TABLE 1.8

Rates of Use of Certain Words by Hamilton and Madison

	Frequency per 1000 words	
Word	*Hamilton*	*Madison*
Upon	3.24	0.23
Also	0.32	0.67
An	5.95	4.58
By	7.32	11.43
Of	64.51	57.89
On	3.38	7.75
There	3.20	1.33
This	7.77	6.00
To	40.79	35.21

source: Reprinted by special permission from Mosteller and Wallace, *Inference and Disputed Authorship: The Federalist*, 1964, Addison–Wesley, Reading, Mass.

1.2   Basic Concepts of Statistics

Experimental and Nonexperimental Research

The six examples in Section 1.1 illustrate a number of important principles in conducting statistical research. Surveys and comparisons between self-selected populations (e.g., smokers and nonsmokers) leave questions about the causal connections among the variables unanswered. For example, if smoking and cancer rates were assessed simultaneously, the possibilities must be considered that (a) smoking causes lung cancer, (b) lung cancer may cause smoking (admittedly not sensible in this instance), or that (c) lung cancer and smoking are simultaneous effects of other causal factors such as alcohol consumption. Data may be collected in such a way that certain explanations are not possible; for example, assessing smoking levels for years prior to the occurrence of cancer precludes explanation (b), but possibilities like explanation (c) remain.

In contrast, the polio vaccine trial and the group pressure experiments conducted by Asch are good examples of *experimental* research. Experiments have two features in common. Differences between the groups being compared are created by the researcher, and groups are established in such a way that there are no systematic differences between them prior to the experimental intervention. As a result, differences at the conclusion of the research, i.e., polio rates or number of conforming responses, can only be explained by the initial experimental manipulation. Since much of science attempts to identify cause-and-effect relationships among variables, experimentation is highly desirable. At the same time, experiments are not always feasible, either for reasons of ethics, the well-being of the subjects being observed, or for practical considerations. It would not be possible to force two groups to smoke or not smoke cigarettes, respectively, to investigate the effects on cancer; the magnitude of the Head Start project, including a large number of communities and individuals, precluded the establishment of near-identical control groups. In each of these situations, nonexperimental research confirms the basic associations between the most important variables, and follow-up investigations are needed to clarify the causal mechanisms.

Populations and Samples

The set of individual persons or objects in which an investigator is primarily interested is the *population* of relevance for that study. Sometimes values for all individuals in the population of relevance are obtained, but often only a set of individuals which can be considered as representatives of that population are observed; such a set of individuals constitutes a *sample*.

Finite Populations   In many cases the population under consideration is one which could be physically listed. The books in a library constitute such a population, and the card catalog provides the list. For *The Literary Digest* poll and other political polls the population of relevance consists of all potential voters; these people are registered on lists held by boards of elections.

The population of the United States as of noon yesterday could be listed, at least in theory.

Hypothetical Populations   If we wish to know if hamburger-producing equipment is correctly putting four ounces of food product in each "quarter pounder," our concern is with the equipment in general, or over the long run. The concern in the polio vaccine trial is the prevention of the disease among all youngsters in a particular age bracket; in the study of smoking and health, it is with all adults; in Asch's experiments, the effects of group pressure on individuals in other locations as well as those at Swarthmore. These larger groups constitute *hypothetical* populations in that their members cannot all be identified or listed. The population may even include objects or persons that do not exist at the present time, e.g., quarter-pounders yet to be produced or children not yet born at the time of testing. Generally speaking, hypothetical populations are comprised of all objects or individuals identified by a list of shared characteristics, e.g., age, health, actual or potential exposure to an experimental intervention, or being produced by the same production equipment.

Samples   In each instance, the *sample* is the subgroup that is actually studied, observed, and measured (e.g., a once-a-day sample of quarter pounders, the individuals actually queried in a political poll, the 401,974 youngsters tabulated in Table 1.4). Members of the sample are called *observational units* or, for short, *observations*. The observations may be individual objects or persons, or they may be groups of people such as families, households, cities, universities, or nations. Also, observations may be events, such as strikes, wars, or volcanic eruptions. In the process of observation we record some particular characteristic(s) of each observational unit; these values constitute the *data*.

Whether we pose questions about industrial equipment, voting preferences, about the effects of a polio vaccine or an educational intervention, or about the effects of group pressure on an individual's judgment, the questions usually pertain to the entire population. By definition, it is impossible to observe all members of a hypothetical population. In any case, it is often too costly or otherwise impractical to enumerate all members of a large population, even if they could be identified. Fortunately, a sample that represents the population well provides an excellent basis for drawing conclusions.

A sample should reflect as accurately as possible the relevant characteristics of the population; it should in a sense be representative. In practice we have to settle for a "fair" sample rather than insisting on perfect representation of the population. For, without complete knowledge of the population (with regard to the characteristics of interest), it is impossible to determine a sample that will be a miniature facsimile of the population.

In order to obtain a sample that is "fair," we sample randomly. If the population is finite, we figuratively put the names of all the members of the population into a bowl, shake the bowl, and draw out as many names as we need for the sample. (Actually this physical procedure is not a good way to draw a random sample; better ways involve the use of random numbers, to be described in later chapters.) A hand dealt in bridge is an example of a random sample of 13 cards from a population of 52 cards; the 10 birthdays with the highest priority numbers in the annual drawing for Selective Service was supposed to be a random sample from the population of birthdays.

The term "random" implies that every individual in the population has the same chance of being included in the sample. This feature helps to make it likely that the sample chosen will contain the salient features of the population in their proper proportions. It is *possible* to obtain a sample on a random basis which, by bad luck, is far from representative, but the probability of doing so is small if the sample is large enough. Random selection is used, therefore, to ensure that the sample is not biased. It is possible to assess the chances that the results obtained from a random sample are misleading. Other methods of selecting members of the population, such as selection by an "expert" or selecting just those individuals who volunteer to participate in a study may result in biased samples. As a result, it may be impossible to evaluate the sampling error.

Descriptive and Inferential Statistics

A statistical data set consists of one or more measurements, scores, or values for each of a number of individuals, objects or events (*observations*). For example, this may be a list of stated preferences of voters in a political poll; or a list of smoking habits coded into categories, accompanied by a 1 or 0 indicating whether or not the individual died from lung cancer before age 70; or a list of school readiness scores for pupils who did or did not participate in Project Head Start; or a list of hamburger weights to the nearest one-tenth of an ounce. Samples of data sometimes have only a small number of observations (say, 2 to 20) and other times have many (e.g., 187,783 men in the smoking-lung cancer study).

The methodology for organizing and summarizing the data for the *sample* is called *descriptive statistics*. The types of descriptive statistics needed depend on what questions the study is designed to answer. Frequently descriptive statistics include charts and graphs that allow the statistician to "see" the data visually and to look for exceptional or erroneous observations, coding or scoring errors, and the like. Usually descriptive statistics also include some numerical summaries of the data, for example, averages, measures of variability or "dispersion" in the data, or measures of relationship between several variables.

When we attempt to use these summaries to draw conclusions about an entire population, we employ the methodology of *statistical inference*. Sometimes the same numerical values serve both purposes. For example, the percentage of voters reported by *Literary Digest* to prefer Franklin Roosevelt as president (42.9%) was both a description of the voters actually sampled and an estimate of the proportion of all voters thought to favor FDR. In this poll, it was clear that the sample did not represent the population particularly well. Even with the most representative of samples, however, there are uncertainties inherent in the process of generalizing from the few to the many. Statistical inference address these uncertainties; inferential techniques perform two general functions:

Estimation: Using Sample Characteristics to Infer Population Characteristics. The second paragraph of Section 1.1 reports that 49% of a national sample of 1380 adults said, when asked, that they approved of the way the president was handling his job in early 1972. We *estimate* the percentage of the voting population who would have answered "approve" to the same question if posed at the same time and the same way to be 49%. How accurate is this estimate? Probability theory[3] tells us that when we take random samples of 1380 persons, for about 95 samples out of 100 the sample percentage for a given variable will not differ from the population percentage for that variable by more than 2.6%. Thus, assuming that the sampling procedure had inherent variability no greater than for random sampling, we can feel confident that 49% is a fairly good approximation of the percentage of voters who would have approved of the President's policy.

Hypothesis Testing: Using Sample Information to Answer Questions About the Population. In the group pressure experiment the first question was whether pressure to conform would affect behavior. An enormous difference between the performance of persons under pressure in the experimental situation and persons not under pressure in the control situation was observed. It seems clear that this difference could not be due to chance in the selection of subjects and in their behavior.

In another study it might not be so clear that an observed difference between the responses of individuals in an experimental group and

---

[3]Details of this theory are discussed in later chapters.

those in a control group reflects a difference which holds also for finite or hypothetical populations of relevance. We test the *hypothesis* that the vaccine has no effect; that is, we ask the question whether or not inoculation by an inert liquid has the same effect on immunity to polio as the Salk vaccine. In the study reported, there was very little chance that the observed difference in rates of incidence could be due to the random assignment of inoculation to children.

Many questions that statistical data may answer are questions that can in principle be answered Yes or No; frequently such a question can be stated in terms of the truth of a hypothesis. It can happen, of course, that the investigator is led to an incorrect answer by chance in obtaining the sample, that is, by chance a sample that is quite different from the population may be drawn. One of the goals of this book is to show how to control the probability that this happens.

Planning Statistical Investigations

Tournaments, run according to specified plans, can be considered as "experiments" for determining who the best players are. The plan, or design, of a tournament should be such as to reduce to a minimum the chance that the player who is best in some overall sense may not win the tournament. This design will include, for example, the method of scoring, the length of play, and the number of tries.

In laboratory experiments investigators often systematically manipulate factors which may cause the results to vary. These factors are then assessed in terms of their effects on the variables under consideration.

A metallurgist trying to determine the best composition for an alloy may try several combinations of concentrations of copper and manganese, concentrations which would have a reasonable cost if adopted for large-scale production. The result may be that increasing the amount of copper included in the alloy increases the strength, but increasing the amount of manganese over the feasible range of cost has little effect on the strength.

When a survey of public opinion is to be made, whether the purpose is to determine voting intentions, preferences of consumers for various products, or social attitudes, the investigators must obtain considerable information while keeping the costs of the survey within the amount of money allocated. Above all, the procedures they use for drawing the samples and obtaining the data must be designed so that it is possible to use statistical methods to assess the precision of the results.

Quantification   General research topics must be formulated in terms of specific questions before scientific investigation can proceed. A study must then be planned in which objective phenomena can be measured, counted, and tabulated. To study the general problem of whether group pressure influences behavior, Asch planned an experimental situation to provide specific evidence. The general problem became a specific question of whether subjects would incorrectly compare the lengths of line segments when others in the group did so before them. The evaluation of the success of Project Head Start was made specific by posing such questions as, will children who have participated in Head Start score better on tests than their counterparts who have not participated? Can such tests measure the effectiveness of the program? If so, what tests are appropriate? In some of the other studies presented in Section 1.1 the relevant quantities were the frequencies of the occurrence of certain phenomena, such as cases of paralytic polio or deaths from lung cancer.

It is sometimes difficult to put observations into numerical form. Such psychological quantities as pain and pleasure, for example, are hard to quantify. A physician, Dr. Harry Daniell, having studied 1104 people over a period of a year, reported [*Time* (1971)] a definite relationship between early and heavy face wrinkling and habitual cigarette smoking. He quantified wrinkling by studying each patient's face to arrive at a "wrinkle score." Then, knowing that age and amount of exposure to the sun are factors in causing wrinkles, he had his patients fill out a questionnaire about their smoking habits, exposure to the sun, and medical history. He found that in each age group cigarette smokers with indoor occupations had more wrinkles on the average than nonsmokers who worked outside.

Explaining Variability

A primary purpose of statistical analysis is to assess the factors that cause a set of measurements to vary. If the heights of a number of children are recorded reasonably accurately, one can expect that not all of the heights will be equal. They will vary. This variability may be "explained" by virtue of the fact that the children's ages are different, the heights of their mothers and fathers vary, their diets are different, etc. When a researcher conducts a study in order to discover to what extent the amount of a vitamin in the diet is related to the height of the children, he or she must take these other factors into account in the analysis.

Likewise, the study of smoking and health sought to explain why some individuals have lung cancer and others do not. Here the variation was limited to two possibilities, i.e., yes or no. Nevertheless, many possible variables were examined to explain this dichotomy, including gender, alcohol consumption, air pollution, and smoking habits.

The task of identifying the reasons for variability in statistical data is made more difficult by the fact that a number of contaminating factors may be present. For example, imprecise measures may give spuriously high or low values to some of the observations. (How accurate is the "wrinkle score," after all?) Also, observations recorded at a particular point in time might give different results on another occasion. The complexities in the data that arise as a function of how or when measurements are obtained are called *measurement error*. Also, there is *sampling error*. Since a sample is only part of the population, differences among observations may be smaller or larger than in some other sample, and may be smaller or larger than differences in the total population. In designing a statistical study, steps must be taken to reduce these extraneous sources of variability if the research is going to be able to identify the more important explanatory factors.

Summary

Statistics is a methodology for collecting, analyzing, interpreting, and drawing conclusions from numerical information. The relevant *population* is the set of individuals, objects, or events in which the investigator is interested; it may be finite and concrete or it may be hypothetical. A *sample*, which is a segment of the relevant population, permits generalizing to the population. The members of the sample are called *observational units*; they may be persons, groups of persons such as families or cities, or they may be objects or events. When samples of data are summarized into tables or numerical indices, the methodology is called *descriptive statistics*; when summary values are used to make generalizations about the population, the methodology is called *inferential statistics*. In order to be confident that statistical inferences are correct, a sample must be representative of the parent population. Sampling randomly protects against systematic sample *bias*. The *experimental method* for conducting empirical research has the advantage over other data collection paradigms that stronger causal connections may be confirmed among the variables being studied.

Exercises

1.1 Find a newspaper or magazine article or advertisement using statistics in a misleading way.

1.2 Find a newspaper or magazine article or advertisement using statistics in a helpful or illuminating way.

1.3   Each of the following studies may be regarded as an example of sampling. Discuss what an appropriate population might be for each.

(a) Collection of prices (at various stores) for a certain brand of cereal

(b) Collection of student ratings of a certain professor

(c) Measurement of blood pressure reduction due to a certain medication in 100 patients of a physician.

1.4   In the examples in Exercise 1.3, discuss some factors other than sampling which might cause the sample to be unrepresentative.

1.5   Give examples of hypothetical populations. What steps must be taken to determine if a given individual, object, or event is a member of a hypothetical population?

1.6   Give examples of finite populations and indicate how to list each one.

1.7   In each of the following situations, give a brief description of the intended population and then indicate how the sample that is chosen may be *biased*.

(a) A psychologist wishes to study the relationship between the aggressiveness of college students and the students' popularity. She places an advertisement in the campus newspaper, requesting student volunteers to participate in the study.

(b) A marketing researcher is interested in understanding the relative roles of husbands and wives in making decisions to buy new family cars. He chooses six large automobile agencies in the suburbs of Los Angeles that sell cars in different price ranges from modest to expensive. He interviews random samples of married couples that have decided to buy cars at each agency.

(c) A sociologist asks whether the incidence of crimes committed by adolescents decreases when school is in session, that is, in the months between September and June. With the cooperation of the police departments in Minneapolis (Minnesota), Chicago (Illinois), and Buffalo (New York) he collects monthly statistics on the crimes committed by adolescents in those cities during the preceding three years.

1.8   (continuation) In each of the situations in the preceding exercise, explain how you would choose a sample so that the bias is reduced or eliminated.

1.9   A professor believes that her accounting textbook is truly outstanding and that students who read the most pages will learn the basic principles of accounting best. She asks a random sample of accounting students to record the number of pages of the textbook they read (assuming that they told the truth, of course), and compares the reports to grades on the final exam at the end of the semester.

(a) The research described was not an experiment because it did not involve an intervention or manipulation on the part of the professor. Thus, if she finds a relationship between pages read and test scores, several explanations for this association may be plausible *besides* the excellence of the textbook. List and explain three of these.

(b) How could you conduct an experiment to determine if reading more pages of this textbook actually causes students to benefit in terms of grades on the final exam?

1.10   Suppose that you were interested in comparing the effectiveness of two toothpastes ("C" and "P") in reducing cavities. Your study entails following random samples of adults for 12 months who use toothpaste "C" or toothpaste "P" on a daily basis. At the end of the year, a dental researcher counts the number of cavities in each participant's mouth.

How would you conduct this study so that it is an *experiment?* How could you conduct this study as *non-experimental* research? What would be the advantages and disadvantages of each approach?

II

Descriptive Statistics

2

Organization of Data

Introduction

In Chapter 2 we start with the statistical information as it is obtained by the investigator; this information might be an instructor's list of students and their grades, a record of the tax rates of counties in Florida, or the prices of Grade A large eggs in each of ten Chicago grocery stores averaged over the past 36 months. We refer to such statistical information as *data*, the recorded results of observation. After the collection of data, the next step in a statistical study is to organize the information in meaningful ways, often in the form of tables and graphs or charts. These displays or summaries are descriptions which help the investigator, as well as the eventual reader of the study, to understand the implications of the collected information. In later chapters we shall develop numerical descriptions that are more succinct than these tables and charts.

In considering methods for organizing, summarizing, and analyzing statistical data, it is necessary to distinguish two types of data. Data that consist of numerical measurement, such as weight, distance, or time, or counts, such as numbers of children or numbers of errors, are called *numerical*. On the other hand, *categorical* data result from the observation of characteristics, e.g., gender, occupations, species, or physical states, defined by categories, such as male and female; or clerical, professional, and laborer; or canine, feline, and bovine; or gas, liquid, and solid. These distinctions among kinds of data are discussed further in Section 2.1.

In Section 2.2 we outline the steps involved in the process of organizing categorical data. These organized data may be presented in a table or a graph.

Numerical data, treated in Section 2.3, permit more flexibility and detail in their organization and summarization. At the stage of first obtaining data there may arise the question of how accurately the information is to be recorded; for example, how well can or should height be measured? We shall see what insight into statistical problems can be obtained by tabulating numerical data and graphing them suitably.

2.1   Kinds of Variables: Scales

A large number of methods are described in this book for organizing and analyzing statistical data. In any particular situation the most appropriate statistical procedure to apply depends on the answers to two primary questions: First, what is the data analyst trying to accomplish, that is, what are the *purposes* or *objectives* of the statistical analysis? And second, what *types of data* are to be analyzed? In answering this question, the distinction between categorical and numeric data is especially important.

Categorical Variables

People may be described as being blond, brunette, or redhead. Employed persons may be classified as managers, clerical workers, skilled laborers, or unskilled laborers. Plants may be classed as vines, trees, shrubs, etc. Rocks are grouped as sedimentary, igneous, or metamorphic. In each of these examples, the members of the population under scrutiny (people, plants, rocks) are placed into *categories*.

A *variable* is any characteristic that varies from one individual member of the (statistical) population to another. Hair color, occupation, and plant type are examples of *categorical variables*; such a variable is defined by the classes or categories into which it falls. The categories may be natural, as in the case of sex, or rather arbitrary, as in the case of occupational classification or an experimental condition to which an observation is assigned.

Frequently the members of separate categories differ on many specific characteristics. For example, managers, clerical workers, and laborers may differ on a host of individual measurable traits including years of education, salary, number of persons supervised, and so on; the different occupations are qualitatively distinct. Thus, categorical variables are also referred to as *qualitative* variables.

The simplest categorical variables have only two possible values, such as male and female or experimental condition and control condition. The categorization may also be the presence or absence of a given quality; for example, a person may or may not be employed, may or may not use a certain product; a plant may or may not bear flowers. In these instances, the information recorded is only whether an individual does or does not have the quality in question, that is, employment, using the soap powder, or bearing flowers. The variable generated is thus of a Yes-No form and may be termed *dichotomous*. The literal meaning of this word, "divided into two parts," becomes especially appropriate when we have reduced a *quantitative* variable such as income to the two categories "low" and "high."

Numerical Variables

Sometimes the characteristic of interest can only be represented correctly by a number. This may arise either from a *counting* process or a *measurement* process. Some variables, such as the number of children in families, the numbers of accidents in a factory on different days, or the number of clubs dealt in a hand of thirteen cards from a deck of playing cards are the results of counting; these are called *discrete* variables.

Typically, a discrete variable is a variable whose possible values are some of the ordinary counting numbers $1, 2, 3, \ldots$ or the integers[1] $0, 1, 2, \ldots$. The number of clubs dealt in a hand is a discrete variable having the possible values $0, 1, 2, 3, 4, 5, 6, 7, 8, 9, 10, 11, 12, 13$. A shorter way of writing this set of values is $0, 1, 2, \ldots, 12, 13$; here the three dots signify the integers omitted, namely, $3, 4, 5, 6, 7, 8, 9, 10, 11$. When it is not difficult for the reader to fill in the missing elements, we shall use these three dots (called an "ellipsis").

As a definition, we can say that a variable is *discrete* if it has only a countable number of distinct possible values. That is, a variable is discrete if it can assume only a finite number of values or as many values as there are integers.

---

[1]We cannot write down all of these numbers because there is no end to them; the three dots in $1, 2, 3, \ldots$ mean that the sequence continues indefinitely.

In contrast, quantities such as length, weight, or temperature can in principle be measured arbitrarily accurately. There is no indivisible unit. Weight may be measured to the nearest gram, but it could be measured more accurately, say to the tenth of a gram; the gram is divisible, as is the tenth of a gram. Such a variable is called *continuous*. The process of actually measuring a continuous variable, however, results in a scale that is not continuous; even if weight were recorded to the nearest milligram, for example, this means that fractions of milligrams are being ignored or attributed to rounding error. Many statistical techniques that are applied to numerical variables work just as well with both discrete and continuous data. The distinction between categorical and numerical variables is more important in choosing a statistical approach.

Scales*

Besides being described as either categorical or numerical, variables are typed according to the *scale* on which they are defined. (See Figure 2.1.) This typology of variables involves the amount of structure and meaning that variables have. The distinctions among measurement scales are sometimes important in selecting a statistical technique as well.

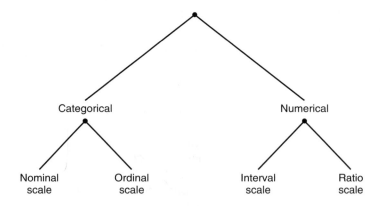

Figure 2.1 *Classification of variables.*

Scales for Categorical Variables. The categories into which a variable falls may or may not have a natural ordering. Occupational categories have no natural ordering. A variable, the categories of which are not

---

*This section may be omitted without affecting the understanding of subsequent sections.

ordered, is said to be defined on a *nominal* scale, the word "nominal" referring to the fact that the categories are merely names.

If the categories can be put in order, the scale is called an *ordinal scale*. Examples of ordinal variables are "socioeconomic status" (classified as low, medium, and high) and "strength of opinion" on some proposal (classified according to whether the individual favors the proposal, is indifferent toward it, or opposes it), and position at the end of a race (first, second, etc.). For some variables, the classification may be somewhat arbitrary: socioeconomic status could as well be categorized simply as low and high, instead of low, medium, and high; strength of opinion could be categorized as strongly in favor, mildly in favor, indifferent, mildly opposed, strongly opposed. Several statistical techniques are discussed in this book that are especially useful for analyzing ordinal data.

Scales for Numerical Variables.   Numerical variables, either discrete or continuous, may be classified as to whether there is a natural, meaningful zero. If I say I have zero cents in my pocket, that statement makes sense; the statement that Mary has twice as many cents as Alice is also meaningful. On the other hand, 0° Fahrenheit is an artificial zero point that does not indicate the complete absence of molecular activity. Thus, to say that an object whose temperature is 32° F is twice as hot as one whose temperature is 16° F is meaningless. However, the numbers can be used in other meaningful ways. An object that is half way in temperature between freezing and boiling (of water) is 50° in the Centrigrade system [(0° C + 100° C) ÷ 2] and is 122° in the Fahrenheit system [(32° F + 212° F) ÷ 2]. In either system the difference between freezing temperature and this temperature is equal to the difference between this temperature and boiling temperature. (The physical meaning is that it takes as much fuel to heat a liter of water from 0° C to 50° C as to heat it from 50° C to 100° C.) Thus, temperature is said to be measured on an *interval scale:* differences but not ratios can be compared. Money and weight are said to be measured on a *ratio scale* because we can sensibly consider ratios of values. For instance, 200 pounds is twice 100 pounds; that is, when weighed, two 100-pound objects will give the same value as one 200-pound object. A ratio scale is an interval scale with a meaningful zero.

A psychologist or a sociologist who wishes to evaluate a conceptual variable such as intelligence, strength of opinion, or degree of belief that cannot be measured directly, may devise a test or questionnaire to measure the variable indirectly, the scores on the test or questionnaire supposedly being related to the variable of real interest. Regardless of the scale of the nonobservable variable, the investigator can choose what sort of scale the measurements of the observable variable will have. Even physical variables may be measured on different scales at different times. A two-point ordinal scale of height may be made by seeing who can walk under a bar without ducking; or an ordinal scaling of height can be made by having a set of individuals line up in order of increasing height. The value for each individual is then his or her position (rank) in line. Any variable that can be measured on an interval or ratio scale could also be measured on the less stringent ordinal scale.

The choice of measurement scale is not the same as deciding what numerals are to be assigned to the observations, however. We may use the integers 1 and 2, for example, to represent males and females (a nominal scale), the tallest and second tallest individuals in a group (an ordinal scale), the scores of two individuals on an achievement test (an interval scale), or the length of time two individuals wait before receiving a phone call (a ratio scale). Thus, it is the measurement process—the way in which numbers are assigned to the observations—that determines if a scale is nominal, ordinal, interval, or ratio. An authoritative discussion of different measurement scales, with further detail, is given by Stevens (1951).

2.2   Organization of Categorical Data

Frequencies

Data can always be arranged in a table having two columns, the first column giving an identification of the observational unit and the second giving the datum for that unit. Table 2.1 is a hypothetical list of occupations of primary householders collected by a student who interviewed householders in a particular metropolitan area. The household is identified by its address, and the occupational category of the primary householder is recorded. The variable observed is occupation; the observational units are households.

Perhaps the simplest kind of summary for this type of data is made by providing a list showing the number of times each category occurs in the list, that is, the *frequency* of each category.

This is done by first writing down a list of the categories. Then as we come to each household in Table 2.1 we enter a tally for the category to which the householder belongs. The result is shown in Table 2.2. In passing from the original list to the table of frequencies we ignore all information except the categories and how often they were observed.

TABLE 2.1

Occupations of Primary Householders

Household identification (address)	Occupational category of primary householder
515 Main Street	Clerical
314 Wilkins Avenue	Professional
212 Shady Avenue	Professional
519 Shady Avenue	Professional
917 Shady Avenue, Apt. 3	Clerical
812 Denniston Street	Professional
423 Denniston Street	Professional
1024 Beechwood Boulevard	Sales
531 Northumberland Street	Laborer
1313 Aylesboro Road	Sales
1057 Aylesboro Road	Sales
1016 Marlborough Drive	Sales
914 Marlborough Drive	Sales
1123 Beeler Street	Sales
13 Solway Place	Professional
597 Maple Street	Clerical
1212 Jones Street, Apt. 5	Sales
612 Smith Street	Clerical
1549 Mary Street	Sales
415 Edna Street	Laborer

Hypothetical data.

TABLE 2.2

Tally Sheet Showing Frequency of Occupational Categories

Category	Tally	Number of primary householders in category (frequency)
Professional	JHT I	6
Sales	JHT III	8
Clerical	IIII	4
Laborer	II	2
		20

source: Table 2.1.

Any two households for which the primary householders have the same occupation are now treated as equivalent. From the table of frequencies we know there are two households whose householders are laborers; we would no longer know that one is from 531 Northumberland Street and the other is from 415 Edna Street. This abstraction is a simplification because information which is regarded as irrelevant for considering the set as a whole is eliminated.

Frequency Distributions.   A table like Table 2.2, or a graph representing the same information, is called *frequency distribution,* because it shows how the individuals are "distributed" over the categories, that is, how many individuals are associated with each category. When we focus attention on the variable observed (occupation here) we may simply say, "distribution of *occupation*" instead of the "distribution of *individuals* in occupational categories."

Percentages.   The *relative frequency* of a category, or the *proportion* of times that category has occurred, is the frequency of that category divided by the total number of observations. The relative frequencies in Table 2.3 are the frequencies in Table 2.2 divided by 20, the number of households observed. The sum of the relative frequencies in any table is 1, except for a possible discrepancy caused by rounding off the figures (for example, recording 1/3 as 0.33). Often we express the relative frequencies as *percentages* by multiplying the relative frequencies by 100. Table 2.3, like Table 2.2, is called a "frequency distribution."

TABLE 2.3
Relative Frequency of Various Occupations

Category	Relative frequency	Percentage
Professional	0.3	30
Sales	0.4	40
Clerical	0.2	20
Laborer	0.1	10
	1.0	100

source: Table 2.2.

Comparing Distributions   Table 2.4 gives the numbers of persons employed in different service occupations for 1970 and 1992. Of interest is the question of whether the composition of the labor force dedicated to various services has

TABLE 2.4

Persons Employed in Various Service Occupations in 1970 and 1992

	Numbers of Persons (in Thousands)	
	1992	*1970*
Business and repair services	6,553	1,403
Personal services	4,400	4,276
Entertainment and recreation	1,957	717
Professional and related services	27,677	12,904
Public administration	5,620	4,476
Total	46,207	23,776

source: *Statistical Abstract of the United States* (1993), p. 409.

TABLE 2.5

Percentage of Persons Employed in Various Service Occupations in 1970 and 1992

	1992	*1970*
Business and repair services	14.2	5.9
Personal services	9.5	18.0
Entertainment and recreation	4.2	3.0
Professional and related services	59.9	54.3
Public administration	12.2	18.8
Total	100.0	100.0
(Thousands of persons)	(46,207)	(23,776)

source: Table 2.4.

changed much over the 22-year span. It is difficult to make the necessary comparison from Table 2.4 because there were more service employees *in total* in 1992 than in 1970. The percentages in Table 2.5 tell the story better.

There are relatively more persons in business and repair services in 1992 and relatively fewer persons in personal services and public administration. The percentage of workers in professional and related services (including health, education, and social services) has increased somewhat from 54.3% of service workers in 1970 to 59.9% in 1992, although Table 2.4 indicates that the *number* of persons in these fields has increased dramatically.

Frequency Graphs

A bar graph can be used to represent frequency or relative frequency. For example, each percentage in Table 2.3 is represented in Figure 2.2 as the length of a horizontal bar; the thicknesses of the bars are the same. Note that in Table 2.2 the tallies themselves are similar to a bar graph. The numbers of persons in the United States belonging to various religions are shown in Figure 2.3 with heights representing fre-

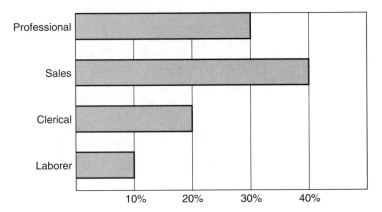

Figure 2.2   *Percentages of persons in various occupational categories (Table 2.3).*

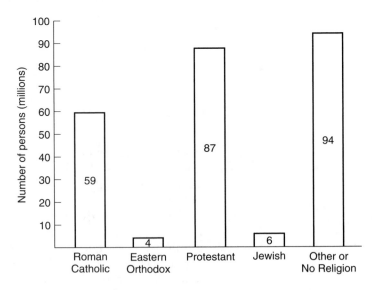

Figure 2.3   *1990 U.S. memberships in various religions, in millions. 1990 U.S. population: 250 million. (Statistical Abstract of the United States (1993), p. 67.)*

quencies. The bar graph indicates that Protestantism is most prevalent, with Catholicism next. The numbers of Eastern Orthodox and Jewish adherents are small, but not negligible.

The purpose of such a visual presentation is to enable the reader to compare the different frequencies easily. Since occupational categories and religions form nominal scales, the orders of labels in Figures 2.2 and 2.3 are arbitrary. In the case of an ordinal scale, the categories are graphed in their intrinsic order.

Another graphical device for representing frequencies or percentages, particularly appropriate for purposes of comparison, is shown in Figure 2.4. The height of each segment of the left-hand bar is proportional to

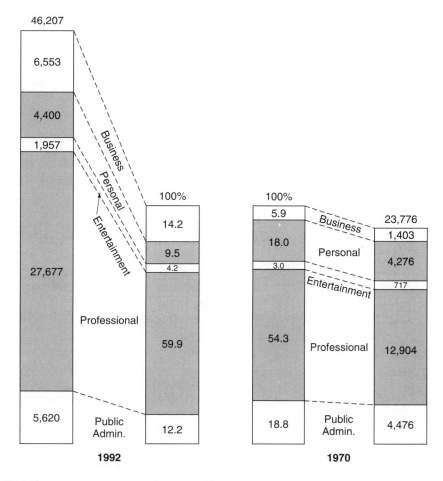

Figure 2.4   *Numbers (in thousands) and percentages of persons employed in various service occupations, illustrating the change from numbers to percentages for comparative purposes (Tables 2.4 and 2.5).*

the frequency in that occupational category in 1992; the right-hand bar represents frequencies in 1970.

The two middle bars represent the percentages in the categories in the two years. Equalizing the base of both number columns to 100 and reducing the figures proportionately makes direct comparison easy. We can see that the professional service portion increased only slightly (from 54.3% to 59.9%) even though the number of professional service employees more than doubled. At the same time, the percentage of public administration employees decreased from 18.8% to 12.2% while the number of public administration employees increased from 4,476 to 5,620.

Another context in which percentaging in this way is interesting is in business, where companies are interested not only in the change in gross sales, but also in the change in their *share of the market*, defined as a percentage of total sales. An automobile manufacturer might compare sales of Chrysler, Ford, General Motors, and various foreign companies in successive years. He or she might want to study share-of-the-market figures in different metropolitan areas, different age groups, or different ethnic groups.

2.3   Organization of Numerical Data

Discrete Data

Frequency Distributions for Discrete Data.   Quantitative data usually are recorded in a list, in a manner similar to the record for qualitative data. Suppose that for each household in Table 2.1 the number of children was recorded as in Table 2.6.

As is the case with categorical data, the information can be summarized in a table of frequencies, ignoring all knowledge except what numerical values were obtained and how often each value was observed. Table 2.7 gives the frequencies derived from Table 2.6, as well as the relative frequencies.

Frequencies can be displayed on a graph (Figure 2.5) by using the horizontal axis to denote the possible values of the variable and the vertical axis to indicate the relative frequency. In Figure 2.5 the dot above 1 and to the right of 0.40 shows that the relative frequency of families with one child is 0.40. Often the graph used to show frequencies is a *bar graph* (Figure 2.6); such a graph is called a *histogram*. The bars are centered at the values, in this case 0, 1, 2, 3, or 4 children per family.

TABLE 2.6

Numbers of Children in 20 Households

Household identification (address)	Number of Children
515 Main Street	2
314 Wilkins Avenue	0
212 Shady Avenue	1
519 Shady Avenue	0
917 Shady Avenue, Apt. 3	0
812 Denniston Street	1
423 Denniston Street	1
1024 Beechwood Boulevard	1
531 Northumberland Street	4
1313 Aylesboro Road	1
1057 Aylesboro Road	3
1016 Marlborough Drive	2
914 Marlborough Drive	2
1123 Beeler Street	1
13 Solway Place	1
597 Maple Street	2
1212 Jones Street, Apt. 5	0
612 Hopkins Road	3
1549 Youngs Road	1
415 Van Ness Street	4

Hypothetical data.

TABLE 2.7

Tally Sheet Showing Frequency of Numbers of Children

Number of children	Tally	Frequency	Relative frequency	Cumulative frequency	Cumulative relative frequency
0	||||	4	0.2	4	0.2
1	ЖН |||	8	0.4	12	0.6
2	||||	4	0.2	16	0.8
3	||	2	0.1	18	0.9
4	||	2	0.1	20	1.0
		20	1.0		

source: Table 2.6.

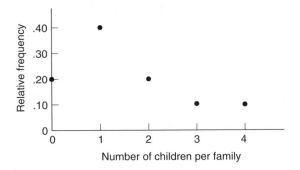

Figure 2.5 *Frequency distribution (dot graph) of number of children (Table 2.7).*

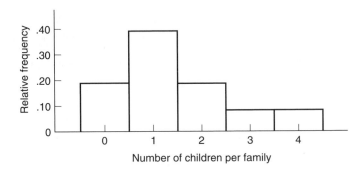

Figure 2.6 *Frequency distribution (bar graph or histogram) of number of children per family (Table 2.7).*

Cumulative Frequency.   The idea of *cumulative* frequency is relevant here because we are dealing with numerical data. The cumulative frequency at a given value is the sum of the frequencies of values less than and equal to that value; it is the number of observational units for which a value was obtained that is less than or equal to the given value. For example, Table 2.7 shows that 16 families had 2 or fewer than 2 children. *Cumulative relative frequency* of a value is the sum of the relative frequencies of values less than and equal to that value; this is equal to cumulative frequency divided by the total number of observational units. In the example, the cumulative relative frequency may be useful because one can read the relative frequency of small families, where "small" can be defined, for example, as families with 0, 1, or 2 children.

Comparing Distributions

Such visual presentations enable the investigator to receive an impression of some features of the data, such as their center and variability, that is, how scattered they are. They also help in comparing different sets of data; the investigator can see whether one set has generally larger values than another.

Consider again the example from the study by Asch discussed in Section 1.1. The frequencies and relative frequencies are given in Table 2.8. The maximum possible number of errors is 12 because there were 12 trials out of the 18 in which the "unanimous majority" matched lines incorrectly. The frequencies show the numbers of errors made by each group separately; the relative frequencies, adjusted for the number of subjects in each group, are better for making comparisons between groups.

TABLE 2.8

Distribution of Errors Made by Last Person in Each Set of Trials

	Frequency		Relative Frequency	
Number of errors	*Pressure group*	*No-pressure group*	*Pressure group*	*No-pressure group*
0	13	35	0.260	0.946
1	4	1	0.080	0.027
2	5	1	0.100	0.027
3	6	0	0.120	0.000
4	3	0	0.060	0.000
5	4	0	0.080	0.000
6	1	0	0.020	0.000
7	2	0	0.040	0.000
8	5	0	0.100	0.000
9	3	0	0.060	0.000
10	3	0	0.060	0.000
11	1	0	0.020	0.000
12	0	0	0.000	0.000
	50	37	1.000	1.000

source: Table 1.7.

Figures 2.7 and 2.8 show the same information in graphical form. It can be seen from either graph that persons under group pressure tend to make more errors than persons not under pressure; that is, most persons in the control groups made no errors, but for the groups with persons under pressure as many as 11 errors were made.

Again, using relative frequency adjusts for the fact that there were different numbers of subjects in the two groups; Figure 2.8 makes the comparisons of the two groups clearer.

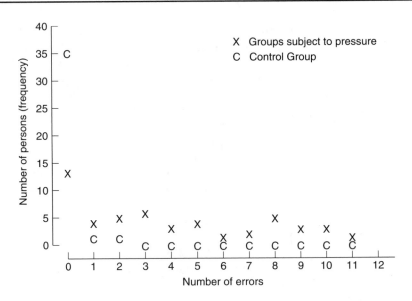

Figure 2.7   *Distributions of number of errors in control group and in group of persons subject to pressure (Table 2.8).*

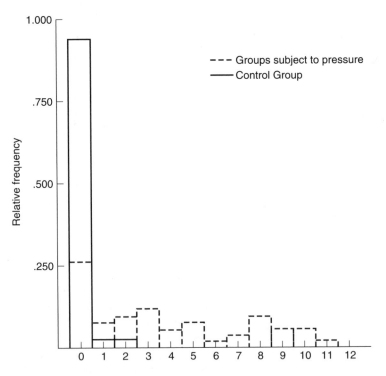

Figure 2.8   *Bar graphs for relative frequency of errors (Table 2.8).*

Grouped Frequency Distributions

When a variable has many discrete values, counting "sets" of values may make the job of summarizing data more efficient. For example, in Table 2.8 we might combine people who made no errors or one error, two or three errors, and so on. The resulting distribution is shown in Table 2.9.

TABLE 2.9

Distribution of Errors Made by Last Person in Each Set of Trials

Number of errors	Frequency Pressure group	Frequency No-pressure group	Relative Frequency Pressure group	Relative Frequency No-pressure group
0– 1	17	36	0.340	0.973
2– 3	11	1	0.220	0.027
4– 5	7	0	0.140	0.000
6– 7	3	0	0.060	0.000
8– 9	8	0	0.160	0.000
10–11	4	0	0.080	0.000
	50	37	1.000	1.000

source: Table 1.7.

The numbers of errors are now counted in *intervals*, and the frequencies and relative frequencies apply to the interval rather than to an exact number of errors. While we can't tell from Table 2.9 exactly how many persons made, say, 0 error or 3 errors, the smaller table makes an overview comparison of the two distributions easier to see, with little loss in detail. It is important, of course, that every interval have exactly the same *width* (two errors in Table 2.9) to avoid distorting the data.

Outliers

We would be very surprised to find a subject had completed the Asch experiment with 12 errors if he or she was a member of the no-pressure group. In fact, if we recorded such a result, we would probably double-check that the errors had been counted correctly, that the individual was truly an experimental subject and not one of the experimenter's "allies," that the individual was indeed in the no-pressure group, and perhaps even check that the individual's eyesight was adequate to see the stimuli.

An observation falling far out of the range of most of the sample is called an *outlier*. It may be a value far above or below the bulk of the sample, such as a family in Table 2.7 with 8 children, or it may be a value that is extreme *given* the particular characteristics of a subgroup. That is, 12 errors is highly unusual only for the no-pressure condition. Outliers are readily apparent in a histogram because they result in bars that are substantially removed from the rest of the graph. Sometimes a data set contains several outliers, either with similar extreme score values or with quite distinct values.

The identification of outliers is an important part of descriptive statistics, especially because extreme values can have a substantial impact on summary statistics. Frequently—especially in large data sets—outliers are data that have been scored or recorded incorrectly. These must be corrected. On occasion, outliers represent populations other than those the researcher intended to study. For example, suppose that Table 2.6 is a list of numbers of children in suburban households. A family with eight children might have resided in a rural area but was living in the suburban area temporarily. Since they are not long-term residents of the suburban community, and do not provide information about the usual residence pattern in the suburbs, this family may justifiably be eliminated from the data set. In an experimental setting, an outlier might be an individual who did not understand or refused to follow the instructions; likewise, this subject can be excluded from any further data analysis.

Outliers also occur because some observations simply do not operate through the same mechanisms as others; these differences require further investigation. For example, during the course of the Asch experiment, members of the research team may have noticed that some subjects in the no-pressure condition paid very little attention to the experiment; they may have been distracted by other individuals or the noise around them. If these individuals had unusually high error rates, it may have been because they did not focus closely on the experimental stimuli. This explanation represents a more detailed understanding of the process of making correct or erroneous judgments. Future experimenters would be well-advised to assess an additional variable—whether the subject is "attending" or "distracted."

Continuous Data

Accuracy.   In theory, continuous variables can be measured as accurately as desired. In practice, the accuracy of measurements is limited by the accuracy of measuring devices. In any particular case we may require only a certain level of accuracy. We might, for example, have a scale accurate enough to determine that a person's weight is 161.4376 pounds, but we would often be content to record it as 161 pounds. The extra accuracy may not be of real interest to our investigation since a person's weight can fluctuate several pounds during the course of the day and it is affected by the amount of clothing worn.

Rounding.   Reducing the accuracy of data measurement involves rounding. The rule for rounding to the nearest pound is to record the weight as 161 if the actual weight is more than $160\frac{1}{2}$ and less than $161\frac{1}{2}$ pounds, as 162 if the actual weight is between $161\frac{1}{2}$ and $162\frac{1}{2}$ pounds, etc. In effect, we are replacing any measurement within the interval $160\frac{1}{2}$ to $161\frac{1}{2}$ by the *midpoint* of the interval, that is, 161, etc.

If we wish to round a measurement of $161\frac{1}{2}$ to the nearest pound, how do we decide whether to round it to 161 or 162? We could flip a coin, which is ambiguous and seems a foolish waste of time and effort, or we could always raise to the higher integer, which would result in an upward bias of our figures. The standard procedure is to round to the nearest *even* integer; this procedure is unambiguous and in the long run, results in rounding up as often as rounding down.

Sometimes we might wish to round to the nearest ten pounds, 160, 170, etc., or to the nearest tenth of a pound, 160.1, 160.2, etc. The same principles apply.

Rounding is not always done accurately in practice. In most countries it is the custom to give one's age, for example, as 28 years until the 29th birthday. (This is an example of rounding *down*.) Thus one's "age" is not *really* one's age, but rather the number of full years of life one has completed since being born. (This is reflected in some languages; in Spanish, for example, one does not ask "How old are you?" but rather "How many years have you completed?") If instead we rounded the years according to the rules of rounding—as is done for life insurance policies—a person having a birthday on November 25th would, during the 29th year, say he or she is 28 years old until May 25th and 29 thereafter. The ages of race horses are changed on January 1st, regardless of the actual birthday.

Frequency Distributions for Continuous Data

As with other statistical information, the data for continuous variables will originally be reported on a listing of individuals and the observed measurements. A list of "after" weights of individuals who participated in a particular diet plan is given in Table 2.10. Continuous data such as these can be displayed along a *continuum* or *axis* (Figure 2.9). The data of Table 2.10 are recorded to the nearest pound; in Figure 2.9 we mark a dot above the axis for each datum. The density of dots shows how the data cluster. Here one can see that there are especially many weights between 145 and 150 pounds.

Even when we record the weights to the nearest pound, there are still not very many individuals at any one recorded weight. In order to obtain a table of frequencies which gives a good summary of the data, we group observations into *classes* according to weights.

TABLE 2.10
Weights of 25 Dieters

Dieter	Weight (pounds)	Dieter	Weight (pounds)
Adams	151	Martin	145
Allan	141	Newcombe	154
Brammer	166	Newsome	144
Brown	147	Peterson	162
Coles	172	Raines	157
Davidson	155	Richards	146
Jackson	153	Samson	149
Johnson	149	Smith	152
Jones	147	Tsu	142
Lee	148	Tucker	138
Levy	150	Uhlman	161
Lopez	143	West	167
Margolis	160		

Hypothetical data.

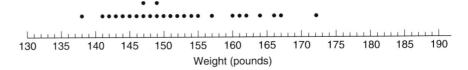

Figure 2.9   *Weights of 25 dieters (Table 2.10).*

The smallest weight is 138 and the largest is 172; the *range* is 172 − 138 = 34. We divide the interval containing all recorded values into seven separate *class intervals*, each 5 pounds wide, choosing the intervals so that their midpoints are convenient numbers, 140, 145, 150, etc. The choice of interval length and midpoints determines the *class limits* (the boundaries of the intervals) since they must lie halfway between the successive equally spaced midpoints. The class limits are thus 137.5, 142.5, 147.5, etc. Then, using a list of the class intervals, for each observation in Table 2.10 we enter a tally for the interval in which that observation lies. Thus, for 151 pounds, we enter a tally for the interval 147.5–152.5, for 141 pounds, we enter a tally in the interval 137.5–142.5, etc. The result is shown in Table 2.11.

Note that tabulating the data into these class intervals is equivalent to simply rounding each figure to the nearest five pounds.

TABLE 2.11
Tally Sheet and Frequency Distribution of Weights

Class interval	Midpoint	Tally	Number of observations in interval (frequency)
137.5–142.5	140	|||	3
142.5–147.5	145	⊬|	6
147.5–152.5	150	⊬|	6
152.5–157.5	155	||||	4
157.5–162.5	160	|||	3
162.5–167.5	165	||	2
167.5–172.5	170			1
			25

source: Table 2.10.

The pattern of tallies provides a picture of how the data cluster. After all the tallies have been made, the tallies for each interval are counted. The number of observations in an interval is the frequency of that interval. The frequencies are given in the last column of Table 2.11 and are shown in the histogram of Figure 2.10. (The break in the horizontal axis of the figure indicates that the axis is not continuous from 0 to 140.) The pattern of weights is more easily discernible from Figure 2.10 than from Figure 2.9. Note that the tally column itself is a histogram when viewed from the side.

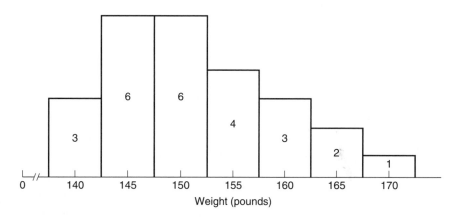

Figure 2.10 *Histogram of weights (Table 2.11).*

There are some guidelines to follow in forming frequency distributions. As a rule of thumb, it is generally satisfactory to use 5 to 15 class intervals. A smaller number of intervals is used if the number of measurements is relatively small. For example, if a data set consists of 20 values, placing them in 10 intervals would give little opportunity to see where the data are concentrated and might leave a number of intervals with no observations at all. In this situation 5 or 6 intervals might be better. If the number of measurements is large, for example, over 200 observations, 12 to 15 intervals or even more will give a clear picture of the distribution.

Second, note that the class limits all have an extra decimal digit 5. As a result, no observation can be equal to a class limit. Each observation lies strictly *within* one class interval, never on the boundary between two intervals. Thus class limits should be chosen with one-half unit more accuracy than the measurements themselves.

The data can also be represented by a dot graph (similar to Figure 2.5 for discrete data). Figure 2.11 represents the frequencies in Table 2.11. When the variable is continuous, sometimes the dots are connected by line segments in order to lead the eye from one point to another; then the graph is a *frequency polygon*. The dots in Figure 2.11 are connected to yield Figure 2.12. The histogram (Figure 2.10) is usually preferred to the line graph (Figure 2.12); the histogram appropriately suggests that the observed measurements are spread out beyond the first and last midpoints, and a single large frequency is given its appropriate weight. In general it is desirable to have all classes the same width, including the first and last; it is better that the first interval have a left-hand endpoint and the last interval a right-hand endpoint. (The right-hand class interval should not be "167.5 and more," for example.)

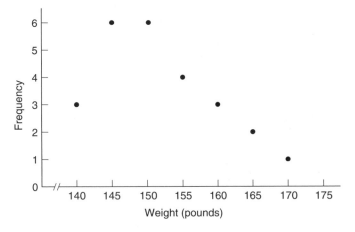

Figure 2.11   *Dot graph of frequency distribution of weights (Table 2.11).*

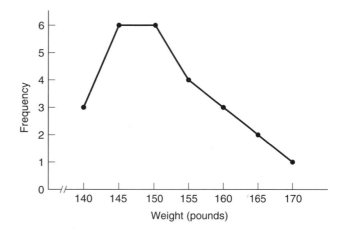

Figure 2.12 *Line graph of frequency distribution of weights (Table 2.11).*

The histogram can be interpreted as representing frequency by *area.* This interpretation is important when we consider changing the width of the class intervals; we want to graph in such a way that the picture is affected as little as possible.

Table 2.12 reports the heights of 1293 11-year-old boys in one-inch intervals. These frequencies are represented by the solid bars in Figure 2.13. In this histogram area represents frequency. For example, the height of the bar for the interval 51.5–52.5 is proportional to the frequency 192, and the area is the width of the interval (namely 1) times the height (namely 192).

Table 2.13 reports the heights in 5 inch intervals.[2] These frequencies are represented by the dashed line in Figure 2.13. The height of the dashed bar over the interval 47.5–52.5 is 109.8 and the area is 5 × 109.8 = 549, which is equal to the area of the 5 bars over the one-inch intervals 47.5–48.5, 48.5–49.5, 49.5–50.5, 50.5–51.5, and 51.5–52.5.

Note that the smaller intervals produce a histogram that is fairly smooth. If the intervals were smaller than one inch, the histogram would be still smoother. Figure 2.14 has a smooth curve superimposed on the histogram. This is the shape we would expect if the interval were very small. It represents the frequency distribution of the continuous variable "height." In pictorial representation we shall often use smooth curves instead of histograms.

---

[2]Of course, this table does not give us very much detail about the distribution of height because the number of intervals is small while the number of observations is quite large.

TABLE 2.12

Heights of 11-Year-Old Boys

Class interval (inches)	Midpoint	Frequency	Relative frequency
44.5–45.5	45	2	0.0015
45.5–46.5	46	8	0.0062
46.5–47.5	47	6	0.0046
47.5–48.5	48	25	0.0193
48.5–49.5	49	50	0.0387
49.5–50.5	50	119	0.0920
50.5–51.5	51	163	0.1261
51.5–52.5	52	192	0.1485
52.5–53.5	53	204	0.1578
53.5–54.5	54	187	0.1446
54.5–55.5	55	156	0.1206
55.5–56.5	56	96	0.0742
56.5–57.5	57	41	0.0317
57.5–58.5	58	23	0.0178
58.5–59.5	59	12	0.0092
59.5–60.5	60	5	0.0039
60.5–61.5	61	1	0.0008
61.5–62.5	62	1	0.0008
62.5–63.5	63	2	0.0015
		1293	0.9998

source: Bowditch (1877).

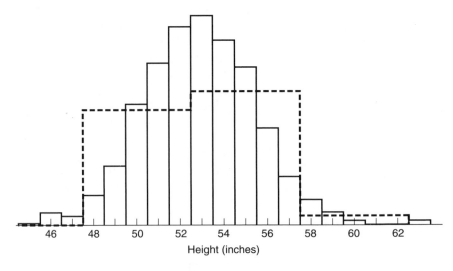

Figure 2.13   *Histograms of heights of 11-year-old boys, with class-interval widths of 5 inches (dashed) and 1 inch (solid) (Tables 2.12 and 2.13).*

TABLE 2.13
Heights of 11-Year-Old Boys

Class interval (inches)	Midpoint	Frequency	Relative frequency
42.5–47.5	45	16	0.0124
47.5–52.5	50	549	0.4246
52.5–57.5	55	684	0.5290
57.5–62.5	60	42	0.0325
62.5–67.5	65	2	0.0015
		1293	1.0000

source: Table 2.11.

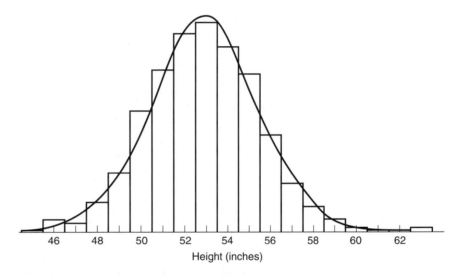

Figure 2.14 *Histogram of heights of 11-year-old boys, with corresponding continuous frequency distribution (Table 2.12).*

Stem and Leaf Display

An alternative form of tabulating frequencies according to a set of intervals uses the intervals implied by the recorded digits themselves. Each numerical observation is split into two parts, the left-hand digit(s) that defines the interval, and the right-hand digit(s) that is listed exactly as it appears in the data. For example, Table 2.14 lists the years of teaching experience reported in a 1988 survey of elementary school teachers in rural schools in Tennessee; the values in the table are already ordered from lowest to highest.

TABLE 2.14

Years Teaching Experience of Tennessee Elementary School Teachers

2	7	12	18
3	7	13	18
3	8	13	19
4	9	15	19
6	10	16	19
6	10	17	21
7	11	17	30
7	11	18	36

source: Tennessee State Education Department (1988).

Since the numbers range from 2 to 36 years, we could split each value into two parts, the "tens" digit and the "ones" digit, respectively. The "tens" digits are listed to the left of the vertical line in the stem-and-leaf display and make up the *stem*. All accompanying ones digits, listed to the right of the vertical line, are called the *leaves*. This is illustrated in Figure 2.15a.

0	233466777789		0	2334
1	00112335677888999		0	66777789
2	1		1	0011233
3	06		1	5677888999
			2	1
			2	
			3	0
			3	6
	(a)			(b)

Figure 2.15   *Stem-and-leaf displays of teaching experience data (Table 2.14).*

All of the observed values from 10 through 19 are recorded in the second line of Figure 2.15(a). The "1" to the left of the vertical line is the tens digit of each measurement. The second digits of the measurements from 10 through 19 are listed to the right of the vertical line in ascending order. Similarly, the first line of Figure 2.15(a) lists observed values 2 through 9, which are assumed to have a tens digit of "0."

Because the stem-and-leaf display shows the data in complete detail, it is particularly useful for showing the range and concentration of data values, the shape of the distribution, specific values that were *not* found in the data, and values that are extreme relative to the rest of the data (outliers). Figure 2.15 shows that rural Tennessee teachers are fairly evenly spread over a range of years of experience, up to 20 years. At that point there is an abrupt cutoff; no teacher in this sample has between 22 and 29 years of experience, while two apparent outliers have 30 and 36 years of experience, respectively. This raises questions about the 22 to 29 year range as well as the process by which these older individuals maintained their employment. This illustration also makes it clear that a stem-and-leaf display is impractical if a sample is very large (perhaps 200 observations or more).

Numerical values may be split in any way that is necessary. For example, scores on the Scholastic Assessment Tests (SAT), which generally range from 400 to 800, may be tabulated with two digits to the left of the vertical line and one to the right. If SAT scores were displayed with one digit to the left and two-digit pairs on the right, the stem-and-leaf display would probably be too large to provide a clear picture of the distribution. On the other hand, any stem-and-leaf display may be further subdivided. For example, the teaching experience data could be displayed in intervals of 0–4 years, 5–9 years, 10–14 years, and so on, as shown in Figure 2.15(b).

Stem-and-leaf displays are also useful for comparing distributions in two separate samples. Figure 2.16 shows "back-to-back" displays of the distribution of a physiological measure obtained from a sample of daughters and from their mothers. The Georgetown University Laboratory for Children's Health Promotion regularly collects health-related data from school children in Washington, DC, and their parents. The samples represented in Figure 2.16 consist of 71 8- to 12-year-old girls and their mothers. One of the many variables studied by the Laboratory is the "Body Mass Index," or BMI. The index is the ratio of weight, in kilograms, to the square of height, in meters.[3] Very high ratios (35 and above) indicate a health risk due to obesity.

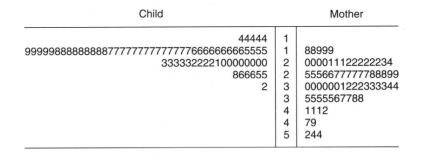

Figure 2.16 *Stem-and-leaf displays of BMI index for 71 young children and their mothers. (Georgetown University Laboratory for Children's Health Promotion.)*

---

[3]The ratios have been rounded to the nearest whole number for this illustration.

The display makes it easy to compare the distributions for younger females and for older, related women. For girls from 8 to 12 years, the most noteworthy characteristic of the distribution is its limited range. All of the values range from 14 to 32 and most are in the interval 15–19. The lower values are fairly evenly distributed; there are 4 or more individuals observed at each value from 14 to 23, with the single exception of 21. Two observations are somewhat outside the range of the others—the values 28 and 32—but not far enough out of the range to be considered outliers. None of the measurements reaches the health risk level (35).

The BMI measures for these youngsters' mothers have quite a different distribution. It is more dispersed, with values ranging from 18 to 54. None is as low as the values of 14, or even 15, 16, or 17 obtained from their daughters. Most of the scores are in the range 18 to 38, but some reach into the low 40's and others are substantially higher, from 47 to 54. All of the scores in the intervals 35–39, 40–44, 45–49, and 50–54 indicate a risk for health problems associated with obesity. The comparative stem-and-leaf display gives us a clear picture of the differences between the two distributions in terms of their shapes, their locations on the scale indicated by the stem, their spread or dispersions, and the occurrence of extreme values.

Summary

Statistical *data* are the recorded results of observation. Data arise from the observation of one or more characteristics of various observational units. The characteristics are called *variables*.

Variables are classified as being *categorical* or *numerical*.

The categories or classes of a categorical variable may be natural, as in the case of sex (male or female), or may be defined on other bases, such as occupational classes or by experimental intervention. A categorical variable with only two categories is called *dichotomous*.

Variables may be characterized according to the *scale* on which they are defined; this typology involves the structure and meaning of the variables. Categorical variables are defined on a *nominal* or an *ordinal* scale. Numerical variables are defined on an *interval* or *ratio* scale. A variable that can be measured on a given scale could also be measured on any less stringent scale.

Categorical data are summarized in *frequency* or *relative frequency* tables and displayed in bar graphs, called *histograms*. Numerical data may be similarly summarized and displayed. In addition, numerical data may be summarized in terms of *cumulative frequencies* and *cumulative relative frequencies*; these are the sums of frequencies and relative frequencies, respectively. Numerical data may also be combined into *class intervals*, and frequencies are calculated for each interval. *Stem-and-leaf displays* are very useful for examining details of a data set that is not extremely large.

The primary purposes of these summaries are to provide an overview of a sample of data, to examine the shape of the distribution and to locate extreme values, called *outliers*.

Exercises

2.1   Indicate for each of the following variables on which scale (nominal, ordinal, interval, or ratio) it is usually measured.

(a) strength of opinion

(b) area

(c) color

(d) shade of red

(e) Fahrenheit temperature

(f) typing speed (words per minute).

2.2   Indicate which of the following data are discrete and which are continuous.

(a) sizes of the student populations on the University of California campuses

(b) Dow-Jones weekly averages over the past year

(c) gross national products (GNP) of 15 countries

(d) number of murders in 1994 in each of 20 large U.S. cities

(e) number of murders in San Francisco for each of the last 20 years

(f) lengths of wheel-base of 100 models of automobiles

(g) lifetimes of lightbulbs.

2.3   Classify the members of your class according to their major subject in college. Find the distribution of majors by sex. Compare the two distributions. (For example, is there a higher proportion of political science majors among the men or among the women?)

2.4   Make a histogram of the heights of the males in your class. Do the same for the females. What is the proportion of males having heights greater than 5'6"? Of the females? What is the proportion of females having heights less than 5'4"? Of the males?

2.5   (continuation) Using the data obtained for Exercise 2.4, make a histogram of the heights of males and females together. Compare it with the separate histograms for males and females.

2.6   The weights and heights of the members of the 1989 Stanford football team are given in Table 2.15.

(a) Mark the weights on a (single) horizontal axis.

(b) Tabulate the weights in a set of classes of weight. What considerations led you to your choice of class intervals?

(continued ...)

TABLE 2.15

Weights and Heights of Stanford Football Players, 1989

Name	Weight (pounds)	Height (inches)	Name	Weight (pounds)	Height (inches)
Archambeau	260	77	Milburn	170	69
Baird	270	76	Nash	260	75
Batson	175	70	Nichols	190	72
Baur	236	77	Palmbush	255	76
Booker	230	75	Palumbus	205	75
Borkowski	240	75	Papathanassiou	265	76
Burton	210	72	Pelluer	215	75
Carpenter	270	77	Pierce	230	73
Daniels	225	75	Pinckney	175	74
Davis	200	72	Price	240	76
Englehardt	195	73	Quigley	185	74
Eschelman	227	74	Richardson	195	72
George	210	74	Roberts	220	72
Gillingham	275	76	Robinson	265	74
Grant	187	70	Roggeman	255	72
Hanson	250	75	Saran	175	73
Hawkins	240	75	Scott	182	71
Hinckley	240	77	Shane	265	74
Hopkins	185	75	Shea	250	76
Johnson	195	74	Sullens	200	75
Lang	225	75	Taylor	197	71
Lawler	215	70	Trousset	195	74
Le	185	72	Tunney	235	74
Machen	185	73	Valdovinos	195	71
McCaffrey	220	78	Vardell	228	74
McGroarty	185	73	Volpe	195	67
Mescher	230	76	Walsh	180	73
Miccichi	235	74	Young	205	73

source: Official program, freshmen players excluded.

(c) Calculate the relative frequency of each interval.

(d) Graph the frequencies in a histogram.

(e) Summarize the major features of the weight distribution in words.

(f) Do you think there is a tendency to report weight in round numbers ending in 5 or 0? Support your opinion.

2.7   Table 2.16 lists the percentage of minority students enrolled in a random sample of 30 of the nation's largest school districts (each district has over 10,000 students) in 1978. Each has been rounded to the nearest whole-number percentage.

(a) Calculate the frequency and relative frequency of minority enrollments in class intervals containing percentages 0% to 4%, 5% to 9%, and so on. Make a table with these results.

(b) Construct a histogram of the data using the same intervals.

(c) Construct a stem-and-leaf display of the data.

(d) Describe the distribution verbally. Is there any particular "concentration" of minority enrollments? Are there districts that should be considered as outliers? The percentage of students in the nation's public schools who were minority in 1978 was 24.7% for the entire country. How did enrollment in the largest districts compare to this?

TABLE 2.16
Percentage of School Districts' Enrollments that is Minority

18	13	5	5	8	29
16	12	15	63	11	19
44	36	10	18	12	23
17	8	11	9	12	4
24	40	33	6	10	12

source: U.S. Office of Civil Rights (1978).

2.8   The ages at death of English kings and queens are given in Table 2.17.

(a) Calculate the frequencies and relative frequencies of ages at death in class intervals of 10 to 19 years, 20 to 29 years, etc. Make a table of the frequencies and relative frequencies.

(b) Construct a histogram of the data using the same intervals.

(c) Construct a stem-and-leaf display of the data.

(d) What observations do you consider as outliers? What are possible reasons for such outliers?

(e) What additional interesting features do these data have?

TABLE 2.17

Ages at Which English Rulers Died

William I	60	Henry VI	49	James II	68
William II	43	Edward IV	41	William III	51
Henry I	67	Edward V	13	Mary II	33
Stephen	50	Richard III	35	Anne	49
Henry II	56	Henry VII	53	George I	67
Richard I	42	Henry VIII	56	George II	77
John	50	Edward VI	16	George III	81
Henry III	65	Mary I	43	George IV	67
Edward I	68	Elizabeth I	69	William IV	71
Edward II	43	James I	59	Victoria	81
Edward III	65	Charles I	48	Edward VII	68
Richard II	34	Cromwell 'I'	59	George V	70
Henry IV	47	Cromwell 'II'	86	Edward VIII	77
Henry V	34	Charles II	55	George VI	56

source: Gebski, V., Leung, O., McNeil, D. R., & Lunn, A. D. (1992). Reprinted in Hand *et al.* (1994), p. 304.

2.9 For the data of Table 2.18

(a) Construct a table of frequencies and relative frequencies of the tax amounts in the intervals 4.00–4.99¢, 5.00–5.99¢, 6.00–6.99¢, etc. (continued . . .)

TABLE 2.18

Federal and State Gasoline Taxes

Alabama	11	Louisiana	20	Ohio	21
Alaska	8	Maine	19	Oklahoma	17
Arizona	18	Maryland	18.5	Oregon	20
Arkansas	18.5	Massachusetts	21	Pennsylvania	12
California	15	Michigan	15	Rhode Island	21
Colorado	22	Minnesota	20	South Carolina	16
Connecticut	25	Mississippi	18	South Dakota	18
Delaware	19	Missouri	11	Tennessee	20
Florida	4	Montana	20	Texas	15
Georgia	7.5	Nebraska	23.9	Utah	19
Hawaii	16	Nevada	16.25	Vermont	15
Idaho	22	New Hampshire	18	Virginia	17.5
Illinois	19	New Jersey	10.5	Washington	23
Indiana	15	New Mexico	16.2	West Virginia	15.5
Iowa	20	New York	8	Wisconsin	22.2
Kansas	17	North Carolina	22.3	Wyoming	9
Kentucky	15	North Dakota	17		

source: *Statistical Abstract of the United States* (1993), p. 305. Note: Tax rates, in cents per gallon, in effect Sept. 1, 1991.

TABLE 2.19 (See Exercise 2.10)

Land Areas, by State

State	Total area (sq mi)	Farm area[a] (1,000,000 acres)	State	Total area (sq mi)	Farm area[a] (1,000,000 acres)
New England			*East South Central*		
Maine	35,387	1.3	Kentucky	40,411	14.0
New Hampshire	9,351	0.4	Tennessee	42,146	11.7
Vermont	9,615	1.4	Alabama	52,423	9.1
Massachusetts	10,555	0.6	Mississippi	48,434	10.7
Rhode Island	1,545	0.1			
Connecticut	5,544	0.4	*West South Central*		
			Arkansas	53,182	14.4
Middle Atlantic			Louisiana	51,843	8.0
New York	54,556	8.4	Oklahoma	69,903	31.5
New Jersey	8,722	0.9	Texas	268,601	130.5
Pennsylvania	46,058	7.9			
			Mountain		
South Atlantic			Montana	147,046	60.2
Delaware	2,489	0.6	Idaho	83,574	13.9
Maryland	12,407	2.4	Wyoming	97,818	33.6
Virginia	42,777	8.7	Colorado	104,100	34.0
West Virginia	24,231	3.4	New Mexico	121,598	46.0
North Carolina	53,821	9.4	Arizona	114,006	36.3
South Carolina	32,008	4.8	Utah	84,904	10.0
Georgia	59,441	10.7	Nevada	110,567	10.0
Florida	65,758	11.2			
			Pacific		
East North Central			Washington	71,302	16.1
Ohio	44,828	15.0	Oregon	98,386	17.8
Indiana	36,420	16.2	California	163,707	30.6
Michigan	96,705	10.3	Alaska	656,424	1.0
Illinois	57,918	28.5	Hawaii	10,932	1.7
Wisconsin	65,499	16.6			
West North Central					
Minnesota	86,943	26.6			
Iowa	55,276	31.6			
Missouri	69,709	29.2			
North Dakota	70,704	40.3			
South Dakota	77,121	44.2			
Nebraska	77,358	45.3			
Kansas	82,282	46.6			

source: *Statistical Abstract of the United States* (1993), pp. 217 and 654.
[a]1987.

(b) Tabulate the cumulative frequency and cumulative relative frequency for each interval.

(c) What tax amount is in the "middle" in the sense that half of the states have that amount or more and half have that amount or less?

2.10   The total areas of the 50 states are given in Table 2.19.

(a) Mark the total areas on a horizontal axis.

(b) Tabulate the total areas in a set of classes. What considerations led to your choice of class intervals?

(c) Graph the frequencies in a histogram.

2.11   The farm areas of the 50 states are given in Table 2.19.

(a) Mark the farm areas on a horizontal axis.

(b) Tabulate the farm areas in a set of classes. What considerations led to your choice of class intervals?

(c) Graph the frequencies in a histogram.

2.12   Refer to Table 2.20.

(a) How many families had incomes less than $2,000?

(b) What percent of all families had incomes less than $2,000?

(c) How many families had incomes of at least $6,000?

(d) What percent of all families had incomes of at least $6,000?

(e) How many families had incomes between $3,000 and $5,999?

(f) What percent of all families had incomes between $3,000 and $5,999?

TABLE 2.20
Family Income in the United States: 1959

Family income	Total number of families	Percent of families
Under $1,000	2,512,668	5.6
$1,000 to $1,999	3,373,813	7.5
$2,000 to $2,999	3,763,758	8.3
$3,000 to $3,999	4,282,945	9.5
$4,000 to $4,999	4,957,534	11.0
$5,000 to $5,999	5,563,516	12.3
$6,000 to $6,999	4,826,563	10.7
$7,000 to $9,999	9,053,220	20.1
$10,000 and over	6,794,380	15.1
Total	45,128,397	100.1

source: *U.S. Census of Population: 1960.* Vol. 1, *Characteristics of the Population.*

2.13   The U.S. Census Bureau has reported projections of population characteristics for selected years between 1980 and 2080. Figures 2.17 and 2.18, reproduced from Pallas, Natriello, and McDill (1989), summarize the projected size of the school-aged population through 2020.

(a) Assuming that these figures are drawn accurately, approximate the *number* of school-aged children projected for each year given.

(b) Assuming that these figures are drawn accurately, approximate the *percent* of children of each racial/ethnic group projected for each year given.

(continued . . .)

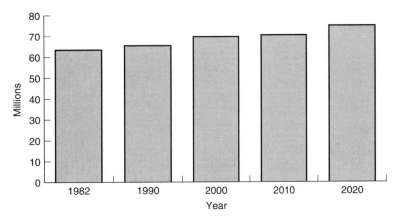

Figure 2.17   *Projected Population, 0- to 17-Year-Olds U.S. Total, 1982–2020. (Copyright 1989 by the American Educational Research Association. Reprinted by permission of the publisher.)*

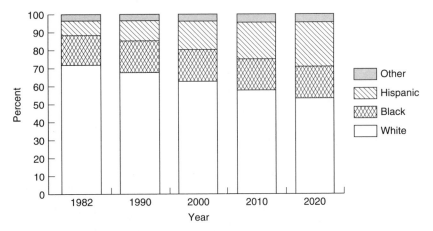

Figure 2.18   *Projections of the 0–17 Population by Race/Ethnicity U.S. Total, 1982–2020. (Copyright 1989 by the American Educational Research Association. Reprinted by permission of the publisher.)*

(c) Approximate the *number* of school-aged children of each racial/ethnic group projected for each year given.

(d) Summarize the change in the racial/ethnic composition of the school-aged population from 1982 to 2020. Which groups are likely to increase or decrease in absolute size? In relative size? Which groups are likely to have the greatest change? The least change?

2.14   Data on students in a graduate course on statistical methods are shown in Table 2.21.

TABLE 2.21

Data for 24 Students in a Statistics Course

Student	Major[a]	Logic[b]	Sets[c]	Sem. stat.[d]	Sem. math. analysis[e]	Final grade[f]
1	EE	Yes	No	0	2	284
2	CE	Yes	Yes	0	3	292
3	S	Yes	Yes	1	6	237
4	P	Yes	Yes	0	5	260
5	PA	No	Yes	1	5	210
6	CE	Yes	No	1	3	238
7	PA	Yes	Yes	0	3	200
8	S	Yes	Yes	3	6	300
9	A	No	Yes	0	1	174
10	PA	Yes	Yes	1	4	266
11	PA	Yes	Yes	1	2	244
12	PA	Yes	Yes	0	5	207
13	PA	Yes	Yes	2	7	261
14	PA	Yes	Yes	2	2	293
15	PA	No	Yes	1	2	213
16	EE	Yes	Yes	1	3	262
17	CE	No	Yes	0	4	272
18	PA	Yes	Yes	0	4	196
19	PA	Yes	Yes	1	1	210
20	IA	Yes	Yes	2	5	267
21	A	Yes	No	1	1	226
22	PA	Yes	Yes	3	0	300
23	CE	No	No	0	4	288
24	M	Yes	Yes	0	2	274

source: Records of a class.

[a] *Major.* A: Architecture; CE: Chemical Engineering; EE: Electrical Engineering; IA: Industrial Administration; M: Materials Science; P: Psychology; PA: Public Administration; S: Statistics.

[b] *Logic.* "Are you familiar with the notions of elementary logic: logical 'and,' logical 'or,' 'implication,' and 'truth tables'?"

[c] *Sets.* "Are you familiar with the notions of 'union' and 'intersection' of sets?"

[d] *Sem. Stat.* "How many semesters of probability or statistics have you taken?"

[e] *Sem. Math Analysis.* "How many semesters of analysis (calculus, advanced calculus, intermediate analysis, etc.) have you taken?"

[f] *Final Grade.* Sum of scores on three exams.

(a) Make a frequency table of final grades that has two sets of columns, one for those who had taken no course in probability or statistics before, the other for those who had taken at least one course in probability or statistics before. Choose an appropriate set of intervals, and give the frequency and relative frequency for each group of students who score in each class interval.

(b) How are the distributions of final grades for students with and without prior coursework similar? Different?

(c) Do previous courses in statistics or probability appear to be important in explaining the final grade?

(d) Mark the semesters of mathematical analysis on a horizontal axis. Describe the main features of this distribution.

2.15 To qualify for a job as firefighter in a large city in the northeastern United States, candidates must take a test of physical agility and a written ability test. The agility measure is obtained by timing candidates as they attempt three tasks: A stair climb, a "body drag," and an obstacle course. Their times (in seconds) are combined into a "scaled" agility total score; of course, lower scores are better on this measure. Data for 28 candidates who participated in the 1987 testing are given in Table 2.22.

(a) Make a frequency table, using appropriate intervals, of scaled agility scores that has columns for males and females separately. For each sex group, give the frequency and relative frequency for each class interval.

(b) Make a histogram, like Figure 2.8, that shows the relative frequencies for each sex group, so that the two may be compared. Label the histogram with the midpoints of the class intervals chosen in part (a).

(c) Summarize the differences in the performance of males and females on the agility test verbally.

2.16 Repeat the procedures of Exercise 2.15, using the Written Test Scores in Table 2.22.

2.17 The data in Table 2.23 constitute a "time series." The numbers are arranged according to time. Plot the data for Massachusetts and California on the same graph, plotting 1.783 and 0.865 above 1880, 2.239 and 1.213 above 1890, etc. Use one symbol for California and another for Massachusetts. (Connect the symbols, making a line graph for each state.)

2.18 In a study of children's abilities to learn and recall words, Gentile, Voelkl, Monaco, and Mt. Pleasant (1995) asked a group of fourth- and fifth-grade pupils to read a 67-word poem repeatedly until they could recite it with at least 75% accuracy. The 27 children who learned the poem most quickly were labeled "fast learners" and the 26 children who took the longest to learn the poem were labeled "slow learners."

TABLE 2.22
Data for 28 Firefighter Applicants

Candidate number	Sex (1 = M 2 = F)	Race (1 = W 2 = M)	Task Times			Scaled agility score	Written test score[a]	Composite score[b]
			Stair climb	Body drag	Obstacle course			
4877	2	1	17.60	29.20	116.50	4.92	71.00	58.93
3592	2	1	20.50	30.80	105.00	6.14	75.00	58.00
4554	2	2	17.70	28.80	162.20	7.20	73.00	55.34
4897	2	2	20.60	30.10	148.50	8.22	74.00	53.76
6297	2	2	20.50	28.90	165.60	8.68	70.00	51.54
2014	2	1	23.70	31.70	178.80	11.69	80.00	49.29
8213	2	1	29.10	39.60	143.60	14.77	76.00	42.15
9766	2	1	16.10	24.60	100.70	2.02	93.00	71.86
5490	2	1	16.20	22.60	115.80	2.22	86.00	69.10
3704	2	2	15.10	21.90	122.40	1.85	78.00	67.07
6014	2	2	15.50	23.50	111.60	1.96	72.00	64.83
3293	2	1	17.40	30.50	124.00	5.62	88.00	63.40
6942	2	1	16.10	27.60	134.40	4.67	78.00	61.78
8387	2	2	17.20	27.80	131.20	5.07	75.00	60.01
1532	1	1	12.30	15.90	84.90	−3.20	98.00	83.36
3475	1	1	13.80	17.10	90.50	−1.85	95.00	79.80
2618	1	1	12.60	18.80	90.50	−1.88	90.00	78.15
8966	1	2	14.00	18.20	102.00	−0.84	92.00	76.88
5922	1	1	11.70	18.50	84.50	−2.69	79.00	75.93
3566	1	2	14.00	20.60	84.20	−1.02	86.00	75.18
2505	1	1	14.90	21.00	82.30	−0.58	86.00	74.36
7093	1	2	12.50	16.10	90.00	−2.79	71.00	73.40
5933	1	1	14.50	20.80	94.90	−0.18	83.00	72.59
4399	1	1	15.00	19.10	99.90	−0.22	80.00	71.64
9047	1	2	13.70	19.40	112.30	−0.08	78.00	70.69
3041	1	1	13.50	23.40	99.50	0.40	77.00	69.45
0722	1	1	15.10	23.20	103.00	1.25	76.00	67.52
5298	1	2	14.60	26.20	128.00	3.23	70.00	61.76

source: Data from Buffalo, New York.
[a]Higher scores on the written test are better.
[b]Individuals with low agility scores and high written scores will obtain the highest values on the composite, while individuals with higher agility scores and/or lower written performance will have lower values on the composite.

TABLE 2.23
Populations of California and Massachusetts (millions)

Year	1880	1890	1900	1910	1920	1930	1940	1950	1960	1970	1980	1990
Massachusetts	1.783	2.239	2.805	3.366	3.852	4.250	4.317	4.691	5.149	5.689	5.737	6.016
California	0.865	1.213	1.485	2.378	3.427	5.677	6.907	10.586	15.717	19.971	23.668	29.760

source: *Historical Statistics of the United States, Colonial Times to 1957*, p. 12, and *Statistical Abstract of the United States*, 1993, p. 28.

One week later, these children were asked to recall the poem and then to relearn the poem to 75% accuracy. After an additional 21 days, the children were asked to recall the poem. The main purpose of the study was to see if fast and slow learners, having twice attained equal accuracy, would remember the same number of words over time. The numbers of words recalled correctly on the third occasion are given in Table 2.24.

TABLE 2.24

Numbers of Words Recalled by Fast- and Slow-Learning Children

Speed of learning	Numbers of words recalled
Fast Learners	36, 47, 36, 52, 50, 43, 49, 51, 47, 45, 52, 47, 51, 55, 59, 46, 38, 47, 54, 55, 50, 50, 41, 50, 60, 53, 44
Slow Learners	47, 38, 58, 35, 55, 19, 51, 46, 40, 36, 43, 36, 40, 59, 38, 44, 41, 41, 33, 43, 48, 49, 30, 36, 31, 37

source: Gentile *et al.* (1995).

Compare the distributions for fast and slow learners by constructing a comparative stem-and-leaf display like Figure 2.16. Summarize the differences between the two distributions in a brief written report. In particular, comment on any obvious differences in the *locations* of the distributions on the scale (the stem), differences in spread or *variability*, differences in the *shape* of the distributions, and differences in the occurrence of *extreme values*. From these distributions would you judge that fast and slow learners, even after attaining a common level of accuracy, have the same distributions of recall scores three weeks later?

2.19 Figure 2.16 gives distributions of the Body Mass Index (BMI) for 71 8- to 12-year-old girls and their mothers. These families were chosen for study because each mother also had a younger daughter between 2 and 5 years of age. The frequency distribution of BMI ratings for the 71 younger children is given in Table 2.25.

Construct a stem-and-leaf display of these values using the same intervals as in Figure 2.16. Write a brief report comparing the distribution of BMI values for the younger children with those of their older sisters. Be sure to compare the distributions in terms of their *locations* on the BMI scale (the stem), the spread or *variability* of the distributions, the *clustering* of scores in particular intervals, *gaps* in the distributions (scores that do not occur), and the occurrence of *extreme values*.

TABLE 2.25

Frequency Distribution of Body Mass Index for 71 Children Between 2 and 5 Years of Age

BMI	Frequency
13	2
14	7
15	12
16	21
17	10
18	9
19	4
20	2
21	3
22	1
Total	71

source: Georgetown University Laboratory for Children's Health Promotion.

3
Measures of Location

Introduction

After a set of data has been collected, it must be organized and condensed or categorized for purposes of analysis. In addition to graphical summaries, numerical indices can be computed that summarize the primary features of the data set. One is an indicator of *location* or *central tendency* that specifies where the set of measurements is "located" on the number line; it is a single number that designates the center of a set of measurements. In this chapter we consider several indices of location and show how each of them tells us about a central point in the data.

If the data values are clustered close to the center point, then a measure of location is an especially good summary statistic. If, on the other hand, values range very widely (e.g., salaries ranging from $4,000 to $100,000) then a measure of location does not convey much information about the entire data set. For this reason, indices of *variability*, or spread of the data, are also informative. These are discussed in Chapter 4.

3.1 The Mode

Definition and Interpretation of the Mode

The *mode* of a categorical or a discrete numerical variable is that category or value which occurs with the greatest frequency. The mode or *modal category* in Table 2.2 is "sales." The mode does not necessarily describe "most" (that is, more than 50%) of the individuals. Note, for example, only 40% of the respondents (Table 2.3) fall into the modal category. The modal occupational category in both 1970 and 1992 (Table 2.4) is "professional service." The modal religious group in the United States (Figure 2.3) is Protestant (among individuals with religious affiliation).

Table 3.1 reports the numbers of children in a sample of 100 families. A family of 2 children occurs most frequently—62 times. The mode, therefore, is 2 children. We emphasize that the mode is a value of the variable, and the frequency of that value suggests its statistical importance.

TABLE 3.1

Number of Children in 100 Families: I

Number	Frequency
0	14
1	18
2	62
3	3
4	1
5	2
	100

Hypothetical data.

The mode is a good measure of location if many of the observations have the modal value as in Table 3.1. In comparison, in Table 3.2 the number of families with 2 children is 30 instead of 62. If the other 32 families are divided evenly with 0, 1, 3, and 4 children, then the mode—still 2—does not summarize the distribution as well. There are almost as many families with no children or one child in Table 3.2, and the lion's share of families have values other than 2.

TABLE 3.2

Number of Children in 100 Families: II

Number	Frequency
0	22
1	26
2	30
3	11
4	9
5	2

Hypothetical data.

The distribution in Table 3.3 has no mode because every value of "number of children" has the same frequency. This would be very nearly the case if the frequencies were, say, 18, 21, 19, 22, and 20. Some measure other than the mode would be needed to tell us about the location of the distribution.

TABLE 3.3

Number of Children in 100 Families: III

Number	Frequency
0	20
1	20
2	20
3	20
4	20

Hypothetical data.

For another 100-family sample, shown in Table 3.4, the frequency increases as the number of children increases to 2, then decreases, then increases again for 4 children, and finally decreases. When the pattern of frequencies rises and falls twice like this, we say that the distribution is *bimodal* because there are two values (here, 2 and 4) which are more frequent than neighboring values. A bimodal distribution often suggests that the data consist of two unimodal distributions. For example, if the families of Table 3.4 could be identified as urban or rural the distributions might resemble those of Table 3.5. We can see that urban families are typified by a mode of 2 children per family, and rural families by a mode of 4 children per family. But neither 2 nor 4 is very useful for describing the entire set of 100 families.

TABLE 3.4

Number of Children in 100 Families: IV

Number	Frequency
0	5
1	11
2	40
3	7
4	32
5	5
	100

Hypothetical data.

TABLE 3.5

Frequency Distributions of Number of Children in 100 Urban and Rural Families: IV

Number	Frequency		
of children	Urban	Rural	Total
0	5	0	5
1	11	0	11
2	35	5	40
3	2	5	7
4	2	30	32
5	1	4	5
	56	44	100

Hypothetical data.

The mode is a good measure of location for both categorical and numerical variables when one value is far more common than other values. It is the *primary* measure of location for variables on a nominal scale for which arithmetic operations such as summing the values are not generally meaningful.

Mode of Grouped Data

When the variable being summarized has a numerical scale, the data can be grouped into class intervals and the mode can then be defined in terms of class frequencies. With grouped quantitative data, the *modal class* is the class interval with the highest frequency.

Data on income may be summarized by frequencies of class intervals as in Table 3.6. The *modal class* in Table 3.6 is $12,000–15,999, since more incomes are in this interval than in any other interval. Of course, when data have been grouped, the modal interval depends to some extent upon the grouping used. Suppose, for example, that in summarizing the income data we had used intervals such as $11,000–14,999 and $15,000–18,999. If there were many families with incomes of exactly $15,000, the modal class would be $15,000–$18,999, which is different from the interval obtained when the data are grouped as in Table 3.6.

TABLE 3.6
Income of 64 Families

Income[a]	Midpoint	Frequency
$ 0–3,999	$ 2,000	4
4,000–7,999	6,000	6
8,000–11,999	10,000	15
12,000–15,999	14,000	28
16,000–19,999	18,000	7
20,000–23,999	22,000	2
24,000–27,999	26,000	2
		64

Hypothetical data.
[a]The interval $4,000–7,999 is meant to include all incomes of at least $4,000 and less than $8,000; for instance $7,999.81 is included. Sometimes, however, incomes are only reported to the nearest dollar. The U.S. Census reports data this way. (See Table 2.20.)

It was suggested in Chapter 2 that class intervals should be chosen so that the midpoints are convenient numbers and there are 5 to 15 classes. In order that the mode be reasonably defined, the class interval should be wide enough to contain many cases, but narrow enough to display several different frequencies. A frequency table with many small frequencies would not ordinarily be informative.

The mode conveys more meaning in the case of ordinal or numerical variables than it does with nominal variables because the existence of an ordering makes possible a grouping around the mode. Moreover, the definition of categories is often arbitrary. In Table 2.1, for example, "sales" could be divided into "salespeople" and "marketing."

The measures of central tendency discussed below are appropriate only for measurements made on ordinal, interval, or ratio scales. The median, to be discussed next, can be used with ordinal data; the mean requires an interval scale.

3.2   The Median and Other Percentiles

The Median

The median ("divider") of an interstate highway is a strip that divides the road in half. The median of a set of observations is a value that divides the set of observations in half, so that the observations in one half are less than or equal to the median value and the observations in the other half are greater than or equal to the median value.

In finding the median of a set of data it is often convenient to put the observations in order, so that the smallest is first, the next smallest is second, and so on. A set of observations arranged in order is called an *ordered sample.* If the number of observations is odd, the median is the middle observation. For example, if the ages of five customers receiving a senior citizen discount are $69\frac{1}{2}$, 64, 65, 67, and $62\frac{1}{2}$ years, the ordered sample is $62\frac{1}{2}$, 64, 65, 67, $69\frac{1}{2}$; and the median is 65 years, since two of the measurements are larger than 65 and two are smaller than 65. Figure 3.1 shows how 65 divides the set of observations into halves; the median is the middle observation.

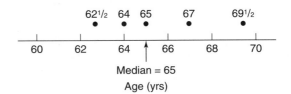

Figure 3.1   *The median of a set of observations.*

If there were 4 measurements instead of 5, say $62\frac{1}{2}$, 64, 65, and $66\frac{1}{2}$, there would not be a middle observation. Here, any number between 64 and 65 could serve as a median; but it is desirable to use a specific number for the median and we usually take this number to be halfway between the two middle measurements, for example $(64+65)/2 = 64\frac{1}{2}$. The median is still a "central" value—there are as many values greater than it as there are less than it.

When a sample contains the same value several times, they can still be put in order. If, for example, the numbers of children in five families are 2, 1, 5, 2, 2, the ordered sample is 1, 2, 2, 2, 5, and the median number of children is the middle observation, which is 2. If there were 4 observations, say 1, 2, 3, 5 children, 1 and 2 are the lower half and 3 and 5 the upper half of the set; any number between 2 and 3 divides the halves. To eliminate ambiguity, we take $2\frac{1}{2}$ as the median, even though $2\frac{1}{2}$ is not a possible value of the variable.

The median is a useful measure of location for ordinal, interval, or ratio data. It is the *primary* measure of location for variables measured on ordinal scales because it indicates which observation is central without attention to how far above or below the median the other observations fall. If data have been grouped, the median can be approximated by interpolation. This procedure is explained in Appendix 3A.

Quartiles

The median divides the set of observations into *halves*. Quartiles divide the set into *quarters*.

Suppose the ages of 8 senior citizens are arranged in order, $60\frac{1}{2}$, 61, 62, 64, 65, 66, 67, and 69 years. The median is $64\frac{1}{2}$; half (four) of the eight observations are less than $64\frac{1}{2}$, and half are greater than $64\frac{1}{2}$.

The quartiles are indicated on Figure 3.2. The value $61\frac{1}{2}$ inches is the *first quartile* of the ages: one-fourth (2) of the 8 observations are less than $61\frac{1}{2}$ and three-fourths (6) of the 8 observations are more than $61\frac{1}{2}$. The value $66\frac{1}{2}$ is the *third quartile* because three-fourths (6) of the observations are less than $66\frac{1}{2}$ and one-fourth (2) of the observations are greater than $66\frac{1}{2}$. The *second quartile* is the median, which is $64\frac{1}{2}$ here.

If the number of observations is a multiple of 4 and the observations are different, the quartiles are particularly easy to find. For example, if there are 8 observations, the first quartile is halfway between the second-ranking and the third-ranking observation, the second quartile is halfway between the fourth- and fifth-ranking observations, and the third quartile is halfway between the sixth- and seventh-ranking observations.

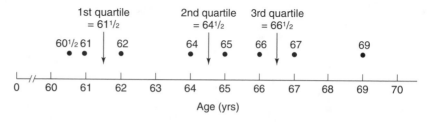

Figure 3.2   *Quartiles of a continuous variable: Ages of 8 senior citizens.*

It should be noted that "quartile" refers to a *value*, not to an individual or a set of individuals. It is a single value of the variable; it is not a set or range of values. We say, "John's test score is in the upper quarter," meaning his score exceeds the third (upper) quartile.

A general procedure for finding quartiles is carried out most easily when data are summarized into cumulative relative frequencies. Table 3.7 is a frequency distribution of the number of children in 100 families. We look for the points at which cumulative relative frequencies of 0.25, 0.50, and 0.75 are achieved. For example, fewer than one-fourth of the observations have 0 or 1 child (0.22 of the families, to be exact). More than one-fourth of the observations have 0, 1, or 2 children (0.47 to be exact). Thus, if all 100 families had been listed in ascending order, the families that "split" the list into two parts with one part having one-fourth of the families and the other part having the remaining three-fourths, would have 2 children. Try it.

In short, the cumulative relative frequency for 1-child families is less than 0.25; for 2-child families it is greater than 0.25. Thus the first quartile is 2. The second quartile is 3 because the cumulative relative frequency corresponding to 2 children is below 0.50 and to 3 children is above 0.50. The third quartile is 4 because the range of cumulative relative frequency corresponding to 4 goes from below 0.75 to above 0.75. If, by coincidence, a cumulative relative frequency is exactly 0.25, 0.50, or 0.75, then the quartile is halfway between that value of the variable and the next higher value (e.g., $2\frac{1}{2}$ or $3\frac{1}{2}$ children).

The quartiles 2, 3, and 4 divide the 100 families into quarters, in that we can assign 25 of the 100 families to the first quarter with 0, 1, or 2 children per family, 25 to the second quarter with 2 or 3 children per family, 25 to the third quarter with 3 or 4 children per family, and 25 to the fourth quarter with 4, 5, or 6 children per family.

TABLE 3.7

Number of Children in 100 Families

Number of children	Frequency	Relative frequency	Cumulative relative frequency
0	3	0.03	0.03
1	19	0.19	0.22
2	25	0.25	0.47
3	27	0.27	0.74
4	21	0.21	0.95
5	4	0.04	0.99
6	1	0.01	1.00
	100	1.00	

Hypothetical data.

Deciles, Percentiles, and Other Quantiles

Deciles, which divide the set of ordered observations into tenths, and *percentiles*, which divide it into hundredths, are defined analogously to quartiles.

If we want to compare colleges according to how "bright" their leading students are, we might use the 9th decile (90th percentile) of scores on a Scholastic Assessment Tests. Telephone company scientists might compare systems of equipment for handling telephone calls by comparing the 90th percentiles of distributions of customers' waiting times under the various systems.

The 90th percentile in Table 3.7 is 4 children because the range of cumulative relative frequency corresponding to 4 children is 0.74 to 0.95, which includes 0.90.

Quantile is a general term that includes quartiles, deciles, and percentiles. The 0.999 quantile (99.9th percentile) is a value exceeded by only one-tenth of one percent (0.1%) of the individuals. Any quantile can be computed by reference to cumulative relative frequencies in the manner already explained for quartiles, deciles, and percentiles.

3.3   The Mean

Definition and Interpretation of the Mean

Definition.   The *mean* of a set of numbers is the familiar *arithmetic average*. There are other kinds of "means," but the term "mean" by itself is understood as denoting the arithmetic average. It is the measure of central tendency most used for numerical variables.

The mean is the sum of observations divided by the number of observations. For example, the mean of the 5 observations 15, 12, 14, 17, and 19 is

$$\frac{15 + 12 + 14 + 17 + 19}{5}$$

which is 77/5, or 15.4. Let these numbers represent the dollars that each of five persons has to spend on an outing; then, if the group pooled its resources and shared them equally, $15.40 is the average amount that people would spend.

The mean summarizes all of the units in every observed value. This is in contrast to the mode, which only indicates the most frequently observed value, and the median, which only considers one or two middle values out of all those recorded.

The mean is like a "center of gravity" for a data set. For example, 10 families are labeled $1, 2, \ldots, 10$, and the number of children in each is recorded in Table 3.8. We can compute the mean number of children per family thus

$$\frac{2 + 4 + 4 + 3 + 4 + 3 + 3 + 3 + 6 + 3}{10} = 3.5.$$

TABLE 3.8
Number of Children in 10 Families

Family	Number of children
1	2
2	4
3	4
4	3
5	4
6	3
7	3
8	3
9	6
10	3
	35

Hypothetical data.

The data in Table 3.8 are plotted on an axis in Figure 3.3 and their mean is indicated. Think of the axis as a board and the data points as one-pound weights. If we place a wedge under the board, it will balance when the wedge is located at the mean, as in Figure 3.4.

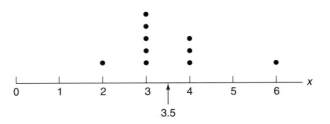

Figure 3.3   *Plotted observations (Table 3.8).*

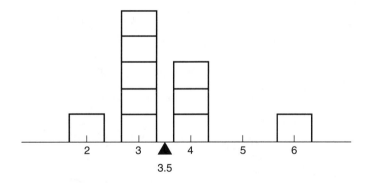

Figure 3.4   *The mean as the center of gravity (Table 3.8).*

The mean has an advantage over other measures of location because it summarizes all of the observed values. This property can also be a drawback, however, since the mean is significantly affected by one or two extreme values. If one more large family with 10 children were added to Table 3.8, the mean of the distribution would be about 4.1 rather than 3.5. This is higher than the number of children in all the families but two, and certainly does not indicate the "center" of the distribution very well.

In contrast, the median is not affected as readily by extreme values. For example, the median number of children in Table 3.8 is 3. If the additional family with 10 children were added to the distribution, the median would still be 3. Even if several observations were added at one end of the distribution or the other, the median would change only slightly; it would remain somewhere among the middle values in the distribution. Thus, before relying on the mean as a summary statistic, it is advisable to examine the distribution for outliers; if they are present at either end of the distribution (and are correctly recorded values), the median may be a better location indicator than the mean.

The mean involves arithmetic operations on the observed values— namely, summation and division. Appendix 3D gives guidelines for the number of significant digits that should be maintained when carrying out these operations. The following section explains a notation to represent the operation of summation.

Use of Notation

Understanding of arithmetic and other mathematical operations will be made considerably easier by the development of an appropriate system of notation. We denote the numbers in a set of five observations by

x_1, x_2, x_3, x_4, and x_5. Their sum is written as

$$x_1 + x_2 + x_3 + x_4 + x_5, \tag{3.1}$$

and their mean is

$$\frac{x_1 + x_2 + x_3 + x_4 + x_5}{5}. \tag{3.2}$$

In the first example of Section 3.3, $x_1 = 15$, $x_2 = 12$, $x_3 = 14$, $x_4 = 17$, and $x_5 = 19$. The phrase "Take the five numbers as listed, add them, and divide by 5" is replaced by (3.2).

We do not always deal with *five* measurements, however, and must be able to express sums and means of an arbitrary number of observations. The number of observations is denoted by the letter n (or some other letter). Since n can be any integer, we do not write all the observations explicitly. We write the n observations as $x_1, x_2, \ldots, x_n$. The three dots (ellipsis) indicate the omitted values. The sum is then written as

$$x_1 + x_2 + \cdots + x_n,$$

and the mean, which we denote by $\bar{x}$, as

$$\bar{x} = \frac{x_1 + x_2 + \cdots + x_n}{n}.$$

The symbol for the arithmetic average $\bar{x}$ is read "x bar." The mean of 2 observations ($n = 2$) is

$$\frac{x_1 + x_2}{2}; \tag{3.3}$$

the mean of 5 observations ($n = 5$) is given in (3.2).

The symbol x_i denotes the observation corresponding to the ith individual in the sample. If there are three individuals in the sample, the index (subscript) i can be 1, 2, or 3; we write $i = 1, 2, 3$. If there are n individuals, we write $i = 1, 2, \ldots, n$.

A more efficient way of indicating a sum is to use the "summation sign," which is $\sum$, the capital Greek letter sigma. The sum of the x's (3.1) is then written

$$\sum_{i=1}^{5} x_i. \tag{3.4}$$

This expression is read "the sum of x_i from i equals 1 to i equals 5" or "the sum of the x_i's from x_1 to x_5." The notation $i = 1$ under $\sum$ says to begin the summation with x_1, and the 5 (short for $i = 5$) above $\sum$ says to end the summation with x_5. It is understood that the sum is over consecutive integer values of i.

The notations "$i = 1$" and "5" tell us *which* values are to be summed. For example, to indicate the sum $x_4 + x_5 + x_6$, we would write

$$\sum_{i=4}^{6} x_i.$$

This is read as the sum of the x_i's from x_4 to x_6. However, in applications in this book, we will always sum *all* of the observations, x_1 through x_n. Thus, we will simplify the notation and write (3.4) as

$$\sum x_i. \tag{3.4a}$$

If we use this form, then (3.2) is written

$$\bar{x} = \frac{\sum x_i}{5}.$$

The general formula for the mean of n observations is

$$\bar{x} = \frac{\sum x_i}{n}.$$

Other Uses of $\sum$

The summation notation may be used to indicate sums of terms other than just observed values x_i. For example, if each of the five observations above were multiplied by 2, the new values would be $2x_1 = 30$, $2x_2 = 24$, $2x_3 = 28$, $2x_4 = 34$, and $2x_5 = 38$. Then

$$\sum 2x_i = 30 + 24 + 28 + 34 + 38 = 154.$$

Likewise, if each of the x_i were reduced by 8 and the result squared, the values would be 49, 16, 36, 81, and 121. Their sum is

$$\sum (x_i - 8)^2 = 303.$$

We note some important principles about the operations used to obtain these results. First, remember that x_i represents *just one* data value at a time—not all of them simultaneously, and not their sum. Second, in evaluating sums that are functions of the x_i's, we follow the usual conventions regarding the order of mathematical operations, namely,

- Expressions inside parentheses are evaluated before all else;
- all powers and roots are evaluated next, before any other operations;
- all multiplications and divisions are evaluated next;
- additions and subtractions are evaluated last.

Thus, in evaluating $\sum 2x_i$, the multiplication of *each* x_i by 2 occurs before the terms are summed. In evaluating $\sum (x_i - 8)^2$ the subtraction of 8 from *each* x_i occurs first because it is contained in parentheses;

each $(x_i - 8)$ is squared next because powers precede addition; the resulting five values are summed last.

A note of caution should be observed by the student to whom this notation is new. The expressions $(\sum x_i)^2$ and $\sum x_i^2$ occur throughout statistical formulas, but indicate two distinct results. In the former, the x_i's are summed first because the summation operation is enclosed in parentheses; the total is squared. In the latter, *each* of the x_i's is squared first, because powers have precedence over sums; the squares are then added. That is,

$$\left(\sum x_i\right)^2 = (15 + 12 + 14 + 17 + 19)^2 = 77^2 = 5929$$

and

$$\sum x_i^2 = 15^2 + 12^2 + 14^2 + 17^2 + 19^2 = 1215.$$

Some rules that make working with summations more efficient are given in Appendix 3B.

Calculating the Mean from a Frequency Distribution

When a set of observations contains repeated values, as do the data in Table 3.8, computation of the mean can be simplified. Let x_i be the number of children in the ith family. Then the computation of the mean is represented by the formula

$$\bar{x} = \frac{\sum x_i}{10}.$$

We can instead compute $\bar{x}$ from the frequency distribution given in Table 3.9. The total number of children is

$$2 + (3 + 3 + 3 + 3 + 3) + (4 + 4 + 4) + 6 = 35.$$

Since there is 1 family with 2 children, 5 families with 3 children, 3 with 4 and 1 with 6, the same result can be obtained by multiplying $(1 \times 2) + (5 \times 3) + (3 \times 4) + (1 \times 6)$. The mean is this sum divided by $n = 10$.

Symbolic notation for these calculations is as follows: We see from the frequency distribution that among the *observations* $x_1, x_2, \ldots, x_{10}$ there are only four distinct *values*. Call these $v_1 = 2$, $v_2 = 3$, $v_3 = 4$, and $v_4 = 6$; the frequencies of these values are $f_1 = 1$, $f_2 = 5$, $f_3 = 3$, and $f_4 = 1$, respectively. Symbolically, then, the above sum is computed as

$$f_1 v_1 + f_2 v_2 + f_3 v_3 + f_4 v_4.$$

TABLE 3.9
Number of Children in 10 Families

j	Number of children v_j	Number of families having v_j number of children f_j
1	2	1
2	3	5
3	4	3
4	6	1
		$n = \overline{10}$

source: Table 3.8.

The mean is

$$\bar{x} = \frac{f_1 v_1 + f_2 v_2 + f_3 v_3 + f_4 v_4}{10}. \tag{3.5}$$

The general rule is that if there are m distinct values $v_1, v_2, \ldots, v_m$ with frequencies $f_1, f_2, \ldots, f_m$, respectively, then we can compute the mean as

$$\bar{x} = \frac{\sum f_j v_j}{n}. \tag{3.6}$$

Note that the number of observations is also equal to the sum of the frequencies $n = \sum f_j$. In these expressions, the subscript j is used instead of i to avoid giving the impression that there are $n = 10$ terms being summed in (3.5) or (3.6); there are only $m = 4$ terms, and j represents the values $1, 2, 3, 4$.

Comparing Distributions

In 1988 the United States Department of Education surveyed a random sample of 15,057 eighth grade students in public schools around the country. Among other questions, each student was asked "During the school year, how many hours a day do you USUALLY watch TV?" Separate answers were requested for weekdays and weekends. The response categories were "Don't watch TV," "Less than 1 hour a day," "1–2 hours," "2–3 hours," "3–4 hours," "4–5 hours," and "Over 5 hours a day." Although some accuracy is lost by using intervals and more is lost by the open-ended interval "over 5 hours," we will use the data to show how measures of location allow us to compare several distributions.

The results of the survey are given in Table 3.10 for students with and without a VCR at home. The mode for each column is the hours of TV viewing with the greatest frequency and greatest percentage of respondents, that is, the "most popular" amount of TV viewing. In this particular table all viewing times over 5 hours were combined into

TABLE 3.10

Hours of Television Viewing by Eighth-Grade Students

	Weekdays		Weekends	
Hours	*No VCR*	*VCR*	*No VCR*	*VCR*
0	99 (4.1%)	297 (2.4%)	132 (5.4%)	474 (3.8%)
0–1	167 (6.9%)	942 (7.5%)	153 (6.3%)	795 (6.3%)
1–2	469 (19.2%)	2790 (22.1%)	279 (11.5%)	1556 (12.3%)
2–3	517 (21.2%)	2999 (23.8%)	344 (14.1%)	2170 (17.2%)
3–4	434 (17.8%)	2366 (18.7%)	389 (16.0%)	2201 (17.4%)
4–5	329 (13.5%)	1543 (12.2%)	368 (15.1%)	2044 (16.2%)
>5	421 (17.3%)	1684 (13.3%)	771 (31.6%)	3381 (26.8%)
Total *n*	2436	12,621	2436	12,621
Mode	2–3	2–3	3–4	3–4
Mean	3.04	2.90	3.54	3.46

source: U.S. Department of Education (1989).

one interval. It may be reasonable to assume that the 771 and 3381 individuals who responded "over 5 hours" on weekends are distributed over several intervals, for example, 5–6 hours, 6–7 hours, and so on.[1] Any one of these intervals would probably have a lower percentage of respondents, and thus the mode is taken as 3–4 hours. The mean was obtained from the frequencies by (3.6) using the midpoints of the measurement intervals, that is, 0, 0.5, 1.5, . . . , 5.5. These summary statistics make it easy to see patterns in four complex distributions.

First, it is clear that eighth grade students watch more TV on weekends than on weekdays. The means for both no-VCR and VCR owners increase by at least one-half hour on weekends. The actual difference is probably greater than this because the many students who watch 5–6 hours, 6–7 hours, 7–8 hours, and so on, on weekends were all combined into a single interval with a midpoint of 5.5. Nevertheless, *individual students* may watch the same amount of television on weekends or even less than they do on weekdays.

The modal amount of TV viewing increases from 2–3 hours on weekdays to 3–4 hours on weekends. The modes are not clearly distinct, however. Among VCR owners, for example, 23.8% watch 2–3 hours of TV per weekday; another 22.1% watch 1–2 hours per day and yet another 18.7% watch 3–4 hours per weekday. That is, there is not a single amount of TV viewing time that is very popular in comparison to others. Likewise the percentage of students who watch 2–3 hours of TV on weekends is almost as great as those who watch 3–4 hours.

---

[1]The errors that may have occurred by combining all responses of "5 hours or more" into one interval are considered in Exercise 3.22.

Second, owning a VCR has a small or negligible effect on the amount of television viewing. The mean viewing time is slightly less among owners of VCRs both on weekdays and weekends, and the mode is no different. If the difference in means represents a real and replicable effect, it may be due to other circumstances such as youngsters from families that cannot afford a VCR tending to watch more television. Since the difference is so small, however, it may even be an artifact of the response categories that were used.

The Proportion as a Mean

Suppose that for a sample of observations a variable is coded 1 if a particular characteristic is present and 0 if it is absent. Any characteristic that can be scored as a simple dichotomy may be coded this way, such as "agree" or "disagree" with a political statement, "alive" or "dead," "pass" or "fail," product quality "acceptable" or "unacceptable," and so on. The sum of the 1's and 0's for the sample is simply the number of observations having the characteristic, and the mean is the *proportion* of the sample in which the characteristic is present.

Suppose that each of the 20 respondents in Table 2.1 is coded 1 if his/her occupational category is "professional" and 0 otherwise. Then $\sum x_i = 6$ is the sum of the coded values, and

$$\bar{x} = 6/20 = 0.30$$

is the proportion of professionals (i.e., 30%) in the sample. This equivalence allows some statistical procedures to be used both for means and proportions.

Other Properties of the Mean

The Mean and the Sum.   The mean is computed by $\bar{x} = \sum x_i/n$. Multiplying both sides of this expression by n gives

$$\sum x_i = n\bar{x}.$$

That is, the sum of the observed values is n times the mean.

This simple equality is a convenience in many applications. Suppose, for example, that an automobile company sold the numbers of cars listed in Table 3.11. To find the average dollar amount of a sale by the company, we need the company's total dollar income for the year. For Guzzlers alone, the total is the product of n (3000) and $\bar{x}$ ($7840), or $23,520,000. For Varooms, the total is 1500 × $12,100 = $18,000,000. For GV-GTs, the total is $8,650,000. Thus the company's total sales for

TABLE 3.11

Auto Sales and Prices in 1988

Model	No. of cars sold	Average selling price
Guzzler	3000	$7,840
Varoom	1500	$12,100
GV-GT	500	$17,300

Hypothetical data.

the year is $50,170,000. Dividing by 5000, the average sale amount is $10,034.

This result is a *weighted average* of the selling prices of the three models. The figure $10,034 was obtained by

$$\frac{3000 \times \$7{,}840 + 1500 \times \$12{,}100 + 500 \times \$17{,}300}{5000}$$

or, equivalently,

$$\frac{3000}{5000} \times \$7{,}840 + \frac{1500}{5000} \times \$12{,}100 + \frac{500}{5000} \times \$17{,}300$$

$$= 0.6 \times \$7{,}840 + 0.3 \times \$12{,}100 + 0.1 \times \$17{,}300.$$

The *weights* $(0.6, 0.3, 0.1)$ are the relative frequencies of the three car models. We could use this procedure, for example, to compare the average sales of different car companies or the same company from one year to another.

Deviations from the Mean.   The mean of the five observations in the first example in Section 3.3 is $\bar{x} = 15.4$. The deviation of a particular observed value from the mean, $x_i - \bar{x}$, is the distance between those two points if they are plotted on a horizontal axis like Figure 3.3. For these five observations, the deviations are

$$15 - 15.4 = -0.4,$$
$$12 - 15.4 = -3.4,$$
$$14 - 15.4 = -1.4,$$
$$17 - 15.4 = 1.6,$$
$$19 - 15.4 = 3.6.$$

The five "distances" sum to zero. This is a numerical counterpart to the idea of the mean as a center of gravity. It is an important property of the mean that there are exactly as many scale units above $\bar{x}$ as below $\bar{x}$ so that the deviations from the mean always sum to zero.

Algebraically, this property is represented $\sum(x_i - \bar{x}) = 0$. The reader will undoubtedly wish to prove this.

Effects of Change of Scale

Adding (or Subtracting) a Constant. The 10 families listed in Table 3.8 have an average of 3.5 children. If each of the families were to have one more child, what would be the effect on the mean? It is obvious that the average would also go up by 1 to 4.5. Computationally, the mean increased family size would be

$$(3 + 5 + 5 + 4 + 5 + 4 + 4 + 4 + 7 + 4)/10$$
$$= [(2 + 1) + (4 + 1) + \cdots + (3 + 1)]/10.$$

Adding 1 to each observation increases the total by 10 which, when divided by 10, adds 1 to the mean. This is a general rule:

RULE 1

Adding (or subtracting) the same constant to every observation increases (or decreases, respectively) the mean by that constant.

In symbols, if variable y is obtained from variable x by $y_i = x_i \pm a$, then the mean of y is $\bar{y} = \bar{x} \pm a$.

Multiplying by a Constant. Suppose that the income tax savings for every child is \$1000. What is the average savings for a family? Computationally, this is

$$(\$2000 + \$4000 + \$4000 + \$3000 + \$4000 + \$3000$$
$$+ \$3000 + \$3000 + \$6000 + \$3000)/10 = \$3500.$$

In this instance, each x_i has been multiplied by the same constant, \$1000. The result was exactly \$1000 times $\bar{x}$. As a general rule

RULE 2

Multiplying every observation by the same constant value multiplies the mean by that constant.

Of course the constant may be positive or negative, and whole numbers or fractions. In symbols, if variable y is obtained from variable x by $y_i = ax_i$, then the mean of y is $\bar{y} = a\bar{x}$.

These properties are proven algebraically in Appendix 3C.

3.4   Choosing Among Measures of Location

A data analyst may choose to use one or more measures of location to summarize a particular data set. One factor to be considered is the measurement scale. The mode is the only measure of location useful for nominal scales but may also be used with ordinal, interval, or ratio data. The median of a sample is the value of the middle observation once the observations have been ordered. The median requires only that the data have ordinal properties (that is, an ordinal scale) but is also a useful measure of location for interval or ratio data. The mean, because it involves arithmetic calculations such as additions and division, is appropriate only for numerical variables, that is, variables on interval or ratio scales.

A second consideration is the nature of the information conveyed by the location measure. The mode identifies the most common value(s). It may be a good measure of location if one value or several adjacent values occur in the sample much more often than others. At the same time, the mode conveys the least amount of information about the distribution as a whole. In some samples the mode may be in the middle of the distribution but in other samples it may be a value at one end of the distribution or the other.

The median indicates the center of the distribution. Equal numbers of observations lie below and above the median regardless of *how far* above or how far below the median the actual values may be. Thus the median has the advantage that it is unlikely to be affected very much by outliers at one end of the distribution or the other. The mean is the only location measure that summarizes the actual values of all the observations. This feature also makes it the most sensitive to extreme values. In some distributions the mean may be very different from the median. (See Exercise 3.14.)

A third consideration in choosing a measure of location is the shape of the distribution; this is discussed separately below. Of all considerations, however, the most important is the research question, that is *What do we need to know from the data*? The following example illustrates the significance of this question.

Telephone Delay.   When you dial a phone number, there is a short delay before you hear either a ring or a busy signal. This delay varies from call to call, that is, it has a distribution. The shape of this distribution depends upon the pattern of telephone traffic through the exchange. It also depends upon the design of the equipment used at the exchange— different systems give different waiting-time distributions. Because it is annoying if the delay is more than momentary, telephone company engineers have paid some attention to evaluating systems on the basis of their associated waiting-time distributions.

Figures 3.5 and 3.6 represent waiting-time distributions under two alternative telephone systems. Statistics summarizing the two distributions are given in Table 3.12. Which statistics are most appropriate for comparing these distributions?

If human nature is such that a silence of one or one-and-a-half seconds is not noticeable, but a longer wait is noticeable and maybe even

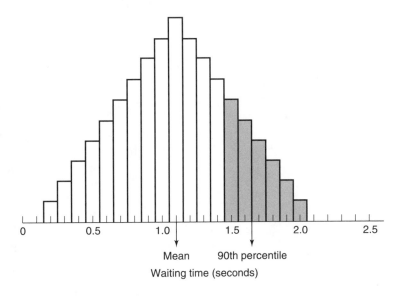

Figure 3.5   *Waiting-time distribution (18% of the calls wait more than 1.45 seconds).*

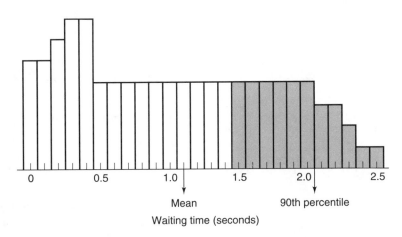

Figure 3.6   *Waiting-time distribution (32% of the calls wait more than 1.45 seconds).*

TABLE 3.12

Summary of Waiting-Time Distributions

	System 1 (Figure 3.5)	System 2 (Figure 3.6)
Proportion of calls delayed more than 1.45 seconds	18%	32%
90th percentile	1.65 seconds	2.05 seconds
Mean	1.10 seconds	1.10 seconds
Median	1.10 seconds	1.05 seconds

source: Figures 3.5 and 3.6. (hypothetical data)

annoying, then it might be appropriate to evaluate a telephone system on the basis of *how many callers* have to wait more than 1.5 seconds—that is, on the basis of what proportion of the waiting-time distribution lies to the right of 1.5 seconds.

A similar idea is to compare the 90th percentiles of the waiting-time distributions, focusing attention on the 10% of calls that are delayed the longest.

Means of the distributions are not an adequate basis for comparison here because a distribution could have a relatively small mean, even though many customers might have to wait a long time. The distributions shown in Figures 3.5 and 3.6 have the same mean, 1.10 seconds, but the 90th percentile is 1.65 seconds in Figure 3.5, while it is 2.05 seconds in Figure 3.6. Obviously the system in which 10% of the calls were delayed more than 1.65 seconds is more satisfactory than the one in which 10% were delayed more than 2.05 seconds. Comparison based on the proportion of calls delayed more than 1.45 seconds leads to the same conclusion. The distribution in Figure 3.6 is less satisfactory because 32% of the calls wait more than 1.45 seconds, compared to only 18% in Figure 3.5.

On the other hand, if the *total delay* experienced by *all callers* combined is felt to be of more importance, it would be appropriate to compare the means of the distributions because

total delay of all calls = mean delay × number of calls.

Shape of Distributions

When the practical problem at hand does not dictate the choice of measure of location, the shape of the distributions may be of some help. In particular, if a distribution is *bimodal* or *skewed*, the mean may not be the best indicator of location.

Bimodality. If a firm employs two kinds of workers, 100 at a salary of $6000 and 100 at a salary or $12,000 per year no one earns the mean salary of $9000 per year. When the distribution is bimodal, as is the case with these payroll data and with the distribution of number of children in Table 3.4, no one measure of location conveys useful information. Both modes should be stated. The median, on the other hand, at least tells us at what point the set of observations is divided in half.

Skewness. Another situation in which the mean may not be descriptive is when the distribution is *skewed*; that is, when there are a few very high or a few very low values. A distribution of income, for example, may have a few very high incomes which can cause the mean to be much too large to be called "typical" or "central." Consider the five incomes $12,000, $14,000, $15,000, $17,000, and $1,000,000. The mean, $211,600, is not particularly close to any one of the incomes.

The median of these five incomes is $15,000. This figure is fairly close to most (4 out of 5) of the incomes. In general the median is a better measure of location than the mean when the distribution is very skewed. The shape of a skewed distribution is shown in Figure 3.7, in which the location of the mean, median, and mode are indicated. A smooth curve is intended to be an idealization of a histogram with very small class intervals. (Alternatively, the reader can think of the smooth curve as an artist's rendering.)

The distribution in Figure 3.7 is said to be *skewed to the right* or *positively skewed* because the extreme values are very high, creating the "tail" on the right side of the curve. Since the high value(s) affect the mean more than the median, it is characteristic of a positively skewed distribution that $\bar{x} >$ median. If the skewness were caused by several

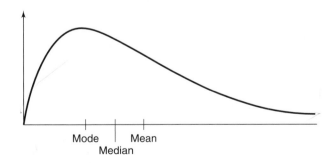

Figure 3.7 *Right-skewed distribution.*

extremely low values, the curve would be the mirror image of Figure 3.7, and the distribution *skewed to the left* or *negatively skewed.* In this case, $\bar{x}$ < median and, again, the median is a more representative central value.

Symmetric Distributions.   A symmetric distribution is a distribution that looks the same in a mirror. If it is drawn on translucent paper, the distribution or histogram will look the same from the back. Two symmetric distributions are sketched in Figures 3.8 and 3.9. The mean and median are equal for symmetric distributions. For symmetric *unimodal* distributions the mean, median, and mode all coincide.

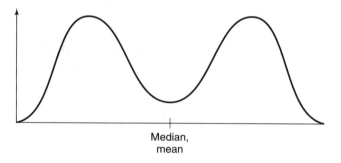

Figure 3.8   *Symmetric bimodal distribution.*

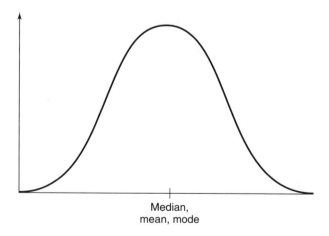

Figure 3.9   *Symmetric unimodal distribution.*

Summary

The *mode* of a distribution of observations is the value with the highest frequency and is valid for data defined on any scale, nominal, ordinal, interval, or ratio; continuous data must be grouped before a mode can be computed. *Bimodal* distributions can arise by mixing two unimodal distributions.

The *median* and other percentiles are valid for data defined on an ordinal, interval, or ratio scale. The median divides the ordered sample in half. *Quartiles* divide the ordered sample into quarters; the median is the second quartile. The median is a good measure of location for distributions that are skewed.

The *mean* (arithmetic average) is valid for data defined on a numerical (interval or ratio) scale and is the only measure of the three that summarizes every unit in every observed value.

Appendices

Appendix 3A   Computing the Median and Other Quantiles of Grouped Continuous Data

When there are many observations we may find it convenient to group data instead of working with the raw data. Consider the frequency distribution of profits of nine businesses (in 1000's of dollars), shown in Table 3.13. (Since there are only 9 observations here, it would be easy to handle the original data. We purposely chose a small example to

TABLE 3.13
Profits

Profit (1000s of dollars)	Frequency	Cumulative frequency
200–250	1	1
250–300	2	3
300–350	5	8
350–400	1	9
	9	

Hypothetical data.

prevent the graphs below from being unwieldy.) The median profit is that of the firm which ranks 5th. The median therefore lies in the class interval 300–350. Within this interval lie the profits of 5 firms. Since we do not know the individual values for the 5 firms in this class, we proceed on the basis that they are spread uniformly over the interval (Figure 3.10). We allot to each of these 5 firms 1/5 of the interval width of 50, which is 10, and we assume that each of the firms falls at the midpoint of its interval. This gives a well-defined median, in this case 315.

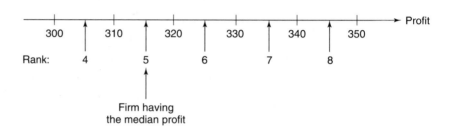

Figure 3.10   *Notion of the median of grouped data (Table 3.13).*

If we make a graph showing the whole range of profits (Figure 3.11), we see that, in terms of the numbers 0 to 9 (on top of the axis), we can define the median as the profit which would correspond to the number $9/2 = 4.5$. It is this reasoning which gives rise to the formula for the median of grouped data as

$$l + \frac{\frac{n}{2} - b}{f}w, \tag{3A.1}$$

where

n = number of observations,

l = lower limit of interval containing the median,

f = frequency of interval containing the median,

b = sum of frequencies of intervals below that containing the median,

w = class interval width.

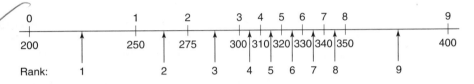

Figure 3.11 *Profits and their ranks (Table 3.11).*

Applying (3A.1) to the distribution of profits, we approximate the median as

$$300 + \frac{\frac{9}{2} - 3}{5} \times 50 = 300 + \frac{1.5}{5} \times 50$$

$$= 300 + 15$$

$$= 315.$$

Other quantiles are computed analogously. For example, the first quartile, third quartile, and ninth decile may be computed by replacing $n/2$ in (3A.1) by $n/4$, $3n/4$, and $9n/10$, respectively.

Appendix 3B   Rules for Summation

The properties of the $\sum$ operator listed below can make certain summations more efficient, and also allow us to work algebraically with more complex expressions containing $\sum$.

RULE 1

> The sum of a single quantity a given number of times is the product of the quantity and the number of times. In symbols
>
> $$\sum a = na.$$

ILLUSTRATION

As an example, suppose that every one of 6 students contributes $2.50 for a present for their professor. The total contribution may be found by

$$\sum \$2.50 = \$2.50 + \$2.50 + \$2.50 + \$2.50 + \$2.50 + \$2.50$$

$$= \$15.00$$

Or, alternatively, $2.50 may be multiplied by 6 to get the same result. Here a is a *constant* that doesn't vary from one observation to another and so a subscript is not needed. It is necessary to define n explicitly as the number of observations, since it enters into the formula na.

RULE 2

The sum of a set of observations, each of which has been multiplied by the same number, is the sum of the original observations multiplied by that number. In symbols

$$\sum cx_i = c \sum x_i .$$

ILLUSTRATION

As an example, suppose that we purchase four items in a department store located in a community where the sales tax is 8%. The item prices are $x_1 = \$6.00$, $x_2 = \$8.50$, $x_3 = \$20.00$, and $x_4 = \$2.00$, with $\sum x_i = \$36.50$. The prices of the items with tax included are

$$1.08x_1 = \$6.48, \quad 1.08x_2 = \$9.18, \quad 1.08x_3 = \$21.60, \quad 1.08x_4 = \$2.16.$$

The total bill is $\sum 1.08x_i = \$39.42$. Or, alternatively (as the sales clerk and cash register well know) $\sum x_i = \$36.50$ may be computed first and only multiplied by 1.08 once to get the same total bill.

RULE 3

Given a set of pairs of observations, adding the two members of each pair and then adding these sums produces the same result as adding the first members of the pairs, adding the second members of the pairs, and adding these two sums. In symbols

$$\sum (x_i + y_i) = \sum x_i + \sum y_i .$$

ILLUSTRATION

As an example, suppose that variable x is years of employment with company Q and 4 employees have $x_1 = 7$, $x_2 = 6$, $x_3 = 2$, and $x_4 = 9$. Then the total years with the company is $\sum x_1 = 24$. Assume that variable y is years of employment prior to working for company Q, and the same 4 individuals have $y_1 = 2$, $y_2 = 8$, $y_3 = 0$, and $y_4 = 1$. The total prior employment is $\sum y_1 = 11$ years.

   Rule 3 indicates that the total years worked altogether, 35, can be obtained in either of two ways. The total can be obtained for each employee

$$x_1 + y_1 = 9, \quad x_2 + y_2 = 14, \quad x_3 + y_3 = 2, \quad x_4 + y_4 = 10.$$

Summing these, $\sum(x_i + y_i) = 35$. Or, alternatively, $\sum x_i$ and $\sum y_i$ may be obtained as in the preceding paragraph, and added to get the same result.

RULE 4

> The sum of a set of observations, to each of which has been added the same quantity, is the sum of the original observations plus the product of the quantity and the number of summands. In symbols
>
> $$\sum(x_i + a) = \sum x_i + na.$$

ILLUSTRATION

As an example, suppose that each company Q employee worked for the firm for three additional years. The number of years worked for the firm can be obtained for each employee

$$x_1 + 3 = 10, \quad x_2 + 3 = 9, \quad x_3 + 3 = 5, \quad x_4 + 3 = 12.$$

Summing these, $\sum(x_i + 3) = 36$. Or, alternatively, $\sum x_i = 24$ may be obtained, and $\sum 3 = 4 \times 3 = 12$ obtained separately, and added to get the same result.

Discussion.   Rules 3 and 4 illustrate a general *distributive* principle for $\sum$ with regard to terms separated by addition (+) or subtraction (−). For example, if variable x is years of employment with company Q, variable y years of prior employment, variable z years of sick or vacation leave, and a an additional number of years of work anticipated for every employee, then total years of on-duty employment is

$$\sum(x_i + y_i - z_i + a) = \sum x_i + \sum y_i - \sum z_i + na.$$

If the terms in the expression are not combined by addition or subtraction, however, the distributive principle cannot be applied directly. For example, $\sum(x_i - 3)^2$ is *not* the same as any simple expression involving only $\sum x_i$ and $\sum 3$ because of the exponent. In this instance we may rewrite the expression in a form that has only sums and differences. Recall that, for two values a and b,

$$(a + b)^2 = a^2 + 2ab + b^2$$

and

$$(a - b)^2 = a^2 - 2ab + b^2.$$

Applying the latter, $(x_i - 3)^2 = x_i^2 - 6x_i + 9$. Now the $\sum$ may be distributed

$$\sum(x_i^2 - 6x_i + 9) = \sum x_i^2 - 6\sum x_i + 9n.$$

The middle and last terms were obtained using rules 2 and 1, respectively.

Appendix 3C   Change of Scale

The effects on the mean of adding a constant to every value of a variable and multiplying every value by a constant are summarized in two rules in Section 3.3. Algebraic proofs of the rules are given below.

RULE 1

If a constant a is added to every value of variable x, then the mean is increased by a. In symbols, if $y_i = x_i + a$, then $\bar{y} = \bar{x} + a$.

Proof

The mean of the resulting variable is $\bar{y} = \sum y_i/n = \sum(x_i + a)/n$. Applying the distributive rule to the numerator,

$$\frac{\sum(x_i + a)}{n} = \frac{\sum x_i + na}{n}.$$

Simplifying, this is

$$\frac{\sum x_i}{n} + a = \bar{x} + a.$$

Comment: If the constant is negative then the mean is decreased by the corresponding amount.

RULE 2

If every value of variable x is multiplied by a constant a, then the mean is multiplied by a. In symbols, if $y_i = ax_i$, then $\bar{y} = a\bar{x}$.

Proof

The mean of the resulting variable is $\bar{y} = \sum y_i/n = \sum ax_i/n$. Applying summation Rule 2, this is

$$\frac{\sum ax_i}{n} = \frac{a\sum x_i}{n} = a\frac{\sum x_i}{n} = a\bar{x}.$$

Appendix 3D Significant Digits

A measurement is usually recorded as accurately as it is measured. Thus, if a ruler can be read to the nearest tenth of a centimeter, one would record a reading to tenths of a centimeter, for example, 3.7 centimeters. Equivalently, this is 37 millimeters and .037 meter. In each case we say there are two *significant digits* (or two *significant figures*) in the number. That means that the measurement is considered valid to 1 part in 100. In .037 the 0 is for the purpose of indicting the decimal place and is not counted as a significant digit. If a measurement is recorded as 4.0 centimeters, the 0 after the decimal point indicates the reading is accurate to 0.1 centimeter; in this case, also, the number of significant digits is two.

The *significant digits* of a number include (a) all the nonzero digits, (b) the zeros that are included between nonzero digits, and (c) zeros that are on the right to signify accuracy.

In reporting the final results of a calculation, one should give only the *accurate significant digits*. If 1.2 and 1.23 are two recorded measurements, then their product, $1.2 \times 1.23 = 1.476$, should be rounded to 1.5, since the factor 1.2 contains only two significant digits.

On the other hand, the *mean* of a set of numbers is generally more precise than any one of the numbers alone. (This is explained in Chapter 9.) Hence it is reasonable to report the mean of 90, 100, 80, 72, and 109, as 90.2 rather than 90.

Exercises

3.1 The 25 families of Little Town live on 4 blocks:

Block	A	B	C	D
Number of persons living on the block	20	50	20	10

(a) Compute the mean number of persons (arithmetic average) per block.

(b) Compute the mean number of persons (arithmetic average) per family.

(c) Compute the mean number of families (arithmetic average) per block.

3.2 In the "postural sway" test a blindfolded subject standing erect is told he or she is moving forward. The actual sway in reaction to this "information" is measured. These measurements, in inches, and rearranged in order, are as follows (a minus sign indicating a backward movement): −2.0, −1.5, −1.4, −1.0, −0.75, −0.60, −0.40, −0.35, −0.20, −0.05, −0.05, +0.10, +0.15, +0.15, +0.20, +0.30, +0.40, +0.40, +0.75, +1.00, +1.50, +2.00, +3.00, +3.25, +4.5, +5.5, +6.0, +8.2.

(a) Make a histogram of these measurements.

(b) Compute the mean, median, and mode.

(c) What proportion of the subjects moved backwards?

3.3   There are 30 students in each of the four sections of Statistic I at Stanu College. The sections are taught by different teachers. In order to learn the students' names (and to help ensure that the students come to class), the teachers call the roll for the first six class periods. For the second through sixth class periods each teacher records the number of names not yet learned. The data are given in Table 3.14.

TABLE 3.14
Number of Names Not Yet Learned

		Teacher		
Trial	*A*	*B*	*C*	*D*
1	21	18	25	20
2	18	16	15	10
3	16	15	12	9
4	12	13	10	7
5	9	10	8	6

Hypothetical data.

(a) Plot the data using the horizontal axis for trials and the vertical axis for number of names not yet learned. Use *A*'s to represent the points for Teacher A, *B*'s for Teacher B, etc. Connect the five points for each teacher.

(b) Compute the mean for each trial.

(c) Plot the means against the trial numbers. Connect the five points.

3.4   Compute the mean, first decile, first quartile, and median for the weights of the Stanford football players given in Table 2.15. What positions do you think those in the lower tenth would be most likely to play?

3.5   The average monthly snowfall for November through May in a hypothetical city in the northeast is 2.2 inches, 16.1 inches, 18.6 inches, 13.1 inches, 9.3 inches, 2.3 inches and 1.1 inch.

(a) Compute the mean and median amount of snowfall for these seven months. Is the distribution skewed to the right or to the left?

(b) Assume that no snow falls in the other five months. Add five zeros to the distribution, and recalculate the mean and median. Is the distribution skewed to the right or to the left?

(c) Which summary statistics—(a) or (b)—gives a clearer picture of the *amount of snow* that falls in this city? Which gives a clearer picture of the *typical* weather in this city?

3.6 Find the median of the distribution of incomes of families shown in Table 3.6. Use the method of interpolation described in Appendix 3A.

3.7 Find the median of the distribution of weights of the 25 dieters in Table 2.10. Recalculate the median from the frequency distribution given in Table 2.11, using the method of interpolation described in Appendix 3A. How similar are the two values? Would they be more similar or less similar if the frequency distribution had only four intervals?

3.8 Ten individuals participated in a study on the effectiveness of two sedatives, A and B. Each individual was given A on some nights and B on other nights. The average number of hours slept after taking the first sedative is compared with the normal amount of sleep; a similar comparison is made with the second drug. Table 3.15 gives the increase in sleep due to each sedative for each individual. (A negative value indicates a decrease in sleep.)

TABLE 3.15
Number of Hours Increase in Sleep Due to Use of Two Different Drugs, for Each of 10 Patients

Patient	Drug A	Drug B
1	1.9	0.7
2	0.8	1.6
3	1.1	−0.2
4	0.1	−1.2
5	−0.1	−0.1
6	4.4	3.4
7	5.5	3.7
8	1.6	0.8
9	4.6	0.0
10	3.4	2.0

source: Student (1908). Printed with permission of the Biometrika Trustees.

(a) Compute the mean increase for drug A and the mean increase for drug B.

(b) For each individual, compute the difference (increase for drug A minus increase for drug B).

(c) Compute the mean of these differences.

(d) Verify that the mean of the differences is equal to the difference between the means.

3.9   Compute the mean and median Scaled Agility Score, separately for males and females using the data in Table 2.22. Compare the performance of the two groups.

3.10   For each of the following hypothetical data sets state whether the mean, median, or mode is the most descriptive measure of location, give its numerical value, and tell why you chose that measure.

(a) Numbers of children in 12 families:

$$3, 2, 2, 2, 3, 2, 1, 2, 2, 2, 2, 2.$$

(b) Incomes of 10 families (dollars per year):

$$8400, 8300, 8600, 7400, 7300, 9700, 8100, 17100, 9100, 9300.$$

(c) Heights of 10 boys (inches):

$$39, 40, 39, 37, 38, 39, 38, 40, 41, 40.$$

(d) Lifetimes of 10 lightbulbs (hours):

$$150, 110, 441, 2100, 1503, 1305, 257, 279, 215, 2536.$$

(Plot the data to judge the amount of skewness.)

(e) Number of cars owned by each of 10 families in the city:

$$1, 1, 2, 1, 1, 1, 1, 1, 2, 1.$$

(f) Number of cars owned by each of 12 families in the suburbs:

$$2, 2, 1, 3, 2, 2, 3, 1, 3, 2, 2, 3.$$

3.11   (From the *San Francisco Gazette*) "The 'average' Bay Area resident is a married, 28-year-old white woman, living with two others in a family-owned home worth $196,200."

What kind of "averages" are referred to in this statement?

3.12   In the United States, although the median duration of marriages ending in divorce is about 7 years, most divorces occur in the first 5 years of marriage or between the 20th and 25th years of marriage.

Sketch a histogram that is consistent with these facts. Can you think of a hypothesis which would explain this bimodal distribution?

3.13   A sample of 25 families living in Amherst Hills owns the number of automobiles given below:

Number of cars	0	1	2	3	4
Number of families having this number of cars	0	2	6	11	6

(a) What is the median number of cars per family?

(b) What is the modal number of cars per family?

(c) What is the mean number of cars per family?

3.14   In 1995 the popular press reported that the mean salary of professional baseball players was approximately $1,100,000 while the median salary was about $500,000. What do these figures tell you about

(a) The skewness of the distribution of salaries?

(b) The salaries of *most* professional baseball players?

(c) The salaries of the lowest-paid and highest paid players in comparison to the mean? In comparison to the median?

3.15   The numbers of books borrowed from the university library during the academic year by a random sample of seniors at two schools are given in Table 3.16.

TABLE 3.16

Numbers of Books Borrowed from Two University Libraries

Number of books	BU	SU
5	1	0
7	2	1
8	12	2
9	6	5
10	17	8
11	14	35
12	35	30
14	8	10
16	3	6
18	2	3

Hypothetical data.

(a) Calculate the mean number of books borrowed by seniors at each university.

(b) Calculate the median number of books borrowed by seniors at each university.

(continued . . .)

(c) Calculate the proportion of seniors at each university who borrowed at least 10 books; who borrowed at least 15 books.

(d) Compare the samples of students at the two universities in terms of these measures.

(e) What other comparisons of the two distributions would be informative?

3.16   In a study of television viewing habits in a community in which 20% of the residents are college graduates, the average number of hours per week which residents spend watching television is to be estimated. Fifty families are sampled. The average number of hours per week for the 30 households whose respondent is a college graduate is 15 hours. For the other 20 families in the sample, the average is 25 hours per week.

(a) Does the sample of 50 seem representative of the population?

(b) Estimate the average number of hours spent watching television by residents in the community. (Hint: Take a weighted average using the weights 0.20 and 0.80.)

3.17   For the data in Exercise 2.14 (Table 2.21) relating to students in a course in statistics, compute the mean final grade of

(a) students who had taken at least one course on probability and statistics,

(b) students who had no previous course on probability and statistics,

(c) all the students.

3.18   Compute the answer to Exercise 3.17(c) as the appropriately weighted average of the answers to Exercise 3.17(a) and (b).

3.19   The mean of variable x for a sample of 200 observations is $\bar{x} = 17.4$. What will the mean be if every observed value is

(a) increased by 2.6?

(b) decreased by 17.4?

(c) multiplied by 2?

(d) multiplied by -0.3?

3.20   The mean of a variable for a large sample of observations (x_i's) is $\bar{x} = 98.6$. If each x_i is modified by computing

$$5(x_i - 32)/9$$

what is the mean of the resulting modified variable? Can you identify a familiar variable that is expressed on two scales that are related by this "transformation"?

3.21   Exercise 3.5 gives the monthly snowfall in a city in the northeastern United States; the values may be viewed as "expected" amounts for the current year. For every inch of snow that falls, it is estimated that heating bills will increase by $8.00 over the base amount of $26.00 per month. What is the "expected" average monthly heating bill for residences in this city? (Don't forget that the average monthly bill is for 12 months, 5 of which have no snowfall.)

3.22   Using the data given in Table 3.10.

(a) Find the percentage of families that own a VCR.

(b) Find the median TV viewing time for each column of data. What conclusions may be drawn about weekdays and weekends and about the effect of a VCR on the median viewing time? Support your conclusions.

(c) Consider the possible effects of recording all amounts over 5 hours per day in a single interval. How might this have affected the three measures of location and the weekday-weekend and VCR–no VCR differences?

(d) What alternative methods of collecting the data might you suggest?

Exercises Using Summation Notation

3.23   Consider a business firm's monthly expenditures over a period of one calendar year. Let x_i represent the amount expended in the ith month of the year ($i = 1, 2, \ldots, 12$). Using this notation and the summation operator, give

(a) the notation for expenditures for March alone,

(b) two notations for December's expenditures,

(c) the notation for the total expenditure for the calendar year.

3.24   Let u_i be the amount expended for salaries in the ith month and v_i be the amount expended for office supplies in the ith month. The year's expenditure for both salaries and office supplies may then be computed either by adding for each month the two kinds of expenditures and then adding these 12 monthly totals, or by finding the year's expenditure for salaries and the year's expenditures for office supplies and then adding these two totals. State this fact symbolically using the u_i's and the v_i's.

3.25 Let

$$
\begin{array}{lll}
x_1 = 10 & y_1 = 100 & z_1 = 1 \\
x_2 = 9 & y_2 = 81 & z_2 = -1 \\
x_3 = 8 & y_3 = 64 & z_3 = 1 \\
x_4 = 7 & y_4 = 49 & z_4 = -1 \\
x_5 = 6 & y_5 = 36 & z_5 = 1 \\
x_6 = 5 & y_6 = 25 & z_6 = -1 \\
x_7 = 4 & y_7 = 16 & z_7 = 1 \\
x_8 = 3 & y_8 = 9 & z_8 = -1 \\
x_9 = 2 & y_9 = 4 & z_9 = 1 \\
x_{10} = 1 & y_{10} = 1 & z_{10} = -1
\end{array}
\qquad (n = 10).
$$

Compute each of the following expressions:

(a) $\sum x_i; \ \sum y_i; \ \sum z_i$,

(b) $\sum x_i^2; \ \sum z_i^2$,

(c) $\frac{1}{2}(\sum x_i)^2; \ (\sum y_i)^2$,

(d) $(\sum \frac{1}{2}x_i)^2; \ \sum \frac{1}{2}z_i^2$,

(e) $\sum(x_i + 1); \ \sum(3y_i - 3); \ \sum(1 - z_i)$,

(f) $\sum 4.4; \ \sum 1$,

(g) $\sum(x_i - 5.5)^2; \ \sum(z_i + 2)^2; \ \sum(z_i^2 + 2)$,

(h) $\sum x_i^2 - 3\sum x_i; \ \sum y_i^2 - (\sum y_i)^2$,

(i) $\sum \frac{1}{n}x_i; \ \sum x_i/n$,

(j) $\sum ix_i$.

3.26 Use the data from Exercise 3.25 to compute the following:

(a) $\sum x_i - \sum y_i; \ \sum(x_i - y_i)$,

(b) $\sum(x_i - y_i)^2; \ [\sum(x_i - y_i)]^2; \ \sum x_i^2 - \sum y_i^2$,

(c) $\sum x_i y_i; \ \sum x_i z_i$,

(d) $\sum x_i \sum y_i; \ \sum x_i \sum z_i$,

(e) $\sum x_i z_i - \sum x_i \sum z_i/n$.

3.27 Use the rules in Appendix 3B to prove algebraically (without substituting numbers for the x_i's) that

$$
\sum \frac{1}{n}x_i - \frac{1}{n}\left(\sum x_i - n\right) = 1.
$$

3.28 Use the rules in Appendix 3B to prove algebraically (without substituting numbers for the x_i's) that

$$
\sum(x_i - \bar{x})^2 = \sum x_i^2 - \left(\sum x_i\right)^2 / n.
$$

3.29 Using the x-values from Exercise 3.25, evaluate *both sides* of the expression in Exercise 3.28 numerically.

4

Measures of Variability

Introduction

Although for some purposes an average may be a sufficient description
of a set of data, usually more information about the data is needed. An
important feature of statistical data is their *variability*—how much the
measurements differ from individual to individual. In this chapter we
discuss the numerical evaluation of variability. A synonym for variability
is *dispersion*, and other terms are sometimes used for the same concept
including "spread" or "scatter."

Unlike the average which is easily understood by most of us, the idea
of dispersion may be a little more elusive. There are three reasons why
the concept plays a central role in statistical analysis.

First, variability may be a "key" feature of a particular data set, and
a measure of dispersion an important descriptive statistic. To give a
simple example, Table 4.1 lists the numbers of minutes that 20 patients
waited to be seen by their physicians. Of these, 10 patients waited for
Doctor A and 10 for Doctor F. Both lists have been ordered from the
shortest waiting time to the longest to make patterns more visible. By
coincidence the average waiting time for the 2 doctors is identical (19.6
minutes), but their behavior is quite different.

TABLE 4.1

Waiting Times of 20 Patients, in Minutes

Doctor A		Doctor F	
7	18	16	20
10	18	17	20
13	19	19	21
16	31	20	21
18	46	20	22

Hypothetical data.

Doctor A sees some patients almost immediately but makes others wait as much as a half or three-quarters of an hour; she is highly inconsistent from one patient to another. Doctor F makes all the patients wait at least a quarter of an hour but at the other extreme none waits more than 20 or 22 minutes; she is highly consistent. Doctor A has greater *variability* than Doctor F and this may be a factor in deciding which doctor we prefer. If we needed to see a doctor during our lunch hour we could probably bring a sandwich, take 15 minutes to eat it, be seen by Doctor F, and return to work with time to spare.

Consistency is especially important when salable merchandise is produced or packaged by weight, volume, or count. If the production of 12-ounce packages of cookies is automated, it is best—both for the producer and the buyer—that the weight of *each* package is very close to the advertised 12 ounces. If the packaging equipment is producing packages that weigh 10 ounces, 14 ounces, 13 ounces, 9 ounces, and so on, we would probably invest in alternate equipment that produced packages of 11.6 ounces, 12.1 ounces, 11.9 ounces, 12.2 ounces, etc. These are more homogeneous package weights—they have less dispersion.

Second, dispersion is a concept that is basic to many statistical methods. As we shall see in later chapters, equal dispersions of the data in several groups is an important condition to assure the validity of certain statistical procedures. Even for one sample, the variability of the data indicates how good a measure of location is as a summary statistic. For example, three heights 71, 72, and 73 inches and three heights 66, 72, and 78 inches have the same mean (72 inches). The mean is a better summary of the first set of heights because the values are closer together and closer to the *representative* value, 72. In statistical work dispersion is often presented in the form of an index that reveals how closely the data are clustered around a measure of location; this is discussed in detail in later chapters.

Third, measures of dispersion are useful for assessing whether two observed values are close together or far apart, or whether a single value is very high, very low, or close to average. For example, if a measure of variability tells us that, in general, people differ from one another by about 3 points on a particular exam, then two people whose scores differ by 10 points must represent extremely different levels of performance. Here, the measure of dispersion is the typical or average difference between pairs of scores (i.e., 3 points) and it is used as a sort of "ruler" to draw conclusions about two particular individuals. As we shall see, this line of reasoning can be formulated into alternative scoring procedures that indicate the relative position of an individual in the distribution, that is, whether an observation is close to average, slightly above or below average, or very much above or below average.

4.1 Ranges

The Range

A student can get an idea of the variability of the heights of men in the class by comparing the tallest man with the shortest, for if the difference is small the heights do not vary much. The difference between the largest measurement in a set (the *maximum*) and the smallest (the *minimum*) is called the *range*.

For example the range of waiting times for Doctor A in Table 4.1 is $46 - 7 = 39$ minutes, and for Doctor F is only 6 minutes. The simplicity of the range makes the concept of dispersion very clear—it indicates how "spread out" the values are. Unfortunately, the range also has a shortcoming that causes us to turn to other indicators of variability instead. The reader should keep in mind, however, that these are just other ways to get at the same simple idea of spread.

The basic shortcoming of the range is that it is entirely dependent on the two extreme values. For example, suppose that all of the males in a large college class are between 69 and 71 inches tall except for one child prodigy and one basketball player. The range would reflect the difference between these two exceptional individuals instead of the typical differences of one or two inches. If the numbers of children in 8 families are 0, 1, 1, 2, 2, 3, 3, and 9, the range is 9, but all of the families except two have 1, 2, or 3 children; again the variability is small though the range is large.

A second disadvantage of using the range to characterize spread is that, because it depends directly on just the largest and smallest measurements, it depends indirectly on the *number* of measurements in the set. For instance, if two families have 1 and 4 children, respectively, the range is $4 - 1 = 3$; if we add a third family, the range continues to be 3 if the number of children in this family is 1, 2, 3, or 4, and it is larger if the number of children is 0 or greater than 4. The point is that the range may increase with the size of the set of observations but it can never decrease, while ideally a measure of variability should be roughly independent of the number of measurements.

The Interquartile Range

A measure of variability that is not sensitive to the sample size and not affected by extremes is the *interquartile range*, which is the difference between the first and third quartiles. For the numbers of children in 8 families, 0, 1, 1, 2, 2, 3, 3, 9 (shown in Figure 4.1), the first quartile is 1 and the third quartile is 3. The interquartile range is $3 - 1 = 2$.

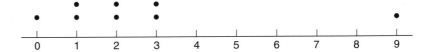

Figure 4.1   *Numbers of children in 8 families: Small variability (range is 9; interquartile range is 2).*

Another set of eight observations with the same range is 0, 1, 1, 2, 3, 6, 6, 9. (See Figure 4.2.) Most of us would feel that this set exhibits more variability than the first set. The range does not reflect this fact, but the interquartile range does. The interquartile range of this new set is $6 - 1 = 5$, instead of 2.

Figure 4.2   *Numbers of children in 8 families: Large variability (range is 9; interquartile range is 5.)*

The quartiles are measures of location, that is, they are positional indicators. The interquartile range may thus be called a *positional measure of variability.*

We have remarked that the range cannot decrease, but can increase, when additional observations are included in the set and that in this sense the range is overly sensitive to the number of observations. The interquartile range does not share this deficiency; it can either increase or decrease when further observations are added to the sample. In the set of observations 1, 2, 3, 4, 5, the range is $5 - 1 = 4$, the first quartile is 2, the third quartile is 4, and the interquartile range is $4 - 2 = 2$. If we obtain the 4 additional observations 6, 3, 3, and 4, the range has increased to $6 - 1 = 5$, but the interquartile range has *decreased* to 1, since the first quartile is now 3 and the third quartile is now 4.

4.2   The Mean Deviation*

The mean deviation is an intuitively appealing measure of variability. Table 4.2 gives the weights at birth of 5 babies. The mean weight is 7 pounds. The deviation of each weight from the mean is given in the third column of Table 4.2. The deviation is the measurement minus the mean. Note that some of the deviations are positive numbers and some are negative. Because we have subtracted the mean, the positive differences balance the negative ones in the sense that the sum (or the mean) of the deviations is 0. (See Chapter 3.)

TABLE 4.2
Weights of 5 Babies

Baby	Weight (pounds)	Deviation	Absolute deviation
Ann	8	1	1
Beth	6	−1	1
Carl	7	0	0
Dean	5	−2	2
Ethel	9	2	2
Total	35	0	6
Arithmetic average	7	0	1.2

Hypothetical data.

The *absolute value* of the deviation, that is, the numerical value without regard to sign, indicates how different a weight is from the mean. The average of these numbers provides a measure of variability. The absolute value of a quantity is indicated by placing the quantity between two vertical bars; for example, $|-2| = 2$. For the five weights in Table 4.2 the mean absolute deviation is

$$\frac{|-2| + |-1| + |0| + |1| + |2|}{5} = \frac{2 + 1 + 0 + 1 + 2}{5} = \frac{6}{5} = 1.2.$$

For a more graphic interpretation of these numbers, consider 5, 6, 7, 8, and 9 as the distances in blocks along a street of 5 stores from a bus transfer point. If one bus stop is to serve all of the stores, it should be located at the middle store (at block 7). (See Figure 4.3.) Then 1.2 is the average number of blocks walked if the same number of shoppers go to each store. If the variability in this sense were less, the shoppers would clearly be pleased.

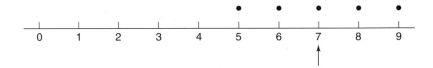

Figure 4.3   *Location of stores.*

In symbols, the mean deviation is

$$\frac{\sum |x_i - \bar{x}|}{n},$$

where $x_1, x_2, \ldots, x_n$ constitute a sample of n and $\bar{x}$ is the sample mean.

This measure of dispersion summarizes the distances of the observed values from the mean. It gives a larger result if the data are more "spread out" and values are farther from the mean, and a smaller result if the data are "close together" and closer to the mean. It is less sensitive to the influence of extreme values than the standard deviation (discussed in the next section) and thus may be particularly useful for skewed distributions. It is used less often than the standard deviation in practice, partially because the algebra of absolute values is particularly complex.

4.3   The Standard Deviation

Definitions

The *standard deviation* is the most frequently used measure of variability, although it is not as simple as the range or mean deviation. Like the mean deviation, it is a kind of average of the distances of the observed values from the mean. If many of the observations are far above or far below the mean, the standard deviation is large; if the observed values are very close to each other, and thus close to their mean as well, the standard deviation is small.

The basic element in the standard deviation is the difference between one observation and the mean, $x_i - \bar{x}$. Since these deviations sum to 0 (Chapter 3), they are first squared to obtain nonnegative numbers. The 5 birthweights in Table 4.2 are 8, 6, 7, 5, and 9 pounds with $\bar{x} = 7$ pounds. The squared deviations are $(8 - 7)^2 = 1^2 = 1$, $(6 - 7)^2 = (-1)^2 = 1$, $(7 - 7)^2 = 0^2 = 0$, $(5 - 7)^2 = (-2)^2 = 4$, and $(9 - 7)^2 = 2^2 = 4$. The squared distances are "averaged" by summing and dividing by $n - 1$. As a last step, the square root is taken in order to put the answer back into the original units (pounds).[1]

The sample standard deviation is defined as

$$s = \sqrt{\frac{\sum(x_i - \bar{x})^2}{n - 1}}, \tag{4.1}$$

where $x_1, x_2, \ldots, x_n$ are the observed sample values. Applying this to the 5 birthweights in Table 4.2,

$$s = \sqrt{\frac{1 + 1 + 0 + 4 + 4}{5 - 1}}$$

$$= \sqrt{10/4} = 1.581 \text{ pounds.}$$

It is obvious that if a distribution has values that are very spread out (e.g., baby weights from 1 to 12 pounds) the deviations from $\bar{x}$ would be larger and the standard deviation would be larger as well. If the weights were all 6 or 7 pounds, the standard deviation would be less than 1.581.

The standard deviation summarizes all of the units in all of the observed values. While it is not affected by extreme values as much as the range is, the standard deviation is more sensitive to outliers than the mean deviation. The difference $x_i - \bar{x}$ is readily inflated by extreme values at either end of the distribution. The operation of squaring increases this impact. It is important to examine the data for outliers that may be erroneous values, and to correct or discard them before calculating any measure of dispersion.

The square of the standard deviation,

$$s^2 = \frac{\sum(x_i - \bar{x})^2}{n - 1},$$

---

[1]The mean has the property that the sum of squared deviations from $\bar{x}$ is smaller than the sum of squared distances from any other point; $\bar{x}$ is called the *least-squares* measure of location.

is the *variance* of the data. It is also a measure of dispersion, but in *squared* units rather than the units of the original x variable. If the standard deviation is large, the variance will be large as well; if the standard deviation is small, the variance will be small. The variance is a concept used often in statistics and is discussed at length in later chapters.

Reasons for Dividing by One Less than the Sample Size

There are two reasons why the sample standard deviation is calculated with $n - 1$ in the denominator instead of n. The first is discussed in more detail in Chapter 10. Namely, if we wish to estimate the standard deviation of the entire *population* from which the sample is drawn, then the sample value is a better estimate if it is calculated with $n - 1$ in the denominator.[2]

The second reason is that the standard deviation is really an average of $n-1$ separate or *independent* pieces of information about dispersion. It is shown in Section 3.3 that the sum of the mean deviations is zero, that is, $\sum(x_i - \bar{x}) = 0$. This means that if we know the value of $x_i - \bar{x}$ for the first $n - 1$ observations, the value of the final $x_n - \bar{x}$ is already determined. For example, with $n = 3$ observations, if the first is 2 units above $\bar{x}$, and the second is 4 units above $\bar{x}$, then the third observation must be 6 units below $\bar{x}$. Although three values $(x_i - \bar{x})^2$ are combined into the numerator of s, the resulting sum $(2^2 + 4^2 + 6^2 = 56)$ is comprised of only two unique pieces of information about dispersion. This count—the number of independent pieces of information contained in a summary statistic—is referred to in later chapters as the number of *degrees of freedom*.

To view this from a slightly different perspective, suppose that a sample consisted of only a single observation, x_1. This value contains some minimal amount of information about location; we could even calculate the sample mean $x_1/1 = \bar{x}$. But the one observation provides no information about spread because there is no other value for comparison. It is the number of observations *beyond* the first one that tells us about dispersion.

With two observations, we can compute the distance between them (the range) or the distance between either value and $\bar{x}$ (half of the range); the other value is the same distance in the other direction. Here

---

[2]It is called an unbiased estimate (Chapter 10). The mean deviation (Section 4.2) is presented in some textbooks with $n - 1$ in the denominator as well, although that does not result in an unbiased estimate of the population mean deviation.

we have one piece of information about dispersion, whether we put it in the form of a simple difference between scores, or in the form of a more complex statistic:

$$ s = \sqrt{\frac{(x_1 - \bar{x})^2 + (x_2 - \bar{x})^2}{2 - 1}} = \sqrt{2(x_1 - \bar{x})^2} \approx 1.4 \times |x_1 - \bar{x}|. $$

With three observations, we have two pieces of information about dispersion, and so on. We divide the resulting total, $\sum(x_i - \bar{x})^2$ by the number of pieces of information it summarizes, and not necessarily by the number of operations it took to get the sum. For the sample standard deviation, this is always $n - 1$.

Interpreting the Standard Deviation

It is important to be able to make meaningful interpretations of the standard deviation of a data set. To begin with, s is just one index of how spread out or close together a set of observed values is. It assesses spread by summarizing distances from a central point, the mean. It is an attempt to answer the question, "how far from the center are typical or average values in this data set?"

A very large or very small standard deviation may be revealing by itself if we have a good intuitive understanding of the measurement scale. For example, suppose that a class is taught 100 new technical terms and then given a test with all 100 words as a final exam. If the standard deviation of test scores is 20, we would expect to find that students in the class vary tremendously in their knowledge of the terms; the students certainly have not benefitted equally from the instruction. If another class has a standard deviation of 3 on the same test, then these students are quite *homogeneous* in their knowledge levels (regardless of whether the mean is high or low).

For the most part, however, standard deviations are used *comparatively*. The standard deviation of waiting times for Doctor A in Table 4.1 would be greater than for Doctor F, because Doctor A is less consistent from one patient to another. Equipment that makes each package of cookies close to the 12-ounce requirement has a smaller standard deviation of weights than equipment that makes some packages much heavier and others much lighter. Three other examples illustrate such comparisons further.

- Two countries having income distributions with the same mean but different standard deviations could have very different patterns of distribution of wealth. One country may have a large middle class; the other, a small middle class or no middle class at all.

- Given a choice between two farm locations where the rainfall had the same average but different variability, a farmer would buy the location with small variability so that he or she could plan for crops with less risk of drought and washouts. Consider, for example, that for the preceding 5 years the rainfall at farm A had been 40, 36, 25, 60, and 54 inches and at farm B it had been 38, 40, 35, 50, and 52 inches. Although the mean, 43, is the same for both locations, the (sample) standard deviation for rainfall at farm A is 14.07, while at farm B it is only 7.55. The relatively large standard deviation for farm A reflects extremely low (25 inches) and extremely high (60 inches) rainfall; farm B would be preferred because there the standard deviation is smaller.

- In Section 3.4 we compared the waiting-time distributions of telephone calls for two alternative systems. To the discussion of these two distributions we now add the fact that the standard deviation for the data in Figure 3.5 is 0.41 second and for the data in Figure 3.6 is 0.69 second. The sizable difference between these standard deviations is a clue that it is not sufficient to compare only the means or medians (as reported in Table 3.12).

It should be noted that location and dispersion are separate descriptions of a data set. In comparing the doctors of Table 4.1, we have two distributions with identical means but different ranges and standard deviations. The 100-item vocabulary test might have a mean of 50 and standard deviation of 3 or a mean of 10 and standard deviation of 3. The mean indicates where the center of the distribution is located and the standard deviation how "spread out" the observations are around that central point. Two distributions can have the same mean but different dispersions, different means but the same dispersion, be identical on both measures or different on both.

Pairwise Distances.   The variance (the square of the standard deviation) has a natural interpretation in terms of *distances* (differences) between pairs of points (observations). The distances between pairs of stores A, B, C, D, and E at blocks 5, 6, 7, 8, and 9 along Center Street are given in Table 4.3. Roughly speaking, the larger the distances between pairs of observations the greater the variability, and a measure of variability should reflect this idea. Each number in Table 4.3 is a quantity of the form $|x_i - x_j|$, where x_i and x_j are pairs from 5, 6, 7, 8, and 9. There is a direct relation between the averages of the *squares* of differences $(x_i - x_j)^2$ and the average of the squares of sample deviations $(x_i - \bar{x})^2$. The mean of the squared distances[3] (given in Table 4.4)

---

[3]Each distance is counted only once; for example, the squared distance between store A and store C (4) is the same as the squared distance between C and A and is only summed once.

is $50/10 = 5$. The sample variance, which is the square of the sample standard deviation, is 2.5. This is one-half of the mean of the squared sample differences. *The variance of a sample is equal to one-half the mean of the squared pairwise distances.* The reader should try other examples to check this statement.

TABLE 4.3
Distances Between Stores

		Store				
		A	B	C	D	E
	A	0	1	2	3	4
	B	1	0	1	2	3
Store	C	2	1	0	1	2
	D	3	2	1	0	1
	E	4	3	2	1	0

TABLE 4.4
Squares of Distances Between Stores (Table 4.3)

		Store				
		A	B	C	D	E
	A	0	1	4	9	16
	B	1	0	1	4	9
Store	C	4	1	0	1	4
	D	9	4	1	0	1
	E	16	9	4	1	0

The interpretation is that the larger the variance the more scattered are the points in the sense of a large average of squared interpoint distances.

The Standard Deviation as a Ruler.   Once the standard deviation of a data set has been calculated, it can be used to assess whether data values are close together, far apart, or close to some specific number. This principle[4] can be stated as a "two standard-deviation rule of thumb":

RULE OF THUMB

In most distributions most of the values are within two standard deviations of the mean.

[4]Sometimes referred to as the "empirical rule."

A more precise form of this principle is discussed in Section 8.5, but even in this loose form, it is a very useful interpretive device.

For example, examining the 5 baby weights given earlier with $\bar{x} = 7$ and $s = 1.581$, two standard deviations below and above the mean, respectively, ranges from

$$7 - 2(1.581) = 3.84 \qquad \text{to} \qquad 7 + 2(1.581) = 10.16$$

pounds. In this case, all five values are within this interval. Furthermore, the weights actually range from 5 to 9 pounds, not very far from the two values above.

We will see that this rule of thumb applies in many situations. It allows us to form a quick mental picture of a data set from just a few summary statistics, and to evaluate individual data points as being near the center of the distribution or at the extremes. For example, an observation that is one standard deviation below average is probably among the more common values in the middle of the distribution, while one that is four standard deviations above average is among the highest, or even an outlier.[5]

4.4   Formulas for the Standard Deviation

Computing Formula

The standard deviation and the variance have been defined in terms of the sum of squares of deviations from the mean. It is possible to calculate these quantities without explicitly obtaining the deviations. This is advantageous when the computations are being done by hand or calculator. When the mean is carried to several significant figures it may be quite troublesome to find the deviations, and also difficult to maintain computational accuracy.

For the observations 5, 6, 7, 8, and 9, the squared deviations from the mean are 4, 1, 0, 1, and 4; these add to 10. An alternative way of obtaining this sum is to square the original numbers to obtain 25, 36, 49, 64, and 81, and add these squares to obtain 255. Then compute the sum of the scores, 35, square it and divide by $n = 5$, obtaining 245. Subtract this from 255, ending up with 10. The rule is

$$\sum(x_i - \bar{x})^2 = \sum x_i^2 - \frac{\left(\sum x_i\right)^2}{n}. \qquad (4.2)$$

---

[5] One important exception to which the principle does not apply is distributions that are highly skewed.

This identity is proved in Appendix 4A. Note that on the right-hand side of (4.2), the first term is the *sum of squares*; the second term involves the *square of the sum*. This formula should be used routinely for finding s^2 and s by hand or with a hand-held calculator.

As another example, consider 7 readings of atmospheric pressure in millibars: 20, 12, 21, 17, 15, 10, and 16. The two sums are

$$\sum x_i = 111 \quad \text{and} \quad \sum x_i^2 = 1{,}855.$$

The average reading is $\bar{x} = 111/7 = 15.86$. Using (4.2), the numerator of the variance is

$$1{,}855 - \frac{(111)^2}{7} = 94.86.$$

Dividing by $n - 1 = 6$, we obtain the variance,

$$s^2 = 15.81.$$

The standard deviation is $s = \sqrt{15.81} = 3.98$. If the two-standard-deviation rule of thumb applies, most of the data values should fall within $2 \times 3.98 = 7.96$ millibars of the mean, or between 7.90 and 23.82; indeed, all 7 values fall neatly within this range.

Care must be taken to carry enough digits to obtain a correct final result. (See Appendix 4B for suggestions for maintaining accuracy.)

Calculating the Standard Deviation from a Frequency Distribution

Just as the calculation of the mean can be simplified when some values are repeated, so can the computation of the variance and standard deviation. Table 4.5 again uses the data in Table 3.8, for which the mean was calculated as 3.5. To calculate the standard deviation we want the sum of squares of the numbers. Table 4.5 gives squares of the observations in the third column. Since many values are repeated, it is convenient to use the frequency distribution in Table 4.6. Then the summing of squares indicated in Table 4.5 can be done by using the frequencies. The sum of squared deviations is

$$133 - \frac{35^2}{10} = 133 - 122.5 = 10.5,$$

the sample variance is $10.5/9 = 1.1667$, and the sample standard deviation is $\sqrt{1.1667} = 1.0801$.

TABLE 4.5
Number of Children in 10 Families

Family	Number of children	Squares of number of children
1	2	4
2	4	16
3	4	16
4	3	9
5	4	16
6	3	9
7	3	9
8	3	9
9	6	36
10	3	9
	35	133

source: Table 3.8.

TABLE 4.6
Number of Children in 10 Families

j	Number of children v_j	Number of families having v_j children f_j	Squares of number of children v_j^2	$f_j v_j^2$
1	2	1	4	4
2	3	5	9	45
3	4	3	16	48
4	6	1	36	36
				133

source: Table 4.5.

In symbols, the calculations are

$$\sum x_i = \sum f_j v_j, \qquad \bar{x} = \frac{\sum f_j v_j}{n}, \qquad \sum x_i^2 = \sum f_j v_j^2,$$

and

$$s = \sqrt{\frac{\sum f_j v_j^2 - \dfrac{\left(\sum f_j v_j\right)^2}{n}}{n-1}}.$$

Effects of Change of Scale

Adding (or Subtracting) a Constant.   The ten families listed in Table 4.5 have an average of 3.5 children and standard deviation of 1.0801 children. If each family were to have one more child, the mean would become 4.5. (See Section 3.3.) To compute the new standard deviation, two sums are required from the increased family sizes:

$$\sum(x_i + 1) = 3 + 5 + 5 + 4 + 5 + 4 + 4 + 4 + 7 + 4 = 45,$$

$$\sum(x_i + 1)^2 = 9 + 25 + 25 + 16 + 25 + 16 + 16 + 16 + 49 + 16 = 213.$$

The standard deviation is

$$\sqrt{\frac{213 - 45^2/10}{9}} = \sqrt{1.1667} = 1.0801.$$

This demonstrates the general rule:

RULE 1   Adding (or subtracting) the same constant to every observation does not affect the standard deviation.

This result is intuitively apparent, since the constant increase affects the mean just as much as it affects the individual values, and thus the deviations remain unchanged. In symbols, if variable y is obtained from variable x by $y_i = x_i + a$, then the standard deviation of y is equal to the standard deviation of x, that is, $s_y = s_x$.

Multiplying (or Dividing) by a Constant.   If the income tax savings for every child in the above example is $1000, the average savings for a family is $3500. (See Section 3.3.) The standard deviation of savings requires the sums:

$$\sum(1000x_i) = \$2000 + \$4000 + \$4000 + \$3000 + \$4000 + \$3000$$

$$+ \$3000 + \$3000 + \$6000 + \$3000 = \$35,000,$$

$$\sum(1000x_i)^2 = \$2000^2 + \$4000^2 + \$4000^2 + \$3000^2 + \$4000^2$$

$$+ \$3000^2 + \$3000^2 + \$3000^2 + \$6000^2 + \$3000^2$$

$$= 133,000,000.$$

The standard deviation of savings is

$$\sqrt{\frac{133,000,000 - 35,000^2/10}{9}} = \sqrt{1,166,666.67} = \$1080.1,$$

which is $1000 times the standard deviation of children.

This illustrates the general rule:

RULE 2

> Multiplying (or dividing) every observation by a constant has the effect of multiplying (or dividing) the standard deviation by the numerical value of the constant.

In symbols, if variable y is obtained from variable x by $y_i = ax_i$, then the standard deviation of y is equal to $|a|s_x$. (This also implies that multiplying by a constant multiplies the variance by the square of the constant; that is, $s_y^2 = a^2 s_x^2$.)

Proofs of these properties are given in Appendix 4A.

4.5   Some Uses of Location and Dispersion Measures Together

Standard Scores

Once the mean and standard deviation of a distribution have been computed, it is a simple matter to re-express any observed value *relative to the group statistics*. As an example, consider the numbers of children in the families listed in Table 4.5. The mean and standard deviation are 3.5 children and 1.08 children, respectively.

The first family has 2 children. This is $2 - 3.5 = -1.5$ or 1.5 children below the average. Dividing the deviation of -1.5 by 1.08 yields -1.39. That is, the size of the first family is 1.39 *standard deviations below average*.

The value $x_1 = 2$ has been re-expressed as a *standard score* or *z-score*, defined by

$$z_i = \frac{x_i - \bar{x}}{s}.$$

z_i is the standard score for the ith observation. In contrast, the original x_i is referred to as the *raw score*. The standard score is a re-scaling of the original value, using the summary statistics as reference points. It addresses the question "How many standard deviations above or below average is a particular observed value?" An x value above average results in a positive standard score; an x value equal to $\bar{x}$ results in a z value of 0; an x value below average results in a negative standard score.

Standard scores have many applications in statistical work. For one, they tell us about the relative position of any observation in a sample (or, as shown in later chapters, in a population as well). The two-standard deviation rule of thumb (Section 4.3) asserts that "in most distributions most of the observations are within two standard deviations of the mean." An equivalent interpretation is that in most samples, most of the standard scores are between -2 and $+2$. If we compute z-scores for all 10 families in Table 4.5, we obtain

$$z_1 = (2 - 3.5)/1.08 = -1.39,$$

$$z_2 = (4 - 3.5)/1.08 = \quad 0.46 \qquad (= z_3 = z_5),$$

$$z_4 = (3 - 3.5)/1.08 = -0.46 \qquad (= z_6 = z_7 = z_8 = z_{10}),$$

$$z_9 = (6 - 3.5)/1.08 = \quad 2.31.$$

Indeed, only one value is more than two standard deviations away from the average, family 9 with 6 children (2.31 standard deviations above average).

It takes only a few computations to discover that the 10 standard scores themselves have an average of 0 and standard deviation of 1 (within rounding error). This is a general rule:

RULE 3   If a data set is rescaled into z-scores, the summary statistics for the resulting values are $\bar{z} = 0$ and $s_z = 1$.

These characteristics make standard scores useful in a variety of applications described in later chapters. They are proved algebraically in Appendix 4A.

Box-and-Whisker Plots

The box-and-whisker plot is among a set of techniques known as "exploratory data analysis" (Tukey, 1977). It is a particularly revealing graphical technique for describing one set of data or for comparing two or more distributions. A single box-and-whisker plot portrays the median, interquartile range, and extremes of a distribution. As an example, consider the ages of 8 senior citizens listed in Section 3.2; in order, these are $60\frac{1}{2}$, 61, 62, 64, 65, 66, 67, and 69 years.

The following values are required to construct the "box":

(1) The median age, $64\frac{1}{2}$ years in the example.
(2) The lower and upper quartiles. In the example these are $61\frac{1}{2}$ and $66\frac{1}{2}$ years, respectively.

The box is usually plotted on a vertical axis that has the range of observed values, with the ends of the boxes located at the quartiles, and a line through the box at the median. This is shown in Figure 4.4. (Usually, the observed values are not plotted.) Note that even this simple diagram displays the dispersion of the data (the interquartile range or length of the box), the location (the median), and the skewness in the middle of the age distribution. Skewness is greater when the two rectangles that make up the box are very different in size. There is little or no skewness when the median splits the box into two parts that are approximately equal.

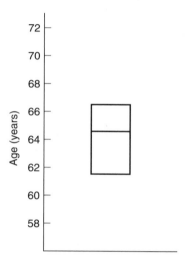

Figure 4.4   *Box plot of age data.*

To complete the diagram, the "whiskers" are added, requiring the minimum value and maximum value observed, respectively. A vertical line segment is drawn from the upper quartile to the maximum and another from the lower quartile to the minimum; these additions to the box are the whiskers. In the example the minimum and maximum are $60\frac{1}{2}$ and 69 years, respectively; these are shown in Figure 4.5.

The addition of whiskers to the box gives more information about skewness. If the whiskers are not equal in length, the distribution is skewed in the direction of the longer one.

Figure 4.6 is a box-and-whisker plot for comparing the weights of Stanford University football players in 1970 and 1989. Over this period of time the median weight has increased considerably as has the interquartile range. Each distribution is skewed in the sense that the upper whisker is greater than the lower. This means that the heavier players vary more in weight than the lighter players.

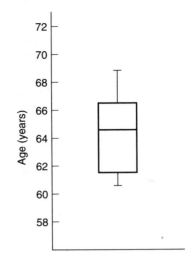

Figure 4.5   *Box-and-whisker plot of age data.*

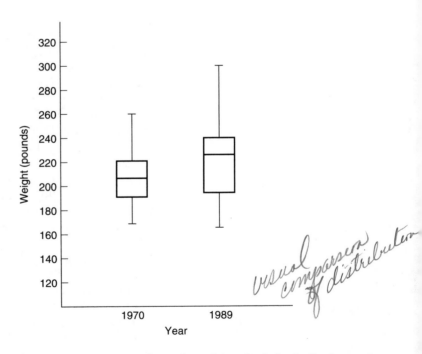

Figure 4.6   *Box-and-whisker plot of weights of Stanford football players in two years. (Source: 1970 data, Anderson and Sclove, 2nd edition. 1989 data, Stanford University football program.)*

The box-and-whisker plot is also useful for identifying outliers. Instead of drawing the whiskers to the minimum and maximum observed values, the length of the whisker is defined as $1\frac{1}{2}$ times the interquartile range. A whisker of this length is drawn upward from the top of the box (the upper quartile) and downward from the bottom of the box (the lower quartile). In this "modified box-and-whisker plot," any observation beyond the endpoint of either whisker is defined as an outlier.[6]

In the example illustrated in Figure 4.5 the interquartile range is 5 years and $1\frac{1}{2}$ times 5 is $7\frac{1}{2}$. In the 8-observation data set, no observation is greater than $66\frac{1}{2} + 7\frac{1}{2} = 74$ years, and no observation is less than $61\frac{1}{2} - 7\frac{1}{2} = 54$ years. If the data set contains outliers, they may be added to the box-and-whisker plot as points beyond the ends of the whiskers.

Summary

Measures of *dispersion* or *variability* provide information about a data set not conveyed by measures of location. In some situations the lack of variability, i.e., consistency, is important in itself. Variability measures also indicate the extent to which an index of location is representative of the whole data set.

Measures of variability include the *range*, the *interquartile range*, the *mean deviation*, and the *standard deviation*. The latter two are both summaries of the deviations of the observed values from the mean. The standard deviation, the most commonly used measure of variability, is the square root of the average *squared* deviations from the mean. This average is computed by dividing the sum of squared deviations by one less than the sample size, a value equal to the number of *independent* deviations in the sample. The sample *variance*, the square of the standard deviation, is $\frac{1}{2}$ of the average squared differences between pairs of observations.

When a constant value is added to, or subtracted from, each observation, the variability does not change. When each observation is multiplied by a constant, the standard deviation is multiplied by the absolute value of that constant.

Standard scores re-express observed values relative to the location and dispersion of a distribution; they indicate how many standard deviations above or below the mean a particular observation lies. In this

---

[6]Roughly speaking, this rule will define about 2% of the observations from a normal distribution as outliers.

application, the standard deviation is being used as a sort of "ruler." According to a "two-standard deviation rule of thumb," most observations in most distributions yield standard scores between -2 and $+2$; values larger than 2 may be outliers.

The box-and-whisker plot is a useful device for summarizing several characteristics of a distribution in an easy-to-read form.

Appendices

Appendix 4A   Proofs of Some Algebraic Principles

Computing Formula for the Standard Deviation

$$s = \sqrt{\frac{\sum(x_i - \bar{x})^2}{n - 1}} = \sqrt{\frac{\sum x_i^2 - \left(\sum x_i\right)^2 / n}{n - 1}}.$$

These two expressions are the same if their numerators are the same. The rules of summation needed for this proof are given in Appendix 3B. The first numerator is

$$\sum(x_i - \bar{x})^2 = \sum(x_i^2 - 2\bar{x}x_i + \bar{x}^2) \qquad \text{[squaring } x_i - \bar{x}\text{]}$$

$$= \sum x_i^2 - \sum 2\bar{x}x_i + \sum \bar{x}^2 \qquad \text{[distributive rule for } \sum\text{]}$$

$$= \sum x_i^2 - 2\bar{x}\sum x_i + n\bar{x}^2 \qquad \text{[Rules 1 and 2 for } \sum\text{]}.$$

Note in applying rules 1 and 2 for $\sum$ that $\bar{x}$ is a constant that does not change from one observation to another, and thus so is $2\bar{x}$ and so is $\bar{x}^2$. With substitution of $\sum x_i/n$ for $\bar{x}$ and combination of similar terms, the expression above becomes

$$\sum x_i^2 - 2\left(\frac{\sum x_i}{n}\right)\sum x_i + n\left(\frac{\sum x_i}{n}\right)^2 = \sum x_i^2 - \frac{\left(\sum x_i\right)^2}{n}.$$

Effects of Change of Scale

RULE 1   If a constant a is added to (or subtracted from) every value of variable x, then the standard deviation does not change.

Proof   The values of the modified variable are $x_i + a$; the mean of the modified variable is $\bar{x} + a$. (See Appendix 3C, Rule 1.) The standard deviation of the modified variable is

$$\sqrt{\frac{\sum[(x_i + a) - (\bar{x} + a)]^2}{n - 1}}.$$

This is simply $\sqrt{\sum(x_i - \bar{x})^2/(n - 1)} = s_x$ since the a's in the numerator cancel.

RULE 2   If every value of variable x is multiplied by a constant a, then the standard deviation is multiplied by $|a|$.

Proof   The modified variable has observed values ax_i and mean equal to $a\bar{x}$. (See Appendix 3C, Rule 2.) The modified standard deviation is

$$\sqrt{\frac{\sum[(ax_i) - (a\bar{x})]^2}{n - 1}}.$$

The numerator is

$$\sum[(ax_i) - (a\bar{x})]^2 = \sum[a(x_i - \bar{x})]^2$$
$$= \sum[a^2(x_i - \bar{x})^2]$$
$$= a^2\sum(x_i - \bar{x})^2.$$

Thus the standard deviation is

$$\sqrt{a^2\frac{\sum(x_i - \bar{x})^2}{n - 1}} = \sqrt{a^2}\,s_x = |a|s_x.$$

RULE 3   *The mean of standard scores is 0 and the standard deviation is 1.*

Proof   A single standard score is defined by

$$z_i = \frac{x_i - \bar{x}}{s_x} = \frac{1}{s_x}(x_i - \bar{x}).$$

Averaging these gives

$$\bar{z} = \frac{1}{n}\sum z_i = \frac{1}{n}\sum \frac{1}{s_x}(x_i - \bar{x}).$$

Since s_x is a constant that does not vary from one observation to another, we may use Rule 2 for $\sum$ (Appendix 3C); $\bar{z}$ is

$$\frac{1}{n}\left(\frac{1}{s_x}\right)\sum(x_i - \bar{x}).$$

Since the sum of the mean deviations, $\sum(x_i - \bar{x}) = 0$, $\bar{z} = 0$ as well.

To prove the second part of the rule, note that a single standard score is

$$z_i = \frac{x_i - \bar{x}}{s_x} = \frac{1}{s_x}(x_i - \bar{x})$$

$$= \frac{1}{s_x}x_i - \frac{1}{s_x}\bar{x}.$$

The term $(1/s_x)\bar{x}$ is a constant that does not vary from one observation to another. Subtracting a constant from every observation does not affect the standard deviation (Rule 1). Multiplying each x_i by a constant multiplies the standard deviation by the absolute value of the constant (Rule 2). Thus the standard deviation of z-scores is equal to the standard deviation of x times the (positive) constant $1/s_x$:

$$s_z = \frac{1}{s_x}(s_x) = 1.$$

Appendix 4B   Adjusting Data to Maintain Computational Accuracy

When data values are large, or particularly when their sum is large (four digits or more), the values may be "adjusted" to maintain accuracy when computing the sample variance or standard deviation. For example, suppose that the 7 atmospheric pressure readings in Section 4.4 were 1000 millibars higher than those given, that is, 1020, 1012, 1021, 1017, 1015, 1010, and 1016. As proven in Appendix 4A (Rule 1), these values have the same variance as the lower readings, $s^2 = 94.86$. Using the higher readings we obtain $\sum x_i = 7111$ and $\sum x_i^2 = 7{,}223{,}855$. Then

$$\sum x_i^2 - \frac{\left(\sum x_i\right)^2}{7} = 7{,}223{,}855 - \frac{(7111)^2}{7}$$
$$= 7{,}223{,}855 - 7{,}223{,}760$$
$$= 95.$$

This result has only two significant digits.

The difficulty arises from the fact that all the data are large numbers, and their squares are even larger. An effective means of dealing with this problem is to subtract a "guessed" mean (*working mean*) from each datum and compute the variance from the resulting "adjusted" data. It is not necessary that the working mean be particularly close to the actual mean of the sample; in the above example 1000 could be used. Often it is convenient, especially with computers, to use the first observation, x_1, as the working mean. In the example, $x_1 = 1020$, and the adjusted observations would be

$$0, -8, 1, -3, -5, -10, -4.$$

It is easy to compute with these small numbers, and seven-digit accuracy will be more than sufficient. The sum of squares is 215; the sum is -29; the square of the sum is 841. Again retaining seven digits in the computation, we now have

$$215 - \frac{841}{7} = 215 - 120.1429 = 94.8571.$$

This result has six significant digits, more than sufficient for most purposes. Not only is this result the sum of squared deviations of the adjusted observations, *it is also the sum of squared deviations of the original measurements, because subtracting the same number from each observation changes only the location, not the spread, of the set of observations.*

Exercises

4.1   If sample A is 1, 2, 1, 2, 4 and sample B is 1, 2, 1, 2, 9, which of the following is true?

(a) Samples A and B have equal standard deviations.

(b) Sample A has a larger standard deviation than Sample B.

(c) Sample B has a larger standard deviation than Sample A.

4.2   For a sample of three observations, the deviations of the first two observations from the sample mean are -3 and -1. What is the deviation of the third observation from the sample mean?

4.3   Compute the standard deviation of waiting times for Doctor A and for Doctor F in Table 4.1.

4.4   (continuation) Find the first and third quartiles and the interquartile range of waiting times for Doctor A and for Doctor F in Table 4.1.

4.5   (continuation) Summarize the differences between Doctor A and Doctor F in Table 4.1, by comparing the average, range, interquartile range, and standard deviations of their waiting times. To what extent do you draw different conclusions from the different measures of dispersion?

4.6   Table 3.15 gives the hours increase in sleep due to the use of two different drugs. Compare the variability of sleep increases due to the drugs by calculating the standard deviation of each set of results.

(a) Which drug gives the greater average sleep increase?

(b) Which drug works more consistently from patient to patient?

4.7   Find the sample standard deviation of $n = 10$ observations with $\sum x_i = 20$ and $\sum x_i^2 = 161$.

4.8   Compute the variances of the data represented in Figures 4.1 and 4.2.

4.9   The numbers of personal computers sold by Compute-er, a store in a suburban shopping center, on Monday through Friday are 3, 8, 2, 0, and 2.

(a) Graph the numbers as points on an axis.

(b) Find the mean and standard deviation of the numbers of sales.

(c) Find the mean deviation of the numbers of sales.

4.10   (continuation) The numbers of personal computers sold by Compute-im, a rival store, during the same five-day period are 2, 2, 3, 4, 3.

(a) Graph the numbers as points on an axis.

(b) Find the mean and standard deviation of the numbers of sales.

(c) Find the mean deviation of the numbers of sales.

(d) For which computer store is the mean more representative of the day-to-day sales? Why?

4.11   Use the data of Exercise 4.9 to verify that the variance is one-half the average of the squared differences between pairs of observed values.

4.12   (a) Compute the variance of each of the following samples.

Sample F: 1, 1, 5, 5
Sample G: 1, 3, 3, 5
Sample H: 1, 1, 1, 5.

(b) Comment on the comparative variability of these three samples.

4.13   A cholesterol count was made on a blood sample from each of nine 30-year old males. The numbers of units counted were 273, 189, 253, 149, 201, 153, 249, 214, and 163.

(a) Compute the mean and standard deviation of these cholesterol readings.

(b) To what extent does the "two-standard deviation rule of thumb" apply to this data set?

4.14   Compute the median and interquartile range for time on the stair climb task given to the firefighter applicants in Table 2.22.

4.15   (continuation) Compute the medians and interquartile ranges for time on the body drag and obstacle course tasks given to the firefighter applicants in Table 2.22.

(a) Which of the three tasks requires the highest median amount of time?

(b) Which of the three tasks has greatest variability in amount of time taken, as indicated by the interquartile range?

4.16   Table 3.16 lists the numbers of books borrowed from two university libraries by a sample of students at each school. Compare the variability of the two distributions by computing the range and standard deviation of the numbers of books borrowed from each library. (Note that the location measures were computed in Exercise 3.15.)

4.17   Summarize the data of Table 2.14 by constructing a frequency distribution of years teaching experience of teachers in rural Tennessee schools. Using the frequencies that you obtain, compute the interquartile range and standard deviation of years experience. To what extent does the "two-standard deviation rule of thumb" apply to this distribution?

4.18   Consider the probable distribution of years teaching experience in schools in urban Tennessee communities.

(a) What factors, if any, would tend to make the standard deviation of this distribution greater than in the distribution of Table 2.14?

(b) Would the range be affected in the same way as the standard deviation?

4.19   The length and breadth of the heads of 25 human males are given in Table 4.7. Of course the mean breadth will be less than the mean length. Is the variance of the breadths less than the variance of the lengths?

TABLE 4.7
Head Measurements (millimeters)

Length	Breadth	Length	Breadth
191	155	190	159
195	149	188	151
181	148	163	137
183	153	195	155
176	144	181	153
208	157	186	145
189	150	175	140
197	159	192	154
188	152	174	143
192	150	176	139
179	158	197	167
183	147	190	163
174	150		

source: Frets (1921).

4.20   Compute the mean and standard deviation of the heights of the football players given in Table 2.15.

4.21   Compute the mean and standard deviation of the weights of the football players given in Table 2.15.

4.22   Table 2.21 lists the final statistics grades for a class of 24 students. Make a box-and-whisker plot for those students who have never had a prior statistics course. Draw a plot on the same graph for students who have had one or more prior courses. How do these distributions compare with one another? (You may wish to compare these results with the answers to Exercise 2.14.)

4.23   Table 2.22 gives qualifying examination scores for 28 firefighter applicants. On a single graph, draw four box-and-whisker plots of the final composite scores:

(a) One plot for all male applicants, and one for all female applicants.

(b) One for all white applicants, and one for all minority applicants.

(c) Compare the distributions.

4.24   Table 4.8 lists the number of telephone calls received by one reservations office of a major airline over a three-year period. On a single graph draw three box-and-whisker plots, one for the calls of each year. Place dots on the graph that indicate any outliers.

(a) Is there any tendency for the "typical" number of calls to increase or decrease from one year to the next?

(b) Is there any tendency for the dispersion of the numbers of calls to increase or decrease from one year to the next?

(c) Is there any evidence that a strike by another major airline in mid-1985 affected the distribution of calls received?

TABLE 4.8

Telephone Calls Received by Airline Reservations Office, 1985–1987 (in thousands, rounded to nearest 100 calls)

Month	1985	1986	1987
January	52.2	54.5	52.3
February	49.8	52.7	48.1
March	64.3	60.0	55.1
April	58.6	53.8	53.3
May	95.3	53.6	56.6
June	93.3	48.1	55.2
July	67.1	48.6	56.8
August	61.0	47.7	51.4
September	46.7	44.5	42.5
October	51.2	44.8	47.2
November	49.6	51.9	51.2
December	52.5	54.0	49.9

source: American Airlines (1988).

4.25   The mean and standard deviation of variable x for a sample of 20 observations are 16.20 and 3.30, respectively. What would the mean and standard deviation become if each of the 20 original observations were

(a) Increased by 1.8?

(b) Decreased by 1.8?

(c) Multiplied by 10?

(d) Divided by 2?

(e) Multiplied by 1/4 and then increased by 1/4?

(f) Multiplied by −2 and then decreased by 1.6?

(g) Multiplied by 1.5 and then increased by 3?

4.26   (continuation) What are the standard scores for the observations of Exercise 4.25 that have the following raw scores?

(a) $x_1 = 12.90$,   (d) $x_8 = 11.25$,

(b) $x_4 = 22.80$,   (e) $x_9 = 16.44$,

(c) $x_5 = 16.20$,   (f) $x_{13} = 8.50$.

4.27   The average life-span of a particular model of refrigerator is $\bar{x} = 12.4$ years. A graduate student has used one of these until it was 18.2 years old; this corresponded to a z-score of 1.5. What is the standard deviation of life-spans of this model refrigerator?

4.28   About five times a year, large groups of high school students gather in auditoriums around the country to take the Test of Ability on Saturday (TAS). The test is comprised of 50 items. This month the mean score of all students taking the test was $\bar{x} = 28.4$ and the standard deviation was $s_x = 5.0$. In order to report test scores in a form that is difficult to understand, the publishers (Scientific Test Evaluators, or STE) multiply each raw score by 0.2 and subtract 5.68; that is, they produce a "rescaled" result, $y_i = 0.2x_i - 5.68$.

(a) What is the y-score for a student who answered 34 questions correctly?

(b) What are the mean and standard deviation of the y values for all students who took the test this month?

4.29   (continuation) Because the general public could not interpret the y values reported by STE, the publishers rescaled each score one more time. Each y value was multiplied by 100 and then 500 was added to the product to obtain a *T-score*. That is, the T-value is $T_i = 100y_i + 500$.

(a) What is the T-score for a student who answered 34 questions correctly?

(b) What are the mean and standard deviation of the T-values for all students who took the test this month?

4.30   Use the ages of the 8 senior citizens used in Figures 4.4 and 4.5 as data. What proportion of the observed values fall within the endpoints of the whiskers in the "modified box-and-whisker plot"? How does this compare to the proportion of cases described by the "two-standard-deviation rule of thumb," that is, within 2 standard deviations of the mean?

4.31   Use the weights of the 1989 Stanford football players given in Table 2.15 as data. What proportion of the observed weights fall within the end-points of the whiskers in a "modified box-and-whisker plot"? How does this compare to the proportion of weights within 2 standard deviations of the mean? Within 3 standard deviations of the mean?

5
Summarizing Multivariate Data: Association Between Numerical Scales

Introduction

Statistical data are often used to answer questions about relationships between variables. Chapters 2 through 4 of this book describe ways to summarize data on a single variable. In Chapters 5 and 6 methods are described for summarizing the relationship or *association* between 2 or among 3 or more variables. Chapter 5 considers association among variables measured on numerical scales; Chapter 6 discusses two or more categorical variables.

Consider some questions that might be asked about relationships between variables. For example, "Does studying more help raise scores on the Scholastic Assessment Tests (SAT)?" This question could also be worded as follows: If we recorded the number of hours spent studying for the SAT and the SAT scores for a sample of students, do individuals with higher "hours" also have higher SAT scores and individuals with lower hours have lower SAT scores? That is, are particular values of the variable hours *associated with* particular values of the variable SAT?

We might ask if high sodium intake in one's diet is associated with elevated blood pressure. The question could be worded as follows: If we recorded the monthly sodium intake for each individual in a sample and his/her blood pressure, do individuals with higher sodium consumption also have higher blood pressure readings while those with lower sodium intakes have the lower blood pressure readings? Again, we might expect to find particular ranges of sodium measurements co-occurring with particular blood pressure levels.

A relationship between variables may also be in the opposite direction, that is, high scores on one variable co-occurring with low scores on the other and vice versa. For example, it is likely that faster typists generally make more errors than slower typists. Thus, in a sample of data entry clerks entering a 10,000-character document, larger numbers of errors are associated with lower entry times while smaller number of errors are made by persons who take more time to enter the document.

The observational units may not be people, but institutions, objects, or events. For example, we might ask whether the selling prices of various models of automobiles are related to the numbers of each that are sold in a given period of time or whether the amount of rainfall in a given agricultural area is related to the size of the crop yield.

Each of these examples involves two numerical variables, but questions about categorical variables are equally common. For example, "is inoculation with a polio vaccine related to the occurrence of paralytic polio?" Like the earlier examples, data may be collected on a sample of observations and we ask whether a particular value on one variable (i.e., "vaccinated") co-occurs with a particular value on the other ("polio absent"). Unlike the earlier examples, both of these variables are simple yes/no dichotomies. Association among categorical variables is discussed separately in Chapter 6.

A basic table of data for any of these questions has the form of Table 5.1. In this table, x and y represent the names of the two variables; x_i

TABLE 5.1

Data Layout for Examining Association Between Variables

Observation	x	y
1	x_1	y_1
2	x_2	y_2
⋮	⋮	⋮
n	x_n	y_n

and y_i are the specific values for the ith observation. These are *bivariate* data, since two variables are listed. If variables x and y are numerical, then x_i and y_i are corresponding quantities (e.g., x_i = number of hours studied, y_i = SAT score). If x and y are categorical variables, x_i and y_i are the names of the categories (e.g., vaccinated or not vaccinated). It is absolutely necessary that two values are obtained from the same observational unit in order to study the relationship of x and y; it is not possible to examine association, for example, if study times are obtained from one group of 25 individuals and SAT scores from 25 other individuals.

Statistical data can also be used to answer questions about the relationships among more than two variables. For example, are study time and the number of hours sleep an individual receives the night before the test related to SAT scores? Or, is sodium intake related to systolic and diastolic blood pressure? At the end of Chapters 5 and 6 there is a discussion of methods for summarizing association among three or more variables. The layout for such data would be a table like Table 5.1 but with an extra column for the values of each additional measure.

5.1   Association of Two Numerical Variables

In this section we describe two ways to assess the association of two numerical variables: a graphical technique (the scatter plot) and a formal numerical index of relationship. Scatter plots frequently depict information about the relationship between variables that is not indicated by a single summary statistic. Also, the interpretation of the numerical index is based on graphical ideas. The section on scatter plots is recommended reading before going on to the more mathematical approach in the section on the correlation coefficient.

Scatter Plots

The language and nonlanguage mental maturity scores (two kinds of "IQs") of 23 school children are given at the right of Figure 5.1. These data are represented visually by making a graph on two axes, the horizontal x axis representing language IQ and the vertical y axis representing nonlanguage IQ. Such a graph is called a *scatter plot* (or *scatter diagram*). Each point in such a plot represents one individual.

The distances in the horizontal direction indicate language IQ; the points are above the axis at the respective language IQ scores as in the graphing of a single numerical variable. The distances in the vertical direction indicate nonlanguage IQ; the points are to the right of the

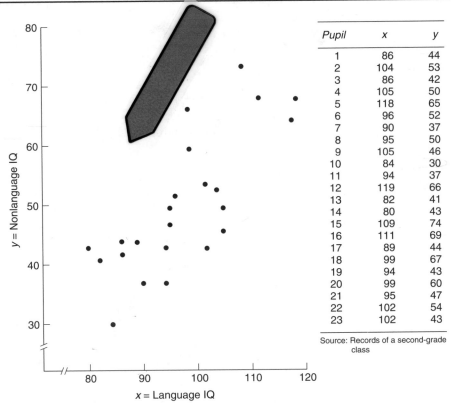

Pupil	x	y
1	86	44
2	104	53
3	86	42
4	105	50
5	118	65
6	96	52
7	90	37
8	95	50
9	105	46
10	84	30
11	94	37
12	119	66
13	82	41
14	80	43
15	109	74
16	111	69
17	89	44
18	99	67
19	94	43
20	99	60
21	95	47
22	102	54
23	102	43

Source: Records of a second-grade class

Figure 5.1 *Scatter plot of language and nonlanguage IQ scores of 23 children: Positive correlation.*

axis at the respective scores. Thus one point represents one pupil by indicating that pupil's two scores. For example, the point for pupil 10 (in the lower left corner of the plot) is placed at 84 on the horizontal axis and 30 on the vertical axis.

When all of the observations are plotted, the diagram conveys information about *direction* and *magnitude* of the association of x and y. The swarm of points in Figure 5.1 goes in a southwest-northeast direction. This indicates a *positive* or *direct* association of x and y. Namely, individuals who have the lower y values are the same people who have the lower values on x; they form a cluster of points in the lower-left portion of the diagram. Individuals who have higher y values are the same individuals who have higher x values; they form a cluster of points in the upper-right portion of the diagram. In general, a positive or direct relationship between two variables has this pattern:

Low x values accompanied by low y values,

Middle x values accompanied by middle y values,

High x values accompanied by high y values.

The strength or magnitude of the association is indicated by the degree to which the points are clustered together around a single line. It is sometimes useful to draw a straight line through the points to give a clearer picture of this, although the line is not a formal part of the scatter diagram; Figure 5.2 has the same data as Figure 5.1, but with a straight line drawn through the middle of the points.

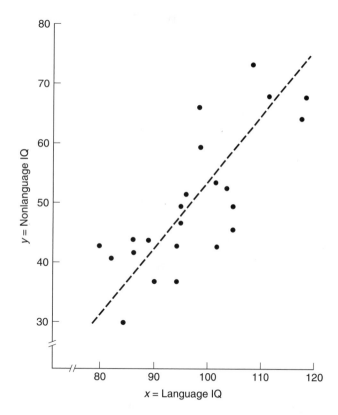

Figure 5.2   *Scatter plot of language and nonlanguage IQ scores of 23 children.*

If all of the points fall exactly on the line, there is a "perfect" association of the two variables. In this case, if we knew an individual's value on variable x, we would be able to compute his/her value on y exactly. Of course, no statistical data attain this degree of association. To the extent that the points in the diagram diverge from a straight line, the association is less than perfect. Since the scatter plot is a non-numerical way of assessing association, adjectives are used to describe the strength of association. Figure 5.1 may be said to depict a "strong" association of x with y, while points dispersed still further from the line would indicate a "moderate" or even "weak" association. The more the points diverge from the line, the less certain we can be about an observation's y value from knowledge of its x value.

Data and a scatterplot of the moisture content and strength of 10 wooden beams are given in Figure 5.3. Variable x is the moisture content of the beams and variable y is beam strength. As in Figure 5.1 each point on the plot represents one observation. For example, beam 1 ($x_1 = 11.1$, $y_1 = 11.14$) is indicated by the point in the lower-right corner of the graph.

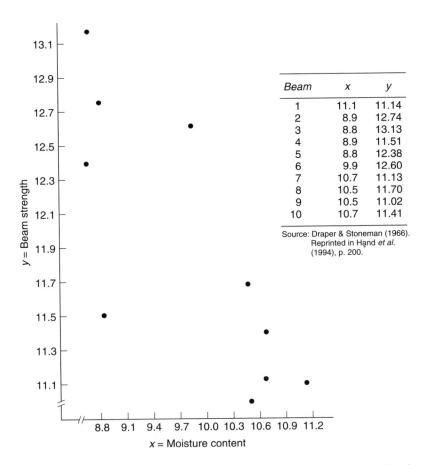

Beam	x	y
1	11.1	11.14
2	8.9	12.74
3	8.8	13.13
4	8.9	11.51
5	8.8	12.38
6	9.9	12.60
7	10.7	11.13
8	10.5	11.70
9	10.5	11.02
10	10.7	11.41

Source: Draper & Stoneman (1966). Reprinted in Hand *et al.* (1994), p. 200.

Figure 5.3   *Scatter plot of moisture content and strength of ten wooden beams.*

In this scatter plot the swarm of points lies in a northwest-southeast direction (i.e., upper left to lower right). There is a *negative* or *inverse* association of x with y. Beams with lower moisture content tend to have greater strength while beams with higher moisture content tend to have less strength. In general, a negative or inverse relationship between two variables has this pattern:

Low x values accompanied by high y values,

Middle x values accompanied by middle y values,

High x values accompanied by low y values.

The magnitude of association of x with y in Figure 5.3 is not quite as great as in Figure 5.1. That is, the points are not clustered quite as closely around a straight line, although it is clear that the best line through the points would be from upper left to lower right; we could characterize the association of moisture content with beam strength as "moderate negative." If the points were closer to a straight line from the upper left to the lower right the association would be "strong negative." If the points in Figure 5.3 fell exactly on a straight line from upper left to lower right, the association would be "perfect negative." It is important to understand that the direction and magnitude of association are two separate aspects of the relationship of two numerical variables.

Data and a scatter plot for two other variables are given in Figure 5.4. In this plot, the points do not indicate a clear straight line in either direction. Or, put another way, if we were to draw a straight line from lower left to upper right or from upper left to lower right, many of the points would be far from the line in either direction; they form more of a circular pattern instead. The association of these two variables is very weak. Lower x values (e.g., $x = 3$ or $x = 4$) are accompanied by low, middle, and high y values; higher x values (e.g., $x = 8$ or $x = 9$) are also accompanied by the entire range of y's. In data such as these knowledge of an observation's x value tells little if anything about the corresponding y value.

Note that the two variables do not have to be measured on the same scale to have a positive (or negative) association. In Figure 5.1, language IQ ranges from 80 to 119, while nonlanguage IQ ranges from 30 to 74. Although the means and standard deviations of the two variables are different, the relationship between the two variables is not a comparison of means or standard deviations. Instead, we are looking for the patterns of association portrayed by individuals' pairs of scores. To make this easy to see in a diagram, it is advisable to spread the *actual observed range* of each variable out along the respective axis, as in Figures 5.1 to 5.4. The axes are marked with slashes (//) to indicate that the first intervals begin close to the lowest observed x and y values, respectively, rather than at zero.

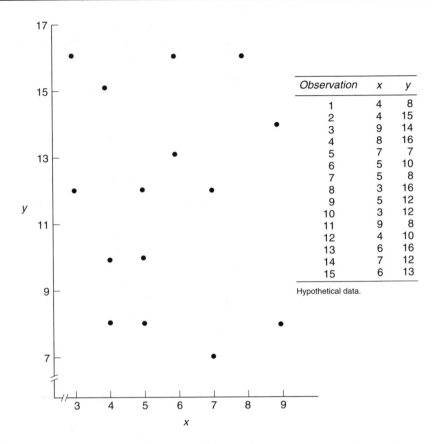

Observation	x	y
1	4	8
2	4	15
3	9	14
4	8	16
5	7	7
6	5	10
7	5	8
8	3	16
9	5	12
10	3	12
11	9	8
12	4	10
13	6	16
14	7	12
15	6	13

Hypothetical data.

Figure 5.4   *Scatter plot of two variables for hypothetical sample of 15 observations.*

Other Information Revealed by Scatter Plots

A formal statistical index of linear association between two variables, the correlation coefficient, is introduced in the next section of this book. It is good practice to construct scatter plots routinely before calculating a correlation since important information can easily be missed in examining a numerical summary statistic. In particular, scatter plots can reveal patterns of association that do not conform to a straight line and can reveal two-variable ("bivariate") outliers.

Nonlinear Association   Pairs of variables are often associated in a clear pattern, but not conforming to a straight line. The four scatter plots in Figure 5.5 show a variety of relationships between x and y. In each of the plots the association is strong, since the points are clustered closely together, but the pattern is *curvilinear*.

Figure 5.5(a) depicts a pattern that is relatively common in the study of human behavior. If the observations are patients who experience pain, variable x might be the amount of aspirin taken orally, in milligrams, and variable y the amount that is absorbed into the bloodstream. Beyond a certain point, additional amounts of ingested aspirin are no longer absorbed, and the amount found in the bloodstream reaches a plateau. Or, in research on human memory, it has been found that initial practice trials are very helpful in increasing the amount of material memorized; after a certain number of trials, however, each additional attempt to memorize has only a small added benefit. Thus, if the observations are students learning a foreign language, Figure 5.5(a) could depict the relationship between x, the number of 25-minute periods devoted to studying vocabulary, and y, the number of words memorized.

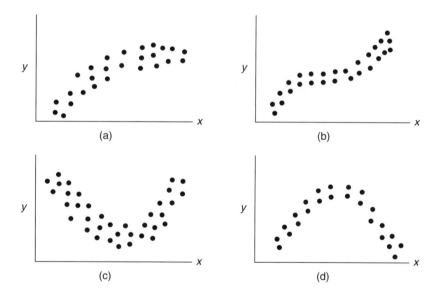

Figure 5.5   *Scatter plots showing curvilinear relationships between two variables.*

A scatter plot like Figure 5.5(b) might be found, for example, in examining the effects of advertising on sales. If observations are branches of a large chain of stores with independent control over expenditures, variable x might be advertising outlays and y the total sales volume in a given period of time.[1] The plot could occur if initial advertising outlays have a substantial effect on sales, additional advertising outlays have little additional impact on sales, but expensive "saturation" advertising again gives a significant boost to sales.

Figure 5.5(c) might depict the relationship between x, the amount of water provided to agricultural plots and y, the proportion of plants on the plot that do not grow to a given size. If the observations are soil plots of equal size, this pattern would suggest that too much as well as too little water is harmful, while moderate amounts of water minimize the plant loss. In contrast, the pattern of Figure 5.5(d) is sometimes found in examining the responses of humans and animals to various kinds of physiological stimulation. Little or no stimulation produces little or no response, while moderate amounts of stimulation produce maximal response. Levels of stimulation that exceed the individual's ability to process the input, however, can result in a partial or complete suppression of the response.

These examples are just a few of the many patterns of nonlinear association that are possible. Scatter plots can also reveal aspects of bivariate data that are more complex than just an association of two variables. For example, a large study of the science productivity of American schools (Finn, 1979) yielded a mean science achievement score for the grade 11 students in each school. Also, a measure of the resources available for teaching science was obtained for each school by combining information on textbooks and library materials, laboratory space and facilities, and characteristics of the teaching staff. The scatter plot, with schools as the observations, looked much like that in Figure 5.6. Lacking a certain "threshold" level of resources, schools do not have high science performance. When that threshold is reached, some schools continue to perform poorly, while others do much better; this is seen in the greater dispersion of y values at the higher ranges of x. It was concluded that the presence of particular levels of resources is necessary but not sufficient in itself to assure that schools will have high science performance. Further research explored the resource threshold in greater detail and differences between high-resource schools that had superior as compared with poorer performance.

Bivariate Outliers   In Chapters 2 and 3 the concept of an outlier is discussed in terms of a single variable, that is, an observation that falls much below or much above most of the data. Scatter plots allow us to refine the process for identifying outliers because they provide two pieces of information (x and y) about each observation.

---

[1] Of course, both variables would need to be adjusted to account for differences in the amount of inventory stocked by each branch and/or the size of the community where the outlet is located.

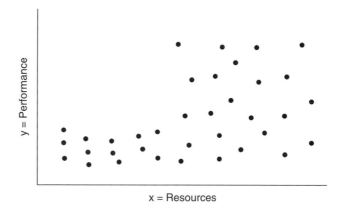

Figure 5.6   *Scatter plot showing relationships between science resources and science achievement in American high schools.*

The data of Figure 5.1 can be used as an example. Suppose that an additional child was tested who obtained $x_{24} = 82$ and $y_{24} = 72$; this observation is indicated by the asterisk ($*$) in Figure 5.7. While the preponderance of points lies in a swarm from lower left to upper right, this observation is far from the rest, falling close to the upper left corner of the plot. Note that 82 is not the lowest x score of all the observations, nor is 72 the highest y value. Both of these are well within the range for the separate variables. It is the *pair* of values—a low x coupled with a high y that causes this observation to fall far from the others. The data should be explored further to determine if these values are recorded correctly and if the child has the same characteristics as the others (e.g., same native language, same age range, not handicapped) before proceeding with the analysis. In Figure 5.3, beam 4 ($x_4 = 8.9$, $y_4 = 11.51$) appears to be an outlier; at very least, the accuracy of the moisture and strength ratings should be re-checked.

A scatter plot may also show the reverse—an apparent outlier on one variable that is not as exceptional when a second variable is brought into the picture. For example, in Figure 5.3, if an additional beam had a strength index of 10.3, we might suspect that an error had been made; certainly this observation is a potential outlier on variable y. But if we also knew that the beam had a moisture content of 11.6, the point on the scatter plot would be directly in line with the rest of the swarm. Although the values should still be checked for accuracy, this appears to be just one more observation indicating that strength is lower when moisture content is greater.

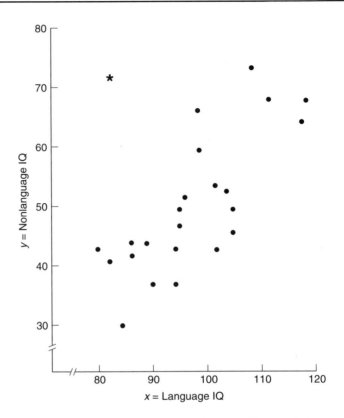

Figure 5.7   *Plot of data from Figure 5.1 with one additional observation (outlier).*

The Correlation Coefficient

A descriptive statistic that indicates the degree of linear association of two numerical variates is the *correlation coefficient*, usually represented by the letter r. The correlation coefficient has a possible range from -1 to $+1$. The sign of r indicates the direction of association—it is positive for a direct association of x and y (like Figure 5.1) and negative for an inverse association (like Figure 5.3). The absolute value of r is exactly 1 for a perfect linear relationship, but lower if the points in the scatter plot diverge from a straight line. A correlation close to zero—either positive or negative—indicates little or no linear association between x and y, as in Figure 5.4.

The question "what is a strong correlation?" has several answers. Statisticians would generally refer to a correlation close to zero as indicating "no correlation"; a correlation between 0 and 0.3 as "weak"; a correlation between 0.3 and 0.6 as "moderate"; a correlation between 0.6 and 1.0 as "strong"; and a correlation of 1.0 as "perfect." Note that the sign of the correlation is not considered in using these descriptors; the strength of association between two variables is indicated by its absolute value. The sign of the correlation indicates only whether the relationship is direct (+) or inverse (−). Thus, a correlation of +0.9 and −0.9 are equally "strong."

The question "what is a strong correlation?" also depends on the particular application, however. Two laboratory technicians counting impurities in the same water samples should have very high agreement— the correlation between their counts may be 0.95 or better. On the other hand, two different human characteristics rarely have such a high correlation. The correlation of height and weight is generally in the neighborhood of 0.8; of scores on the Scholastic Assessment Tests with college freshman grade average about 0.6; of measured intelligence with socioeconomic status about 0.4; of heart rate with blood pressure about 0.2. Likewise, with objects, events, or institutions as the observational units, correlations of two different measures are rarely as high as 0.8 or 0.9. For example, hypothesizing that schools with a larger teaching staff yield better learning outcomes, researchers find that the correlations of pupil:teacher ratios with indicators of academic performance are about −0.2 (negative because lower pupil:teacher ratios are associated with higher performance). In this instance, school practices that correlate with performance more than 0.2 are "stronger" explanations of academic outcomes.[2]

Definition of the Correlation Coefficient.   The correlation coefficient, represented by r, is

$$r = \frac{s_{xy}}{s_x s_y},$$

where

$$s_x = \sqrt{\frac{\sum(x_i - \bar{x})^2}{n - 1}}$$

and

$$s_y = \sqrt{\frac{\sum(y_i - \bar{y})^2}{n - 1}},$$

---

[2]Regardless of whether they are greater than +0.2 or less than −0.2.

are simply the sample standard deviations of x and y. The numerator of r,

$$s_{xy} = \frac{\sum(x_i - \bar{x})(y_i - \bar{y})}{n-1},$$

is the *sample covariance* between x and y.

As an example, consider the data collected by Nanji and French (1985) to examine the relationship of alcohol consumption with mortality due to cirrhosis of the liver in the 10 Canadian provinces. Variable x in Table 5.2 is an index of the amount of alcohol consumed in the province in one year (1978) and variable y is the number of individuals who died from cirrhosis of the liver per 100,000 residents. A scatter plot of the data is shown in Figure 5.8. The means of the two variables are $\bar{x} = 10.99$ and $\bar{y} = 15.29$; these are also indicated in this scatter plot.

Deviations from the means are the basic elements in the numerator and denominator of the formula for r, that is, both for the covariance and for the standard deviations. Thus the columns of deviations, $x_i - \bar{x}$ and $y_i - \bar{y}$, are computed first and added to Table 5.2. For the first observation (Prince Edward Island) these are $x_1 - \bar{x} = 11.00 - 10.99 = 0.01$ and $y_1 - \bar{y} = 6.50 - 15.29 = -8.79$. The covariance requires the product of these deviations, given in the right-hand column of the table. For the first observation this is $(x_1 - \bar{x})(y_1 - \bar{y}) = (0.01)(-8.79) = -0.0879$.

TABLE 5.2

Data for Correlation of Alcohol Consumption with Cirrhosis Mortality for 10 Canadian Provinces

	Alcohol	Mortality			
Observation	x_i	y_i	$x_i - \bar{x}$	$y_i - \bar{y}$	$(x_i - \bar{x})(y_i - \bar{y})$
1 Pr. Edward Is.	11.00	6.50	0.01	−8.79	−0.088
2 Newfoundland	10.68	10.20	−0.31	−5.09	1.578
3 Nova Scotia	10.32	10.60	−0.67	−4.69	3.142
4 Saskatchewan	10.14	13.40	−0.85	−1.89	1.606
5 New Brunswick	9.23	14.50	−1.76	−0.79	1.390
6 Alberta	13.05	16.40	2.06	1.11	2.287
7 Manitoba	10.68	16.60	−0.31	1.31	−0.406
8 Ontario	11.50	18.20	0.51	2.91	1.484
9 Quebec	10.46	19.00	−0.53	3.71	−1.966
10 Brit. Columbia	12.82	27.50	1.83	12.21	22.344

source: Nanji & French (1985). Data reprinted in Hand *et al.* (1994), pp. 91–92.

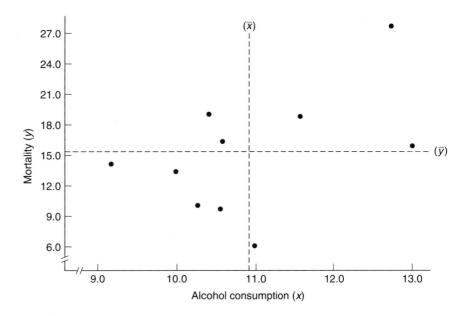

Figure 5.8 *Scatter plot of data in Table 5.2, with $\bar{x}$ and $\bar{y}$ indicated.*

The numerators of the standard deviations are the sums of squared deviations from the mean. For x this is

$$\sum(x_i - \bar{x})^2 = (0.01)^2 + (-0.31)^2 + \cdots + (1.83)^2 = 12.595.$$

For y the result is $\sum(y_i - \bar{y})^2 = 303.629$. The standard deviations are

$$s_x = \sqrt{\frac{12.595}{9}} = 1.183 \quad \text{and} \quad s_y = \sqrt{\frac{303.629}{9}} = 5.808.$$

The covariance requires the sum of the cross-products of the deviations, $\sum(x_i - \bar{x})(y_i - \bar{y})$. This is the sum of the values in the right-hand column of Table 5.2, that is, $(-0.088) + (1.578) + \cdots + (22.344) = 31.371$. Note that the individual cross-product for each observation must be computed *before* summing. The covariance is

$$s_{xy} = \frac{31.371}{9} = 3.486.$$

The correlation of x with y is

$$r = \frac{3.486}{1.183 \times 5.808} = 0.51,$$

rounded to two decimal places. Both the scatter plot and computed value indicate a moderate positive association of these two variables.

The numerator of r (the covariance s_{xy}) conveys the essential information about association. The standard deviations serve to scale the correlation so that it is always in the range -1 to $+1$. The larger the covariance, the larger the correlation, and vice versa. The magnitude of the covariance is primarily a function of the size of the cross-product terms $(x_i - \bar{x})(y_i - \bar{y})$. This gives us an easy way to see why the correlation works as it does.

The covariance in the example has a numerator of 31.371. This is the sum of 10 terms, the largest of which are 3.142 for Nova Scotia and 22.344 for British Columbia. The 22.344 is as large as this because British Columbia is in the upper-right corner of the scatter plot. It is much above the mean on both variables x and y, giving two positive deviations whose product is relatively large. In contrast, Nova Scotia is in the lower-left quadrant of the scatter plot, somewhat below the mean on both variables. It has two negative deviations, whose product is relatively large and also positive.

The principle illustrated by these two observations is true in general. Observations close to the "best" straight line, especially those at the extremes, increase the magnitude of r. If any observation is moved farther from the line either horizontally or vertically, the magnitude of r is reduced. The same principle holds for both direct and inverse relationships. For inverse relationships, points close to the straight line, especially those at the extremes, have relatively large *negative* cross-products. The more points that lie in the upper-left and lower-right portions of the scatter plot, the more strongly negative r will be.

Since the correlation coefficient is particularly sensitive to observations with large x or y deviations from the mean, it is especially important to identify and to verify, correct, or eliminate outliers. The scatter plot is especially valuable for locating such observations.

Computing the Correlation Coefficient.   The computation of r involves calculating two standard deviations and the covariance of x and y. A computational formula for the standard deviation, given in Section 4.4, is

$$s_x = \sqrt{\frac{\sum x_i^2 - \left(\sum x_i\right)^2 / n}{n - 1}} \quad \text{and} \quad s_y = \sqrt{\frac{\sum y_i^2 - \left(\sum y_i\right)^2 / n}{n - 1}}.$$

A parallel formula for computing the covariance is

$$s_{xy} = \frac{\sum x_i y_i - \left(\sum x_i\right)\left(\sum y_i\right) / n}{n - 1}.$$

The proof that this is equivalent to the earlier formula is given in Appendix 5A.

The correlation is the ratio of s_{xy} to the product $s_x s_y$. Note that s_{xy} has $n - 1$ in the denominator, while both s_x and s_y have $\sqrt{n - 1}$ in the denominator. Thus the $n - 1$ in the covariance and the product of the $\sqrt{n - 1}$'s in the standard deviations "cancel" each other in the formula for r, leaving

$$r = \frac{\sum x_i y_i - (\sum x_i)(\sum y_i)/n}{\sqrt{\left[\sum x_i^2 - (\sum x_i)^2/n\right]\left[\sum y_i^2 - (\sum y_i)^2/n\right]}}. \tag{5.1}$$

For the data of Table 5.2 the components of the formula are

$$\sum x_i = 109.88, \qquad \sum y_i = 152.90,$$

$$\sum x_i^2 = 1219.956, \qquad \sum y_i^2 = 2641.470,$$

$$\sum x_i y_i = (11.00 \times 6.50) + (10.68 \times 10.20) + \cdots + (12.82 \times 27.50)$$

$$= 1711.437$$

and

$$r = \frac{1711.437 - 109.88 \times 152.90/10}{\sqrt{[1219.956 - 109.88^2/10][2641.470 - 152.90^2/10]}}$$

$$= \frac{31.372}{\sqrt{12.595 \times 303.629}} = 0.51.$$

Expression (5.1) is particularly simple to apply when the observations are whole numbers. If data values on one or both variables are very large, a constant may be subtracted from each x value or each y value, or both, without affecting the resulting correlation coefficient.

Interpreting Association

The correlation coefficient is useful for summarizing the direction and magnitude of association between two variables. It is a widely-used statistic, cited frequently in both scientific and popular reports. Nevertheless, the informed data analyst must be aware of two limitations to the meaning of any particular correlation.

First, the correlation coefficient described in this chapter reveals only the *straight line* (linear) association between x and y. If r were to be computed from the data sets plotted in Figure 5.5 or 5.6, for example, the values that result could be misleading. Both Figures 5.5(a) and 5.6 show definite patterns of positive linear association. The scatter plots reveal that a straight line is not the whole story of the relationship of x to y in either case. In Figures 5.5(c) and 5.5(d), the linear correlation would be close to 0, while the scatter plot reveals important curvilinear patterns. It is obvious that a scatter plot should always be examined when a correlation coefficient is to be computed or interpreted.

Second, even a strong correlation between two variables does not imply that one *causes* the other. When a statistical analysis reveals association between two variables it is generally desirable to know more about the association. Does it persist under different conditions? Does one factor "cause" the other? Is there a third factor that causes both? Is there another link in the chain, a factor influenced by one variable and in turn influencing the other?

In spite of our desire to find answers to these questions, the correlation indicates *only* that certain pairings of values on x and y occur more frequently than other combinations. For example, if the x and y variables of Table 5.2 were "years on the job" and "job satisfaction ratings" for a sample of employees, the positive correlation between them might indicate that (a) if employees hold their jobs for more years, the work seems to become more satisfying, or (b) if employees are more satisfied, they keep their jobs longer. That is, the causal connection may go in either direction and may also be affected by other intermediary mechanisms. For example, it may be that more senior employees are given subtle or overt rewards that in turn enhance their satisfaction. Thus, the distribution of rewards—referred to as an *intervening variable*—explains the association of years with satisfaction; the correlation itself does not imply a direct cause-and-effect relationship.

Association between two variables may also occur because x and y are both consequences of some third variable that has not been observed. This is seen in the following illustration:

Do Storks Bring Babies? In Scandinavian countries a positive association between the number of storks living in the area and the number of babies born in the area was noticed. Do storks bring the babies? Without shattering the illusions of the incurably romantic, we may suggest the following: Districts with large populations have a large number of births and also have many buildings, in the chimneys of which storks can nest. Figure 5.9 is a diagram representing the idea that the population factor explains both the number of births and the frequency with which storks are sighted. The three variables to study are populations of districts, numbers of births in districts, and numbers of storks seen in the districts.

Although the stork example is only used to illustrate the point, the association between two variables is often dependent on a third. Even the established relationship between smoking and lung cancer was at one time thought to be attributable to other factors including alcohol consumption; that is, if alcohol "causes" cancer and causes people to smoke more, the two consequences would be seen as related. Sometimes the phrase "spurious association" is used to describe this situation.

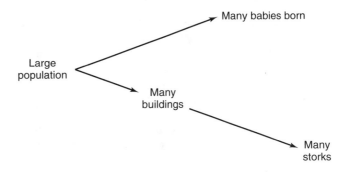

Figure 5.9   *Explanatory diagram indicating that high population accounts for many babies being born and for many storks being sighted.*

It is important to note that it is not the associations which are spurious; rather it is the *interpretations* people place on the associations. There is a real association here between birthrate and the number of storks. What would be "spurious" in this case is an interpretation that involves an assumption of some intrinsic or causal relationship between them.

Two conclusions can be drawn from this discussion. First, the discovery of an association between two variables does not in itself demonstrate that one "causes" the other. Caution must be exercised in making or interpreting causal arguments, including any suggestion that changing an observation's value on x by whatever means will necessarily result in a change in y. Second, in attempting to understand the relationship of two measures, it is usually necessary to consider other variables as well that help to explain how the two are connected. This is the topic of Section 5.2.

Effect of Change of Scale

In Chapters 3 and 4 it is demonstrated that adding the same constant to every observation adds the same constant to the mean but does not change the standard deviation. Also, multiplying every observation by a constant multiplies the mean by that constant and the standard deviation by the absolute value of the constant. Since the correlation is the ratio of the covariance to the product of standard deviations, it is easy to see the effect on the correlation of either rescaling by focusing on the covariance.

In Table 5.2 suppose that x, alcohol consumption, was one unit higher in each province, that is, every x_i is increased by 1. What would the correlation of the new alcohol consumption variable be with y? To answer this, we may examine the sample covariance, given by

$$s_{xy} = \frac{\sum(x_i - \bar{x})(y_i - \bar{y})}{n - 1}.$$

Adding 1 to each x value increases the x_i, but also increases $\bar{x}$ by the same amount. The difference $(x_i+1)-(\bar{x}+1)$ is identical to $x_i-\bar{x}$. Thus, like the standard deviation, the covariance is not affected by adding a constant, and the correlation remains unchanged. This can be stated as a general rule:

RULE 1   Adding (or subtracting) a constant to every observation, on one or both variables, does not affect the correlation between them.

If the mortality rate for each province was *twice* as high as that given in Table 5.2, the new y values would be those in the table multiplied by 2, that is, 13.0, 20.4, ..., 55.0. What would the correlation of alcohol consumption with the new mortality rates be?

To answer this, we return to the covariance formula. In place of each value y_i would be the new value $2y_i$. From Chapter 3 we know that the mean of the new mortality index is $2\bar{y}$. Thus the deviations from the mean are twice as large as before, that is, $(2y_i - 2\bar{y}) = 2(y_i - \bar{y})$ and the standard deviation of the new mortality index is twice that of the old (Chapter 4). Similarly, the covariance s_{xy} is twice as large as it was when the index was just y. The correlation has the increased covariance in the numerator and the increased standard deviation in the denominator; the ratio (the correlation) remains the same. This can be stated as a general rule:

RULE 2   Multiplying (or dividing) every observation by a constant, on one or both variables, does not affect the magnitude of their correlation. Multiplication by a negative constant changes the sign.

The correlation coefficient is a remarkably stable measure of association. Proofs of these properties are given in Appendix 5B. However, the reader may wish to demonstrate this constancy by adding 1 to each x value in Table 5.2, multiplying each y value by 2, and calculating the correlation of the revised columns of data; they will still have a correlation of 0.51.

Rank Correlation

The rank correlation is the correlation coefficient defined in terms of the ranks of the two variables. It is useful when data on interval or ratio scales are given ranks instead of their measured values, or when data are defined only on an ordinal scale. An ordinal scale is created by assigning the numeral 1 to the "first" or "best" observation, 2 to the "second best," and so on.

Instead of x_i and y_i, the symbol s_i is used to indicate the rank of observation i on one variable, and t_i the rank of the same observation on the second variable. Substitution of these in the computational formula for r yields

$$r_s = \frac{\sum s_i t_i - (\sum s_i)(\sum t_i)/n}{\sqrt{\sum s_i^2 - (\sum s_i)^2/n}\sqrt{\sum t_i^2 - (\sum t_i)^2/n}}. \qquad (5.2)$$

The subscript s is used to acknowledge that Charles Spearman is responsible for developing this application of correlation.

Because ranks always range from 1 to n on each variable, the formula can be simplified. The sum of integer values from 1 to n is

$$\sum s_i = \sum t_i = \frac{n(n+1)}{2}$$

and the sum of squared integers from 1 to n is

$$\sum s_i^2 = \sum t_i^2 = \frac{n(n+1)(2n+1)}{6}.$$

(Try these, if you like.) A fair amount of additional algebra shows that the rank correlation is

$$r_s = 1 - \frac{6\sum d_i^2}{n(n^2-1)}, \qquad (5.3)$$

where d_i is the difference of the two ranks for observation i, that is, $d_i = s_i - t_i$.

The data in Table 5.3 were gathered to investigate the possibility that radioactive contamination in areas where nuclear materials are produced poses a health hazard to the nearby residents. Since World War II, plutonium has been produced at the Hanford, Washington, facility of the Atomic Energy Commission. Over the years, appreciable quantities of radioactive wastes have leaked from their open-pit storage areas into the nearby Columbia River, which flows through parts of Oregon to the Pacific. As part of the assessment of the consequences of this contamination on human health, investigators calculated, for each of the nine Oregon counties having frontage on either the Columbia River or the Pacific Ocean, an "index of exposure." This index of exposure was based on several factors, including distance from Hanford and average distance of the population from water frontage. The cancer mortality rate, cancer mortalities per 100,000 person-years (1959–1964), was also

TABLE 5.3

Radioactive Contamination and Cancer Mortality

County	Index of exposure	Cancer mortality	Exposure rank (s_i)	Mortality rank (t_i)	Difference ($s_i - t_i$)
Clatsup	8.34	210.3	8	9	−1
Columbia	6.41	177.9	7	7	0
Gilliam	3.41	129.9	5	2	3
Hood River	3.83	162.3	6	6	0
Morrow	2.57	130.1	4	3	1
Portland	11.64	207.5	9	8	1
Sherman	1.25	113.5	1	1	0
Umatilla	2.49	147.1	3	5	−2
Wasco	1.62	137.5	2	4	−2

source: Fadeley (1965).

determined for each of these nine counties. Table 5.3 shows the index of exposure and the cancer mortality rate for the nine counties.

To illustrate the rank correlation, the exposure index was converted to ranks; for example, Sherman County with the lowest index was ranked 1. Likewise, the 9 counties were ranked on mortalities from lowest (Sherman County) to highest (Clatsup County). The sum of squared differences in ranks is $\sum(s_i - t_i)^2 = 20$, and the rank correlation between the variables is

$$r_s = 1 - \frac{6 \times 20}{9 \times (81 - 1)}$$

$$= 0.833.$$

Although the sample of counties is small, a noteworthy association is found between exposure and cancer mortality. In this example, a more precise correlation value could be obtained from the original unranked data. However, the rank correlation is useful for many variables that can only be assessed as rankings, including multifaceted quality ratings of persons, objects, products or institutions, or events.

At times ties may occur in the rankings. Two or more competitors in an athletic event may have performances that are indistinguishable; two or more artistic products may appear equally good to a panel of judges; even the *New York Times'* weekly listing of best-sellers marks those whose ranks are indistinguishable from the next most popular book. When this occurs, an appropriate procedure is to assign a value to each of the tied observations that is the median of their rankings had they not been tied. If, in Table 5.3, both Gilliam County and Morrow County had exposure indexes of 3.41, they would both be assigned 4.5

as ranked values instead of 4 and 5 before r_s is computed. Hood River remains rank 6 because there are still 5 counties with larger exposure indexes. Likewise, if Sherman, Gilliam, and Morrow Counties all had the same mortality rate, each would be assigned a t value of 2, instead of 1, 2, and 3. When ties occur, expression (5.2) should be used to compute r_s from the median-assigned ranks instead of (5.3).

5.2   More than Two Variables

When a data set consists of two numerical variables, an ordinary scatter plot is used to represent individuals on a plane; the linear association is summarized by a single correlation coefficient. If there are three variables, we could represent the data in three-dimensional space, say as points in a room, and calculate the correlation of every pair of variables. Since the problem of representing data becomes more complex as the number of variables increases, statisticians have adopted some conventions for displaying the results.

The following sections describe a graphical technique for displaying *multivariate* data, that is, two or more variables, and introduce the *correlation matrix*, a table that facilitates viewing correlations among 3 or more measures.

Profiles

It is often desirable to be able to visualize data for more than three variables simultaneously. A simple method is the use of *profiles*. A profile is obtained by plotting the values of the variables for a given person vertically, but at different horizontal positions; the points are connected by lines. The scores of firefighter candidates 4877, 9766, and 3475 from Table 2.22 are listed in Table 5.4 and the corresponding profiles are shown in Figure 5.10. The four variables are scaled for plotting on the same picture. The shape of the profile is dependent upon the scaling and the ordering of the variables along the axis.

It is evident from the profiles that, except for the stair climb on which all 3 times were about the same, candidate 4877 was the slowest. Also, candidate 4877 stands out by a very low score on the written test. The other two applicants have similar, higher written test results. The profiles suggest similar performance in general between candidates 9766 and 3475, while candidate 4877 appears to be the least qualified of the three.

TABLE 5.4

Task Completion Times for Three Firefighter Applicants (in seconds[a])

Candidate	Stair climb	Body drag	Obstacle course	Written test
4877	97.6	109.2	116.5	71
9766	96.1	104.6	100.7	93
3475	93.8	97.1	90.5	95

source: Table 2.22.

[a]The stair climb and body drag times were increased by 80 seconds so that they would fit on the same graph with the other measures. Note that adding the constant value does not change the shape of the profile.

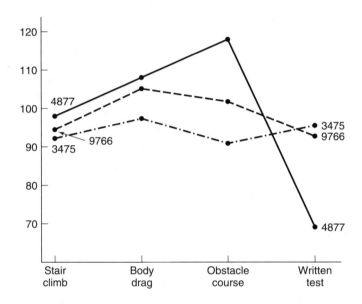

Figure 5.10   *Profiles of Three Firefighter Candidates (Table 2.21).*

Correlation Matrix

When more than two variables are examined, it is common statistical practice to compute the correlation of each pair of measures and place them in a table, or *matrix*, that has one row for each variable and one column for each variable. For example, if we wished to ask about the factors that make some employees remain in their jobs longer than others, we might speculate that one important antecedent is the employee's satisfaction with the work conditions. Suppose that for a sample of employees variable x is "years on the job," variable y is "satisfaction," and the correlation between them is $r_{xy} = 0.54$. (See footnote 3.)

However, other important factors may include salary level and the quality of the employee's relationship with an immediate supervisor. If we call these variables a and b, respectively, then we may wish to calculate the correlation of "years" with each of these as well, i.e., r_{xa} and r_{xb}. Unfortunately, there is no short method for computing any of these. The data must be listed as in Table 5.1 or 5.2, and the computational formulas applied to each pair of variables separately; there is some slight savings in work since the sums of the x values and the squared x values do not change from one correlation to the next.[4]

Once these three correlations are computed, we can compare the direction and magnitude of association of each of three antecedents with x. Also, we may be curious about the associations among y, a, and b themselves. Again, the basic formula for correlation may be applied three more times to obtain r_{ya}, r_{yb}, and r_{ab}.

A common form for displaying the six correlation coefficients is the correlation matrix, as shown in Table 5.5. In practice, actual numerical results are substituted for the symbols in the table. Suppose we found (hypothetically) that $r_{xa} = 0.61$, $r_{xb} = 0.08$, $r_{ya} = 0.22$, $r_{yb} = 0.25$, and $r_{ab} = -0.17$. Then the actual matrix would be the one shown in Table 5.6.

The matrix makes it easy to scan a single row or single column to see the stronger and weaker correlates of any particular variable; here we might be most interested in the first column in particular. It also highlights patterns in the direction of association. For example, the result $r_{ab} = -0.17$ stands out from all the others because of its sign. We might wish to ask a follow-up question such as "why is there some

TABLE 5.5

Correlation Matrix for Four Measures

Variable	x	y	a	b
x	r_{xx}	r_{xy}	r_{xa}	r_{xb}
y	r_{yx}	r_{yy}	r_{ya}	r_{yb}
a	r_{ax}	r_{ay}	r_{aa}	r_{ab}
b	r_{bx}	r_{by}	r_{ba}	r_{bb}

---

[3]The subscripts are added to r to distinguish the correlation of variables x and y from other correlations, discussed in paragraphs that follow.

[4]Computer programs such as MINITAB can compute many correlations rapidly, but there are no shortcuts to the number of arithmetic operations required.

TABLE 5.6

Correlation Matrix for Hypothetical Data

Variable	Years	Satisfaction	Salary	Relationship
Years (*x*)	1.00	0.54	0.61	0.08
Satisfaction (*y*)	0.54	1.00	0.22	0.25
Salary (*a*)	0.61	0.22	1.00	−0.17
Relationship (*b*)	0.08	0.25	−0.17	1.00

(slight) tendency of higher-paid employees to have poorer relationships with their supervisors?"

The matrix in Table 5.6 also demonstrates some mathematical properties of the correlation coefficient. First, it is a "symmetric" index, that is, the same result is obtained regardless of which variable is called x and which is y. This can be seen in any of the correlation formulas, in which multiplication is the only operation by which x and y values are combined. Multiplication, of course, is commutative. As a result, each correlation appears twice in Table 5.6 (e.g., $r_{yx} = r_{xy} = 0.54$), and the matrix itself is said to be symmetric. Sometimes, as in the example to follow, only half of the matrix is given to eliminate the redundancy.

Second, the correlation of each measure with itself (e.g., r_{xx} or r_{yy}) is exactly 1. This is obvious intuitively if one imagines a data table like Table 5.1 in which the same x values were placed in both columns. A scatter plot of the data would have all of the points exactly on a straight line since each value in the left column would be identical to the corresponding value in the right-hand column.[5] Thus, the 1's on the diagonal of the matrix provide no useful information about association, and are sometimes omitted as well.

The matrix in Table 5.7 contains the correlations among the heights of 126 children measured annually from 3 days after birth to age 5. The matrix has some interesting properties often found in studies of human development as well as in other contexts.

The lowest correlations are between the 3-day measures and all other time points. This is attributable to factors that affect an infant's prenatal development, including the length of the gestational period. By age one, however, height becomes relatively stable. This does not mean that children reach their full height by one or two years of age (of course). But they do maintain their *relative* heights to a large extent between ages 1 and 2 ($r = 0.79$), more so between ages 2 and 3 ($r = 0.82$), and the correlation is quite high between ages 4 and 5 ($r = 0.89$).

---

[5]The algebraic proof of this property is saved for a chapter exercise.

TABLE 5.7

Correlations among Heights of 66 Males and 60 Females at 6 Ages

Age	3 days	1 year	2 years	3 years	4 years	5 years
3 days	1.00					
1 year	0.41	1.00				
2 years	0.27	0.79	1.00			
3 years	0.28	0.74	0.82	1.00		
4 years	0.28	0.67	0.76	0.85	1.00	
5 years	0.30	0.66	0.75	0.80	0.89	1.00

source: Finn (1969).

Also, the matrix exhibits a *simplex* pattern in which close time points tend to be more highly correlated than more separate times. This is seen, for example, by comparing the values in the column headed "1 year." The correlations with 2-year, 3-year, 4-year, and 5-year measures are 0.79, 0.74, 0.67, and 0.66, respectively. The same pattern is evident in each of the other columns except for the 3-day measure, which is affected by prenatal and perinatal circumstances.

Summary

Bivariate numerical data can be represented in a *scatter plot* that reveals the shape and magnitude of association between the variables. A visual inspection of *multivariate* data may be gained by *profiles* or by other pictorial representations.

The *correlation coefficient* is a numerical index of the linear association of two variables. It has a range from -1 to $+1$. The sign is negative if the swarm of points falls largely in the upper-left and lower-right portions of the scatter plot; it is positive if the swarm of points falls largely in the lower-left and upper-right section of the plot. The magnitude indicates how close the points are to a straight line. If the points fall exactly on such a line, there is a *perfect* association of the two variables, and the correlation has an absolute value of 1. Even a strong correlation, however, does not mean that one variable "causes" the other. A scatter plot should always be examined, to identify features that the correlation coefficient alone does not indicate.

There is a special form of the correlation coefficient when both variables are rankings (measured on an ordinal scale); this is Spearman's *rank correlation*.

When there are 3 or more variables in a data set, the correlation may be computed separately for each pair of variables and entered into a *correlation matrix*.

Appendices

Appendix 5A   Computational Form for the Covariance

The covariance of x and y is

$$s_{xy} = \frac{\sum(x_i - \bar{x})(y_i - \bar{y})}{n-1} = \frac{\sum x_i y_i - (\sum x_i)(\sum y_i)/n}{n-1}.$$

These two formulas are the same if their numerators are the same. The rules of summation needed for this proof are given in Appendix 3B. The numerator is

$$\sum(x_i - \bar{x})(y_i - \bar{y})$$

$$= \sum(x_i y_i - \bar{x}y_i - \bar{y}x_i + \bar{x}\bar{y}) \qquad \text{(multiplying)}$$

$$= \sum x_i y_i - \sum \bar{x}y_i - \sum \bar{y}x_i + \sum \bar{x}\bar{y} \qquad \text{(distributive rule)}$$

$$= \sum x_i y_i - \bar{x}\sum y_i - \bar{y}\sum x_i + n\bar{x}\bar{y} \qquad \text{(Rule 2).}$$

Substituting $\sum x_i/n$ for $\bar{x}$ and $\sum y_i/n$ for $\bar{y}$ and combining similar terms shows that the expression above is

$$\sum x_i y_i - \frac{\sum x_i}{n}\left(\sum y_i\right) - \frac{\sum y_i}{n}\left(\sum x_i\right) + \frac{\sum x_i}{n}\left(\sum y_i\right)$$

$$= \sum x_i y_i - \frac{(\sum x_i)(\sum y_i)}{n}.$$

Appendix 5B   Change of Scale

RULE 1   If a constant is added to (subtracted from) every value of variable x and/or variable y, the correlation r_{xy} is not affected.

Proof   The correlation is the ratio $s_{xy}/s_x s_y$. It is proven in Appendix 4A that the standard deviations are not affected by adding or subtracting a constant.

Likewise, the covariance is not affected. If the constant value is a, and each x value is increased by a, the covariance is

$$\frac{\sum[(x_i + a) - (\bar{x} + a)](y_i - \bar{y})}{n - 1}$$

since the mean, $\bar{x}$, is also increased by a units. (See Appendix 3C.) Since the value of a is both added and subtracted in the numerator of this expression, it becomes just

$$\frac{\sum(x_i - \bar{x})(y_i - \bar{y})}{n - 1} = s_{xy}.$$

The same result obtains for variable y, or for both x and y at the same time.

Rule 1 may be stated in symbols as

$$r_{xy} = r_{(x+a)y} = r_{x(y+b)} = r_{(x+a)(y+b)},$$

where a and b are constants.

RULE 2

If every value of variable x and/or variable y is multiplied by a constant, the magnitude of the correlation r_{xy} is not affected. Multiplication by a negative constant changes the sign.

Proof

The correlation is the ratio $s_{xy}/s_x s_y$. It is proven in Appendix 4A that multiplying every value of x by a constant multiplies the standard deviation s_x by the absolute value of the constant. It is also the case that multiplying every value of x by the constant a multiplies the covariance of x with any other variable y by the same constant. The modified covariance is

$$\frac{\sum(ax_i - a\bar{x})(y_i - \bar{y})}{n - 1},$$

since $\bar{x}$ is also multiplied by a. (See Appendix 3C.) This is

$$\frac{\sum a(x_i - \bar{x})(y_i - \bar{y})}{n - 1} = \frac{a\sum(x_i - \bar{x})(y_i - \bar{y})}{n - 1} = as_{xy}.$$

The correlation is

$$\frac{as_{xy}}{|a| s_x s_y} = \frac{a}{|a|} \times \frac{s_{xy}}{s_x s_y} = \frac{a}{|a|} r_{xy}.$$

This is equal to the original correlation r_{xy} if the constant is positive, and the original correlation *with the opposite sign* if the constant is negative. The magnitude of the correlation is unchanged in either case. The same result occurs if variable y is multiplied by a constant or if both x and y are multiplied by different constants.

The fact that the magnitude of the correlation is unchanged by multiplying by constants may be stated in symbols as

$$|r_{xy}| = |r_{(ax)y}| = |r_{x(by)}| = |r_{(ax)(by)}|,$$

where a and b are the constant multipliers.

Exercises

5.1   For each of the following examples, state from your general knowledge whether the correlation between the two variables is positive, negative, or zero.

(a) Severity of earthquakes and amplitude of change in a recording pen on a seismograph,

(b) Golf scores and number of hours spent practicing,

(c) The length of dogs' tails and their owners' waist measurements,

(d) Heights of fathers and their sons,

(e) Bowling scores and number of hours spent practicing.

5.2   For each of the following examples, state from your general knowledge whether the correlation between the two variables is positive, negative, or zero.

(a) Height and weight,

(b) Amount of traffic and time needed to reach destination,

(c) Education and income,

(d) Price of personal computers in the United States and price of tea in China,

(e) Grade in physics and grade in mathematics.

5.3   Table 5.8 gives five (x, y) pairs. Treating these data as a sample, calculate

(a) The variance of x,

(b) The standard deviation of x,

(c) The variance of y,

(d) The standard deviation of y,

(e) The covariance of x and y,

(f) The coefficient of correlation between x and y.

TABLE 5.8
Data Set

x	1	2	3	4	5
y	5	4	3	1	2

5.4   Consider the data of Table 5.9.

(a) Make a scatter plot of the data.

(b) Note that the four points fall on a straight line. Verify that the value of the correlation coefficient is 1.

TABLE 5.9
Four Collinear Points

x	1	2	3	4
y	4	7	10	13

5.5   Scores of 20 husbands and wives on a test of conformity are given in Table 5.10.

TABLE 5.10
Scores of 20 Husbands and Wives on a Test of Conformity

Couple	Husband's score, x	Wife's score, y
1	9	8
2	19	31
3	13	18
4	8	7
5	6	10
6	25	28
7	6	8
8	15	16
9	4	8
10	3	4
11	12	21
12	22	26
13	14	12
14	13	17
15	12	9
16	20	16
17	12	12
18	18	18
19	7	6
20	15	24

Hypothetical data.

(a) Make a scatter plot of the data. What does the plot indicate about the magnitude (high, medium, low) and direction of the correlation?

(b) Compute the value of r.

5.6 Compute the correlation between the language and nonlanguage IQ scores in Figure 5.1.

5.7 Compute the correlation between moisture content and beam strength for the data in Figure 5.3. Compare both the direction and magnitude of this result to the correlation obtained from Figure 5.1.

5.8 In a traffic study the average speed of cars (in miles per hour) and the "acceleration noise" level were measured for each of 30 sections of roadway. In Table 5.11 the sections are listed in order of increasing average speed.

(a) Plot noise level y against average speed x.

(b) Calculate the correlation of noise level with speed.

(c) What sort of policy on speed limits should be encouraged to decrease noise? (Consider lower and upper limits.) Does your antinoise policy agree with or conflict with safety practices?

TABLE 5.11
Average Speed and Noise Level in 30 Highway Sections

Section	x = Average speed (mph)	y = Noise level	Section	x = Average speed (mph)	y = Noise level
1	10.0	1.60	16	40.5	0.00
2	11.5	1.45	17	41.5	0.00
3	12.0	1.10	18	42.0	0.00
4	18.0	1.05	19	42.5	0.00
5	19.5	1.25	20	43.0	0.00
6	20.0	0.50	21	43.9	0.40
7	28.0	0.70	22	44.0	0.20
8	31.0	0.45	23	44.1	0.06
9	32.0	0.40	24	45.0	0.09
10	33.0	0.10	25	47.9	0.35
11	33.5	0.18	26	48.0	0.38
12	35.0	0.11	27	50.0	0.37
13	36.0	0.00	28	50.2	0.10
14	39.0	0.00	29	50.5	0.12
15	40.0	0.00	30	51.0	0.25

source: Hypothetical data suggested by Drew and Dudek (1965), Figure 20, p. 52.

5.9 (continuation) Compute the values of x^2 from Table 5.11. Plot y against this new variable. Try to draw some conclusion about the mathematical relationship of y to x.

5.10 Blau (1955) compared scores of competitiveness and productivity of interviewers in two sections of a state employment agency. (See Table 5.12.) Circumstances are such that an interviewer who wants to make more job placements than other interviewers in the section is forced to monopolize the available openings so that the other interviewers are prevented from filling them. In Section A the norms of the group favor competition. The more competitive interviewers, then, hoard openings. In Section B the norms of the group favor cooperation, so that all job openings coming to the section are shared by the five interviewers in that section.

(a) Graph these data using the horizontal axis to represent competitiveness and the vertical axis to represent productivity. Mark each point representing an interviewer in Section A with the symbol A and each point representing one in Section B with B.

(b) Do the two sections seem to differ with respect to competitiveness? If so, how?

(c) Do the two sections seem to differ with respect to productivity? If so, how?

(d) Does there seem to be a relationship between competitiveness and productivity in Section A? If so, is it positive or negative?
(continued . . .)

TABLE 5.12

Competitiveness and Productivity in Two Sections of a State Employment Agency

	Section A		Section B	
Competitiveness[a] (x)	*Productivity*[b] (y)	*Competitiveness*[a] (x)	*Productivity*[b] (y)	
3.9	0.70	2.2	0.53	
3.1	0.49	1.6	0.71	
4.9	0.97	1.5	0.75	
3.2	0.71	2.1	0.55	
1.8	0.45	2.1	0.97	
2.9	0.61			
2.1	0.39			

SOURCE: Blau (1955), Table 2, p. 53.
[a]Based on the degree to which each interviewer made more than the expected number of referrals of his or her own clients to job openings received personally, that is, the degree to which the interviewer hoarded job openings as they came in.
[b]The proportion: actual placements made (client actually gets job) divided by the number of openings per interviewer per section.

(e) Does there seem to be a relationship between competitiveness and productivity in Section B? If so, is it positive or negative?

(f) What conclusions do you draw from this analysis? (Answer this question on the assumption that the differences observed are typical of such interviewers.)

5.11   Assume that the sociologist of Exercise 5.10 collected data from a third section of the same agency (Section C) and obtained the results in Table 5.13.

(a) Make a scatter plot of these data.

(b) Examine the data for outliers. If any data points seem to be erroneously recorded, eliminate them.

(c) Describe the relationship of competitiveness with productivity.

(d) Calculate the correlation coefficient. Is it consistent with what you see in the scatter plot? If not, why not?

TABLE 5.13

Competitiveness and Productivity in a Hypothetical Work Setting

Competitiveness (x)	Productivity (y)
3.1	0.72
1.3	0.53
4.4	0.51
1.4	1.02
4.8	0.52
3.6	0.77
2.1	0.51
1.5	0.48
5.1	0.52
4.0	0.66
2.8	0.63
3.2	0.81

Hypothetical data.

5.12   Draw the scatter plot for the heights and weights given in Table 2.15. Are any outliers apparent from the plot?

5.13   For the data for firefighter applicants in Table 2.22, do the following:

(a) Make a scatter plot of agility and written test scores for males, using the letter "M" to denote each point. Does there appear to be an association between the two variables? If so, describe it.

(continued . . .)

(b) On the same graph, make a scatter plot of agility and written test scores for females, using the letter "F" to denote each point. Does there appear to be an association between the two variables for females? If so, describe it.

(c) How would you describe the differences that you find between the distributions for males and females?

5.14 Demonstrate algebraically (without data) that the correlation of a variable with itself is +1.0.

5.15 Table 5.14 gives data on domestic flights made by 10 airlines during the fourth quarter of 1988. Variable x, "available seat miles," is an index of the size of the airline in terms of the number of miles flown; it is reported in billions. Variable y, the "load factor," is an index of efficiency—the percent of seats occupied by revenue-paying passengers.

(a) Compute ranks for each airline on both variables.

(b) Calculate the rank correlation of the two variables.

(c) Describe the association of the size of airlines, in terms of miles flown, with efficiency.

TABLE 5.14
Domestic Flights During October–December, 1988

Airline	Available seat miles	Load factor
American	23.8	61.0
Delta	20.1	56.6
United	20.6	64.9
Northwest	10.1	57.3
Continental	12.4	57.9
Eastern	8.9	58.7
TWA (not the author)	8.7	55.4
Pan American	2.6	55.7
Piedmont	5.6	56.3
USAir	7.7	58.8

source: American Airlines.

5.16 Having wondered for years if the students who were first to finish an exam were the most knowledgeable or simply could not answer the questions, Professor F piled last semester's papers in the order in which they were returned. Papers were given a numerical grade and were also ranked from the best paper to the poorest. The ranks are given in Table

TABLE 5.15
Examination Data from Last Semester's Class

Student (actual initials used)	Order of completion	Ranked performance
LB	1	3
AR	2	7
DR	3	1
JF	4	8
DJ	5	2
MY	6	4.5
SC	7	4.5
LS	8	13
MS	9	9
CK	10	6
KF	11	16
RR	12	10
BR	13	14
GB	14	11
DH	15	15
JG	16	17
MJ	17	12

source: Last semester's class.

5.15. Calculate a rank correlation between the variables, and describe the relationship between the order of completion and performance on the test.

5.17   Calculate the correlation of the actual exposure index with cancer mortality for the data in Table 5.3. How does this result compare with the rank correlation? How is the difference between the two correlations explained?

5.18   For every x_i value in Table 5.8, substitute a new value by computing $x_i + 10$. For example, replace $x_1 = 1$ by 11 instead. Recalculate the standard deviation of x and the correlation of x with y. How do these compare with the results obtained in Exercise 5.3?

5.19   For every y_i value in Table 5.8, substitute a new value by computing $y_i/2$. For example, replace $y_1 = 5$ by 2.5. Recalculate the standard deviation of y and the correlation of y with the x values as *modified for Exercise 5.18*. How do these compare with the results obtained in Exercise 5.3?

5.20   Two variables, x and y, are summarized in Table 5.16. Assume that two other variables are created by transforming each x_i value to $x_i - 6$, and each y_i value to $3y_i + 1.5$. Construct a table like Table 5.16, giving the summary statistics for the two transformed variables.

TABLE 5.16

Summary Statistics for Two Hypothetical Variables

x	y
$\bar{x} = 7.42$	$\bar{y} = 19.62$
$s_x^2 = 1.69$	$s_y^2 = 4.41$
$s_x = 1.30$	$s_y = 2.10$
$s_{xy} = -1.150$	
$r_{xy} = -0.421$	

5.21   The mean and standard deviation of variable x are $\bar{x}$ and s_x, respectively, and its correlation with variable y is r. Variable v is created by transforming each x value to

$$v_i = ax_i + b,$$

where $a = 1/s_x$ and $b = \bar{x}/s_x$. In their simplest form, what are $\bar{v}$, s_v, and the correlation of variable v with variable y?

5.22   The data on alcohol consumption given in Table 5.2 were reported originally by Nanji and French in 1985. The study also included a measure of pork consumption for each Canadian province. For the 10 provinces in the order listed in Table 5.2, the pork consumption measure was 5.8, 6.8, 3.6, 4.3, 4.4, 5.7, 6.9, 7.2, 14.9, and 8.4.

(a) Make a scatter plot that shows the relationship of pork consumption with mortality due to cirrhosis of the liver, that is, with variable y of Table 5.2.

(b) Compute the correlation coefficient between pork consumption and mortality due to cirrhosis of the liver for the 10 provinces.

(c) How does this correlation compare to the correlation of mortality with alcohol consumption?

5.23   (continuation) Compute the correlation between alcohol consumption as given in Table 5.2 and pork consumption as given in Exercise 5.22. Arrange this correlation, the correlation between alcohol consumption and mortality, and the correlation between pork consumption and mortality into a correlation matrix with appropriate row and column labels. What conclusions can be drawn from the elements of this matrix?

5.24   In a study of factors that influence school performance among adolescents, Wise and Cramer (1988) obtained the correlations among measures of achievement in school subjects and two abilities thought to underlie school performance. These are general aptitude, as indicated by an IQ test, and the specific ability to approach problems analytically. The latter was assessed by giving each student the "Embedded Figures Test" (Witkin, 1971). The subjects for the study were 840 junior-high-school students. The correlations among the variables are given in Table 5.17.

(a) What is the correlation between the ability to locate embedded figures and IQ?

(b) What is the correlation between language performance and mathematics performance in school?

(c) Of the variables listed, what is the strongest correlate of mathematics performance? Of reading performance?

(d) Which of the two ability indicators has the higher correlation with achievement in each of the three academic subjects?

(e) One reader of these results noticed that of the three academic subjects, mathematics had the highest correlation with both IQ and embedded figures. Based on these correlations, she recommended "to increase individuals' mathematics performance, we should design exercises to raise their scores on both of the aptitude tests." What is wrong with this recommendation?

TABLE 5.17

Correlations among Measures of School Performance and Two Ability Indicators

Variable	IQ	Embedded figures	Reading	Language	Mathematics
IQ	1.00				
Embedded figures	0.50	1.00			
Reading	0.69	0.39	1.00		
Language	0.74	0.40	0.72	1.00	
Mathematics	0.77	0.47	0.67	0.15	1.00

Reproduced with permission of authors and publisher from: Wise, P.S., & Cramer, S.H. Correlates of empathy and cognitive style in early adolescence. *Psychological Reports*, 1988, *63*, 179–192. © Psychological Reports 1988.

5.25   As part of a study of cardiovascular health, researchers at Georgetown University Medical School collected information on 66 mothers, each with a daughter 10 to 12 years old and another daughter 4 to 5 years old. Each mother and daughter provided four pieces of information: heart rate in beats per minute (HR), systolic blood pressure (SBP), diastolic blood pressure (DBP), and a general indicator of physical health, the body mass index (BMI). Systolic and diastolic blood pressure are measured in millimeters elevation in a column of mercury when the heart is pumping and when it is at rest, respectively. BMI is the ratio of weight to the square of height, where weight is measured in kilograms and height in meters. Correlations among these measures are given in Table 5.18.

(a) What is the correlation between a child's heart rate and her older sister's heart rate? Between a child's heart rate and her mother's heart rate? Between an adolescent's heart rate and her mother's heart rate? Draw a general conclusion from these three values.

(b) What is the correlation between a child's BMI and her older sister's BMI? Between a child's BMI and her mother's BMI? Between an adolescent's BMI and her mother's BMI? Draw a general conclusion from these three values.

(c) What is the correlation between SBP and DBP among young children in the study? Among adolescents? Among adults? Draw a general conclusion from these three values.

(continued . . .)

TABLE 5.18

Correlations among Cardiovascular Measures for 66 Families

Variable	Child				Adolescent				Mother			
	HR	SBP	DBP	BMI	HR	SBP	DBP	BMI	HR	SBP	DBP	BMI
Child												
HR	1.00											
SBP	0.12	1.00										
DBP	0.21	0.58	1.00									
BMI	0.03	0.19	−0.03	1.00								
Adolescent												
HR	−0.06	−0.14	−0.14	0.23	1.00							
SBP	0.05	0.01	0.17	0.02	0.14	1.00						
DBP	0.04	0.16	0.27	0.24	0.16	0.52	1.00					
BMI	−0.07	−0.10	−0.13	0.46	0.32	0.30	0.16	1.00				
Mother												
HR	0.03	0.12	0.08	0.17	0.00	0.06	0.09	0.17	1.00			
SBP	−0.09	0.42	0.37	0.02	−0.07	0.10	0.29	−0.09	0.09	1.00		
DBP	0.00	0.41	0.46	−0.02	−0.24	0.16	0.28	−0.21	−0.04	0.72	1.00	
BMI	−0.03	0.14	0.06	0.21	0.28	0.15	0.10	0.40	0.18	0.40	0.08	1.00

source: Laboratory for Children's Health Promotion, Georgetown University Medical School.

(d) How strongly are heart rate and blood pressure correlated with physical well being (BMI) among children? Among adolescents? Among adults? How would you explain the lower correlations with BMI among the younger children?

5.26   (continuation) Using the data of Table 5.18, write a brief summary that answers each of the following:

(a) Examine the six correlations among the health measures (HR, SBP, DBP, BMI) for children only, the six for adolescents only, and the six for mothers only. Which of the correlations tend to *increase* with age (either from childhood to adolescence or from adolescence to adulthood, or both)?

(b) Which of the correlations tend to *decrease* with age?

(c) What are the most noteworthy similarities between the characteristics of young girls and their older sisters, between young girls and their mothers, and between adolescent girls and their mothers?

5.27   Use SPSS or another computer program to calculate the matrix of correlations among the three task times and written test scores for the data of Table 2.22.

(a) What conclusions can be drawn about the relationship of the agility scores (task times) with the written test?

(b) What conclusions can be drawn about the relationships among the times on the three tasks?

Summarizing Multivariate Data: Association Between Categorical Variables

Introduction

Statistical data are used frequently to answer questions about the association of two or more variables. When the variables have numerical scales, association may be examined through scatter plots and the correlational techniques discussed in Chapter 5. In this chapter we discuss methods for examining relationships between and among categorical variables.

The idea of association is basically the same as in Chapter 5, that is, do certain values of one variable tend to occur more frequently with certain values of another? The values of categorical variables, however, may not have a range from lower to higher quantities, but may represent qualitatively distinct conditions, such as a condition being "present" or "absent." Summary statistics such as the mean and standard deviation are not applicable. Instead, observations are simply counted; for example, how many observations have both condition A present and condition B present? Counts such as these are entered into *frequency tables* that make patterns of association apparent.

A frequency table for two variables is a two-way table. The special but important case of two dichotomous variables is treated in Section 6.1. Section 6.2 treats larger two-way tables.

As with numerical variables, an association of two categorical variables does not imply that one *causes* the other. The direction of causation may be reciprocal, with each variable affecting the other; causation may be mediated by one or more intervening third variable(s); or both variables may be consequences of additional variables not included in the data causing the two outcomes to occur together. To understand these effects more completely, it is often necessary to examine the association of three or more variables.

The tabulation of data for three categorical variables results in a three-way frequency table (Section 6.3). In addition to describing the relationship of each pair of variables, three-way tables make it possible to explore the effects of a third variable on the pattern of association of the other two. The possible effects of a third variable are discussed in Section 6.4.

6.1   Two-by-Two Frequency Tables

Organization of Data into Two-by-Two Tables

Bivariate Categorical Data.   *Bivariate* categorical data result from the observation of two categorical variables for each individual. Such data might result, for example, by classifying the 25 students in a statistics course by sex and university standing (graduate or undergraduate). Table 6.1 illustrates such data as originally recorded.

Table 6.2 is a tally sheet and summarization of the data listed in Table 6.1. (For example, "Adams M G" is recorded as the first tally in the lower left-hand box in Table 6.2) This kind of summary is called a *cross-tabulation* or *cross-classification* of individuals, because they are categorized using two classifications simultaneously. The table indicates that 10 persons are male undergraduates, 4 are female undergraduates, 8 are female graduates, and 3 are male graduates.

TABLE 6.1

Data with Two Categorical Variables

	Name	Sex	University division		Name	Sex	University division
1	Adams	M	G	14	Sanchez	F	G
2	Brown	M	U	15	Schwartz	M	G
3	Clark	F	G	16	Thomas	M	U
4	Davis	M	G	17	Torres	M	U
5	Franz	M	U	18	Tucker	M	U
6	Humphreys	M	U	19	Tyler	M	U
7	Jones	M	U	20	Waters	F	G
8	Keller	M	U	21	York	F	U
9	Mann	M	U	22	Young	F	U
10	Martin	F	U	23	Youngman	F	G
11	Morgan	F	G	24	Yule	F	U
12	Rogers	F	G	25	Ziegler	F	G
13	Ruiz	F	G				

Hypothetical data.
U = undergraduate
G = graduate

TABLE 6.2

2×2 *Frequency Table: Cross-Tabulation of 25 Students*

		University Division	
		G	U
Sex	F	ⅢⅢⅢ 8	ⅢⅢ 4
	M	Ⅲ 3	ⅢⅢⅢⅢ 10

source: Table 6.1.

A variable with two categories, such as male/female or graduate/undergraduate, is called a *dichotomous* variable. A pair of dichotomous variables is called a *double dichotomy*. Because each variable has just two values, the table that results is a *two-by-two (2 × 2) frequency table*.

Double dichotomies arise, for example, when each of a number of persons is asked a pair of yes-no questions. For example, Company X may ask each of 600 men whether or not they use Brand X razors and whether or not they use Brand X blades. A physician may classify patients according to whether or not they have been innoculated against a disease and whether or not they contracted the disease. Of course, the observations do not have to be persons, but may be objects, institutions, or events. Businesses of a certain type, for example, might be classified based on whether or not they provide "day-care" facilities and whether or not they provide maternity leave for pregnant employees.

The questions that are addressed by these counts can be seen in Table 6.2. It is obvious that the total number of females in the course is 12 (i.e., 8 + 4); the total number of males is 13. Also, summing the observations in each column, 11 graduate students and 14 undergraduates took the course. While these totals may be interesting, they are *univariate* summaries of the data (i.e., one variable at a time).

Association in Two-by-Two Tables

Association between the pair of variables is seen by examining the patterns of frequencies in the individual cells. For example, among females in the class, 8 are graduate students and 4 are undergraduates; the ratio is 2:1, or 67% of females are graduate students. Among males, the pattern is the reverse. Three are graduate students and 10 are undergraduates; the ratio is about 1:3.3, and only about 23% of males are graduate students. Put another way, there is a substantial tendency in Table 6.2 for "femaleness" to be associated with "graduate" and for "maleness" to be associated with "undergraduate" status. The very different graduate:undergraduate ratios for females and males and, equivalently, the clear pattern of "pile ups" of individuals in the two cells in opposite corners of the table, indicate an association between sex and university division.

If the proportion of graduate students is nearly the same for females and males, then there is little or no association of gender with university division. For example, suppose that the course had the numbers of students shown in Table 6.3. There are more females than males in the course and more graduate students than undergraduates. However, the graduate-to-undergraduate ratio is 2:1 and 67% of the students are graduate students in both sex groups. Stated in terms used for numerical variables in Chapter 5, knowledge of an individual's gender does not provide any information about the likelihood of the person being a graduate or undergraduate student. Even if the proportions were not

TABLE 6.3

2 × 2 Frequency Table: Cross-Tabulation of 30 Hypothetical Students

		University Division	
		G	U
Sex	F	12	6
	M	8	4

identical for males and females, but were very close, we would still conclude that there is no noteworthy association of gender with university status.

Because many statistical studies involve 2 × 2 tables and because many concepts of statistics can be presented in this form, we shall consider such tables in some detail. This discussion is simplified by the notation displayed in Table 6.4. The letters *a*, *b*, *c*, and *d* represent *frequencies*, that is, numbers of persons falling into the four possible categories:

a is the number of persons answering "yes" to both questions.

b is the number of persons answering "yes" to Question 1 and "no" to Question 2.

c is the number of persons answering "no" to Question 1 and "yes" to Question 2.

d is the number of persons answering "no" to both questions.

We denote by *n* the total number of persons included in the table:

$$n = a + b + c + d.$$

TABLE 6.4

2 × 2 *Table: General Notation for Frequencies*

		Question 2 Yes	No
Question 1	Yes	a	b
	No	c	d

Other Kinds of 2 × 2 *Tables*

Dichotomous variables sometimes have categories that are *ordered*, so that one value reflects more of some characteristic than the other. For example, income might be classified simply as above or below poverty level, education as having completed fewer than 12 years of schooling or 12 years or more, or dosage of a medication as high or low. In Table 6.5, for example, *b* represents the number of persons with poverty-level incomes who completed 12 years or more of high school, *c* represent the number of persons with incomes above poverty level who did not complete 12 years of schooling, and so on.

Still another type of double dichotomy is shown by the two-by-two tabulation of a dichotomous variable for the *same individuals* at *two different* times. Suppose in August we asked a number of people which of two presidential candidates they favored and then asked the same

TABLE 6.5
2 × 2 Table for Ordinal Variables

		Education	
		Less than 12 years	*12 years or more*
Income	*Below poverty*	*a*	*b*
	Above poverty	*c*	*d*

people the same question in September. We could tabulate the results as in Table 6.6. Here the number *c* would be the number of persons who switched from Democratic candidate Clinton to Republican candidate Bush between August and September 1992. These data are change-in-time data and the table is called a "turnover" table. (Persons favoring candidates other than Bush or Clinton are omitted from this tabulation.)

TABLE 6.6
2 × 2 Table for Change-in-Time Data ("Turnover" Table)

		September	
		Bush	*Clinton*
August	*Bush*	*a*	*b*
	Clinton	*c*	*d*

Turnover tables are useful to address questions about change. For example, we may ask whether the number of people who kept their original preference (cells *a* and *d*) is substantially larger than the number who changed (cells *b* and *c*). Or, we might ask whether the proportion of voters who changed from Bush to Clinton is greater than, the same, or less than the proportion of Clinton's original supporters who changed to Bush.

Calculation of Percentages

The interpretation of association in a frequency table is complicated by the fact that the total number of observations in the first row may differ from the total in the second row; likewise the columns may have different totals. To examine tables for association, it is often better to express all of the cell counts as proportions or percentages. The denominators may be either the row totals or the column totals, whichever is more relevant to the question being posed. In Table 6.5 for example, if we are interested in knowing whether individuals from poverty-level homes complete less education than persons from homes with higher incomes, it would make sense to express each count in the first row (*a* and *b*) as a percent of the number of persons having below-poverty incomes, and each count in the second row (*c* and *d*) as a percent of the number of persons with higher incomes. If, on the other hand, we are interested in the income levels of people who have completed more or less schooling, it would be more informative to express each frequency as a percent of the number of people in the corresponding column.

Marginal Totals. From Table 6.2 we see that $8 + 4 = 12$ of the 25 students are females, while $3 + 10 = 13$ are males. Also, $8 + 3 = 11$ are graduates, while $4 + 10 = 14$ are undergraduates. These totals are appended to the margins of Table 6.2 to obtain Table 6.7 and therefore are called *marginal totals*, or marginals.

TABLE 6.7

2×2 *Frequency Table, with Marginals*

		University Division		Total
		Graduate	*Undergraduate*	*Total*
Sex	*Female*	8	4	12
	Male	3	10	13
	Total	11	14	25

source: Table 6.2.

Percentages Based on Row Totals. Table 6.8 shows what percentage of the female students are graduates ($100\% \times 8/12 = 67\%$), what percentage of the female students are undergraduates (33%), what percentage of the male students are graduates ($100\% \times 3/13 = 23\%$), and what percentage of the male students are undergraduates (77%). Here the *row* totals, 12 and 13, have been used as *bases* for the percentages. (A *row* is a horizontal line of the table.)

Percentages Based on Column Totals. Table 6.9 shows what frequency and what percentage of the graduate students are females ($100\% \times 8/11 = 73\%$), what frequency and what percentage of the graduate students are males (27%), what frequency and what percentage of the

TABLE 6.8

Percent Graduates and Percent Undergraduates, by Sex

		University Division		Total
		Graduate	*Undergraduate*	*Total*
Sex	*Female*	67%	33%	100%
	Male	23%	77%	100%
	Both sexes	44%	56%	100%

source: Table 6.7.

TABLE 6.9

2×2 *Frequency Table, with Percentages Based on Column Totals*

		University Division		Both divisions
		Graduate	*Undergraduate*	*Both divisions*
Sex	*Female*	8 (73%)	4 (29%)	12 (48%)
	Male	3 (27%)	10 (71%)	13 (52%)
	Total	11 (100%)	14 (100%)	25 (100%)

source: Tables 6.7 and 6.8.

undergraduates are females ($100\% \times 4/14 = 29\%$), and what frequency and what percentage of the undergraduates are males (71%). (A *column* is a vertical line in the table.) An alternative presentation of the information in Table 6.9 is shown in Table 6.10.

Percentages Based on the Grand Total.   There is a total of 25 students in the class. Table 6.11 shows that 32% of the students in the class are female graduate students, 16% are female undergraduates, 12% are male graduates, and 40% are male undergraduates. The percentages in the margins show that 48% of all the students in the class are females, 52% are males, 44% are graduate students, and 56% are undergraduate students.

TABLE 6.10

Percent Females, by University Division

	Graduate	*Undergraduate*	*Both divisions*
Percent females	73%	29%	48%
(Number of students)	(11)	(14)	(25)

source: Table 6.9.

TABLE 6.11
Percentages Resulting from Cross-Classification of Students by Sex and University Division

| | | University Division | | |
		Graduate	Undergraduate	Total
Sex	Female	32%	16%	48%
	Male	12%	40%	52%
	Total	44%	56%	100%

source: Table 6.7.

Interpretation of Frequencies

Is there a tendency for those who use Brand X razors also to use Brand X blades? Company X conducted a market survey, interviewing a sample of 600 men. Each man was asked whether he uses the Brand X razor and whether he uses Brand X blades.

Such a large data set would usually not be processed by hand, but by computer instead. Cross-tabulation by making tallies manually as in Table 6.2 would be tedious and possibly error producing.

From Table 6.12 it can be seen that

- Of the 600 men, 245, or 41%, use Brand X blades.
- Of the 279 men using Brand X razors, 186, or 67%, use Brand X blades.
- Of the 321 men not using Brand X razors, 59, or 18% use Brand X blades.

TABLE 6.12
Results of the X Company's Survey

	Use Brand X blades	Do not use Brand X blades	Total
Use Brand X razor	186 (67%)	93 (33%)	279 (100%)
Do not use Brand X razor	59 (18%)	262 (82%)	321 (100%)
Total	245 (41%)	355 (59%)	600 (100%)

Hypothetical data.

Association   Most (67%) of the men using Brand X razors also use Brand X blades; only a few (18%) of the men not using Brand X razors use Brand X blades. Because the percentage of men using Brand X razors who use Brand X blades differs from that percentage among men not using Brand X razors, we say there is an association[1] between using the Brand X razor and using Brand X blades. Thus, from Table 6.13 we can say that men who use the Brand X razor are more likely to use Brand X blades than are men who do not use the razor. This difference may suggest that an advertising campaign for blades should be directed to men not using Brand X razors.

TABLE 6.13
Percent Using Brand X Blades, Among Those Who Use Brand X Razor and Among Those Who Do Not

	Percent using Brand X blades	(Number of men)
Men in sample who use Brand X razor	67%	(279)
Men in sample who do not use Brand X razor	18%	(321)
All men in sample	41%	(600)

source: Table 6.12.

Independence   Table 6.14 is a cross tabulation of 350 persons according to whether they are Democrats or Republicans and whether they intend to vote in the next election. The proportion intending to vote in the next election is 80% among both the Democrats and the Republicans. We say, therefore, that intending to vote in the next election is independent of whether

TABLE 6.14
2×2 *Table of Vote Intentions*

		Democrat	*Republican*	*Both groups*
Intend to vote in next election	*Yes*	160 (80%)	120 (80%)	280 (80%)
	No	40 (20%)	30 (20%)	70 (20%)
	Total	200 (100%)	150 (100%)	350 (100%)

Hypothetical data.

---

[1]Many authors use the term "correlation." We prefer to reserve that term for relationships among *quantitative* variables.

one is a Democrat or a Republican. The percentages in the two groups are the same. Such a lack of association is called *independence.*

Since the percentage intending to vote is 80% in both columns of the table it is also 80% in the column labeled "Both Groups." That is, the marginal proportion 280/350 must also be equal to 0.8. In fact, it is an algebraic identity that if in a row the percentages in the two columns are equal, the same percentage holds for the groups combined. We demonstrate this arithmetically for Table 6.14. For "Both Groups" the denominator is the sum of the column totals $200 + 150 = 350$. The entries in the first row of the Democrat and Republican columns are $0.8 \times 200 = 160$ and $0.8 \times 150 = 120$, respectively. The entry in the first row of "Both Groups" is the sum $0.8 \times 200 + 0.8 \times 150 = 0.8 \times (200 + 150) = 0.8 \times 350$. Thus the ratio of the first entry to the total for "Both Groups" is $0.8 \times 350/350 = 0.8$.

The property of independence still holds when we use the *row* totals as bases for percentages, which results in Table 6.15. The percentage who are Democrats is 57.1%, whether one considers those persons who intend to vote, those who do not, or the entire group. The conclusion, that there is no association between the two variables, is the same. In fact, it is always the case that if the percentages based on column totals exhibit independence, then the percentages based on row totals also exhibit independence.

There is a third, very important, property of a table that exhibits independence, and that is that *any* frequency in the body of the table is equal to the product of the overall row proportion and the column total ($160 = 0.8 \times 200$, $120 = 0.8 \times 150$, $40 = 0.2 \times 200$, $30 = 0.2 \times 150$), and to the product of the overall column proportion and the row total ($160 = 0.571 \times 280$, $40 = 0.571 \times 70$, $120 = 0.429 \times 280$, $30 = 0.429 \times 70$).

TABLE 6.15
2×2 *Table: Percentages Relative to Row Totals*

		Democrat	Republican	Total
Intend to vote in	*Yes*	57.1%	42.9%	100%
next election	*No*	57.1%	42.9%	100%
	Both groups	57.1%	42.9%	100%

source: Table 6.14.

Algebraic Condition for Independence.   These properties can be stated using a simple algebraic notation. The general notation for 2×2 tables with marginal totals is given in Table 6.16 where A_1 and A_2 denote the two values of one dichotomous variable, and B_1 and B_2 denote the two values of another dichotomous variable. The individual frequencies are represented by the letters *a, b, c,* and *d,* as in Table 6.4. The total number of observations in the first row (A_1) is the sum $a + b$, and the total in the second row (A_2) is $c + d$. Likewise, the total numbers of observations in the two columns are $a + c$ and $b + d$, respectively.

TABLE 6.16

2 × 2 Frequency Table: General Notation for Cell Frequencies, Marginal Totals, and Variables

		B		
		B_1	B_2	Total
A	A_1	a	b	$a + b$
	A_2	c	d	$c + d$
	Total	$a + c$	$b + d$	$n = a + b + c + d$

The condition of independence is that the proportion of observations in any one column is the same in the two rows. That is, *a* is the same proportion of the total $a + b$ as *c* is of $c + d$, or

$$\frac{a}{a + b} = \frac{c}{c + d}.$$

This expression can be put in a form that is simpler still. By cross-multiplying, we obtain

$$a(c + d) = c(a + b)$$

or

$$ac + ad = ac + bc,$$

which simplifies to $ad = bc$.

Exactly the same result is obtained if we begin with columns. The condition of independence stipulates that *a* is the same proportion of $a + c$ as *b* is of $b + d$, or

$$\frac{a}{a + c} = \frac{b}{b + d}.$$

By cross-multiplying, we obtain

$$ab + ad = ab + bc,$$

which also simplifies to $ad = bc$. For example, in Table 6.14, $ad = 160 \times 30 = 4800$ and $bc = 120 \times 40 = 4800$. Political party and intention to vote are independent.

To summarize, a 2 × 2 table (written in the notation of Table 6.16) *exhibits independence* if and only if *ad = bc*. This equation provides an easy check for independence, namely, comparison of the product of frequencies on one diagonal with the product of frequencies on the other.

We have discussed the notion of independence in idealized terms; in our hypothetical numerical examples of independence corresponding percentages have been identical. In real-life situations one cannot expect to have percentages exactly the same. (In fact, the row or column totals may be such integers as 7 and 13 so that the possible equal percentages are only 0% and 100%.) In actual data two variables would for practical purpose be considered independent if the percentages differed by very small amounts.

The Amount of Association.   Suppose the 2 × 2 table looked like Table 6.17. Among all 350 persons the proportion intending to vote in the next election is 280/350, or 80%. However, of those who voted in the last election, 90% intend to vote in the next election, while only two-thirds (67%) of those who did not vote in the last election intend to vote in the next one. Thus it appears that there is a noticeable association between having voted in the past election and intending to vote in the coming election. If we were to select a person at random from among the 350, that person is more likely to vote in the coming election if he or she voted in the last election than if he or she did not.

TABLE 6.17

2 × 2 *Table: Moderate Association*

		Voted in last election		
		Yes	No	Total
Intend to vote in	Yes	180 (90%)	100 (67%)	280 (80%)
next election	No	20 (10%)	50 (33%)	70 (20%)
	Total	200 (100%)	150 (100%)	350 (100%)

Hypothetical data.

Suppose the data had appeared as in Table 6.18. Now the percentages intending to vote are 75% and 67%, compared to 90% and 67% before, so there is less of a difference between the two groups than before.

We need numerical measures of the *degree* of association in order to compare different tables and also to gauge how far the relationship in a given table differs from independence; one such measure is discussed below.

TABLE 6.18

2 × 2 Table: Less Association Than in Table 6.17

| | | Voted in last election | | Total |
		Yes	No	
Intend to vote in next election	*Yes*	150 (75%)	100 (67%)	250 (71%)
	No	50 (25%)	50 (33%)	100 (29%)
	Total	200 (100%)	150 (100%)	350 (100%)

Hypothetical data.

Direction of Association.   For variables with categories that have no in-herent order (e.g., male/female; voted/did not vote; agree/disagree; de-fective/not defective) we can only determine the *degree* of association. On the other hand, for variables with ordered categories, there is also a notion of *direction* of association. Table 6.19 is a cross-classification of 250 persons according to income levels and years of education. In this table, both variables have ordered categories, that is, greater or lesser amounts of income and more or fewer years of schooling. To help with interpretation, each frequency has been expressed as a percentage of the corresponding column total. It is evident that higher income and more years of education go together, and lower income and fewer years of education go together; 80% of individuals who did not complete 12 years of schooling have low incomes, and 60% of those who completed 12 years of schooling have higher incomes. In this case—as with the correlation coefficient in Chapter 5—we would say that the association between education and income is *positive*.

TABLE 6.19

2 × 2 Table: Positive Association

| | | Education | | Total |
		Less than 12 years	*12 years or more*	
Income	*Below poverty*	100 (80%)	50 (40%)	150 (60%)
	Above poverty	25 (20%)	75 (60%)	100 (40%)
	Total	125 (100%)	125 (100%)	250 (100%)

Hypothetical data.

Consider, on the other hand, the hypothetical data summarized in Table 6.20 regarding the relationship between college grade-point aver-age and the number of semesters of statistics courses taken. Four-fifths

TABLE 6.20

2 × 2 *Table: Negative Association*

| | | Number of semesters of statistics | | |
		Low (0 or 1)	High (More than 1)	Total
Grade-point average	Low (0.0–2.9)	200 (44%)	40 (80%)	240 (48%)
	High (3.0–4.0)	250 (56%)	10 (20%)	260 (52%)
	Total	450 (100%)	50 (100%)	500 (100%)

Hypothetical data.

(80%) of those students having more than one semester of statistics have low grade-point averages, and most (56% = 100% × 250 ÷ 450) of those having taken only a little statistics have *high* grade-point averages. A *high* grade-point average is associated with a *low* number of semesters of statistics; we say that the association between the two variables, "grade-point average" and "number of semesters of statistics," is *negative*.

Faced with such data, we would want to know whether the cause of the negative association is that the grades in statistics courses are low or whether the students who take statistics courses have lower averages to begin with. Association does not imply causality; if two variables are associated, either *may* "cause" the other, or they may both be associated with a third variable, a "common cause." This topic is examined in more detail in Section 6.4.

Index of Association for 2 × 2 Tables[2]   A numerical indicator of association between two dichotomous variables can be based on the quantity $ad - bc$. Note that a and d are the upper-left and lower-right entries in Table 6.16, and b and c are the upper-right and lower-left entries. If the difference $ad - bc$ is zero, it indicates independence. If the difference is positive, it indicates that A_1 occurs more frequently with B_1, and A_2 with B_2, than the other way around. If the difference is negative, it indicates that A_1 occurs more frequently with B_2, and A_2 with B_1, than the other way around. The difference $ad - bc$ is divided by a quantity that keeps the final index of association between −1 and +1, like the correlation coefficient for numerical scales.

---

[2]Other measures of association that can be computed for larger tables, but which also apply to 2 × 2 tables, are discussed in Chapter 14 (Section 14.3) of this book.

The index of association for a 2×2 table is called ϕ (lower case Greek phi) or the phi-coefficient:

$$\phi = \frac{ad - bc}{\sqrt{(a+b)(c+d)(a+c)(b+d)}}.$$

For Table 6.17

$$\phi = \frac{180 \times 50 - 100 \times 20}{\sqrt{280 \times 70 \times 200 \times 150}}$$

$$= \frac{7,000}{\sqrt{588,000,000}} = \frac{7,000}{24,249} = 0.289,$$

and for Table 6.18

$$\phi = \frac{150 \times 50 - 100 \times 50}{\sqrt{250 \times 100 \times 200 \times 150}}$$

$$= \frac{2,500}{\sqrt{750,000,000}} = \frac{2,500}{27,386} = 0.091.$$

The definition of ϕ does not depend on whether the rows of the table correspond to variable A and the columns to B or vice versa.

When the categories have no inherent order but just represent qualitatively different states, a positive ϕ coefficient indicates only that the preponderance of the observations are the quantities represented by a and d; a negative ϕ indicates that the preponderance of observations are those represented by b and c. In either case, it is necessary to return to the table and see what values of each variable are represented by a, b, c, and d to understand the meaning of the sign of ϕ.

When the categories of the two variables are ordered, so that A_1 and B_1 are both the lower values of their respective variables, and A_2 and B_2 are the two higher values (as in Tables 6.19 and 6.20) then the *direction* of association has the same interpretation as a correlation coefficient. That is, a positive ϕ coefficient indicates that high values on one variable are associated with high on the other, and low values of one variable are associated with low on the other; a negative ϕ coefficient indicates that high values on either variable are associated with low values of the other.

When the categories are ordered, it may make sense to assign numerical values to them, for example, assign -1 to observations that are low on A and 1 to observations high on A; likewise, we can assign -1 to observations that are low on B and 1 to observations high on B. Then each of the n individuals is associated with one of the four pairs of numbers $(1, 1)$, $(1, -1)$, $(-1, 1)$, or $(-1, -1)$. The coefficient of correlation r (Section 5.1) between variables A and B then turns out to be $r = \phi$. That is, for a 2×2 table created from two variables with ordered categories, the ϕ coefficient of association is the same as the ordinary correlation coefficient between 2 two-valued numerical scales.[3]

Perfect Association.   Association is complete or perfect if a value of one variable implies the value of the other. In Table 6.21 every individual in category A_1 is in category B_1 and every individual in category A_2 is in category B_2. The combinations of A_1 and B_2 and of A_2 and B_1 never occur; that is, b and c are zero. We have

$$\phi = \frac{ad - 0}{\sqrt{adad}} = \frac{ad}{ad} = +1.$$

This is the case of *perfect positive association*. Likewise, if all of the observations fell into the categories represented by A_1, B_2 and A_2, B_1, and the entries represented by a and d were zero, the ϕ coefficient would be exactly -1. Like the correlation coefficient r, the extent to which ϕ approaches 1 reflects the degree of association between the two variables.

TABLE 6.21
Perfect Positive Association: $\phi = +1$

		B		
		B_1	B_2	Total
A	A_1	a	0	a
	A_2	0	d	d
	Total	a	d	$a + d$

6.2   Larger Two-Way Frequency Tables

Organization of Data for Two Categorical Variables

Larger Two-Way Tables.   Two-way frequency tables are the result of classification of a number of individuals according to two sets of *categories* simultaneously. They may be 2×2 tables, 2×3 tables, 3×3 tables,

---

[3]In fact, the value of r is the same for all choices of values assigned to the categories of A and B, provided only that observations high on A are assigned a larger number than observations low on A and observations high on B are assigned a larger number than observations low on B.

etc. Thus a two-way frequency table is the *joint frequency distribution of two variables*, that is, a *bivariate distribution*. It contains information about each of the separate variables as the row and column margins, and information about the joint distribution of the two variables as the entries within the table. In general, we refer to a table like this as an $r \times c$ table, where r is the number of rows and c is the number of columns.

Table 6.22 is a 5×3 cross-classification of 36,892 criminal cases that were tried in U. S. district courts in 1992. The total number of cases that appeared in which the defendant was not convicted was 5408, in which the defendant was put on probation was 8810, and in which the defendant was sentenced to a prison term was 22,674. Convictions of one type or the other outnumber non-convictions, in total, by about 6 to 1, with prison sentences being more common than probation. The row margins indicate that drug-related cases are the most common, followed by embezzlement cases, and then by larceny, robbery, and forgery.

TABLE 6.22

Cross-Classification of Cases that Appeared in U.S. District Courts in 1992

| | | Outcome | | |
		Not convicted	*Convicted: probation*	*Convicted: prison*	*All outcomes*
	Robbery	132	32	1459	1623
	Larceny-theft	886	1792	1082	3760
Offense	*Embezzlement and fraud*	1189	4443	4457	10,089
	Forgery	176	589	644	1409
	Drug abuse	3025	1954	<u>15,032</u>	20,011
	All offenses	5408	8810	22,674	36,892

source: Adapted from *Statistical Abstract of the United States*, 1993, p. 206.

The number of robbery defendants who appeared in district courts but were not convicted was 132; this is the *joint frequency* of occurrence of "robbery" and "not convicted." These values tell us which offenses occur with particular outcomes most often and least often. For example, the most common combination is drug abuse cases in which the defendant is sentenced to a prison term, this is the *mode of the bivariate distribution*, with a joint frequency of 15,032 (underlined in table). The joint frequencies provide more detailed information than either the row or column totals alone. In this table they show that drug offenders sent to prison are by far the most common occurrence among cases heard by district court judges.

The percentage of cases in each offense-outcome category is given in Table 6.23. These percentages are called *joint relative frequencies*. For example, the percentage corresponding to the modal category is $100\% \times 15{,}032/36{,}892 = 40.75\%$. The percentage of district court cases that are embezzlement and fraud and result in imprisonment is 12.08%. Less than $\frac{1}{2}$ of a percent of district court cases are robbery defendants or forgery defendants who are not convicted.

TABLE 6.23

Percentages of Cases that Appeared in U. S. District Courts in 1992

| | | Outcome | | | |
		Not convicted	*Convicted: probation*	*Convicted: prison*	*All outcomes*
	Robbery	0.36	0.09	3.95	4.40
	Larceny-theft	2.40	4.86	2.93	10.19
Offense	*Embezzlement and fraud*	3.22	12.04	12.08	27.35
	Forgery	0.48	1.60	1.75	3.82
	Drug abuse	8.20	5.30	<u>40.75</u>	54.24
	All offenses	14.66	23.88	61.46	100.00

source: Table 6.22.

Interpretation of Frequencies

The primary use of a two-way table is to understand the type and degree of association between the variables that define the table's rows and columns. The following examples illustrate how *differences between the distributions* in the rows (or columns) of a two-way table indicate association between the variables.

A 2 × 3 Table.   Table 6.24 is a 2 × 3 frequency table based on a 1990 survey of academic libraries (Williams, 1992). The rows of the table classify the colleges and universities that responded according to the length of the programs they offer, that is, "less-than-4-year institutions" and "4-years-and-greater institutions," respectively. The columns classify the numbers of volumes (books, serials, and documents) in the libraries' holdings into 3 categories: fewer than 30,000 volumes, 30,000 to 99,999 volumes, and 100,000 volumes or more. Since there is a different number of institutions of each type (as well as a different number of libraries with each size classification) it is best to divide by either the row or column totals in order to examine the table for association. In Table 6.24 each joint frequency has been expressed as a percentage of the corresponding row total.

TABLE 6.24
Length of Programs Offered and Size of Library Collection

		Number of volumes in library			Total
		0–29,999	*30,000–99,999*	*100,000 or more*	*Total*
Length of programs	*Less than 4 years*	555 (44%)	656 (51%)	66 (5%)	1277 (100%)
	4 years or more	244 (12%)	530 (27%)	1223 (61%)	1997 (100%)
	Total	799 (24%)	1186 (36%)	1289 (39%)	3274 (100%)

source: Williams (1992), pp. 28–29.

The two variables, length of programs offered and number of volumes, are seen to be associated because the percentage distributions are not the same in the two rows. For example, 44% of the less-than-4-year institutions have libraries with fewer than 30,000 volumes, compared to only 12% of the 4-year-or-more schools. And 61% of the 4-year-or-more schools have libraries with 100,000 volumes or more compared to 5% of the less-than-4-year schools.

Both of the variables in this table are ordinal—the length of programs offered are classified as shorter or longer and the numbers of volumes are ordered from least to most. In such a case, as with correlations, the association may be designated as *positive* or *negative*. The association between length of programs offered and number of volumes in the library is positive because there is a larger percentage of small libraries (0–29,999 volumes) in schools with shorter programs (less-than-4-years) and a larger percentage of big libraries (100,000 volumes or more) in schools with longer programs. That is, the lower value of one variable tends to be coupled with the lower value of the other and the higher value of one variable with high on the other. A negative association would have the reverse pattern, that is, low with high and high with low.

3 × 3 *Tables.* The 3×3 cross-classification shown in Table 6.25 is from a study by Columbia University's Bureau of Applied Social Research. This study dealt with an international student exchange organization which sent young people abroad for a summer of living with a foreign family.

Each group of ten students was accompanied by a leader. Part of the research focused on ways of evaluating participants in the program, and two possible ways of evaluation are shown in the cross-classification in Table 6.25. The columns of the table categorize group members according to the ratings given them by their leader in response to the following:

> Considering that one of the purposes of the Experiment is international good will, how would you rate the members of your group as ambassadors of America abroad? Please name the two who you think were the best and the two you think were the least successful as ambassadors.

Thus each leader designated two students from his or her group as "best ambassadors" and two as "worst ambassadors." The rows of the table distinguish group members according to the degree of change (that is, increase) in their "understanding of some of the simple cultural practices of their host family" as measured by a series of questions asked them both before and after the summer program.

TABLE 6.25

3×3 *Table: Evaluation in Exchange Program*

		Leader's rating			
		Best ambassadors	*Not mentioned*	*Worst ambassadors*	*Total*
Degree of	*Large*	8	16	2	26
change in	*Small*	7	30	9	46
understanding	*None*	2	8	4	14
	Total	17	54	15	86

source: Somers (1959).

The data shown in Table 6.25 are for a total of 86 group members (not a multiple of ten) because some groups actually had less than ten members, and in some of those cases the leader named only one "best ambassador" or one "worst ambassador." The column totals were nevertheless more or less fixed by the plan of the study.

The column totals 17, 54, and 15, and the row totals 26, 46, and 14 are shown in Table 6.25. We can see, for example, that 46 students showed a small change in understanding. Seven of these 46 were mentioned as "best ambassador," 30 were not mentioned, and 9 were mentioned as "worst ambassador."

In Table 6.26 each frequency has been expressed as a percentage of the corresponding column total. The percentages show a definite association between the variables. They decrease from left to right across the top row and increase from left to right across the middle and bottom rows:

- 47% of those mentioned as best ambassador showed a large change in understanding, compared to only 30% for those not mentioned and 13% for those mentioned as worst ambassador.
- Of those mentioned as best ambassador, only 41% showed a small change in understanding, compared to 56% for those not mentioned and 60% for those mentioned as worst ambassador.
- Of those mentioned as best ambassador, only 12% showed no change in understanding, while 15% of those not mentioned showed no change in understanding, and 27% of those mentioned as worst ambassador showed no change in understanding.

The data provide strong evidence of a positive relationship between degree of change in understanding of cultural practices and rated effectiveness as an ambassador of America abroad.

TABLE 6.26

Distribution of Change in Understanding by Leader's Rating

		Leader's rating		
		Best ambassadors	*Not mentioned*	*Worst ambassadors*
Degree of	*Large*	47%	30%	13%
change in	*Small*	41%	56%	60%
understanding	*None*	12%	15%	27%
		100%	101%	100%
	(Number of persons)	(17)	(54)	(15)

source: Table 6.25.

Notice what happens when we calculate percentages from Table 6.25 by using the *row* totals as bases, to obtain Table 6.27. This table shows, for example, that among those who showed a large change in understanding, 31% were mentioned as best ambassadors and 8% were mentioned as worst ambassadors. Those who showed a large change in understanding tended to be good or fair ambassadors but the 8% indicates two "surprises." The second and third rows are studied in the same way.

TABLE 6.27

Best and Worst Ambassadors, by Degree of Change in Understanding

		Leader's rating			
		Best ambassadors	*Not mentioned*	*Worst ambassadors*	*Total*
Degree of	*Large*	31%	62%	8%	101% (26)
change in	*Small*	15%	65%	20%	100% (46)
understanding	*None*	14%	57%	29%	100% (14)

source: Table 6.25.

Association.   Association between the two variables is discovered by using either the column totals or the row totals to calculate percentages. However, one set of percentages may lead to more interesting or relevant interpretations than the other. For example, the percentages in Table 6.26 may suggest that the students who tried to be "good ambassadors" gained more understanding, while Table 6.27 might indicate that those who were affected more by the experience showed up better than fellow students.

The tables give a good summary of the information, but summaries necessarily hide some information. For example, these tables are based only on the *change* in understanding. One wonders, for example, if the two persons among the best ambassadors who had no change in understanding already had a high degree of understanding, which would certainly reduce the significance of the fact that their understanding did not increase. As a general rule, an analysis of *changes*, or *gains*, should include a simultaneous analysis of the initial scores, as well as a study of the relationship between the initial scores and the gains.

Independence.   When the joint frequencies for two variables have no association, they are said to be independent. For example, suppose communities were cross-classified according to average income and crime rate, as in Table 6.28. (In this table the units of observation are *communities*.) The table shows that for each income level most communities (70%) have a medium crime rate. In fact, the entire distribution of crime rates is the same for each level of average income; that is, crime rate is *independent* of level of average income. In this case, knowledge of a community's average income level provides no information about its crime rate. The percentages of districts with low, medium, and high crime rates are 10%, 70%, and 20% *regardless* of the average income. Note that, since there is independence, the marginal distribution of crime rate has to be the same as the distribution for each income level.

TABLE 6.28

Distribution of Crime Rates, by Level of Average Income, for 80 Communities

		Average income level			All income levels
		Low	*Medium*	*High*	
Crime rate	*Low*	1 (10%)	4 (10%)	3 (10%)	8 (10%)
	Medium	7 (70%)	28 (70%)	21 (70%)	56 (70%)
	High	2 (20%)	8 (20%)	6 (20%)	16 (20%)
	Total	10 (100%)	40 (100%)	30 (100%)	80 (100%)

Hypothetical data.

Now let us calculate percentages in the other direction, calculating the percentage of communities of different income levels within each category of crime rate to obtain Table 6.29. Note that the distribution of income levels is the same for all levels of crime rate and is equal to the marginal distribution. Thus, income level is *independent* of crime rate.

TABLE 6.29

Distribution of Income Levels, by Crime Rate

		Average income level			Total
		Low	*Medium*	*High*	
Crime rate	*Low*	1 (12%)	4 (50%)	3 (38%)	8 (100%)
	Medium	7 (12%)	28 (50%)	21 (38%)	56 (100%)
	High	2 (12%)	8 (50%)	6 (38%)	16 (100%)
	All crime levels	10 (12%)	40 (50%)	30 (38%)	80 (100%)

source: Table 6.28.

So it works both ways: income is independent of crime rate, and crime rate is independent of income level. In fact, it is *always* the case that if within the categories of the variable x the distributions of the variable y are the same, then it is also true that within the categories of the variable y the distributions of the variable x are the same. Because of this, we simply say, "x and y are independent," or, in the present example, "crime rate and income level are independent."

It two variables are not independent they are said to be *dependent*, or *associated*, as noted in the previous sections.

The algebraic condition for independence in 2 × 2 tables is given in Section 6.1. Independence is exhibited if the proportion of observations in any one column is the same in the two rows. The example above illustrates that this condition generalizes to larger two-way tables.[4] In many tables, however, the distributions in the rows (or columns) are *almost* the same and the percentages exhibit independence approximately. In fact, exact independence may sometimes be impossible for a specific set of marginal totals because the frequencies are always integers.

6.3 Three Categorical Variables

Organization of Data for Three Yes-No Variables

The Three-Way Table.   We now consider situations in which interest centers on analyzing the interrelationships among three categorical variables.

In the simplest case each variable is dichotomous. The interviewers sent out by Company X to investigate razor and blade use also asked each of 600 men in the sample if he had seen the Brand X television commercial. The full set of data appeared as in Table 6.30.

TABLE 6.30
Data for Three Yes-No Variables

Person	Do you use a Brand X razor?	Do you use Brand X blades?	Have you seen the Brand X television commercial?
1 Adams	No	Yes	Yes
2 Balch	No	No	No
⋮	⋮	⋮	⋮
600 Zuckerman	Yes	Yes	Yes

Hypothetical data.

Data for three dichotomous variables can be summarized as in Table 6.31. The rows indicate whether or not the Brand X razor is used. The columns are paired according to use of the blades, the first two columns for those who have seen the commercial, the second two for those who

---

[4]The generalized algebraic condition is given in Chapter 14.

TABLE 6.31

$2 \times 2 \times 2$ *Table: Cross-Tabulation of 600 Men, by Use of Blades, Use of Razor, and Viewing of Commercial*

	Have seen commercial		Have not seen commercial	
	Use Brand X blades	*Do not use Brand X blades*	*Use Brand X blades*	*Do not use Brand X blades*
Use Brand X razor	86	38	100	55
Do not use Brand X razor	9	17	50	245

Hypothetical data.

have not. For example, there were 86 men who answered Yes to all three of the questions: they have seen the commercial and use Brand X products, the blades and the razor.

The three-way table may be viewed as a set of two 2×2 tables, one for those who have seen the commercial and the other for those who have not. The effectiveness of the television commercial is assessed by comparing these two tables.

By thinking of one of these 2×2 subtables as being placed above the other, we can visualize this $2 \times 2 \times 2$ table as a three-dimensional array of numbers. The table is represented in this way in Figure 6.1. Those who have not seen the commercial are represented at the top of the cube.

How to Examine Three-Way Tables

There are two general approaches that can be taken to understand the associations among three categorical variables. The first is to explore association between each pair of variables in turn. We sum over the categories of one variable (in effect ignoring the categories of that variable) and examine association between the other two; this is called an examination of the *marginal* association between the pair of variables. Through this summing, we obtain the three frequency tables for the three variable pairs shown in Table 6.32.

As a second alternative, we can examine the association between a pair of variables for each category of the third variable separately, that is, the association of buying Brand X razors and Brand X blades for only those individuals who saw the commercial and for only those who did not. This is called the *conditional* association of two variables, because it is examined under one condition—a specific value of the third variable—at a time.

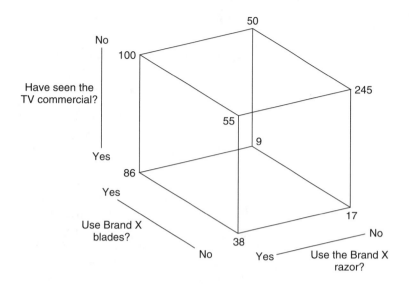

Figure 6.1   *A three-dimensional representation of a three-way table (Table 6.31).*

TABLE 6.32
2 × 2 Frequency Tables for Three Pairs of Variables

		Use Brand X blades?		Total
		Yes	*No*	
Use Brand X razor?	*Yes*	186	93	279
	No	59	262	321
	Total	245	355	600

		Seen TV commercial?		Total
		Yes	*No*	
Use Brand X razor?	*Yes*	124	155	279
	No	26	295	321
	Total	150	450	600

		Use Brand X blades?		Total
		Yes	*No*	
Seen TV commercial?	*Yes*	95	55	150
	No	150	300	450
	Total	245	355	600

source: Table 6.31.

Interpretation.   Of the 600 men in Table 6.32, 150 (or 25%) have seen the Brand X television commercial. Of these 150 men who have seen the commercial, 124 (83%) use the razor.

Table 6.32 can be used to study the marginal associations among the three variables. For example, the top table shows the relationship between just the use of Brand X razors and Brand X blades (discussed in Section 6.1). Likewise the second and third tables allow us to examine the relationship of viewing the commercial with use of Brand X razors and Brand X blades, respectively.

We can also discover, for example, the relationship between popularity of the razor and the blades among those who have and who have not seen the commercial; the data needed for this are in Table 6.31. Among those who have seen the commercial, 124 men (86 + 38) use Brand X razor and 86 of them use the blades. This is 100% × (86/124) = 69%; the remaining 31% do not. Also, of the 26 men who do not use the razor, 9 (35%) use the blades and 17 (65%) do not. We could calculate similar figures for individuals who have not seen the commercial and then compare the tables for the two groups.

Since use of the blades is a simple dichotomy, the percentage of individuals who do not use the blades is exactly 100% minus the percentage of individuals who do use the blades; thus for a dichotomy the data may be condensed somewhat further. Table 6.33 shows just the percentage using the blades, in each of the four groups. (The base for each percentage is indicated below it in parentheses.) The percentages using the blades are higher for those who use the razor, both among those who have seen the TV commercial and among those who have not. Thus the association between using Brand X blades and using the Brand X razor exhibits itself in both groups.

TABLE 6.33

Percent Using Brand X Blades, by Use of Razor and Viewing of Commercial

		Have seen commercial	*Have not seen commercial*
Use brand X razor	*Percent*	69%	65%
	(Number)	(124)	(155)
Do not use brand X razor	*Percent*	35%	17%
	(Number)	(26)	(295)

source: Tables 6.31 and 6.32.

Furthermore, we see that the third variable, viewing the commercial, has its own, separate effect: the percentage using the blades is higher among those who have seen the commercial, both among those who use the Brand X razor and among those who do not.

Various possible effects of a third variable are studied further in Section 6.4. Exercises 6.15 and 6.16 involve further analysis of the effects of Company X's television commercial.

Larger Three-Way Frequency Tables

Each dichotomous variable in the $2 \times 2 \times 2$ table could be replaced by a variable with more categories. We may refer to more general three-way tables as $r \times c \times t$ tables. Such a table can be considered as a set of t tables, each $r \times c$.

Table 6.34 is a $3 \times 3 \times 2$ table that was obtained by Lazarsfeld nad Thielens (1958) in their study on the relationships between age, productivity, and party vote in 1952 for a sample of social scientists. Each of a number of social scientists was classified according to age, productivity score (an index constructed to measure professional productivity), and party vote in 1952.

TABLE 6.34

Classification of Social Scientists by Age, Productivity Score, and Party Vote in 1952

		Low		Medium		High	
		Democrats	Others	Democrats	Others	Democrats	Others
	40 or younger	260	118	226	60	224	60
Age	*41 to 50*	60	60	78	46	231	91
	51 or older	43	60	59	60	206	175

PRODUCTIVITY SCORE

source: Lazarsfeld and Thielens (1958), Figure 1.7, p. 17.

These are data resulting from the observation of three variables for each social scientist. The three variables are categorical: age with the three categories, 40 or younger, 41 to 50, 51 or older; productivity score with the three categories Low, Medium, and High; and Party Vote with the two categories, Democrat and Others. This table can be considered as a series of three two-way tables for age and party, one for each productivity level.

The interrelationships among the three variables can be conveniently summarized by computing the percent Democratic in each of the nine age-productivity categories. This is part of Exercise 6.14.

6.4   Effects of a Third Variable

Association and Interpretation

As with numerical data, an association between two categorical variables is generally clarified further by studying additional variables. It is important to address such questions as: Does the association persist across different values of a third variable? Does a third variable "explain" the association of the other two, perhaps as an underlying cause of both, or perhaps as an intervening variable? Or, does a third variable *hide* the association between two others?

When three variables are categorical, the interrelations are found from three-way tables. In this section we shall compare the overall association between two categorical variables with the association between them in the two-way tables defined by various values of a third variable. That is, we compare the *marginal* association of two variables with their *conditional* association. If two variables have a certain relationship overall, that association may be sustained in the component tables defined by values of the third variable, may be decreased, may be obliterated, or may be reversed. We examine all of these possibilities.

Independence in Subtables

It frequently happens that two variables which appear to be associated are not when a third variable is taken into account. This idea can be illustrated most easily by a hypothetical example displaying *exact* independence.

Table 6.35 is a cross-classification of 6000 persons living in two communities according to income and residence (Community A or Community B). There is an association between residence and income level; 34% of the residents of Community A are in the low-income bracket,

TABLE 6.35
Income Distributions in Two Communities

		Income level			
		Low	*Medium*	*High*	*Total*
Community	*A*	1350 (34%)	1400 (35%)	1250 (31%)	4000 (100%)
	B	475 (24%)	775 (39%)	750 (37%)	2000 (100%)

Hypothetical data.

compared to only 24% for Community B, and 37% of the inhabitants of Community B are in the high-income bracket, compared to only 31% for Community A.

This study can be expanded by considering other factors. Residence and income may both be related to occupation. Table 6.36 is a three-way table. For each of three occupational groups, there is a 2 × 3 table of frequencies according to residence and income level. In each such sub-table residence and income are independent, that is, the income distributions are the same for communities A and B (though they differ from one occupational group to another).

TABLE 6.36

Three-Way Frequency Table: Income and Occupation in Communities A and B

| Occupational group | Community | Income level | | | Total |
		Low	*Medium*	*High*	
Agricultural	*A*	1000 (50%)	500 (25%)	500 (25%)	2000 (100%)
	B	250 (50%)	125 (25%)	125 (25%)	500 (100%)
Blue Collar	*A*	250 (25%)	500 (50%)	250 (25%)	1000 (100%)
	B	125 (25%)	250 (50%)	125 (25%)	500 (100%)
White Collar	*A*	100 (10%)	400 (40%)	500 (50%)	1000 (100%)
	B	100 (10%)	400 (40%)	500 (50%)	1000 (100%)

Hypothetical data.

How can it happen that income and residence are associated when all persons are considered, but are not associated within more homogeneous occupational groups? This can occur because occupation is related to both residence and income.

Table 6.37 shows a relationship between residence and occupation. (The modal occupational categories (with frequencies underlined) are agricultural for Community A and White Collar for Community B.) A little investigation reveals the reason: Community A is close to the farms where the agricultural workers are employed.

TABLE 6.37

Occupation Distributions in Communities A and B

| | | Occupational group | | | Total |
		Agricultural	*Blue collar*	*White collar*	
Community	*A*	2000 (50%)	1000 (25%)	1000 (25%)	4000 (100%)
	B	500 (25%)	500 (25%)	1000 (50%)	2000 (100%)

source: Table 6.36.

Table 6.38 is a cross-classification by occupation and income level. The modal income level is underlined for each occupational category. There is an association between income and occupation.

TABLE 6.38
Cross-Classification of Persons in Communities A and B by Occupation

| | | Occupational group | | |
		Agricultural	*Blue collar*	*White collar*
Income level	*Low*	1250 (50%)	375 (25%)	200 (10%)
	Medium	625 (25%)	750 (25%)	800 (40%)
	High	625 (25%)	375 (25%)	1000 (50%)
	Total	2500 (100%)	1500 (100%)	2000 (100%)

source: Table 6.36.

The association between income level and residence is due to the fact that each of these variables is related to occupation. Figure 6.2 is a *casual diagram* representing this.

If we had restricted ourselves to observing only the association between income and residence, we might have suggested that persons with low incomes would prefer Community A because of certain public facilities (for example, more public parks, child-care centers, etc.). If this were the case, then within occupational categories, we would expect the percent of residents with low income to be greater in A than in B. Such is not the case, however; the association between income and residence does not persist after occupation is taken into account.

When two variables are associated, it is not necessarily true that one factor causes the other. Considering other factors, the investigator may find one which explains the relationship between the other two; it may cause them both.

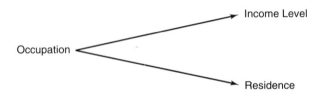

Figure 6.2 *Explanatory diagram: Occupation as a common cause, explaining both income level and residence.*

Education, Intent to Vote, and Political Interest.   Table 6.39 exhibits positive association between level of education and intent to vote, since 92% of those with at least some high school education intended to vote, and only 86% of those with no high school education intend to vote. Table 6.40 is a three-way table in which the third factor, "level of political interest" (in terms of a scale based on the responses to certain questions), is introduced. Table 6.41 shows that there is a positive association between level of education and level of political interest, while

TABLE 6.39

Level of Education and Intention to Vote

| | | Intend to vote? | | |
		Yes	*No*	*Total*
Level of education	*No high school*	1026 (86%)	173 (14%)	1199 (100%)
	At least some high school	1481 (92%)	132 (8%)	1613 (100%)

source: Lazarsfeld, Berelson, and Gaudet (1968). See Table 6.40.

TABLE 6.40

Level of Education, Intent to Vote, and Political Interest

| | GREAT POLITICAL INTEREST | | | MODERATE POLITICAL INTEREST | | | NO POLITICAL INTEREST | | |
| | Intent to vote | | | Intent to vote | | | Intent to vote | | |
	Yes	*No*	*Total*	*Yes*	*No*	*Total*	*Yes*	*No*	*Total*
No high school	279	6	285	602	67	669	145	100	245
At least some high school	490	5	495	917	69	986	74	58	132
Total	769	11	780	1519	136	1655	219	158	377

source: Lazarsfeld, Berelson, and Gaudet (1968), p. 47.

TABLE 6.41

Distribution of Political Interest, by Education

| | Level of political interest | | | *Number of* |
	Great	*Moderate*	*No interest*	*persons*
No high school	24%	56%	20%	(1199)
At least some high school	31%	61%	8%	(1613)

source: Table 6.40.

Table 6.42 indicates that there is a strong association between level of political interest and intent to vote: 99% of those having great political interest intend to vote, compared to 92% for those with moderate political interest and only 58% for those with no political interest.

TABLE 6.42

Percent Intending to Vote, by Level of Political Interest

| | Level of political interest | | |
	Great	Moderate	No interest
Percent intending to vote	99%	92%	58%
(Number of persons)	(780)	(1655)	(377)

source: Table 6.40.

Table 6.43 shows that, within level-of-interest categories, there are only small differences in the percentages intending to vote. The differences between 99% and 98%, 93% and 90%, and 56% and 59% are small compared to the differences manifested in Table 6.39. We conclude that, within level-of-interest categories, there is essentially no association between education and intent to vote. The apparent association between these two variables in Table 6.39 is explained by their mutual associations with level of political interest. Figure 6.3 is a causal diagram consistent with this finding. It indicates that education increases political awareness, which in turn increases the likelihood of voting.

TABLE 6.43

Percent Intending to Vote, by Level of Education and Level of Political Interest

| | | Level of political interest | | |
		Great	Moderate	No interest
Level of	*At least some high school*	99%	93%	56%
education	*No high school*	98%	90%	59%

source: Table 6.40.

In the case where the third variable can be interpreted as a link in a chain from one variable to another, we call that third variable an *intervening variable.*

More Education ⟶ Greater Political Interest ⟶ Intent to Vote

Figure 6.3 *Causal diagram: Intervening variable.*

Summary. In both of these examples, the relationship between two variables was "explained" by a third variable. That is, the conditional association, at particular values of a third variable, were small or nonexistent. Two causal diagrams are drawn to demonstrate the mechanisms through which the third variable operates. In Figure 6.2 the third variable is shown as preceding both of the others; in Figure 6.3 it is shown as intervening between the other two. Since nonexperimental data can never demonstrate causation definitively, it must be remembered that the placement of the third variable in the "causal chain" is based on the data, but also on logic and theory. In short, it makes considerable sense to view occupation as preceding income level and residence, and political interest as intervening between education and intentions to vote. Other orders of events do not make as much sense. Like all explanations of data, however, further research may explain these relationships in still more detail. To note just one possibility, all three variables in either example may be simultaneous outcomes of still other causes, for example, the desire for more education, high political interest and intending to vote may all result from being influenced by parents or peers who view all of these activities as very important.

Similar Association in Subtables

When an association between two variables persists even after other variables are introduced, we consider the interpretation that there is an intrinsic relationship to be better established. The associations in the subtables may be less than that in the overall table, but still in the same direction.

Population Density and Social Pathology. Galle, Gove, and McPherson (1972) studied the relationship, if any, between crowding (as measured by population density) and various "social pathologies" (high rates of juvenile delinquency, high admission rates to mental hospitals, high mortality ratios, etc.) in the 75 community areas of Chicago. We shall phrase our discussion in terms of one social pathology, juvenile delinquency. Table 6.44 shows a marked positive relationship between population density and juvenile delinquency.

It is known that social pathologies are related to socioeconomic status (SES) and that SES is a variable which should be considered when studying the relation between population density and social pathologies. Table 6.45 is a cross-classification of 75 districts according to the three variables, population density, rate of juvenile delinquency, and SES. Examination of Table 6.45 shows that in each half of the table— the left half corresponding to low SES and the right corresponding to

TABLE 6.44

Population Density and Juvenile Delinquency

		Population density		Total
		Low	*High*	
Rate of juvenile delinquency	*Low*	30 (88%)	5 (12%)	35 (47%)
	High	4 (12%)	36 (88%)	40 (53%)
	Total	34 (100%)	41 (100%)	75 (100%)

source: Hypothetical data suggested by the study by Galle, Gove, and McPherson (1972).

TABLE 6.45

Cross-Classification of 75 Community Areas of a Large City by Population Density, Rate of Juvenile Delinquency, and Socioeconomic Status (SES)

		Low SES			High SES		
		Low population density	*High population density*	*Total*	*Low population density*	*High population density*	*Total*
Rate of juvenile delinquency	*Low*	3	2	5	27	3	30
	High	2	33	35	2	3	5
	Total	5	35	40	29	6	35

source: Hypothetical data suggested by the study of Galle, Gove, and McPherson (1972).

high SES—there is positive association between population density and juvenile delinquency. Table 6.46 gives the percentage of community areas having high rates of juvenile delinquency, in each of the four SES-population-density combinations. In both low-SES districts and high-SES districts, the relationship between density and delinquency persists: for

TABLE 6.46

Percentage of Community Areas Having High Rates of Juvenile Delinquency, by Socioeconomic Status (SES) and Population Density

		SES	
		Low	*High*
Population density	*Low*	40%	7%
	High	94%	50%

source: Table 6.45.

low SES areas, the 94% for high-density areas exceeds the 40% for low-density areas; and similarly for high-SES areas, the 50% for high-density areas exceeds the 7% for low-density areas.

Since the positive association between population density and juvenile delinquency is not so great in the SES subtables as it is in Table 6.44, we next ask whether SES is associated with population density and whether it is associated with juvenile delinquency. The two-way tables relevant to these questions are given in Table 6.47. It will be seen that there is negative association in each table.

TABLE 6.47
Relationships of Population Density and Juvenile Delinquency to Socioeconomic Status

| | | SES | | |
		Low	*High*	*Total*
Population density	*Low*	5	29	34
	High	35	6	41
	Total	40	35	75

| | | SES | | |
		Low	*High*	*Total*
Juvenile delinquency	*Low*	5	30	35
	High	35	5	40
	Total	40	35	75

source: Table 6.45.

We may ask whether SES is a common cause, totally explaining the association between population density and juvenile delinquency, as represented in Figure 6.4. However, this proposition is not consistent with the data of Table 6.45. It seems more plausible that SES affects both population density and juvenile delinquency; and, in addition,

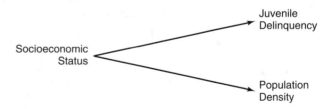

Figure 6.4   *A causal diagram representing low SES as a "common cause" of juvenile delinquency and population density.*

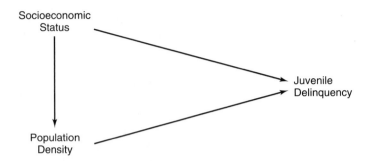

Figure 6.5 *Causal diagram: Density and socioeconomic status each affect delinquency, and socioeconomic status affects density.*

population density has its own, additional effect on delinquency (as represented in Figure 6.5).

The association between population density and juvenile delinquency is "true," in the sense that it persists even after effects of a highly relevant third variable have been taken into account.

Drinking, Smoking, and Health.   Since drinking and smoking are associated (see Table 6.54, for example), alcohol consumption was considered as a possible explanation of the association between smoking and lung cancer. But it was found that among both those who drink a great deal and those who drink little, the rate of lung cancer is higher for smokers than for nonsmokers.

Urban Life, Smoking, and Health.   Air pollution is greater in urban environments than in rural ones. Table 6.48 gives the death rate due to lung cancer in various urban and rural areas. It will be seen that in each type of area the incidence of lung cancer is higher for smokers than for nonsmokers. The relationship between smoking and lung cancer persists when environment is taken into account.

TABLE 6.48

Death Rates (Per 100,000 Person-Years) from Lung Cancer According to Residence and Smoking Habits

	Cities of over 50,000	10,000–50,000	Suburb or town	Rural
Cigarette smokers	85.2	70.9	71.7	65.2
Nonsmokers	14.7	9.3	4.7	0.0

source: Figure 6 from Hammond and Horn, *Journal of the American Medical Association* 166: 1294–1308. Copyright 1958, American Medical Association.

Also, the relationship between environment and lung cancer persists when smoking habits are taken into account: urban smokers are more prone to lung cancer than are rural smokers, and urban non-smokers are more prone to lung cancer than are rural nonsmokers.

The conclusion is that both urban environment and smoking are contributing factors in lung cancer.

Reversal of Association in Subtables

So far we have considered examples in which a third factor "explained away" an association between two variables, and examples in which the association between two variables remains after taking the third factor into account. Now we consider an example in which taking a third factor into account changes the direction of association.

The Relation Between Weather and Crop Yield. This example is suggested by the study of Hooker (1907), who studied the interrelationships among temperature, rainfall, and yield of hay and other crops in part of England. Table 6.49 is a three-way table giving a cross-classification of 40 years, by yield of hay, rainfall, and temperature.

TABLE 6.49
Cross-Classification of 40 Years, by Yield of Hay, Rainfall, and Temperature

YEARS OF LIGHT RAINFALL

| | | Yield | | |
		Low	High	Total
Temperature	Low	4	1	5
	High	11	4	15
	Total	15	5	20

YEARS OF HEAVY RAINFALL

| | | Yield | | |
		Low	High	Total
Temperature	Low	4	11	15
	High	1	4	5
	Total	5	15	20

source: Hypothetical data suggested by the study by Hooker (1907).

The two-by-two tables derived from Table 6.49 are given in Table 6.50. The association between temperature and yield in Table 6.50 is negative, which is somewhat surprising because one would expect a warm summer to make for a good crop. The table suggests that high temperature tends to cause low yield. In contrast, each part of Table 6.49 (for years of light and heavy rainfall, respectively) shows a weak positive association between temperature and yield: that is, in both subtables

$$ad - bc = +5$$

TABLE 6.50

Three 2 × 2 Tables for Rainfall, Temperature, and Yield of Hay in 40 Seasons

		Yield		
		Low	*High*	*Total*
Temperature	*Low*	8	12	20
	High	12	8	20
	Total	20	20	40

		Rainfall		
		Light	*Heavy*	*Total*
Temperature	*Low*	5	15	20
	High	15	5	20
	Total	20	20	40

		Yield		
		Low	*High*	*Total*
Rainfall	*Light*	15	5	20
	Heavy	5	15	20
	Total	20	20	40

source: Table 6.49.

Why is it that these two subtables, when combined, yield an association in the opposite direction? The answer lies in the relationship of temperature with rainfall. From Table 6.50, it is clear that rainfall and temperature are negatively associated. That is, the temperature is apt to be low when there is a good deal of rain. This is seen in Table 6.49 in the overall (marginal) distribution of temperature. In the light-rainfall table, 15 of 20 years had high temperatures; in the heavy-rainfall table, only 5 of 20 years had high temperatures. When the two tables with weak positive associations are combined, the total number of inconsistent years becomes substantially greater than the total number of consistent years.

The conclusion from Table 6.49 is that, among years when the rainfall is light, the yield is apt to be greater when the temperature is high than when it is low, and the same is true among years when the rainfall is heavy. From Table 6.50, we see that the association of rainfall and yield is positive, and the association of rainfall with temperature is negative. A hypothesis consistent with this pattern of association is that both heavy rainfall and high temperature increase yield, but usually the temperature is low when there is a good deal of rain. Figure 6.6 is a "causal diagram" corresponding to this hypothesis.

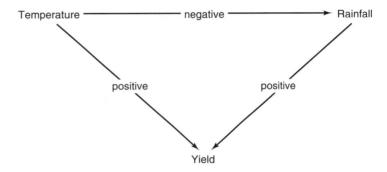

Figure 6.6 *Explanatory diagram: Rainfall as an intervening variable.*

It would be a mistake to take the overall negative association between temperature and yield at face value. In particular, it would be incorrect to deduce from the fact that the association is negative that decreasing the temperature would *cause* larger yield. It is just that low temperature *occurs naturally* with high rainfall, which gives a larger yield. In a greenhouse, we could achieve *both* high temperature *and* high humidity.

In summary: what we have seen in this example is that although the overall association was strongly negative, the associations in the subtables corresponding to fixed levels of rainfall are slightly positive. The overall negative relationship exists between temperature and yield because rainfall is an intervening variable between temperature and yield (Figure 6.5); rainfall is positively related to yield but negatively related to temperature.

Hidden Relationships

In Section 6.4 we saw that when a third variable is taken into account the association between two variables may disappear. The opposite can happen; two variables which appear independent may be found to be associated when the individuals are classified according to a third variable; i.e., the conditional association may be nonzero even if the marginal association is zero. This phenomenon occurs when there is a reversal, that is, when the direction of the association in one group is different from that in the other.

In Table 6.51 there is little association between age and "listening to classical music." The percentages listening to classical music, among those below 40 and those 40 or older are 65% and 64%, respectively.

TABLE 6.51

Percentage Who Listen to Classical Music, by Age

	Age	
	Below 40	*40 and Above*
Percentage who listen to		
classical music	65%	64%
(Number of persons)	(603)	(506)

source: Adapted from Lazarsfeld (1940), p. 98.

But when the additional factor, education, is brought in, Table 6.52 results. Note the reversal: among college-educated persons, 76% of the "oldsters" listen to classical music, while 73% of the "youngsters" do. Among the noncollege persons, *fewer* of those 40 or older listen to classical music: 52% compared to 60% for those below 40.

TABLE 6.52

Percentage Who Listen to Classical Music, by Age and Education

	Age	
	Below 40	*40 and Above*
College	73%	76%
(Number of persons)	(224)	(251)
Below college	60%	52%
(Number of persons)	(379)	(255)

source: Same as Table 6.51.

Within the college group, the association between age and listening to classical music is small, but positive. Within the noncollege group, it is stronger but negative. When the groups are combined, these two opposite effects tend to cancel out, making the overall association small.

Summary

Bivariate categorical data are summarized in *two-way frequency tables*. In the case of two Yes-No (dichotomous) variables, this is a 2×2 *table*. Two-by-two tables arise also from two Low-High variables, and from change-in-time data. A frequency table for change-in-time data is called a *turnover table*.

Larger two-way tables arise when either one or both variables have more than two categories.

Interpreting two-way tables involves examining the percentages in each row and in each column (the *margins*) and observing patterns of *association*. Two variables are associated if the percentages in the body of the table differ from row to row or from column to column. A correlation-like measure of association for 2×2 tables is the *phi coefficient*. When two variables are not associated, they are *independent*. This occurs when the percentage distributions are the same in the rows of the table. In this case, the distributions are also the same in the columns.

Three-way frequency tables arise from the cross-classification of three categorical variables. *Marginal association* may be discovered by examining each pair of variables in isolation; *conditional association* may be found by examining the association of any pair of variables for specific categories of the third variable.

Examination of a third variable can add substantially to understanding the relationship between two others. There may be an overall relationship between two variables that is not present at specific levels of a third variable; there may be a relationship at specific levels of a third variable that is hidden or reversed when levels of the third variable are combined; or a relationship may persist at all levels of a third variable, adding to its generalizability.

Association must be supplemented by further evidence in order to establish causal relationship. Some evidence is supplied when the association persists after relevant third factors have been taken into account. The strongest evidence is supplied by a controlled experiment.

Exercises

6.1   Fill in the following 2 × 2 frequency tables in all possible ways.

(a)
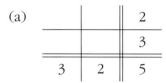

			2
			3
3	2	5	

(b)

2		
5	1	

6.2   Fill in the following 2 × 2 table in all possible ways. Rank the possible tables according to the degree and direction of association exhibited. For each table, compute the phi coefficient of association between variables A and B.

		B		
		Low	*High*	
A	*Low*			4
	High			4
		4	4	8

6.3   Fill in the following 2 × 2 frequency table in all possible ways. Rank the possible tables according to the degree and direction of association exhibited. For each table, compute the phi coefficient of association between variables A and B.

		B		
		Low	*High*	
A	*Low*			5
	High			3
		4	4	8

6.4   The 1000 cars sold by a new-car dealer in a recent year were cross-classified as in Table 6.53. What do these data suggest about the relation between body type and type of transmission purchased?

TABLE 6.53

Cross-Classification of 1000 New Cars Sold in a Recent Year

		Transmission		
		Automatic	*Manual*	*Total*
Body type	*2-door*	250	225	475
	4-door	500	25	525
	Total	750	250	1000

Hypothetical data.

6.5 Table 6.54 gives the joint frequency distribution (in percentages) of white males in a single, long-range study, by smoking and drinking habits.

(a) Compute the proportion of smokers who drink.

(b) Compute the proportion of nonsmokers who drink.

(c) For this group of men, are smoking and drinking associated?

TABLE 6.54

Joint Frequency Distribution of Men by Smoking and Drinking Habits (Percentages)

	Smoker	*Nonsmoker*	*Total*
Drinker	73%	8%	81%
Nondrinker	14%	5%	19%
Total	87%	13%	100%

source: Marvin A. Kastenbaum, Director of Statistics, The Tobacco Institute, Inc., personal communication.

6.6 Summarize Table 6.55 using two different low-high categorizations of the variable, "Highest Proficiency Achieved."

(a) Combine "Average," "Better than Average," and "Extremely High" into the single category "High."

(b) Combine only "Better than Average" and "Extremely High" into the single category "High." Combine "Minimal" and "Average" into a single category "Low."

(continued . . .)

In each case, write down the 2 × 2 frequency table, and compute the percentages using languages after doctoral study in the Low and in the High groups. In each case compute the phi coefficient of association between Proficiency and Use.

TABLE 6.55

Cross-Classification of 247 Professional Geographers by Highest Proficiency Achieved in at Least One Language in Meeting the Language Requirement for the Doctoral Degree and Use of the Language After Doctoral Work

		Use the language after doctoral study?		*Total*
		No	*Yes*	
Highest proficiency	*Minimal*	32	23	55
achieved in meeting	*Average*	51	64	115
the foreign language	*Better than average*	12	37	49
requirement	*Extremely high*	1	27	28
	Total	96	151	247

source: Wiltsey, Robert G., *Doctoral Use of Foreign Languages: A Survey, Part II: Supplementary Tables.* p. 61. Copyright ©1972 by Educational Testing Service. All rights reserved. Adapted and reproduced under license.

6.7   In 1987 Singer, Levine, Rowley and Bazargan reported the results of a survey of students in public and private high schools in western New York state. Table 6.56 gives the results for gender and the question "How often have you damaged or destroyed property that did not belong to you?"

TABLE 6.56

How Often Have You Damaged Property that Did Not Belong to You?

		Response				*Total*
		Never	*1 or 2 times*	*3 to 11 times*	*12 or more times*	
Gender	*Male*	129	162	41	14	346
	Female	259	84	18	1	362
	Total	388	246	59	15	708

source: Singer, Levine, Rowley, and Bazargan, 1987, p. 16.

(a) Compute percentages based on row totals.

(b) Form a summary table having two columns ("Never" and "One Or More Times")

    (i) Compute percentages based on row totals.

    (ii) Describe the association, if any, between gender and property damage.

6.8 Each owner in a random sample of 100 business firms was asked whether he or she favored a new tax proposal. The results are given in Table 6.57.

TABLE 6.57
Poll of Owners of Business Firms

		Size of firm			
		Small	*Medium*	*Large*	*Total*
Opinion	*In favor*	40	10	2	52
	Opposed	30	10	8	48
	Total	70	20	10	100

Hypothetical data.

(a) (i) Compute percentages based on the column totals.

    (ii) Compute percentages based on the row totals.

(b) (i) Form a summary table having two columns: "Small- or medium-size firms" and "Large-size firms."

    (ii) Give percentages based on the column totals.

    (iii) Give percentages based on the row totals.

(c) (i) Record the 2×2 subtable for the small- and medium-size firms only.

    (ii) Compute percentages based on the row totals.

6.9 A study of reading errors made by second-grade pupils was carried out to help decide whether the use of different sorts of drills for pupils of different reading abilities is warranted. Errors were categorized as follows.

DK: Did not know the word at all
C: Substitution of a word of similar *configuration* (e.g., "bad" for "had")
T: Substitution of a synonym suggested by the *context*
OS: Other substitution

(These are combinations of categories that were suggested by Kottmeyer (1959), p. 30.) The children had been clustered into three relatively homogenous reading groups on the basis of (1) their reading achievement scores at the end of first grade, (2) their verbal IQ's, and (3) the opinions of their first-grade teachers. The three groups of children each chose the name of an animal as their group name. It happened that the least able readers chose the name Squirrels; the most able readers chose the name Cats; and the middle group, the name Bears. There were five children in the Squirrels group, nine in the Bears, and eleven in the Cats. The numbers of errors of each type made by each child were added to obtain the group totals given in Table 6.58.

The *numbers* of errors are not directly comparable: the more able readers used more difficult texts and read more. For each group, compute the *relative frequency* of each type of error. Compare the three frequency distributions.

TABLE 6.58

Distribution of Reading Errors in Three Ability Groups

	T	C	OS	DK	Total
Squirrels	5	10	15	53	83
Bears	28	34	72	172	306
Cats	8	10	15	36	69
Total	41	54	102	261	458

source: Anonymous.

6.10   Despite the fact that 90% of the people in a community watch soap operas on television and 90% exercise regularly, watching soap operas on television and exercising regularly may not be associated in that community. Explain.

6.11   A researcher has collected certain data on all the students in a statistics course and claims that the summary in Table 6.59 shows that graduate students do better in the course. The following facts are also selected from the data: exactly 100 students in the course are male; exactly 60 are undergraduate females; exactly 25 of the females did poorly in the course; and exactly 35 undergraduate females did well in the course. Taking all this into account, do you think the graduate/undergraduate dichotomy is an important factor in determining performance in the course?

TABLE 6.59
Student Performance in a Statistics Course

	Graduate student	Undergraduate student
Did well in the course	20	35
Did poorly in the course	40	85

Hypothetical data.

6.12   Table 6.60 gives a cross-classification of pupils according to their scores on a test at the beginning of the semester and another test at the end of the semester.

(a) Decide whether it is more appropriate to express the distributions as percentages of the row totals or the column totals. Justify your decision.

(b) Calculate the corresponding percentages.

(c) Describe the association between the two tests in terms of the row or column distributions you have obtained.

TABLE 6.60
3×3 *Table: Ratings by Initial Test and Final Test Scores*

		Score on final test			
		Low	Medium	High	Total
Score	*Low*	30	15	5	50
initial	*Medium*	4	32	44	80
test	*High*	4	32	4	40
	Total	38	79	53	170

Hypothetical data.

6.13   Summarize the District Court cases in Table 6.22 as follows:

(a) Express each frequency as a percentage of the corresponding row total.

    (i)   What percent of all offenses results in no conviction? In conviction with probation? In conviction with a prison sentence?

(continued . . .)

(ii) What offense has the highest percent of prison sentences? The lowest? What offense has the highest percent of "no conviction" outcomes?

(iii) Is there a tendency for some offenses to be followed by less severe outcomes, and other offenses by more serious outcomes? If so, describe the pattern of association.

(b) Combine all non-drug offenses into one category, and both conviction outcomes into one category ("convicted"), so that a 2 × 2 table results. Express each frequency as a percentage of the corresponding row total.

(i) Describe the association, if any, between drug/non-drug offenses and conviction/no-conviction outcome.

(ii) Compute the phi index of association between the offenses and outcomes. What numerical methods might help you simplify the problems presented by such large numbers?

6.14   Summarize the data given in Table 6.34 from the study, *The Academic Mind*, as follows:

(a) Calculate the indicated percentages.

(i) Find the percentage of social scientists who voted Democratic in 1952.

(ii) Among those social scientists in the 41–50 age group, what percentage are in the medium productivity group?

(iii) Among those social scientists 51 years of age or older having a low productivity score, what percentage voted Democratic in 1952?

(iv) Among those social scientists who did not vote Democratic in 1952, what percentage were 40 years or younger?

(v) Among those social scientists 51 years or older having a low productivity score, what percentage did not vote Democratic in 1952?

(vi) Among those social scientists low on the productivity scale and Democratic in 1952, what percentage are in the 41–50 year age group?

(b) Compute the percent Democratic in each of the nine age-productivity categories. Arrange these percentages into a 3 × 3 table.

(c) Summarize the relationship between age, productivity, and vote for the Democrats in 1952.

6.15   The X Company's commercial (Section 6.3) advertised Brand X products in general ("Brand X, a trusted name in men's toiletries," etc.), not the blades or razor specifically. The advertising department is trying to decide whether it may be worthwhile to advertise specific products separately. Some background study is necessary.

Figure 6.7 shows use of the razor as an intervening variable, relating use of the blades to viewing of the commercial. On the other hand, Figure 6.8 indicates that the commercial affected use of the blades not only indirectly (through increased use of the razor), but also directly. Which causal diagram do you find more consistent with the data in Table 6.32?

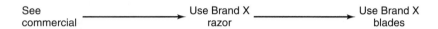

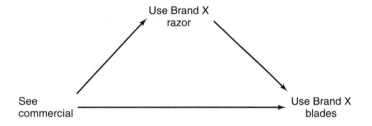

Figure 6.7   *Use of Brand X razor as an intervening variable.*

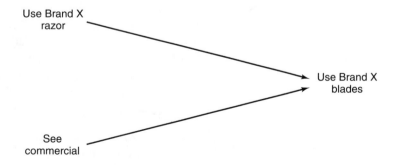

Figure 6.8   *Viewing the commercial has also an independent effect on use of Brand X blades.*

6.16   (continuation) Is Figure 6.9 consistent with the data concerning Brand X products? (Hint: Does taking use of the blades into account as a third factor eliminate the association between use of the razor and viewing of the commerial?)

Figure 6.9   *Another possible causal diagram.*

6.17   Basic to the analysis in Lazarsfeld and Thielens' (1958) study, *The Academic Mind*, was a concept of "apprehensiveness," a mixture of worry and caution about the effects of infringements on academic freedom. The analysis based on Table 6.61 is concerned with the impact of an individual's attitudes on his or her perception of the state of affairs at the college. Thus professors were asked questions designed to gauge their own apprehensiveness and to learn how apprehensive they believed their colleagues were. In addition, colleges themselves were given a rating in terms of apprehensiveness—characterizing them by the (sampled) proportion of "apprehensive" faculty members at the school. It is with the perception of this "true apprehension rate" at the college that the analysis of these tables deals.

(a) From Table 6.61, compute the percent saying apprehension is high at their college, among those at low-apprehension colleges and among those at high-apprehension colleges.

(b) Compute the phi coefficient of association between "true" and perceived apprehension.

TABLE 6.61

True Apprehension and Perceived Apprehension Among Colleagues of Faculty Members

		True rate of apprehension	
		Low	*High*
Perceived	*Low*	463	620
apprehension	*High*	198	574
	Total	661	1194

source: Lazarsfeld and Thielens (1958).

6.18   (continuation) Respondents were classified into those whose own apprehension was high or low. For each group separately, a 2 × 2 table was constructed indicating "true" apprehension and perceived apprehension among the respondents' colleagues.

(a) From Table 6.62 compute the percent saying apprehension is high, in each of the four groups.

(b) Interpret the results.

6.19   (*Cross-tabulation according to two discrete numerical variables.*) Table 6.63 is a cross-tabulation of 6000 household heads by number of children in the household and number of magazines read. Compute the distributions in each row, that is, the distributions of number of mag-

TABLE 6.62

True Apprehension, Perceived Apprehension, and Respondent's Own Level of Apprehension

		RESPONDENTS WITH LOW LEVEL OF APPREHENSION			RESPONDENTS WITH HIGH LEVEL OF APPREHENSION	
		True rate of apprehension			True rate of apprehension	
		Low	*High*		*Low*	*High*
Perceived	*Low*	358	381	*Low*	105	239
apprehension	*High*	84	148	*High*	114	426
	Total	442	529	*Total*	219	665

source: Lazarsfeld and Thielens (1958).

TABLE 6.63

Cross-Tabulation of Household Heads, by Number of Children in the Household and Number of Magazines Read

		Number of magazines read					
		0	*1*	*2*	*3*	*4*	*Total*
	0	754	868	781	162	63	2628
Number of	*1*	297	363	330	87	23	1100
children	*2*	256	358	307	71	32	1024
	3 or more	314	391	382	99	62	1248
	Total	1621	1980	1800	419	180	6000

Hypothetical data.

azines read by number of children. Is the number of magazines read associated with the number of children in the household? If so, is the association strong?

6.20 On the other hand, Table 6.64 is a cross-classification of the same 6000 household heads by number of magazines read and income level. Is there an association between income level and number of magazines read? If so, what is its direction?

6.21 (continuation) Summarize Table 6.64 by computing the mean number of magazines read in each income group.

6.22 Figure 6.10 is a scatter plot consistent with Hooker's (1907) study. Points marked *H* are years in which the rainfall was heavy. Points marked *L* are years in which the rainfall was light.

TABLE 6.64
Cross-Tabulation of Household Heads by Income Level and Number of Magazines Read

		Number of magazines read					
		0	*1*	*2*	*3*	*4*	*Total*
Income	*Less than $8,000*	1005	905	452	126	25	2513
	$8,000 to $16,000	536	930	1139	274	107	2986
	More than $16,000	80	145	209	19	48	501
	Total	1621	1980	1800	419	180	6000

Hypothetical data.

(a) Cross-tabulate the 20 years into the $2 \times 2 \times 2$ frequency table which summarizes Figure 6.10.

(b) Make the three possible 2×2 frequency tables corresponding to the three-way table of item (a).

Note that the table for yield and temperature corresponds to ignoring whether a point in the figure is an *H* or an *L*; that is, to ignoring whether the rainfall was heavy or light.

6.23 Hooker also found a strong association between the size of the crop of peas (harvested in summer) and the warmth of the *subsequent* autumn. How is this possible?

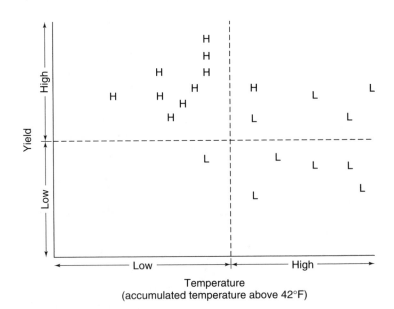

Figure 6.10 *Scatter plot of yield of hay and temperature for 20 years.*

III
Probability

7

Basic Ideas
of Probability

Introduction

Each of us has some intuitive notion of what "probability" is. Everyday conversation is full of references to it: "He'll probably return on Saturday." "Maybe he won't." "The chances are she'll forget." "The odds on winning are small."

What do these vague statements mean? What are the ideas behind them?

In this chapter we discuss some of the basic notions of probability and make them precise. We show how to find probabilities of various events on the basis of certain given information.

7.1 Intuitive Examples of Probability

Physical Devices Which Approximate Randomness

Some of our ideas of probability come from playing games in which chance is an essential part. In many card games a deck of 52 playing cards is used. There are two colors of cards, red and black, and four suits: clubs (black), diamonds (red), hearts (red), and spades (black). Within each suit there are the 13 cards Ace, 2, 3, . . . , 10, Jack, Queen, and King, the Jack, Queen, and King being "face cards" and bearing the letters J, Q, K, respectively, the other cards bearing their respective numbers. The deck is shuffled a few times, hopefully to put the cards in a *random order*—which means that all possible orderings of the cards are equally likely. In particular, we believe that when the cards are distributed (dealt), all of the 52 cards in the deck are equally likely to be the first card dealt. The terms "random" and "equally likely" are meaningful to a card player.

In bridge all of the cards are dealt out to 4 players, each receiving 13 cards. Bridge players have some ideas of the probabilities of receiving various combinations of cards, and their strategy of play involves applying an intuitive knowledge of these probabilities to the unseen combinations held by the other players.

There are some important factors underlying the determination of probabilities in card playing. For one thing, there is a finite, though very large, number of possibilities. Another aspect is that the players themselves produce the randomness; the player shuffling continues to shuffle until he or she feels the order is random. (We assume the players are honest.) The basis of calculating probabilities in card playing is that certain fundamental occurrences are "equally likely."

The shuffling of a deck of cards may be considered as a physical device for approximating randomness. Another physical device useful in describing probability is the flipping of a coin. Many games are played by flipping one or more coins. For a "fair" coin the probability of a head is considered to be $\frac{1}{2}$. In other games one rolls dice (singular: *die*). A die is a cube with 1 dot on 1 face, 2 dots on a 2nd face, etc. Our intuition suggests that for a balanced or fair die the probability of any specified one of the 6 faces coming up is $\frac{1}{6}$.

In some games the player spins a pointer (Figure 7.1). When the pointer stops on blue one move is prescribed, on red another is prescribed, etc. The color areas can be designed so that the various moves have any desired probabilities.

As another example of probability, consider a jar with 1000 beads in it. The beads are spherical and of the same size and weight; 500 are red and 500 are black. The beads are stirred, or the lid is put on the jar and the jar is shaken. A blindfolded person takes out a bead. We would agree that all the beads are equally likely to be drawn and the probability that the bead drawn is red is 500 chances in 1000, or $\frac{1}{2}$, that is, the first bead drawn is as likely to be red as it is to be black.

This example has shortcomings because in actuality it is hard to make the beads identical in size and weight and difficult to shake or stir them into random positions. In a somewhat similar situation slips of paper

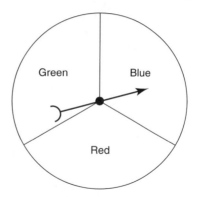

Figure 7.1 *A pointer for spinning, as in a game.*

bearing the names of the 365 days of the year were put into capsules; the capsules were put into a drum which was rotated. The sequence of dates then drawn from the drum was supposed to be in random order, but there was considerable question whether this physical device in fact could come close to achieving randomness.

The Draft Lottery

In 1969 a draft lottery was instituted by the Selective Service System. Men born in 1950 would be called for military service according to a schedule determined by the lottery. Given the personnel requirements of the military services, those having birthdates near the top of the list would almost certainly be drafted and those having birthdates near the end of the list would almost certainly not be drafted. The priority list was determined by drawing capsules containing the dates from a rotated drum.

Was the 1969 draft lottery really a random drawing? That is to say, was the procedure one in which all possible orders were equally likely? A more detailed account indicates that the capsules containing the dates of a particular month were put into one box. The 12 boxes were emptied into the drum one by one, first January, then February, etc., and then the drum was rotated and capsules were drawn out, one by one. Perhaps the capsules put into the drum last were more likely to be drawn first. Table 7.1 gives evidence that this was the case.

The capsules for days in December tended to be drawn out of the drum early in the drawing: 17 December days have low numbers (between 1 and 122). Only 4 December days have high numbers. The strong trend manifests itself in the table in a striking manner when one

TABLE 7.1

1969 Draft Lottery: Month of Birthday and Priority Number

Month	Priority numbers			*Total*
	1 to 122	*123 to 244*	*245 to 366*	
January	9	12	10	31
February	7	12	10	29
March	5	10	16	31
April	8	8	14	30
May	9	7	15	31
June	11	7	12	30
July	12	7	12	31
August	13	7	11	31
September	10	15	5	30
October	9	15	7	31
November	12	12	6	30
December	17	10	4	31
	122	122	122	366

NOTE: Because some men born in a leap year were eligible, February 29 was included, making the total number of days 366.

looks at the clusters of single-digit frequencies. In the first column of figures, these are at the top, in the second column, they are in the middle, in the third column, they are at the bottom.

For the next year's lottery, deciding the order of call for men born in 1951, a procedure was used which could not be criticized so easily. Two drums were used, one for capsules containing the priority ranks 1 to 365, the other for capsules containing the numbers corresponding to birth dates. This time, each drum was loaded in a random order. The final drawing was by hand: a date drawn from one drum was paired with a number drawn from the second drum [Rosenblatt and Filliben (1971)].

Probability and Everyday Life

In the examples so far, the probabilities are fairly obvious. In many everyday situations things are not so simple. One may try to avoid driving an automobile on a holiday weekend because one thinks that the probability of an accident is greater on such a weekend; this idea may come from reading the newspaper reports on the frequency of accidents on weekends. A weather forecast in a newspaper may state that tomorrow the chance of rain is 1 in 10, but for another day it may be 8 in 10; one would not bother with a raincoat on the first day, but would on the second.

7.2   Probability and Statistics

Often in statistical analysis the available data are obtained from individuals who constitute a *sample* from the population of interest.

A *random sample* of n individuals from a population of N is a sample drawn in such a way that all possible sets of n have the same chance of being chosen. One of the consequences of this is that all N members of the population have the same chance of being one of the n included in the sample. In many cases a sample is drawn from a population by a procedure that insures the property of randomness; in other cases there may be a natural physical mechanism operating to give a more or less random sample. In Chapter 1 examples of the use of statistics were given; let us see how notions of sampling and probability enter into some of these.

The procedure followed in political polls approximately achieves a random sample. Enumeration of the responses may lead the pollster to report that 40% of the respondents favor the Democratic candidate. How accurate is this as an estimate of the proportion of the entire population favoring the Democratic candidate? What is the probability that the pollster comes up with a percentage that differs from that of the entire population by as much as 5%? The pollster should be able to answer such questions and plan the sample so that the margin of probable error is sufficiently small.

In the polio vaccine trial the vials of vaccine and placebo were distributed in lots of 50; of the 50, 1/2 were vaccine and 1/2 were placebo. The positions of these vials in the container were random. There is some chance that the placebo would more often be given to children susceptible to paralytic polio than was the vaccine and that therefore the higher incidence of polio among the children receiving the placebo could be due to chance and not to the vaccine. It is essential to be able to evaluate the probability of such an occurrence.

In the study of group pressure discussed in Section 1.1 the sets of measurements obtained are not strictly random samples from populations of direct interest. The persons in the experiment were students at Swarthmore College. The behavior of the students in the experimental situation is considered as the behavior of a sample of persons under group pressure, and the behavior of the students in the control situation as that of a sample of persons from a population not under group pressure. Could the large observed difference in behavior be reasonably expected if group pressure had no effect? What is the *probability* of such a large difference under the assumption of random sampling if in fact these populations are the same? If this probability is very small, then we cannot hold to the idea that the populations are the same.

On the other hand, if the results of the experiment could have arisen by random sampling from identical populations, they do not seem very important. In order to evaluate such statistical studies properly, we must be able to calculate the probabilities of certain results occurring under conditions of sampling from two identical populations. Thus, a study of probability is essential if we are to use samples to make inferences about populations.

There are times when, although samples are not drawn randomly from the corresponding populations, the investigator considers the method of selection close enough to random to justify statistical inference. The methods used in the selection of students for an experiment, the allocation to experimental or control situations, and the assignment of roles within the situation can all contribute to randomness.

7.3   Probability in Terms of Equally Likely Cases

Simple physical devices can be used, at least in principle, to draw a random sample from a given population. To choose at random one of two possibilities we can flip a coin, and to choose one of six possibilities we can roll a die. In this section we introduce the ideas of probability in terms of such simple operations. In the rest of this chapter and in Chapter 8 we extend these ideas to provide a background in probability for the study of statistical inference.

The various possible outcomes of such simple operations as tossing coins, drawing beads from a jar, throwing dice, and dealing cards are thought of as being *equally likely*. This notion of "equally likely" is not defined; it is anticipated that the reader will have an intuitive feeling for it. This notion goes along with what it means for the coin to be balanced or "fair," for the beads to be thoroughly mixed, for the dice to be balanced or "true," for the cards to be thoroughly shuffled.

Consider dealing from a shuffled deck of ordinary playing cards. Since there are 52 different cards in the deck, the first card dealt can be any one of 52 possibilities. We call these possibilities *outcomes*; they are the "atoms" or indivisible units under consideration.

We may be interested in the probability that the first card has some specific property, such as being a spade. The *event* that the first card dealt is a spade corresponds to a set of 13 possible outcomes: the ace of spades, the deuce of spades, the 3 of spades, . . . , the king of spades.

DEFINITION   An *event* is a set of outcomes.

Dealing a card which is a spade is an event. Dealing a card which is not a spade is an event. Dealing a face card is an event. Dealing a black card is an event.

If the individual outcomes are atoms, then events, being collections of outcomes, are molecules. Dealing a *particular* card, say the queen of hearts, is an event consisting of but a single outcome. It is a molecule, but a molecule which consists of a single "atom."

Typically an event is a set of outcomes with some interesting property in common.

What is the "probability" of dealing a spade? Since there are 13 spades and all cards are equally likely to be drawn, it seems reasonable to consider the probability of drawing a spade as $13/52 = 1/4$. This leads to the following definition.

DEFINITION

> If there are n equally likely outcomes and an event consists of m outcomes, the probability of the event is m/n.

In this situation, to find the probability of an event we count the number of possible outcomes to find n and we count the number of outcomes in the event to find m and calculate the ratio. Thus we obtain $13/52 = 1/4$ as the probability of a spade. The probability of an ace is $4/52 = 1/13$. The probability of a black card (spade or club) is $26/52 = 1/2$. The probability of a nonspade is $39/52 = 3/4$.

We want to be able to use probabilities of some events to obtain probabilities of new events. For example, the event of a black card being dealt is made up of the event of a spade being dealt together with the event of a club being dealt. (See Figure 7.2.) We shall show

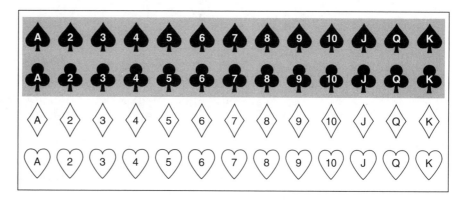

Figure 7.2 *A black card is a spade or a club.*

that we can calculate the probability of a black card from the probability of a spade and the probability of a club. First we note

$$\text{black cards} = \text{spades} + \text{clubs},$$
$$26 \quad = \quad 13 \quad + \quad 13.$$

By definition the probability of an event is the number of outcomes in the event divided by the total number of possible outcomes. If we divide the above equation by 52, the number of cards in the deck, we obtain[1]

$$\frac{\text{\# black cards}}{\text{\# cards}} = \frac{\text{\# spades}}{\text{\# cards}} + \frac{\text{\# clubs}}{\text{\# cards}},$$

$$\frac{26}{52} = \frac{13}{52} + \frac{13}{52},$$

$$\begin{array}{ccc} \text{probability of} \\ \text{a black card} \end{array} = \begin{array}{c} \text{probability of} \\ \text{a spade} \end{array} + \begin{array}{c} \text{probability of} \\ \text{a club}. \end{array}$$

The general idea is that if we have two events with no outcomes in common, the first consisting of l outcomes and the second consisting of m outcomes, then the event which consists of all the outcomes in the two given events consists of $l + m$ outcomes. Similarly, the two events have probabilities l/n and m/n, respectively, and the event of all outcomes has probability $(l + m)/n$. Thus if an event consists of all the outcomes of two events having no outcome in common, the probability of that event is the sum of the probabilities of the two events.

If we have three events with no outcome in more than one of the three events, the probability of an event consisting of all outcomes in the three events is the sum of the probabilities of the three events. For example, if the three events are a spade being dealt, a club being dealt, and a diamond being dealt, then

$$\frac{\begin{array}{c}\text{\# spades, clubs, and}\\ \text{diamonds}\end{array}}{\text{\# cards}} = \frac{\text{\# spades}}{\text{\# cards}} + \frac{\text{\# clubs}}{\text{\# cards}} + \frac{\text{\# diamonds}}{\text{\# cards}},$$

$$\begin{array}{c}\text{probability of a spade,}\\ \text{club, or diamond}\end{array} = \begin{array}{c}\text{probability}\\ \text{of a spade}\end{array} + \begin{array}{c}\text{probability}\\ \text{of a club}\end{array} + \begin{array}{c}\text{probability}\\ \text{of a diamond}.\end{array}$$

What about the probability of an event consisting of all outcomes in two events which may have some outcomes in common? For example, consider the event of the card being *either* a spade *or* a face card (or

---

[1]The symbol # stands for "number of."

both). The cards which are spade face cards are common to the two events. It is convenient to think of the large event as consisting of the outcomes in three events, no two of which have outcomes in common: spades that are face cards, spades that are not face cards, and face cards that are not spades. (See Figure 7.3.) Then we have

$$\frac{\text{\# cards that are either spades or face cards}}{\text{\# cards}} = \frac{\text{\# spades that are face cards}}{\text{\# cards}} + \frac{\text{\# spades that are not face cards}}{\text{\# cards}} + \frac{\text{\# face cards that are not spades}}{\text{\# cards}}$$

$$\frac{22}{52} = \frac{3}{52} + \frac{10}{52} + \frac{9}{52},$$

$$\underset{\substack{\text{probability of a} \\ \text{card that is} \\ \text{either a spade} \\ \text{or a face card}}}{} = \underset{\substack{\text{probability of} \\ \text{a spade that} \\ \text{is a face card}}}{} + \underset{\substack{\text{probability of} \\ \text{a spade that is} \\ \text{not a face card}}}{} + \underset{\substack{\text{probability of} \\ \text{a face card} \\ \text{that is not} \\ \text{a spade.}}}{}$$

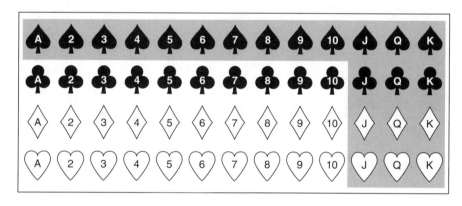

Figure 7.3 *Face cards, together with spades.*

An important event is the set of *all* cards. It has a probability of $52/52 = 1$; it is considered an event that happens for sure. Any event and the event consisting of all outcomes not in that event have no outcome in common; but the event consisting of the outcomes in these two events contains all the outcomes, and has a probability of 1. Hence

$$\underset{\substack{\text{probability of} \\ \text{a given event}}}{} + \underset{\substack{\text{probability of event} \\ \text{consisting of all} \\ \text{outcomes not in} \\ \text{given event}}}{} = 1.$$

It is convenient to define the absence of any outcome as the *empty* event, and it has probability $0/52 = 0$; it is certain not to happen.

Finally we note that if one event contains all the outcomes in another event (and possibly more), the probability of the first event is at least as great as the probability of the second. The probability of dealing a black card is 1/2, which is greater than the probability of dealing a club, which is 1/4.

Equally likely outcomes occur (in principle) in many games of chance. We consider that a head and a tail are equally likely in the toss of a coin. When we toss a penny and a nickel we consider the following outcomes equally likely: heads on both coins, head on the penny and tail on the nickel, tail on the penny and head on the nickel, and tails on both coins. In rolling a die we consider all of the six faces as equally likely.

The hand (set of 13 cards) dealt to one of four bridge players is a sample from a population of 52 cards. There are many possible hands; all are considered equally likely. Likewise, from any specific population there are many possible combinations of individuals. *Random sampling* is designed to make all combinations equally likely. The calculation of many probabilities can be based on the notion of equally likely outcomes. However, there are often situations in which a more general notion of probability is required; we develop this in the next section.

7.4   Events and Probabilities in General Terms

Outcomes, Events, and Probabilities

There are two contexts in which the notion of a definite number of equally likely cases does not apply: (1) where the number of possible outcomes is finite but all outcomes are not equally likely, and (2) where the whole set of outcomes is not finite. If a coin is not fair, the probability of a head may not be 1/2; even if eyes are classified only as blue or brown, the probability of a blue-eyed baby is not 1/2. The possible states of weather are not finite. Outcomes which depend on the value of a continuous variable are not finite in number. For instance, in the spinner portrayed in Figure 7.1, an outcome is the position of the pointer measured by the angle between the pointer and the vertical; that angle can take on any of the infinite number of values between 0° and 360°. The events, red, blue, and green, are made up from these outcomes.

In this section we treat probability in a more general setting. The means used to calculate probabilities from general principles should agree with the methods that obtain for equally likely outcomes; the properties and rules of probability will be the same in the general case as in the special case of equally likely outcomes.

As before, we start with a set of outcomes and consider events, which are collections of outcomes.

DEFINITION

An *event* is a set of outcomes.

The *probability* of each event E is a number denoted as $\Pr(E)$. In Section 7.3, where the space of all outcomes, as well as each event, consists of a finite number of equally likely outcomes, $\Pr(E)$ is the ratio of the number of outcomes in the event to the total number of outcomes. Since probabilities in the general setting must have the same properties as these ratios, any probability is a number between 0 and 1 (inclusive).

PROPERTY 1

$$0 \leq \Pr(E) \leq 1.$$

The set of all outcomes is called the *space* (e.g., the entire deck of cards) and the set containing no outcome is called the *empty event*. In agreement with Section 7.3 we require the following property:

PROPERTY 2

$$\Pr(\text{empty event}) = 0, \qquad \Pr(\text{space}) = 1.$$

Addition of Probabilities of Mutually Exclusive Events

Mutually Exclusive Events.   The events of a spade being dealt and of a club being dealt are incompatible, in the sense that they cannot occur simultaneously; they are *mutually exclusive*. Figure 7.4 illustrates mutually exclusive events.

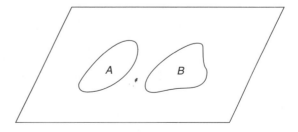

Figure 7.4   *Mutually exclusive events.*

DEFINITION   Two events are *mutually exclusive* if they have no outcome in common.

Probabilities of Mutually Exclusive Events.   In terms of equally likely events we saw that the probability of the event made up of all outcomes of two mutually exclusive events (spades and clubs, for example) was the sum of the probabilities of the two events. We want this fact to be true in general. If A and B are two mutually exclusive events, we denote the event composed of the outcomes from *both* A and B as "A or B" because an outcome of this event is an outcome of A or an outcome of B.

PROPERTY 3   If the events A and B are mutually exclusive, then

$$\Pr(A \text{ or } B) = \Pr(A) + \Pr(B).$$

The probability of an event may be determined by a systematic procedure such as the proportion of "equally likely" outcomes contained in the event, or it may be determined on another basis. (See Section 7.5.) However determined, if the numbers $\Pr(E)$ have Properties 1, 2, and 3, then the mathematics of probability theory can be used.

If the three events A, B, and C are mutually exclusive, then

$$\Pr(A \text{ or } B \text{ or } C) = \Pr(A) + \Pr(B) + \Pr(C).$$

If $A_1, A_2, \ldots, A_m$ are an arbitrary number of mutually exclusive events, then

$$\Pr(A_1 \text{ or } A_2 \text{ or } \ldots \text{ or } A_m) = \Pr(A_1) + \Pr(A_2) + \cdots + \Pr(A_m)$$

$$= \sum \Pr(A_j).$$

DEFINITION   The *complement* of an event is the event consisting of all outcomes not in that event.

If A is the given event we denote the complement of A by $\bar{A}$. By definition, A and $\bar{A}$ are mutually exclusive, but every outcome in the space is either in A or $\bar{A}$. If we substitute $\bar{A}$ for B in Property 3 and use the second part of Property 2, we obtain

$$1 = \Pr(A) + \Pr(\bar{A})$$

or

$$\Pr(\bar{A}) = 1 - \Pr(A).$$

Addition of Probabilities

The Event "A and B."   In Figure 7.5 we indicate the event consisting of outcomes that are in both A and B. We call the event "A and B." Another representation is similar to the 2×2 frequency table (Figure 7.6). The event A consists of the points in the square above the horizontal line; B consists of the points to the left of the vertical line; "A and B" consists of the points in the upper left-hand corner. In obtaining a corresponding 2×2 frequency table one counts the number of outcomes in A and B.

The probabilities of the events indicated in Figure 7.6 are given in Table 7.2. Since the event "A and B" and the event "A and $\bar{B}$" are mutually exclusive (because B and $\bar{B}$ are mutually exclusive),

$$\Pr(A) = \Pr(A \text{ and } B) + \Pr(A \text{ and } \bar{B}).$$

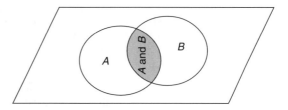

Figure 7.5   *A and B.*

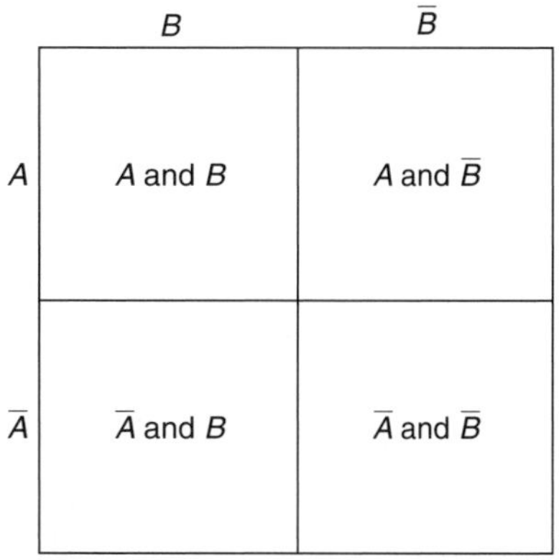

	B	$\bar{B}$
A	A and B	A and $\bar{B}$
$\bar{A}$	$\bar{A}$ and B	$\bar{A}$ and $\bar{B}$

Figure 7.6   *Events based on a two-way classification.*

The other sums indicated in Table 7.2 are derived similarly.

TABLE 7.2
Probabilities Associated with Two Events

	B	$\bar{B}$	
A	$\Pr(A \text{ and } B)$	$\Pr(A \text{ and } \bar{B})$	$\Pr(A)$
$\bar{A}$	$\Pr(\bar{A} \text{ and } B)$	$\Pr(\bar{A} \text{ and } \bar{B})$	$\Pr(\bar{A})$
	$\Pr(B)$	$\Pr(\bar{B})$	1

Probability of the Event "A or B."   Let us find the probability of "A or B," when the events A and B are not necessarily mutually exclusive. We know the three events "A and $\bar{B}$," "$\bar{A}$ and B," and "A and B" are mutually exclusive. Thus

$$\Pr(A \text{ or } B) = \Pr(A \text{ and } \bar{B}) + \Pr(\bar{A} \text{ and } B) + \Pr(A \text{ and } B).$$

This is the sum of all the probabilities in the interior of Table 7.2 except the probability in the lower right, namely, that of "$\bar{A}$ and $\bar{B}$" ("neither A nor B"). We can also see from Table 7.2 that

$$\Pr(A \text{ or } B) = \Pr(A) + \Pr(B) - \Pr(A \text{ and } B).$$

An event B is *contained* in an event A if every outcome of B is in A, as indicated in Figure 7.7. Then

$$\Pr(B) \le \Pr(A)$$

because A is made up of the two mutually exclusive events, B and the event consisting of outcomes in A but not in B.

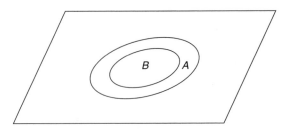

Figure 7.7   *B contained in A.*

Sets.   The notion of set used here is the same as that already encountered by students having some familiarity with the branch of mathematics known as "set theory." We have used the term "event" where "set" is used in set theory. The correspondence between the terms and notations of this book and of set theory are given in Table 7.3.

TABLE 7.3
Correspondence Between Terms and Notations

	Term or Notation
In this book	*In set theory*
Event	Set
Outcome	Element, member, or point
Mutually exclusive	Disjoint
A or B	$A \cup B$ ("A union B")
A and B	$A \cap B$ ("A intersect B")
$\bar{A}$	$\bar{A}$, A^C, or $\tilde{A}$ ("A complement")
empty event	null set

7.5   Interpretation of Probability: Relation to Real Life

Probability theory, as presented in Section 7.4, can be developed as a formal theory; it is a branch of mathematics. How does this model relate to real life?

In Section 7.3 we saw that one way of assigning probabilities to events was to allocate *equal probabilities* to all possible outcomes and then define the probability of an event as the proportion of outcomes in the event. Although this seems a reasonable procedure in appropriate situations, these probabilities may be only approximations. In tossing a coin one side may be more likely to turn up than another; the shuffling of a deck of cards may not be random; the drawing of draft priority numbers from a rotated basket is not conceded to make all orderings equally likely. Another shortcoming of this approach is that in many cases all events cannot be expressed in terms of a finite number of equally likely outcomes. For example, it is hard to see how the probability of rain tomorrow can be given a meaning in that way.

A second, more satisfactory way of relating probabilities to the real world is in terms of *relative frequency*. If a coin was tossed unendingly, one would expect that after a while the ratio of heads to the total number of tosses would stay very near a certain number, which would be considered the probability of a head. In principle, the probability of an event can be thought of as the relative frequency with which it occurs in an indefinitely large number of trials. But the frequency approach to probability also has its problems. For one thing the determination is hypothetical (we spoke of tossing a coin "unendingly"). For another thing the notion of the relative frequency being approximately a constant for a large number of trials needs more careful development.

A third approach is *subjective* or *personal*. A person can assign a probability to an event according to how likely he or she feels the event is to occur. In this assignment he or she may be guided by mathematical analysis as well as other information. An advantage of this approach is that an individual can give probabilities to events when the other approaches cannot. He or she can, for example, provide a probability that extrasensory perception exists. A disadvantage is that different people may assign different subjective probabilities to the same event. Another disadvantage is that a given individual might assign probabilities in an inconsistent manner so that the rules of Section 7.4 would not hold.

For this book we shall adopt the point of view that a probability represents the relative frequency in a large number of trials. We shall, however, often appeal to the reader's intuition and use the "equally likely" approach in many examples.

7.6   Conditional Probability

The probability of one event given that another event has occurred is a *conditional probability*. A conditional probability is defined in terms of the probabilities of the individual events and combinations of them.

Consider a population of 100 individuals who have answered the questions "Have you seen an advertisement of Spotless toothpaste in the last month?" and "Did you buy Spotless toothpaste in the last month?" Their responses are tabulated in Table 7.4.

TABLE 7.4
2×2 *Table of Frequencies*

	Buy	*Not buy*	
Seen ad	20 (50%)	20 (50%)	40 (100%)
Not seen ad	10 (16.7%)	50 (83.3%)	60 (100%)
	30 (30%)	70 (70%)	100 (100%)

Hypothetical data.

If we were to draw one person at random from those who had seen the ad, the probability of obtaining a person who bought the toothpaste is $20/40 = 1/2$; that is, we have considered each of the 40 individuals who have seen the ad as equally likely to be drawn. We could as well have made the calculation from the table of proportions, Table 7.5, as

$$\frac{20/100}{40/100} = \frac{20}{40} = \frac{1}{2}.$$

The entries in Table 7.5 can also be considered probabilities in terms of a random draw.

TABLE 7.5

2×2 *Table of Proportions.*

	B: Buy	$\bar{B}$*: Not buy*	
A: Seen ad	$\Pr(A \text{ and } B) = \frac{20}{100}$	$\Pr(A \text{ and } \bar{B}) = \frac{20}{100}$	$\Pr(A) = \frac{40}{100}$
$\bar{A}$*: Not seen ad*	$\Pr(\bar{A} \text{ and } B) = \frac{10}{100}$	$\Pr(\bar{A} \text{ and } \bar{B}) = \frac{50}{100}$	$\Pr(\bar{A}) = \frac{60}{100}$
	$\Pr(B) = \frac{30}{100}$	$\Pr(\bar{B}) = \frac{70}{100}$	$\frac{100}{100}$

source: Table 7.4.

The probabilities of events A and B and derived events were given in Table 7.2. The conditional probability of B given A, denoted $\Pr(B \mid A)$, is $\Pr(A \text{ and } B)/\Pr(A)$.

Of course, this formula can be used only if $\Pr(A)$, the denominator, is different from zero.[2]

DEFINITION

The *conditional probability* of B given A when $\Pr(A) > 0$ is

$$\Pr(B \mid A) = \frac{\Pr(A \text{ and } B)}{\Pr(A)}.$$

The calculation of $\Pr(B \mid A)$ from Table 7.2 and the calculation of percentages in 2×2 frequency tables in Chapter 6 are the same. The phrase, "given A," says to restrict attention to row A of the bivariate

---

[2]If $\Pr(A) = 0$, then A cannot occur, and the conditional probability of B given A has no meaning. Mathematically, if $\Pr(A) = 0$ then $\Pr(A \text{ and } B) = 0$ and the ratio

$$\Pr(A \text{ and } B)/\Pr(A)$$

is not defined.

table, just as is done when calculating a percentage based on the total for row A. From Table 7.2 we can calculate other conditional probabilities, such as

$$\Pr(\bar{B} \mid A) = \frac{\Pr(A \text{ and } \bar{B})}{\Pr(A)}.$$

The results of these calculations are tabulated in Table 7.6.

TABLE 7.6
Conditional Probabilities of B given A or $\bar{A}$.

	B	$\bar{B}$	
A	$\Pr(B \mid A)$	$\Pr(\bar{B} \mid A)$	1
$\bar{A}$	$\Pr(B \mid \bar{A})$	$\Pr(\bar{B} \mid \bar{A})$	1

The conditional probabilities given one specified event A have all the properties of ordinary probabilities. They are greater than or equal to 0 (because the numerator is nonnegative and the denominator is positive) and they are less than or equal to 1 (because the numerator is less than or equal to the denominator since "A and B" is an event contained in A). If C is another event that is mutually exclusive of B, then "A and B" and "A and C" are mutually exclusive and

$$\Pr(B \text{ or } C \mid A) = \Pr(B \mid A) + \Pr(C \mid A).$$

In particular,

$$\Pr(B \mid A) + \Pr(\bar{B} \mid A) = 1.$$

Conditional probabilities do not need to be calculated in terms of equally likely cases. For instance, at some season in some places the probability of clear skies is 0.6, the probability of cloudy skies without rain is 0.2, and the probability of cloudy skies with rain is 0.2. Then the conditional probability of rain given the skies are cloudy is $0.2/(0.2 + 0.2) = 1/2$.

The equation defining the conditional probability $\Pr(B \mid A)$ can be rewritten

$$\Pr(A \text{ and } B) = \Pr(A) \times \Pr(B \mid A).$$

This says that the probability of the event "A and B" can be calculated by multiplying the probability of A by the conditional probability of B given A.

A cautious man planning to fly to New York asked an actuary friend, "What is the chance that there will be a bomb on the plane?" On hearing that the probability was 1/1000, he became anxious and puzzled. Then he asked his friend, "What is the chance that there will be *two* bombs on the plane?" His friend answered 1/1,000,000. "Aha," said the man, "I know what I'll do; I'll carry one bomb myself." The reader can use a knowledge of conditional probability to find the fallacy in this fictitious story.

An important use of the last equation is to calculate probabilities of events defined by sampling from finite populations. Dealing n cards from a shuffled deck of 52 cards can be thought of as drawing a random sample of n from a population of 52. The deck can be thought of as a model for any finite population of 52, such as the weeks in a year or 52 students in a class. When one deals n cards, the sample consists of n different cards. This is called *sampling without replacement*. Sampling *with replacement* would mean dealing one card, recording it, returning it to the deck and shuffling again; the procedure would be repeated until n cards have been recorded. We shall always consider dealing as done without replacement.

Suppose the event A is dealing the spade ace as the first card and the event B is dealing the spade king as the second card. The probability of A is 1/52. Given that the spade ace has been dealt, there are 51 cards left, and one of these is the spade king. The conditional probability of the spade king given the spade ace has been dealt is 1/51. Thus the probability of dealing the spade ace and then the king is

$$\frac{1}{52} \times \frac{1}{51} = \frac{1}{52 \times 51}$$

This probability can be calculated in another way. There are 52 different cards that can be dealt first and for each of them there are 51 ways of dealing the next card. Thus there are 52×51 ways of dealing a pair of cards when we distinguish the first card dealt from the second. The number of "ordered" pairs is 52×51. Hence, the probability of dealing the particular ordered pair, spade ace followed by spade king, is $1/(52 \times 51)$.

The event of dealing the spade ace and king as the first two cards (regardless of order) is made up of the two mutually exclusive events, spade ace followed by spade king and spade king followed by spade ace, each having probability $1/(52 \times 51)$. Hence, the probability of the spade ace and king appearing as the first two cards in either order is $2/(52 \times 51)$.

As another example, consider the probability of dealing aces as the first two cards. The event A is an ace as the first card, having probability 4/52 and the event B is an ace as the second card. If an ace is dealt as the first card, there are 3 aces left in the remaining 51 cards; hence $\Pr(B \mid A) = 3/51$. The probability of aces as the first two cards dealt is

$$\Pr(A \text{ and } B) = \Pr(A) \times \Pr(B \mid A) = \frac{4}{52} \times \frac{3}{51}.$$

7.7   Independence

Independence of events in terms of probabilities is defined exactly like independence in frequency tables. Table 7.7 resembles Table 6.53 with rows and columns interchanged and percentages replaced by proportions. The responses in Table 6.53 were said to be independent because the proportion of individuals listening to classical music was about the same among those less than 40 and among those over 40.

If we sample one person at random from the 35 million persons enumerated in Table 7.7 we have the probabilities given in Table 7.8: each frequency has been divided by 35. The conditional probability of

TABLE 7.7

2 × 2 Frequency Table

| | | Listen to classical music [Number of Persons (in millions)] | | |
		Yes	*No*	*Total*
Age	*Less than 40*	12 (0.6)	8 (0.4)	20 (1.0)
	40 or above	9 (0.6)	6 (0.4)	15 (1.0)
	Total	21 (0.6)	14 (0.4)	35 (1.0)

Hypothetical data suggested by Table 6.51.

TABLE 7.8

2 × 2 Probability Table.

| | | Listen to classical music | | |
		Yes	*No*	*Total*
Age	*Less than 40*	$\dfrac{12}{35}$	$\dfrac{8}{35}$	$\dfrac{20}{35}$
	40 or above	$\dfrac{9}{35}$	$\dfrac{6}{35}$	$\dfrac{15}{35}$
	Total	$\dfrac{21}{35}$	$\dfrac{14}{35}$	$\dfrac{35}{35}$

source: Table 7.7.

drawing a person listening to classical music given that the person was less than 40 is

$$\frac{12/35}{20/35} = \frac{12}{20} = 0.6,$$

and the conditional probability of drawing a person listening to classical music given that the person was 40 or over is

$$\frac{9/35}{15/35} = \frac{9}{15} = 0.6$$

These two conditional probabilities are of necessity the proportions in the first column of Table 7.7. Since the conditional probability of listening to classical music does not depend on age, we say that listening to classical music is independent of age.

Table 7.2 gives the probabilities associated with two arbitrary events A and B. If we divide the entries in row A by $\Pr(A)$ and in row $\bar{A}$ by $\Pr(\bar{A})$ we obtain Table 7.9. If the corresponding conditional probabilities in the two rows are equal, we say the A and B are independent.

TABLE 7.9
Conditional Probabilities of B Given A or $\bar{A}$.

	B	$\bar{B}$	
A	$\Pr(B \mid A)$	$\Pr(\bar{B} \mid A)$	1
$\bar{A}$	$\Pr(B \mid \bar{A})$	$\Pr(\bar{B} \mid \bar{A})$	1

DEFINITION

The event B is *independent* of the event A if

$$\Pr(B \mid A) = \Pr(B \mid \bar{A}). \tag{7.1}$$

This notion of independence is analogous to the notion of independence in 2×2 frequency tables, as discussed in Chapter 6. There we noted that, if B is independent of A, it is also true that A is independent of B. The same is true in terms of probabilities; (7.1) implies that A is independent of B:

$$\Pr(A \mid B) = \Pr(A \mid \bar{B}). \tag{7.2}$$

Just as the condition of independence in 2×2 tables can be expressed in different ways so can independence in the probability sense.

In Table 7.7 the marginal distribution is $(0.6, 0.4)$ because both of the distributions in the body of the table are $(0.6, 0.4)$. In the general notation of Table 7.9, this says that the common value of the conditional probabilities in (7.1) is $\Pr(B)$:

$$\Pr(B) = \Pr(B \mid A) = \Pr(B \mid \bar{A}). \tag{7.3}$$

From (7.3) and the definition of $\Pr(B \mid A)$, we obtain

$$\Pr(B) = \frac{\Pr(A \text{ and } B)}{\Pr(A)},$$

or

$$\Pr(A \text{ and } B) = \Pr(A) \times \Pr(B).$$

Thus the definition of independence is equivalent to the following:
The events A and B are independent if

$$\Pr(A \text{ and } B) = \Pr(A) \times \Pr(B). \tag{7.4}$$

It follows from (7.4) that every entry in Table 7.2 is the product of the corresponding row and column probabilities when A and B are independent. Thus, independence of A and B is equivalent to independence of A and $\bar{B}$, independence of $\bar{A}$ and B, and independence of $\bar{A}$ and $\bar{B}$. In Table 7.8, each entry is the product of corresponding row and column probabilities.

A coin, tossed twice, illustrates the idea of independence in another context. The four possible outcomes, which we assume to be equally likely, can be represented as (H, H), (H, T), (T, H), and (T, T); each outcome has a probability of $1/4$. The probability of getting a head on any one toss is $1/2$ and so is the probability of getting a tail. To check independence of the two tosses, we note that $\Pr[(H, H)] = \Pr(\text{heads on both tosses}) = 1/4$ and $\Pr(\text{head on first toss}) \times \Pr(\text{head on 2nd toss}) = 1/2 \times 1/2 = 1/4$. We can similarly check (H, T), (T, H), and (T, T).

Mutually exclusive events are represented in a diagram by nonoverlapping figures, because then "A and B" is the empty event. Thus, for mutually exclusive events, $\Pr(A \text{ and } B) = \Pr(\text{empty event}) = 0$.

On the other hand, *independent* events are events A and B such that

$$\Pr(A \text{ and } B) = \Pr(A) \times \Pr(B).$$

Independent events are *not* mutually exclusive; in fact, they overlap *just enough*, so that "A and B" has probability equal to the product of the separate probabilities.

7.8 Random Sampling; Random Numbers

Random Devices

We have already described elsewhere a number of devices designed to achieve randomness. These will be reviewed briefly.

We can obtain a random sample from a deck of 52 playing cards by shuffling the deck adequately and dealing a specified number of cards from the top of the deck. Each of the four hands in bridge is a random sample of 13 from the deck of 52. A random sample from the population of 52 weeks in the year can be obtained by identifying each week with a different card and then dealing a sample of the desired size.

A population of 1000 beads in a jar may be numbered from 1 to 1000 or 000, 001, 002, ..., 999. If the beads are otherwise identical, a random sample of 15 may be obtained by mixing the beads in the jar and, without looking at the numbers, taking out 15 beads. A random sample can be drawn from any arbitrary finite population by numbering (or otherwise identifying) each individual in the population, placing in the jar beads labelled with those identities, and drawing a sample of beads from the jar. The sample consists of those individuals whose identities were drawn.

Flipping a fair coin yields a probability of 1/2 for a tail and for a head. Thus, if a tail denotes 0 and a head 1, coin-tossing provides a mechanism for obtaining the integers 0 and 1 with a probability of 1/2 each. Rolling a balanced die yields a probability of 1/6 for each integer between 1 and 6. There is also a "die" that will provide the integers 0, 1, ..., 9 with equal probabilities. It is a long cylinder, the cross-section of which is a regular 10-sided polygon; that is, the cylinder has 10 faces (in addition to the ends) numbered 0, 1, ..., 9. When it is rolled the probability of each integer coming up is 1/10. A spinner (Figure 7.1) over a circle that is divided into n equal segments will provide n equally likely outcomes.

Random Numbers

Unfortunately, these mechanical devices for drawing at random are subject to biases. A deck of cards may be shuffled inadequately, the jar of beads may not be mixed well, and a die may have imperfect edges or be weighted asymmetrically. To avoid these difficulties special numerical methods involving computers have been developed to create processes that generate events with specified probabilities; these methods have essentially no biases. Such a method is used by an expert to generate a series of random events. Since the expert and a computer may not be on hand when the series of random events is needed, the expert records the series in advance. When the series of events is a sequence of integers, the record is a table of random numbers. Appendix VI is an example.

Let us pursue the principle of random numbers and their use in drawing random samples. Given a device which has as equally likely outcomes the *digits* 0, 1, ..., 9, we have what are called *random digits*. We have

$$Pr(i) = \frac{1}{10}, \qquad i = 0, 1, \dots 9.$$

We can determine two random digits, say i and j, each having one of the values $0, 1, \dots, 9$. The two-digit number ij formed from these two random digits can be any one of the numbers $00, 01, \dots, 09, 10, 11, \dots, 99$. Since the two digits i and j are (presumably) independent, the probability of the pair ij is the product of the probability of the first digit i and the probability of the second digit j,

$$Pr(ij) = Pr(i) \times Pr(j) = \frac{1}{10} \times \frac{1}{10} = \frac{1}{100}.$$

Thus all pairs are equally likely. To draw one individual randomly from 100, we number the individuals from 00 to 99, obtain the random pair of digits, and select the corresponding individual.

What if the population size is not a power of 10? For example, suppose we wish to draw randomly from among 4 individuals. Let us identify them as 1, 2, 3, and 4. The probability of the event consisting of the integers 1, 2, 3, 4, from the set of 10 integers, is

$$Pr(1, 2, 3, 4) = \frac{4}{10}.$$

The conditional probability of the integer i $(i = 1, 2, 3, 4)$ given the event 1, 2, 3, 4 is

$$Pr(i \mid 1, 2, 3, 4) = \frac{1/10}{4/10} = \frac{1}{4},$$

which is the probability associated with 4 equally likely outcomes. A procedure for drawing one individual at random from a population of 4 is to take random digits until one obtains one of the digits 1, 2, 3, or 4; this digit is a random selection among 4. Any of the digits $5, 6, \dots,$ 9, 0 is discarded. (A more efficient procedure is to also let 5 correspond to 1, 6 correspond to 2, 7 correspond to 3, and 8 correspond to 4, and discard only 9 and 0.)

If we have a population consisting of N individuals, we may number them from 1 to N. If N is less than 10, we draw random digits singly until we obtain a digit between 1 and N, inclusive. If N is more than 10, but less than 100, we draw pairs of random digits until we obtain a pair between 01 and N, inclusive.

Appendix VI contains "random" digits. This means that the procedure by which they were obtained is supposed to produce random digits. Actually, they are produced on a computer by numerical operations that give equal proportions of the 10 digits.[3] Sometimes they are called "pseudo-random" digits.

To illustrate how to use the table in Appendix VI, suppose we want to choose a random sample of 10 observations from a population of 80 individuals. We pick a starting point in the table by some more or less random device, such as pointing a pencil with our eyes closed. Suppose that the pencil point lands in the extreme left-hand column of figures, 6 rows from the top (i.e., quintuple 02760). We should decide in advance whether to proceed vertically (down the page) or horizontally (across the page).

The first pair of digits is 02 and thus individual 2 is chosen to be included in the sample. If we go horizontally, the next pair of digits is 76 and individual 76 is chosen for the sample. The next digit pair is 07. (0 is the last digit of this quintuple; 7 is the first digit of the quintuple to its right.) Individual 7 is chosen for the sample. The next digit pair is 95 but, since the population contains only 80 individuals, this digit pair is discarded. Likewise, the pairs 00 and 83 are discarded, and individuals 27 and 29 are included in the sample. We continue in this manner until 10 digit pairs between 01 and 80 have been found. If we do not complete the sample using the first row of random digits, we continue with the next row.

If the population consisted of, say, 10,000 individuals, we would use 4 digits for each observation, with "0000" representing individual 10,000. If we started at the same point as in the preceding example, we would select individuals 276 (0276), 795 (0795), 83 (0083), 2729, 8876, and so on until we obtained the needed sample size.

Sampling with Replacement

One procedure for obtaining a sample of n individuals from a population of N individuals is simply to draw n random numbers (between 1

---

[3]One method is to take as the "random number" the remainder obtained when dividing the product of two large numbers by a third large number.

and N) and let these determine the sample. In this procedure, however, there might be some duplications; some individuals might appear in the sample several times. This is called *sampling with replacement* because it corresponds to drawing beads from a jar when one replaces the bead that has been drawn before drawing the next bead.

Sampling without Replacement

Now consider taking a random sample such that all the individuals in the sample are different. If we want a sample of 2 individuals from a population of 4, the first is selected randomly as described above by drawing random digits until 1, 2, 3, or 4 appears. Suppose 3 appears. That leaves 1, 2, and 4 in the population. Then one of these three is drawn randomly. We draw random digits until we obtain 1, 2, or 4; the result determines the second member of the sample. Thus, to draw a sample of n from a population of N, we draw the first member of the sample randomly, then the second member of the sample from the population which is left, etc. This is called *sampling without replacement*.

Sampling without replacement is considered in detail in Sections 9.1 and 10.7.

7.9   Bayes' Theorem*

Bayes' Theorem in a Simplified Case

In Chapter 6 we have seen that the 4 entries in a 2 × 2 table can be added to obtain column totals, row totals, and the overall total; from these, percentages can be computed by using the row totals, the column totals, or the overall total as bases. Since we can construct the original table from the total frequency and certain percentages, all the information is available from the total frequency and these percentages.

We have seen how conditional probabilities are obtained from other probabilities in a similar fashion. Bayes' Theorem in the simplified case is used in obtaining conditional probabilities one way from conditional probabilities the other way.

We consider events A, $\bar{A}$, B, $\bar{B}$, and events formed from them. Suppose we are given $\Pr(B)$, $\Pr(A \mid B)$, and $\Pr(A \mid \bar{B})$ as indicated in

Table 7.10. If we subtract $\Pr(B)$ from 1, we obtain $\Pr(\bar{B})$, as Step 1 of Table 7.11. Then we calculate $\Pr(A \text{ and } B) = \Pr(A \mid B) \times \Pr(B)$ and $\Pr(A \text{ and } \bar{B}) = \Pr(A \mid \bar{B}) \times \Pr(\bar{B})$ as Step 2 of Table 7.11. We add these to obtain $\Pr(A)$ as Step 3. Finally we obtain the conditional probability $\Pr(B \mid A) = \Pr(A \text{ and } B)/\Pr(A)$. The full calculation is Bayes' Theorem.[4]

TABLE 7.10
Given Probabilities

	B	$\bar{B}$				B	$\bar{B}$	
A					A	$\Pr(A \mid B)$	$\Pr(A \mid \bar{B})$	
$\bar{A}$					$\bar{A}$			
	$\Pr(B)$		1			1	1	

TABLE 7.11
Derived Probabilities

Step 1

	B	$\bar{B}$				B	$\bar{B}$	
A					A	$\Pr(A \mid B)$	$\Pr(A \mid \bar{B})$	
$\bar{A}$					$\bar{A}$			
	$\Pr(B)$	$\Pr(\bar{B})$	1			1	1	

Step 2

	B	$\bar{B}$	
A	$\Pr(A \text{ and } B) = \Pr(A \mid B) \times \Pr(B)$	$\Pr(A \text{ and } \bar{B}) = \Pr(A \mid \bar{B}) \times \Pr(\bar{B})$	
$\bar{A}$			
	$\Pr(B)$	$\Pr(\bar{B})$	

Step 3

	B	$\bar{B}$	
A	$\Pr(A \text{ and } B)$	$\Pr(A \text{ and } \bar{B})$	$\Pr(A)$
$\bar{A}$			

Step 4

	B	$\bar{B}$	
A	$\Pr(B \mid A) = \dfrac{\Pr(A \text{ and } B)}{\Pr(A)}$		1
$\bar{A}$			

[4]Due to the Reverend Thomas Bayes (1763).

BAYES' THEOREM

$$\Pr(B \mid A) = \frac{\Pr(B) \times \Pr(A \mid B)}{\Pr(B) \times \Pr(A \mid B) + \Pr(\bar{B}) \times \Pr(A \mid \bar{B})}.$$

Examples

Now let us follow an example to see how Bayes' Theorem can be used. Suppose a new simple and economical test for a certain disease has been invented; the test is good, but not perfect. To validate the test, an experiment is performed. A sample of 1000 persons known to have the disease ("Ill") and a sample of 5000 known not to have the disease ("Not Ill") are tested. These sample sizes are decided by practical budgetary and administrative decisions and are not representative of the incidence of the disease in the overall population. In fact, doctors know that at any one time about 1 in 200 persons in the population have the disease. The results of the experiment are shown in Table 7.12. Of those ill, 99% are detected, while only 5% of those not ill give a positive response. Thus the test seems reasonably good. (Note that Table 7.12 is a 2×2 table in which the column totals were fixed in advance by the experimental plan.)

TABLE 7.12
Summary of Experimental Results

	Ill	Not ill	
+	990 (99%)	250 (5%)	
−	10 (1%)	4750 (95%)	
	1000 (100%)	5000 (100%)	

Hypothetical data.

In terms of Table 7.10 A and $\bar{A}$ represent positive and negative responses, respectively, and B and $\bar{B}$ represent Ill and Not Ill, respectively. We shall extrapolate to the entire population:

$$\Pr(A \mid B) = \Pr(+ \mid \text{Ill}) = 0.99,$$

$$\Pr(A \mid \bar{B}) = \Pr(+ \mid \text{Not Ill}) = 0.05.$$

In the entire population, we have the doctors' knowledge that

$$\Pr(B) = \Pr(\text{Ill}) = \frac{1}{200} = 0.005.$$

Suppose the test is used to detect the disease in members of the general public. What proportion of those having a positive reaction to the test should we expect actually to have the disease? In other words, if a person gives a positive reaction to the test, what is the conditional probability that that person is ill, $\Pr(B \mid A)$?

To answer this question, we imagine the test being given to everyone in the whole population. If the population size were 200,000,000, the figure 0.005 (1 in 200) provided by doctors for the overall incidence of the disease would tell us that the total number of Ill is about $0.005 \times 200,000,000 = 1,000,000$. The number Not Ill is then about 199,000,000 (Step 1). These numbers go into the bottom row of Table 7.13. Now we fill in the rest of the table so that the percentages in each column are as in Table 7.12 (Step 2). After summing across rows we have Table 7.14 (Step 3), and thereby obtain 9.05% as the expected percentage of persons with a positive reaction who are actually ill (Step 4). This process is equivalent to using the formula

$$\Pr(\text{Ill} \mid +) = \frac{\Pr(+ \mid \text{Ill}) \times \Pr(\text{Ill})}{\Pr(+ \mid \text{Ill}) \times \Pr(\text{Ill}) + \Pr(+ \mid \text{Not Ill}) \times \Pr(\text{Not Ill})}$$

with $\Pr(\text{Ill}) = 0.005$ [so that $\Pr(\text{Not Ill}) = 0.995$], and $\Pr(+ \mid \text{Ill}) = 0.99$, $\Pr(+ \mid \text{Not Ill}) = 0.05$. Information about the overall population [$\Pr(\text{Ill}) = 0.005$] is combined with information about the diagnostic test [$\Pr(+ \mid \text{Ill}) = 0.99$, $\Pr(+ \mid \text{Not Ill}) = 0.05$] to yield new information [$\Pr(\text{Ill} \mid +) = 0.0905$].

TABLE 7.13
Calculated Frequencies

	Ill	Not ill	
+	990,000 (99%)	9,950,000 (5%)	
−	10,000 (1%)	189,050,000 (95%)	
	1,000,000 (100%)	199,000,000 (100%)	200,000,000

TABLE 7.14
Table of Percentages

	Ill	Not ill	Total
+	990,000 (9.05%)	9,950,000 (90.95%)	10,940,000 (100%)
−	10,000 (0.005%)	189,050,000 (99.995%)	189,060,000 (100%)
	1,000,000 (0.5%)	199,000,000 (99.5%)	200,000,000 (100%)

In spite of the relatively low rate of positives among those not ill (5%), less than 1 in 10 of those who react positively to the test actually have the disease. This illustrates a difficulty of mass screening programs [Dunn and Greenhouse (1950)]. To be useful, the test would presumably have to have a smaller rate of positives among those not ill.

One more example will serve to indicate how information may be combined using Bayes' Theorem. Suppose that the voters of Ohio can be divided into two groups with regard to their views on foreign affairs. Call these liberal and conservative. Further, suppose that the division is known to be 40% liberal and 60% conservative, so that the probability that a person picked at random would be a conservative would be 0.60. Let us suppose that the two groups respond differently to the question, "Do you think that the U.S. should leave the U.N.?", 1/10 of the liberals answering Yes, while 1/2 of the conservatives respond Yes. We would like to know the probability that an individual chosen at random would be a liberal, *given the information that that individual had responded to this question in a certain way.*

We formulate the problem in terms of symbols:

B = The individual is a liberal.
$\bar{B}$ = The individual is conservative.
A = The individual thinks the U.S. should leave the U.N.
$\bar{A}$ = The individual does not think the U.S. should leave the U.N.

$$\Pr(A \mid B) = 0.10, \qquad \Pr(A \mid \bar{B}) = 0.50$$

$$\Pr(B) \quad = 0.40, \qquad \Pr(\bar{B}) \quad = 0.60.$$

Given these probabilities, we want to find $\Pr(B \mid A)$, the probability that a randomly chosen individual is a liberal, given that the response is Yes to the test question. Applying Bayes' Theorem, we find

$$\Pr(B \mid A) = \frac{\Pr(B) \times \Pr(A \mid B)}{\Pr(B) \times \Pr(A \mid B) + \Pr(\bar{B}) \times \Pr(A \mid \bar{B})}$$

$$= \frac{0.4 \times 0.1}{(0.4 \times 0.1) + (0.6 \times 0.5)} = \frac{0.04}{0.34} = \frac{2}{17} = 0.12.$$

Similarly, the probability that the individual is a liberal given that the answer is No is

$$\Pr(B \mid \bar{A}) = \frac{0.4 \times 0.9}{(0.4 \times 0.9) + (0.6 \times 0.5)} = \frac{0.36}{0.66} = \frac{6}{11} = 0.54.$$

Thus, the observed information (the response to the question) allows us to make a better guess of whether the individual is liberal or conservative. The original probabilities, 0.4 for liberal and 0.6 for conservative, are called the *prior* (or *a priori*) probabilities, while the altered probabilities, given the information, are called the *posterior* (or *a posteriori*) probabilities. If a person responds A ("get out of U.N."), the prior probability of B (liberal) is converted from 0.4 to the posterior probability of 0.12, while if a person responds $\bar{A}$ ("stay in U.N."), the posterior probability of B (liberal) is 0.54. The posterior probabilities are summarized in Table 7.15.

TABLE 7.15
Posterior Probabilities Resulting from Application of Bayes' Theorem

	B: *Liberal*	$\bar{B}$: *Conservative*	*Total*
A: Yes, we should get out of the U.N.	0.12	0.88	1.00
$\bar{A}$: No, we should stay in the U.N.	0.54	0.46	1.00

Use of Subjective Prior Probabilities in Bayes' Theorem

Although the formula that is called Bayes' Theorem is a perfectly valid mathematical rule, it has been the center of considerable discussion and controversy. The controversy has to do with the use of the rule when the *prior probabilities*, $\Pr(B)$ and $\Pr(\bar{B}) = 1 - \Pr(B)$, are subjective. In the medical diagnosis example, the prior probability $\Pr(B)$, the frequency of illness in a population, is really a relative frequency based on the clinic's experience. As such, it is objectively based and readily interpretable. In other examples, the prior probability may be only subjective probabilities. Some people feel that multiplying an ordinary probability by a subjective probability is like adding 4 oranges to 5 apples; they just don't mix.

As an example, consider the following:

B = "The phenomenon known as extra sensory perception exists";

A = "A test displaying some ESP is successful."

For a given test of ESP, the probability of the test being a success if there is no ESP, $\Pr(A \mid \bar{B})$, could perhaps be computed on the basis of randomness; for instance if the test consisted of predicting which face of a fair die will appear on a given toss. We might also be able to agree on what $\Pr(A \mid B)$ should be.

Skeptics, however, might insist that the probability $\Pr(B)$ is very close to zero, while some others might argue that it should be assumed to be 0.5. By applying Bayes' Theorem and working out the conditional probabilities, skeptics and their opponents would obtain quite different results. If, for example, $\Pr(A \mid \bar{B}) = 0.1$ and $\Pr(A \mid B) = 0.2$, then, a person who assigns a prior subjective probability of 0.01 that ESP exists will, when the test is successful, assign a posterior probability of

$$\Pr(B \mid A) = \frac{0.01 \times 0.2}{0.01 \times 0.2 + 0.99 \times 0.1} = \frac{0.002}{0.002 + 0.099} = \frac{2}{101}.$$

However, a person who assigns a prior probability of 0.5 will compute a posterior probability as

$$\Pr(B \mid A) = \frac{0.5 \times 0.2}{0.5 \times 0.2 + 0.5 \times 0.1} = \frac{0.10}{0.15} = \frac{2}{3}.$$

Since the test was successful, in each case the *posterior probability*, $\Pr(B \mid A)$, is higher than the prior probability; but the two posterior probabilities are still very different.

A statement such as "on the basis of the results of experiments the probability that ESP effects exist is 0.99" must depend on one's *a priori* belief concerning the effect. More generally, statements of the form, "it is 99% probable that the hypothesis is true," must depend upon what prior probability was used, and the prior that suits one person may not suit another. Nevertheless, if the experiment results in evidence favorable to the hypothesis, everybody's posterior probability will be larger than his or her prior probability was (even though some may be large and others small).

Summary

Operations relating to two events A and B, and to the corresponding probabilities, can be understood by reference to the 2×2 frequency table.

The event "A or B" denotes the collection of all outcomes contained either in A *or* in B (or in both). The event "A and B" denotes the collection of all outcomes contained in both A *and* in B.

If A and B are *mutually exclusive* events, that is, have no outcome in common, then

$$\Pr(A \text{ or } B) = \Pr(A) + \Pr(B)$$

If A and B are *any* two events,

$$\Pr(A \text{ or } B) = \Pr(A) + \Pr(B) - \Pr(A \text{ and } B).$$

If a set A is *contained* in a set B, then $\Pr(A) \leq \Pr(B)$.

The *conditional probability* of B given A, defined when $\Pr(A) > 0$, is

$$\Pr(B \mid A) = \frac{\Pr(A \text{ and } B)}{\Pr(A)}.$$

The events A and B are *independent* if $\Pr(A \text{ and } B) = \Pr(A) \times \Pr(B)$. In this case,

$$\Pr(B \mid A) = \Pr(B \mid \bar{A}) = \Pr(B)$$

and

$$\Pr(A \mid B) = \Pr(A \mid \bar{B}) = \Pr(A).$$

Class Exercise

7.1 Flip or shake two distinguishable coins (labelled coins 1 and 2 below) in such a way that you feel confident that the outcome for each coin (head or tail) is random. Record the outcomes for the two coins. Repeat this process until the pair of coins has been tossed 25 times.

(a) Record the frequency of outcomes in Table 7.16 below.

TABLE 7.16
Observed Frequencies

		Coin 1	
		Head	Tail
Coin 2	Head		
	Tail		

(b) Record the proportion of outcomes in Table 7.17.

TABLE 7.17
Observed Proportions

		Coin 1	
		Head	Tail
Coin 2	Head		
	Tail		

(c) Assuming the coins are unbiased and are tossed independently, indicate in Table 7.18 the probabilities of various outcomes for the toss of one pair of coins.

Save the results for use in Chapter 9.

TABLE 7.18
Probabilities

		Coin 1	
		Head	*Tail*
Coin 2	*Head*		
	Tail		

Exercises

7.2 Let *A* be the event that a person is a college graduate and *B* the event that a person is wealthy. In each case give in words the event whose probability is represented.

(a) $\Pr(A \text{ or } B)$,

(b) $\Pr(A \text{ and } B)$,

(c) $\Pr(A \text{ or } B) - \Pr(A \text{ and } B)$,

(d) $1 - \Pr(A)$,

(e) $1 - \Pr(B)$,

(f) $1 - \Pr(A \text{ or } B)$.

7.3 Let *A* be the event that a person is a male and *B* the event that a person is more than six feet tall. In each case give in words the event whose probability is represented.

(a) $\Pr(A \text{ or } B)$,

(b) $\Pr(A \text{ and } B)$,

(c) $\Pr(A \text{ or } B) - \Pr(A \text{ and } B)$,

(d) $1 - \Pr(A)$,

(e) $1 - \Pr(B)$,

(f) $1 - \Pr(A \text{ or } B)$.

7.4 Let *C* be the event that a college student belongs to a fraternity or sorority and *D* the event that a college student has a scholastic grade-point average of *B* or better. Express each of the following symbolically.

(a) the probability that a college student belongs to a fraternity or sorority,

(b) the probability that a college student has a grade-point average of B or better,

(c) the probability that a college student has a grade-point average of less than B,

(d) the probability that a college student does not belong to a fraternity or sorority and has a grade-point average of B or better.

7.5 If A and B are mutually exclusive events with $\Pr(A) = 0.3$ and $\Pr(B) = 0.4$, find the following:

(a) $\Pr(A \text{ or } B)$,

(b) $\Pr(A \text{ and } B)$,

(c) $\Pr(\bar{A} \text{ and } \bar{B})$.

7.6 If A and B are independent events with $\Pr(A) = 0.2$ and $\Pr(B) = 0.5$, find the following:

(a) $\Pr(A \text{ or } B)$,

(b) $\Pr(A \text{ and } B)$,

(c) $\Pr(\bar{A} \text{ and } \bar{B})$.

7.7 Which of the following pairs of events A and B are mutually exclusive?

(a) A: being the child of a lawyer
B: being born in Chicago,

(b) A: being under 18 years of age
B: voting legally in a U.S. election,

(c) A: owning a Chevrolet
B: owning a Ford.

7.8 Which of the following pairs of events A and B are independent?

(a) A: getting a six on the first roll of a die
B: getting a six on the second roll of a die,

(b) A: being intoxicated
B: having an auto accident,

(c) A: being on time for class
B: the weather being good.

7.9 By the "odds" against the occurrence of event A we mean the ratio $\Pr(\bar{A})/\Pr(A)$. Give the odds for each of the following:

(a) against the occurrence of A if $\Pr(A) = 1/4$,

(b) against the occurrence of A if $\Pr(A) = 1/3$,

(c) against the occurrence of A if $\Pr(A) = 1/2$,

(continued . . .)

(d) against the occurrence of A if $\Pr(A) = 3/4$,

(e) for the occurrence of A if $\Pr(A) = 3/4$.

7.10   Give the probability of getting an ace and a queen, not necessarily in that order, in two draws from a standard deck of 52 playing cards if

(a) the first card is not replaced before the second card is drawn,

(b) the first card is replaced before the second card is drawn.

7.11   Give the probability of getting three aces in three successive drawings from a deck of cards

(a) if each card is replaced before the next is drawn,

(b) if the successive cards are not replaced.

7.12   In rolling a balanced die, what is the probability that the first 3 will occur on the fourth try?

7.13   In drawing one card at a time without replacement from a deck of cards, what is the probability that the first spade will be the second card drawn?

7.14   From a sample of 2000 persons, each tested for colorblindness, Table 7.19 was obtained.

TABLE 7.19

Frequencies of Colorblindness

	Male	*Female*	*Total*
Normal eyesight	980	936	1916
Colorblind	72	12	84
Total	1052	948	2000

Hypothetical data.

(a) What is the relative frequency of occurrence of colorblindness?

(b) What is the relative frequency of colorblindness among males, that is, the conditional frequency of colorblindness, given that the individuals in question are males?

(c) What is the relative frequency of colorblindness among females?

(d) Assuming these 2000 persons to be a random sample from some given population, relative frequencies obtained from the table are estimates of the corresponding probabilities in the population. For example, the answer to (a) is an estimate of the overall rate of colorblindness in the population. Suppose it is known that the population is half males and half females. (Then our sample is slightly nonrepresentative, in the sense that we obtained 1052/2000, or 52.6% males rather than 50%.) How would you use this information to alter your estimate of the rate of colorblindness in the population?

7.15   Suppose we toss a coin twice. The probability of getting two heads is 1/4. What is the conditional probability of getting two heads, given that at least one head is obtained?

7.16   Suppose we toss a coin twice. What is the conditional probability of the sequence *HT* given that one *H* and one *T* occurred?

7.17   A test for a rare disease is positive 99% of the time in case of disease and is negative 95% of the time in the case of no disease. The rate of occurrence of the disease is 0.01. Compute the conditional probability of having the disease, given that the test is positive. How do you account for the fact that this probability is so low, in spite of the fact that a high percentage of those with the disease test positively and a high percentage of those without the disease test negatively?

7.18   Suppose that we have carried out a study relating voting behavior to social class, and find that the population can be adequately represented by two social classes, High and Low, having relative sizes 1/3 and 2/3, respectively. Of the people in the High social class, 75% voted Republican, while 36% in the Low social class voted Republican. Making use of Bayes' Theorem, find the conditional probability that a person chosen at random will be from the High social class if he or she votes Republican. What is the conditional probability that a person who votes Republican is from the Low social class?

7.19   In the Land of Aridona it rains on one out of ten days; that is, the probability of rain on any given day is 0.1. The conditional probability of rain on any one day, given that it rained on the previous day, is 0.6. Compute the probability that it will rain on two given consecutive days.

7.20   (continuation) Compute the conditional probability that it will rain on two given consecutive days given that it rained on the day before. Assume

$$\Pr(\text{Rain on Day 3} \mid \text{Rain on Days 1 and 2})$$

$$= \Pr(\text{Rain on Day 3} \mid \text{Rain on Day 2}).$$

7.21   In a certain routine medical examination, a common disorder, *D*, is found in 20% of the cases; that is, $\Pr(D) = 0.2$. The disorder is detected by a moderately expensive laboratory analysis. The disorder is, however, related to a symptom *S* which can be determined without a laboratory analysis.

Among those who have the disorder, 80% are found to have symptom S; that is, $\Pr(S \mid D) = 0.8$. Among those without the disorder, only 5% are found to have the symptom; that is $\Pr(S \mid \bar{D}) = 0.05$.

(a) Find the probability that presence of the symptom correctly predicts the disorder, that is, find $\Pr(D \mid S)$.

(b) Find the probability that lack of the symptom incorrectly predicts the disorder; that is, find $\Pr(D \mid \bar{S})$.

(c) Why is it reasonable that $\Pr(D \mid S)$ exceeds $\Pr(D)$?

7.22   You have one of those strings of Christmas tree lights where the whole string lights up only if each bulb is okay. Your string has four bulbs. If the probability is 0.9 that any one bulb is okay, what is the probability

(a) that the whole string lights up?

(b) that the string does not light up?

7.23   A family of three go to a photographic studio to have a picture made. A picture is a success if each of the three persons looks good in the picture; otherwise, it is a failure. Suppose that the probability that any one person looks good in the picture is 0.4. Assume independence.

(a) What is the probability that the picture is a success?

(b) Suppose the photographer makes two pictures. What is the probability that at least one of them is a success?

(c) If the photographer is good, do you think the assumption of independence is warranted?

7.24   (continuation) How many pictures need the photographer make to be 50% sure of getting at least one picture that is a success?

7.25   Eye color is governed by a pair of genes, one of which is contributed by the father, the other by the mother. Let B denote a gene for brown eyes and b a gene for blue eyes. An individual with a BB genotype has brown eyes. And, since B gene dominates the b gene, an individual with a Bb genotype will have brown eyes; only the bb genotype leads to blue eyes. Assume there are only the two genes B and b for eye color and that the father and mother each contribute one of these genes with a probability of $1/2$, independently of one another. What are the probabilities of the various genotypes for the offspring? What are the probabilities of the various eyecolors for the offspring?

7.26   Under the assumptions of Exercise 7.25, and given that a child has brown eyes, what is the conditional probability that the child's genotype is BB (rather than Bb or bB)?

7.27   Under the assumptions of Exercise 7.25, suppose that a mother has blue eyes and the father has brown eyes. What is the conditional probability that the father has the BB genotype, given that the couple has

(a) a child with brown eyes?

(b) two children with brown eyes?

(c) three children with brown eyes?

7.28   Suppose there were a society in which couples had children until they had a boy. Find the probability that the number of children in a family is 1, 2, 3, . . . , respectively.

7.29   Suppose couples had children until they had at least one child of each sex. Find the probability that the number of children in a family is 2, 3, . . . , respectively.

7.30   Fill in the following 2 × 2 tables.

	B	$\bar{B}$	Total
A	100	200	
$\bar{A}$	80		
Total		300	

	B	$\bar{B}$	Total
A			
$\bar{A}$	30	5	
Total	90		100

	B	$\bar{B}$	Total
A			
$\bar{A}$	0.20	0.40	
Total	0.30		1.00

	B	$\bar{B}$	Total
A	0.35	0.25	
$\bar{A}$	0.15		
Total			1.00

7.31   You have a probability of 0.6 for hitting the dart board each time you throw a dart, and successive throws are independent. What is the probability that you hit the dart board at least once in two throws?

7.32   Figure 7.8 represents a system of two components A and B, *in series*, which means the system functions if and only if both components function. Suppose the probability that A functions is 0.9, the probability that B functions is 0.9, and these events are independent. What is the probability that the system functions?

Figure 7.8   *Two components in series.*

7.33   (continuation) Figure 7.9 represents a system with two *parallel* sub-systems, $A - B$ and $C - D$. If one of the subsystems fails, the other can carry on. Thus the system fails only if both subsystems fail. Assuming that C is similar to A, D to B, and that the functioning of the two subsystems is independent, what is the probability that the system functions?

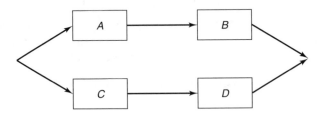

Figure 7.9   *Two parallel subsystems.*

7.34   (continuation) How many parallel subsystems are needed to make the probability that the system functions at least 0.95?

7.35   The application of statistical methods to the problem of deciding the authorship of the Federalist Papers was discussed in Section 1.1.

Suppose that a work, the author of which is known to be either Madison or Hamilton, contains a certain key phrase. Suppose further that this phrase occurs in 60% of the papers known to have been written by Madison, but only 20% of those by Hamilton. Finally, suppose a historian gives subjective probability 0.3 to the event that the author is Madison (and consequently 0.7 to the complementary event that the author is Hamilton). Compute the historian's posterior probability for the event that the author is Madison.

8
Probability Distributions

Introduction

Statistical inference is discussed in Part Four of this book. Inference is the process of drawing conclusions about populations of interest from samples of data. In this chapter we introduce the terminology associated with population distributions. In Chapter 9 we present the theory used as a basis for drawing inferential conclusions. As the reader will see, these two sets of principles are closely related.

8.1 Random Variables

In Chapter 7 we discussed probabilities of events described *categorically*. Categories define the event that a person is a lawyer and the event that a person is a doctor, the event that a person has a certain disease and the event that a person does not have the disease. We are also interested in events described *numerically*, such as the event that a person regularly reads 3 magazines and the event that a person regularly reads 2 magazines, the event that a person has an annual income of $28,000 and the event that a person has an annual income not greater than $36,000, the event that a tree bears 560 apples and the event that a tree bears 72 apples.

Table 8.1 gives the distribution of families by number of children under 18 years of age for the United States in 1990. This table is a frequency distribution for a population. The relative frequency of families with 3 children is 0.07; the relative frequency of families with 2 children is 0.19. If a family is chosen at random, then 0.07 is the probability that this random family will have 3 children. Relative frequencies for a population are probabilities. A (relative) frequency distribution for a population is called a *probability distribution.*

TABLE 8.1

Distribution of Families by Number of Children Under 18 Years of Age (United States, 1990)

Number of children	0	1	2	3	4 or more
Proportion of families	0.51	0.20	0.19	0.07	0.03

source: Statistical Abstract of the United States: 1993, p. 60.

Often in describing a probability distribution we focus on the *variable* and say "distribution of *number of children*" rather than "distribution of *families* by number of children." If a family were chosen at random, the probability that the *number of children* would be 3 is 0.07, the probability that the number of children would be 2 is 0.19, etc. The variable, "number of children," is a *random variable* in the sense that its value depends on which family is chosen.

In general, any variable that can be described by a probability distribution is a random variable. In the case of families, we might describe the distribution of the number of children, of total family size, of family income, of the number of automobiles, or of the number of magazines read. For each of these, there is a certain probability that a randomly-chosen family will have a particular quantity (e.g., 3 autos or 8 magazines). The quantity is not a foregone conclusion because there is more than one possible value and because the value is not assigned to families in any fixed, predetermined way. The quantity can only be known once a family has been chosen and the observation is made. If we were to sample many, many families (perhaps an infinite number of families, if that were possible) then the relative frequency of each quantity would be its probability and the list of relative frequencies for all quantities is the probability distribution.

For notation, let x represent the number of children. Then the notation $x = 3$ represents the *event* that a family has 3 children, and $\Pr(x = 3)$ stands for the probability that a randomly chosen family has 3 children. Thus, $\Pr(x = 3) = 0.07$.

The variables described above are *discrete* random variables because the scales are comprised of distinct values with no quantities in between, e.g., 0, 1, 2, 3, ... children but not $1/2$ or $1\frac{1}{16}$ children. Many other variables may be thought of as existing in all possible quantities, for example, between a 3-minute waiting time and a 4-minute time period there is an infinite number of other possible times. However, the process of observing and recording statistical data always results in a discrete set of values. Even if we were to record time to the nearest nanosecond, we are ignoring values that are fractions of nanoseconds; if we record time to the nearest $1/10$ of a nanosecond, we are not permitting values between $1/10$ and $2/10$ of a nanosecond; and so on.

For theoretical purposes, however, it is important to be able to describe the probability distribution of a *continuous* random variable. As an example, consider the heights of 1293 11-year-old boys; Table 2.12 gives the numbers of boys with heights in one-inch intervals. The distribution of heights of the 1293 boys is approximated by considering all heights within an interval as equal to the midpoint of that interval. The height of a boy picked at random from among the 1293 is a random variable. We can think of Table 2.12 replaced by a frequency table with smaller intervals. If the initial population is very large—and the method of measurement is very accurate—the intervals can be very small. As noted in Chapter 2, if this idea is carried to the extreme—even infinitely small intervals—the result is a smooth curve.

Figure 8.1 shows a histogram and a smooth curve. The relative frequency of heights between 54.5 and 56.5 inches is represented by the area of the two bars at 55 inches and 56 inches, that is, by the area of the histogram from 54.5 inches to 56.5 inches, because the area of the entire histogram is taken to be 1. The area under the histogram is about the same as the area under the smooth curve from 54.5 inches to 56.5 inches. When we sample randomly, the probability of a height between 54.5 and 56.5 is equal to this area. In fact, the probability of obtaining a measurement in any interval is the area over that interval.

8.2   Cumulative Probability

Cumulative probabilities are analogous to the cumulative relative frequencies discussed in Chapter 2. Table 8.3 gives the cumulative probabilities corresponding to the probability distribution of Table 8.2. For

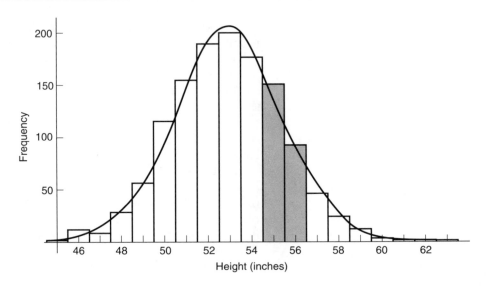

Figure 8.1   *Histogram of heights of 11-year-old boys, with corresponding continuous probability distribution (Table 2.12).*

TABLE 8.2

Probability Distribution of Number of Children

Number of children	0	1	2	3	4	5
Proportion of families	0.2	0.2	0.3	0.1	0.1	0.1

Hypothetical data.

TABLE 8.3

Cumulative Distribution Function

Number of children	0	1	2	3	4	5
Proportion of families having this many children or fewer	0.2	0.4	0.7	0.8	0.9	1.0

source: Table 8.2.

each number of children, the corresponding cumulative probability is simply the proportion of families having that number of children or fewer. Cumulative probabilities can also be obtained for continuous random variables.

8.3 The Mean and Variance of a Probability Distribution

The Mean of a Discrete Random Variable

Since probabilities are relative frequencies for a population, it is possible to find the mean of a probability distribution in the same way as the sample mean can be found from a frequency distribution. In Section 3.3 we note that the mean of a sample of n individuals can be written as

$$\bar{x} = \frac{f_1 v_1 + f_2 v_2 + \cdots + f_m v_m}{n},$$

where $v_1, v_2, \ldots, v_m$ are the *different* observed values of the variable and $f_1, f_2, \ldots, f_m$ are their respective frequencies. Dividing each f_j by n permits this expression to be written as

$$\bar{x} = \frac{f_1}{n} v_1 + \frac{f_2}{n} v_2 + \cdots + \frac{f_m}{n} v_m.$$

This is the same as

$$\bar{x} = p_1 v_1 + p_2 v_2 + \cdots + p_m v_m,$$

where the ratios $p_1 = f_1/n$, $p_2 = f_2/n, \ldots, p_m = f_m/n$ are the *relative frequencies* of the values $v_1, v_2, \ldots, v_m$. Thus $\bar{x}$ is a *weighted average* of the values $v_1, v_2, \ldots, v_m$, the weights being the relative frequencies $p_1, p_2, \ldots, p_m$.

Applying these same arithmetic operations to the hypothetical data in Table 8.2, we have

$$(0.2 \times 0) + (0.2 \times 1) + (0.3 \times 2) + (0.1 \times 3) + (0.1 \times 4) + (0.1 \times 5)$$

$$= 0 + 0.2 + 0.6 + 0.3 + 0.4 + 0.5$$

$$= 2.0 \text{ children per family.}$$

This value 2.0 is the mean of the variable "number of children," or, equivalently, the mean of the corresponding probability distribution.

In general, the mean of a probability distribution of a discrete random variable is

$$p_1 v_1 + p_2 v_2 + \cdots p_m v_m,$$

where $p_1, p_2, \ldots, p_m$ are the probabilities associated with the values $v_1, v_2, \ldots, v_m$. That is, the mean of a probability distribution is a weighted average of the values $v_1, v_2, \ldots, v_m$, where the weights are the probabilities $p_1, p_2, \ldots, p_m$; it is the *probability-weighted average* of the values

$v_1, v_2, \ldots, v_m$. This is also referred to as the *expected value* of the variable under consideration. That is, if the variable is x, its mean is the *expected value of x*, or just $E(x)$, where

$$E(x) = p_1 v_1 + p_2 v_2 + \cdots + p_m v_m = \sum p_j v_j. \qquad (8.1)$$

The mean of a probability distribution is also the same as the population mean of the variable. While the sample mean is represented by $\bar{x}$, the population mean is represented by the Greek symbol μ, so we may also write

$$\mu = E(x).$$

It may seem that there are two symbols for the same concept, but the expected value $E(x)$ is used to denote the *process* of finding the mean of a probability distribution and μ is a symbol for the *result*.[1]

In the special case where a population is finite, the mean can be found by averaging the individual values, just as in a sample. If $y_1, y_2, \ldots, y_N$ are the individual values, then

$$\mu = \frac{\sum y_i}{N} = \sum p_j v_j.$$

The first sum is the sum of the N individual observations (y_i). The second sum is the sum of m *different* values (v_j), each multiplied by its probability.

The Variance of a Discrete Random Variable

The variance of a probability distribution, like the variance of a sample, summarizes the squared distances from the mean. If the mean of a probability distribution is represented by μ, the variance is the probability-weighted average of the squared deviations $(v_1 - \mu)^2$, $(v_2 - \mu)^2, \ldots,$ $(v_m - \mu)^2$, that is,

$$p_1(v_1 - \mu)^2 + p_2(v_2 - \mu)^2 + \cdots + p_m(v_m - \mu)^2$$
$$= \sum p_j(v_j - \mu)^2. \qquad (8.2)$$

For the distribution of the discrete random variable in Table 8.2, we have $\mu = 2.0$. The variance is

---

[1] In later sections, we compute expected values of more complex random variables (not just x but some functions of x); there we use the $E(\ )$ notation with the function given in the parentheses, but the result is no longer represented as μ.

$$0.2 \times (0 - 2.0)^2 + 0.2 \times (1 - 2.0)^2 + 0.3 \times (2 - 2.0)^2$$
$$+ 0.1 \times (3 - 2.0)^2 + 0.1 \times (4 - 2.0)^2 + 0.1 \times (5 - 2.0)^2$$
$$= 0.2 \times (-2.0)^2 + 0.2 \times (-1.0)^2 + 0.3 \times (0.0)^2 + 0.1 \times (1.0)^2$$
$$+ 0.1 \times (2.0)^2 + 0.1 \times (3.0)^2$$
$$= 0.8 + 0.2 + 0 + 0.1 + 0.4 + 0.9$$
$$= 2.4.$$

In the previous section we used the expected-value notation $E(x)$ to denote the probability-weighted average of the values of variable x. The notation can be used to denote the average of squared deviations from the mean as well. Using this notation, the variance formula (8.2) can be written $E[(x - \mu)^2]$, that is, the expected value of the squared deviations from the mean.

The result is the variance of the corresponding population, represented by the Greek symbol σ^2. Thus in summary,

$$E\left[(x - \mu)^2\right] = \sum p_j(v_j - \mu)^2 = \sigma^2. \tag{8.3}$$

The population *standard deviation* is the square root of the variance, that is $\sigma = \sqrt{\sigma^2}$. For the data of Table 8.2, $\sigma = \sqrt{2.4} = 1.55$.

In Chapter 4 a computing formula is given for the sample variance that simplifies the arithmetic operations. Likewise a computing formula for the population variance, equivalent to (8.3), is

$$\sigma^2 = E(x^2) - [E(x)]^2,$$

where $E(x^2) = p_1 v_1^2 + p_2 v_2^2 + \cdots + p_m v_m^2$ and $[E(x)]^2 = \mu^2$. For the variable in Table 8.2, $E(x^2) = 0.2 \times 0^2 + 0.2 \times 1^2 + 0.3 \times 2^2 + 0.1 \times 3^2 + 0.1 \times 4^2 + 0.1 \times 5^2 = 6.4$ and $[E(x)]^2 = 2.0^2 = 4.0$. Thus $\sigma^2 = 6.4 - 4.0 = 2.4$.

In the special case of a finite population, the mean can be obtained by averaging the individual values, as shown above. The variance is

$$\sigma^2 = \frac{\sum (y_i - \mu)^2}{N} = \sum p_j(v_j - \mu)^2.$$

The Mean and Variance of a Continuous Random Variable

The examples in the preceding sections involved discrete random variables having only specific whole-number values (e.g., 0, 1, 2, 3, or 4 children). We also want to be able to discuss the mean and variance of a continuous variable. In Section 8.1 we considered the heights of 11-year-old boys as given in Table 2.12. From this table we can calculate the mean and the variance of the distribution of heights by using the midpoint of each interval as the height of all boys grouped in that interval. There may be small differences between the mean and variance computed this way and the the mean and variance computed on the basis of the 1293 heights if they were measured more accurately than to the nearest inch. We could reduce these differences by using intervals smaller than 1 inch.

If the population is large, the intervals can be very small. The probability distribution of a continuous random variable represented by a smooth curve has a mean and variance which are approximated by the mean and variance of an approximating histogram. Obviously, the smaller the intervals, the better the approximation will be.

If we carry this process to the extreme, something we could only do in theory, there would be an infinite number of infinitely narrow intervals. This is one way to think of a continuous probability distribution. Probabilities, means, and variances would have to be obtained using calculus. The mean may still be thought of as the center of gravity of the distribution. If a flat metal plate were cut to the shape of the distribution, the plate would balance at the mean (Figure 8.2).

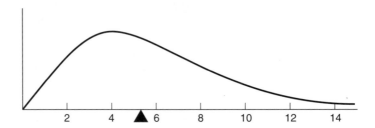

Figure 8.2 *The mean is the center of gravity.*

8.4 Uniform Distributions

The Discrete Uniform Distribution

One of the simplest probability distributions is that of a digit taken at random. In this case the integers 0, 1, 2, ..., 9 are to be equally likely. The probabilities are

$$\Pr(\text{integer chosen} = i) = \frac{1}{10}, \qquad i = 0, 1, 2, \ldots, 9.$$

This probability distribution is called the *uniform distribution* on the integers 0, 1, 2, ..., 9. This distribution provides the *probability model* for "choosing a digit at random."

Similarly, we can consider choosing one of the ten values, 0.0, 0.1, 0.2, ..., 0.9, at random. In this case

$$\Pr\left(\text{value chosen} = \frac{i}{10}\right) = \frac{1}{10}, \qquad i = 0, 1, \ldots, 9.$$

These probabilities are graphed in Figure 8.3.

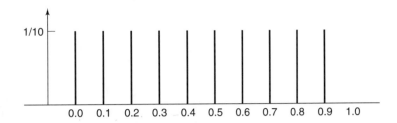

Figure 8.3   *Uniform distribution on tenths.*

The use of *pairs* of random digits leads to the uniform distribution on 0.00, 0.01, 0.02, ..., 0.99, with probabilities

$$\Pr\left(\text{value chosen} = \frac{i}{100}\right) = \frac{1}{100}, \qquad i = 0, 1, 2, \ldots, 99.$$

We can also consider the uniform distribution on the one thousand values 0.000, 0.001, 0.002, ..., 0.999. The graphs of these probability distributions look like Figure 8.3, but with 100 and 1000 bars, respectively.

The Continuous Uniform Distribution

If we continue this process of considering uniform distributions over the numbers between 0 and 1 with more and more decimal places, in the end the resulting distribution is the *continuous uniform distribution* over the interval from 0 to 1. This probability distribution assigns a probability of $\frac{1}{2}$ to the interval 0 to $\frac{1}{2}$, to the interval $\frac{1}{2}$ to 1, to the interval $\frac{1}{4}$ to $\frac{3}{4}$; it assigns a probability of $\frac{1}{4}$ to any interval of length $\frac{1}{4}$. The continuous uniform distribution on the interval 0 to 1 assigns to any subinterval of the interval from 0 to 1 a probability equal to its length.

8.5   The Family of Normal Distributions

The Normal Distributions

Many continuous statistical variables have distributions that look similar. The histograms of such biological measurements as heights and weights, and such psychological measurements as IQ scores have common features. The histogram of Figure 8.1, based on the frequencies of heights to the nearest inch of 1293 Boston schoolboys (Table 2.12), shows these features. The modal class is 52.5–53.5 inches, with a midpoint of 53 inches. In classes on either side of 53 inches the frequencies decrease slowly at first, then rapidly, and then slowly again. The frequency for 54 inches is about the same as that for 52 inches, the frequency for 55 inches is about the same as that for 51 inches, etc.; that is, the histogram is approximately symmetric about 53. If the class intervals had been half as large, the histogram would have been smoother. This is approximated by a smooth curve, as shown in Figure 8.1.

It turns out that in many cases the smooth curve has the shape of a particular theoretical probability distribution known as the *normal* distribution. [It is also known as the *Gaussian* distribution, after the famous German mathematician, K. F. Gauss (1777–1855); in France it is called Laplace's distribution, after P. S. Laplace (1749–1827).] Several graphs of normal distributions are given in Figures 8.4–8.7.

It is not a coincidence that so many variables have probability distributions that look like this bell-shaped curve. A powerful mathematical theorem, the *Central Limit Theorem*, assures that this will be the case. In general terms the Central Limit Theorem states that whenever a variable x is made up of many separate components, each of which can have two or more values, the resulting variable will have approximately a normal probability distribution. Further, the more separate components there are, the more perfectly normal the probability distribution of x will be.

Consider several examples. The height of a single boy in Table 2.12 is undoubtedly the result of a number of separate factors, each of which could have been present to a greater or lesser extent during his lifetime. These might include the the height of his mother, the height of his father, prenatal health care, his exact age, his health status and illnesses that might affect growth since birth, numerous dietary factors, and so on. If we listed these components in sufficient detail, we could easily derive a list of 30 or more variables "behind the scene" that went into determining his present height. Because there are so many components the Central Limit Theorem tells us that the distribution of heights for a population of boys will be *very* close to normal in shape. Likewise, test scores for a population of respondents with similar characteristics are normally distributed if the test is comprised of a number of separate items. The more items to the test, the more normal the resulting distribution of test scores will be. Thus while scores on a 20-item test may have a probability distribution that is close to a normal curve, scores on a 50-item test may be truly indistinguishable from a "perfect" normal distribution. In general, the lion's share of measures of humans and their institutions and complex objects and events have normal-shaped probability curves.

According to the Theorem, the components must also be "separate"; that is, they must each affect the resulting variable in some different way from the other components. Thus, mother's height, father's height, and the average of both parents' heights are not 3 separate components but 2 because the average is just the two individuals' values expressed another way. On the other hand, since mother's height and father's height have somewhat different impacts on the child's height, these 2 components are at least somewhat "separate." Likewise, if a test had two items that covered exactly the same topic so that a respondent would answer one correctly if and only if he or she answered the other correctly, these would constitute one component, not 2. For the Central Limit Theorem to apply, variable x must be comprised of a number of components that are at least somewhat separate.

The degree of separateness does matter. When the components that lie behind a variable x are very different from one another (i.e., almost or completely *independent*[2]) a relatively small number of components—perhaps 12 to 20—will assure that x has a normal distribution. When the components are somewhat interrelated, a larger number is needed before we can be certain that the resulting distribution is normal. Thus, all of the items on a science test or a personality test are likely to have a lot in common with one another because they are all items on the same general topic; certain responses to one item will often be accompanied by certain responses to other items. In such a situation, the instrument may need 30 to 50 items or more before we can be confident that the probability distribution of test scores is normal in shape.

Because so many variables have this many-faceted structure, the normal distribution is frequently used as an ideal or "model" for analyzing statistical data. To decide whether the normal distribution is applicable, the data analyst can examine the composition of the variable(s) being studied to see if the Central Limit Theorem applies or else inspect a histogram of the data in the sample. If the sample is drawn from a normally-distributed population, the sample distribution should

---

[2]Independence of random variables is defined precisely in Chapter 9.

be roughly normal in shape—i.e., central values more common than extreme values and no extreme skew in either direction. The sample distribution does not have to be perfectly normal in order for the normal distribution to be an appropriate model for the parent population, only roughly so. In fact, it may be difficult to see normality in a small sample, even if the population is highly normal. On the other hand, if the sample size is large and the histogram is extremely skewed or non-normal in some other way, an alternate statistical model may be required.

Different Normal Distributions

The histogram of heights of 11-year-old boys (Figure 8.1) shows roughly the typical shape of a normal distribution. The histogram of heights of 12-year-old boys (Exercise 8.20) is similar but is centered at 54.7 inches. The histograms of heights of boys in one-year age brackets will have similar shapes but different means and percentiles, and perhaps different variabilities. Heights of cats would give a histogram which would again approximate the normal distribution, but the location and spread would differ from those for boys.

Thus, there is a whole *family* of normal distributions. The peak can moved to various points and the width can be expanded or contracted. Three of these are shown in Figures 8.4, 8.5, and 8.6.

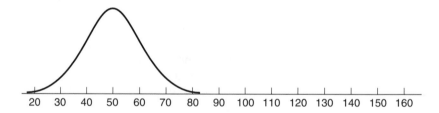

Figure 8.4   *Normal distribution with center at 50.*

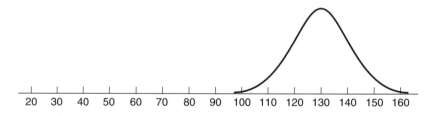

Figure 8.5   *Normal distribution with same spread as that of Figure 8.4, but with center at 130.*

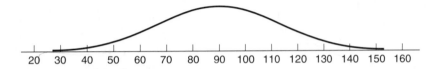

Figure 8.6   *Normal distribution with large spread and center at 90.*

All the normal distributions are symmetric, unimodal, and bell-shaped, and they are all defined by the same mathematical formula. The characteristics that distinguish one from another are the mean and standard deviation. The mean μ gives the location along the horizontal axis, and the standard deviation σ tells how spread out the distribution is; the larger σ is, the more spread out is the distribution. The distribution in Figure 8.4 has $\mu = 50$ and $\sigma = 10$; in Figure 8.4, $\mu = 130$ and $\sigma = 10$; and in Figure 8.6, $\mu = 90$ and $\sigma = 20$. The pair of characteristics μ and σ are the *parameters* of the family of normal distributions.

The formula that relates y to x in these curves is

$$y = \frac{1}{\sigma\sqrt{2\pi}}e^{-(x-\mu)^2/(2\sigma^2)},$$

where x denotes the horizontal axis and y the vertical axis. In theory, in order to use the normal distribution as a model for statistical data we would have to use calculus to integrate this formula over a range of x values to obtain the corresponding probability.

In practice, the job has been simplified for us. Statisticians have created tables that give the probability of x values in particular intervals, if x has a normal distribution. This was possible because all normal distributions have the same defining formula, and because the formula involves just the two parameters μ and σ. As a result, the probability of a value between the mean and one standard deviation above the mean—to pick one of many possible examples—is the same in *all* normal distributions regardless of the particular value of μ or σ.

The Standard Normal Distribution

A table of normal probabilities is given in Appendix I for a variable with mean 0 and standard deviation 1. In Section 4.5 we note that values with this mean and standard deviation are termed *standard scores*, often represented by the letter z. A normally-shaped distribution of standard scores is called the *standard normal* distribution. Although we might never encounter a distribution with mean 0 and standard deviation 1 in real data, we can obtain all of the probabilities we need from this distribution and then convert them to any other distribution that is normal in shape.

Figure 8.7 is a graph of the standard normal distribution. Based on the total area under the curve, the percentages in the figure indicate the area under the curve between various vertical lines. For instance, 34.1% of the total area is between 0 and +1 and the same percentage is between −1 and 0. In probability terms this expresses the idea that the probability of a value between 0 and 1 is 0.341. This corresponds to the interpretation of relative area in a histogram as indicating relative frequency.

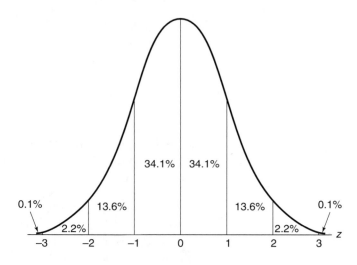

Figure 8.7   *Graph of the standard normal distribution.*

Table of Standard Normal Probabilities

This value is given in Appendix I. The table is entered by locating the desired z value 1.0 in the left-hand column. Then read across the headings at the top of the table until you reach the second decimal digit for z. For $z = 1.00$, this is just the first column; for $z = 1.01$ we would look in the second column of probabilities, and so on. Corresponding to $z = 1.00$, the tabled entry is .3413. This is the proportion of the distribution contained *between* 0 and the z value 1.00. In symbols, this is written $\Pr(0 \le z \le 1) = .3413$. In words, in the standard normal distribution, the probability of a z value between 0.00 and 1.00 is .3413. It is also the case that 34.13% of the values in *any* normal distribution are between the mean and one standard deviation above the mean.

Two additional properties of the table are needed to use it generally. First, the table is constructed so that the *sum* of the areas (probabilities) of the intervals *in the entire distribution* is 1.00. Second, the normal distribution is *symmetric* about the mean. In the standard normal distribution the mean is 0, and symmetry about 0 implies that $\Pr(z \le z_0) = \Pr(z \ge -z_0)$ where z_0 is any specific value of z. In particular, $\Pr(z \le 0) = \Pr(z \ge 0) = .5000$. The property of symmetry can be used to find probabilities for negative z values, which are not tabled.

For example, the probability of a z value between -1 and 0 is the same as that between 0 and 1; $\Pr(-1 \le z \le 0) = .3413$. From this, we can see that the probability of a z value between -1 and $+1$ is $.3413 + .3413 = .6826$. In symbols, $\Pr(-1 \le z \le 1) = .6826$. About two-thirds of the values in the standard normal distribution are between -1 and $+1$. It is also the case that about two-thirds of the values in *any* normal distribution are within one standard deviation of the mean.

Since one-half of the curve is to the right of 0, the probability of a z value greater than or equal to $+1$ is $.5000 - .3413 = .1587$. In symbols, $\Pr(z \ge 1) = .1587$. By symmetry $\Pr(z \le -1)$, in the left-hand tail of the curve, is also $.1587$.

If we enter the table in Appendix I with $z = 2.00$, we see that the proportion of the standard normal distribution between 0 and 2.00 is $.4772$. By symmetry, the probability of a value between -2.00 and 0 is also $.4772$. Combining these, the probability of a value from the standard normal distribution between -2 and $+2$ is $\Pr(-2 \le z \le 2) = .4772 + .4772 = .9544$. About 95% of the values in the standard normal distribution are in the range -2 to $+2$. It is also the case that about 95% of the values in *any* normal distribution are within two standard deviations of the mean.

This result is the origin of the "two-standard-deviation rule of thumb" stated in Chapter 4. Since many, many variables have normal population distributions, *and* since about 95% of the values in a normal distribution are within two standard deviations of the mean, *and* since most samples have distributions that approximate their respective populations, it is a safe bet that in most samples of statistical data most of the observations are also within two standard deviations of the mean.

Additional Examples.   Several more examples illustrate the use of the table. The probability of a z value between 0.7 and 1.7 is indicated by the darker shaded area in Figure 8.8. This is obtained by finding the probability of a z between 0 and 1.7 and subtracting the probability of a z between 0 and 0.7; the shaded area is just the difference between the larger and smaller regions. From the table, $\Pr(0 \le z \le 1.7) = .4554$ and $\Pr(0 \le z \le 0.7) = .2580$. Thus, $\Pr(0.7 \le z \le 1.7) = .4554 - .2580 = .1974$.

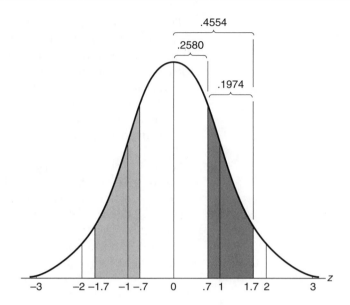

Figure 8.8 *Graph of the standard normal distribution.*

Should the probability of a z value between -1.7 and -0.7 be needed, it can be obtained by "reflecting" the problem to the right side of the distribution. That is, the lighter shaded region and the darker shaded region of Figure 8.8 have the same area. In symbols, $\Pr(-1.7 \le z \le -0.7) = \Pr(0.7 \le z \le 1.7) = .1974$.

The symmetry property can be used to find other probabilities involving negative z values. For example, $\Pr(-2.22 \le z \le 0.34)$ can be obtained by summing two regions, $\Pr(-2.22 \le z \le 0)$ and $\Pr(0 \le z \le 0.34)$. The first of these is identical to $\Pr(0 \le z \le 2.22)$ which is tabled as .4868. The second is tabled directly as .1331, and the sum of the two is the desired probability, .6199.

In general, finding probabilities in the standard normal distribution is easy if you (1) draw a diagram, (2) express the problem in terms of areas between 0 and a particular z value, because this is how the table in Appendix I is constructed, and (3) "reflect" any problem or subproblem involving a negative z because Appendix I gives only probabilities for the right-hand side of the curve.

As has been emphasized, the probability of a value in an interval is the area of that region below the normal curve. What is the probability of a *particular* value, say $z = 1.2$? We can think of the point 1.2 as an interval of 0 length. The area above the point and below the normal curve is the height of the normal curve times the length of the interval; this product is 0. We conclude that the probability of a particular point

$\Pr(z_0 \le z \le z_0) = \Pr(z = z_0) = 0$. Note that $\Pr(z \le z_0) = \Pr(z < z_0) + \Pr(z = z_0) = \Pr(z < z_0)$. Thus for a continuous distribution it does not matter whether an inequality has the symbol "$<$" or "$\le$." The probability is the same.

Finding a z Value for a Given Probability.   In the preceding discussion the table of the standard normal distribution is used to find probabilities associated with a range of z values. It can be used in the opposite direction as well to find a z value that satisfies a particular probability statement. For example, the value that separates the upper 50% of the standard normal distribution from the lower 50% is obviously $z = 0$. In symbols, if the particular z value we seek is represented z_0, then we are looking for z_0 that satisfies $\Pr(z \ge z_0) = .5000$. The probability statement can be read "the proportion of z values above the *particular* z value (z_0) is .5000."

Suppose we require z_0 such that $\Pr(z \ge z_0) = .1587$. This is the particular z value that has .1587 of the standard normal distribution at or above it, indicated by the shaded area in Figure 8.9.

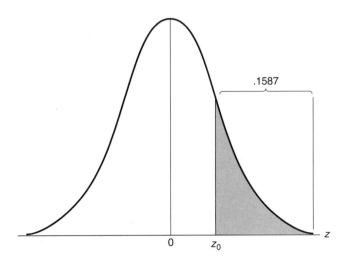

Figure 8.9   *Graph of the standard normal distribution.*

Because the area to the right of 0 is .5000, it can be seen in Figure 8.9 that z_0 is the value of z such that the probability of the interval 0 to z_0 is $.5000 - .1587 = .3413$, that is, .3413 of the standard normal distribution lies between 0 and z_0. The subtraction is necessary because the table in Appendix I gives only the probability *between 0 and a given z value.*

Now we can look through the center part of the table for the probability .3413, which is found about one-third of the way down the first column. Reading outward to the left and top margins, this corresponds to a z value of 1.00 which is the required z_0.

Suppose that we require z_0 such that $\Pr(z \geq z_0) = 0.9$. This is the particular z value that has 90% of the standard normal distribution above it, corresponding to the shaded area of Figure 8.10. Obviously it is to the left of 0 and is a negative z, and so must be reflected to use the table in Appendix I. By symmetry, the z_0 we seek has the same numerical value as the z that has 90% of the distribution below it, but with the opposite sign. This z value has $.9000 - .5000$ or $.4000$ of the distribution between it and 0. Again, this subtraction is necessary in order to conform to the entries in the table.

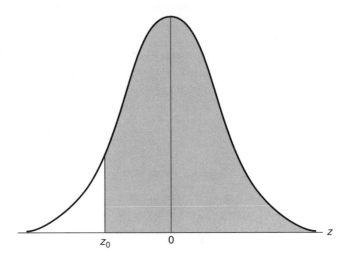

Figure 8.10   *Graph of the standard normal distribution.*

Searching through the table for a probability of .4000, the closest entry is .3997 (sufficiently close for our purposes). If we read out to the corresponding row and column margins, we see that this corresponds to a z value of 1.28. Reflecting back to the left half of the curve, the required z_0 is -1.28. That is, $\Pr(z \geq -1.28) = .8997$.

As an additional example, let us find z_0 that satisfies $\Pr(-z_0 \leq z \leq z_0) = 0.90$; this is a value such that 90% of the normal distribution is contained between $-z_0$ and $+z_0$. Since the same numerical value is required for both ends of the interval, the portion of the curve that is enclosed must be the central 90%, divided evenly at 0 as shown in Figure 8.11. Thus one-half of the interval is between 0 and $+z_0$. The probability of this portion is $0.90/2 = .4500$. The closest probability entries in the table are .4495 corresponding to a z value of 1.64, and .4505 corresponding to $z = 1.65$. For our purposes, either z value is sufficiently close, or we may interpolate to a result half-way between them, $z_0 = 1.645$. This value, with two signs affixed, is the one we seek, that is $\Pr(-1.645 \leq z \leq 1.645) = 0.90$.

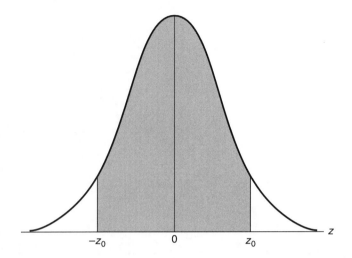

Figure 8.11 *Graph of the standard normal distribution.*

A Notation for Values from the Standard Normal Distribution. Throughout this book there are many statistical applications that require values from the standard normal distribution; thus it is convenient to have a general notation for normally distributed z scores. We shall use $z(p)$ to represent the z value that has the proportion p of the standard normal distribution *to the right of it*, that is, $z(p)$ is the value that satisfies

$$\Pr[z > z(p)] = p. \tag{8.4}$$

For example, $z(0.50)$ is the z value that has proportion 0.50 of the distribution to the right of it, $z(0.50) = 0$. Using the same notation, $z(0.05)$ is the z value that has proportion 0.05 of the normal distribution to the right of it, that is, $z(0.05) = 1.645$, and $z(0.95)$ has proportion 0.95 of the standard normal distribution to the right of it,[3] $z(0.95) = -1.645$. These points and several others are labeled in Figure 8.12.

Some of the examples in the preceding section ("Finding a z value for a given probability") might have used this notation. The example that illustrated how to find z_0 that has 90% of the standard normal dis-

---

[3]Because of the symmetry of the standard normal distribution, $z(0.95)$ is equal to $-z(0.05)$.

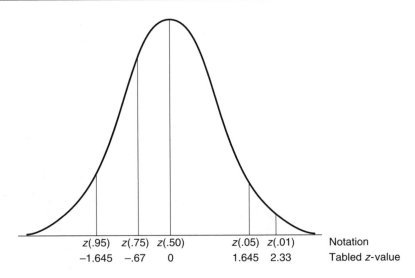

z(.95)	z(.75)	z(.50)		z(.05)	z(.01)	Notation
−1.645	−.67	0		1.645	2.33	Tabled z-value

Figure 8.12   *Graph of standard normal distribution with values of z(p) indicated.*

tribution above it was solving for $z(0.90)$. The example that illustrated how to find z_0 that satisfied $\Pr(-z_0 < z < z_0)$ was solving for $z(0.95)$ and $z(0.05)$. We shall use this notation a few times in the remainder of this chapter and more often in the chapters to come.

Other Normal Distributions

The table of the standard normal distribution can be used for other normally distributed variables because the probability of an interval depends in a simple way on its relationship to the mean and standard deviation. To find probabilities associated with any normally distributed variable x, the x values only need to be converted to the standard score scale and then the table in Appendix I may be applied.

The relationship between a variable and the corresponding standard score is given for samples in Chapter 4. The relationship to the population mean μ and standard deviation σ is similar:

$$z = \frac{x - \mu}{\sigma}.$$

As in the sample, the standard score or *standardized version of x* tells how many standard deviations above or below the mean a particular x value lies. Values of x above the mean give positive z scores and x's below the mean give negative z's. For the entire population the mean of the z values is 0 and the standard deviation is 1.

Suppose that we encountered a normally distributed variable, x, with a mean of 50 and standard deviation 8, and needed to know the proportion of the distribution above 54. It is only necessary to convert 54 to a z value to use the table. For this example,

$$z = \frac{54 - 50}{8} = 0.5.$$

In symbols, $\Pr(x \geq 54) = \Pr(z \geq 0.5)$. The required probability corresponds to the shaded portion of Figure 8.13. From Appendix I the probability of a z value between 0 and 0.5 is .1915. Since one-half of the normal curve is above the mean, the probability of a z value greater than 0.5 is .5000 − .1915 or .3085. This is also the probability of an x value greater than 54.

Other x values and their corresponding z's are labeled along the axis of Figure 8.13 as reference points. For example, if $x = 42$ is entered into the z formula, the result is $z = -1$. This x value is one standard deviation below the mean. This dual labeling of abscissa helps in obtaining any probabilities that are needed.

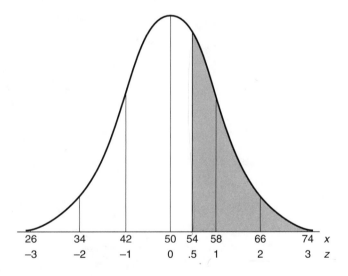

Figure 8.13   *Graph of normal distribution with $\mu = 50$, $\sigma = 8$.*

Suppose that, in the same distribution of x we require the proportion of values between 42 and 62, or $\Pr(42 \leq x \leq 62)$. It can be seen from Figure 8.13 that this interval spans the mean and must be obtained by adding two components. The z score for the left-hand component is −1 and for the right-hand component is between +1 and +2. If the z values are computed exactly, they are $(42 - 50)/8 = -1$ and

$(62 - 50)/8 = 1.5$. Using the table, the probability of a z value between -1 and 0 is .3413 and between 0 and 1.5 is .4332. Thus, $\Pr(-1 \leq z \leq 1.5) = .3413 + .4332 = .7745$. This is also the required probability of x between 42 and 62.

From Probabilities to z Values to x Values.* In practical applications of the normal distribution, the most common uses involve finding some particular x values that meet certain probability conditions. Suppose that the distribution shown in Figure 8.13 represented the dollar value of sales made by a very large retail chain and that gold stars were to be awarded every time a sale is made that falls in the top 20% of the sales distribution. What dollar value distinguishes gold-star sales from the rest?

This application begins with a probability, 0.20, and requires that we obtain an x value as the result; the particular x value needed is represented as x_0. The probability in this instance is the proportion of the curve *above* x_0. This is shown in Figure 8.14. Note that the required value is above the mean since only 20% of the curve has higher x values.

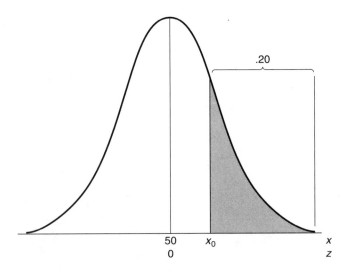

Figure 8.14   *Graph of normal distribution with $\mu = 50$, $\sigma = 8$.*

To use Appendix I, two steps are necessary. First, the problem must be restated in terms of the proportion of the curve between the mean and a given value of the variable; all of the entries in the table are in this form. In this case, we require a value of x that has 0.30 of the distribution between the mean and x_0. Second, the resulting z

value obtained from the table must be converted to the x scale. This is done by substituting z in the standard score formula and solving for x. Algebraically, σ multiplies both sides of the z formula and μ is added to both sides. This yields

$$x = \mu + z\sigma.$$

To complete the example, we are seeking x_0 that satisfies $\Pr(x \geq x_0) = 0.20$. The probability of x between the mean and x_0 is 0.30. The closest probability to 0.30 in the body of the table is .2995; the margins tell us that this corresponds to a z value[4] of 0.84. Using the z-to-x conversion above, $x_0 = 50 + 0.84 \times 8 = 56.72$. Sales above \$56.72 fall in the top 20% of this company's sales distribution.

The relationship of z values with corresponding x values may be used as often as needed, as illustrated in the following example. A professor knows that scores on the exam given in her large undergraduate class are normally distributed with mean $\mu = 79$ and standard deviation $\sigma = 9$. She wishes to grade so that 10% of the students receive A's, 20% B's, 40% C's, 20% D's, and 10% F's, and would like to announce the cut-off point for each grade at the beginning of the semester. (See Figure 8.15.) What should the cutoffs be?

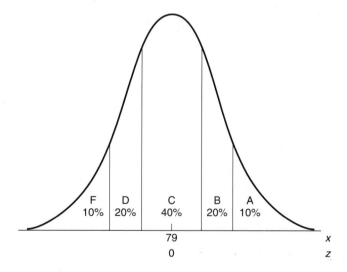

Figure 8.15 *Graph of normal distribution with $\mu = 79$, $\sigma = 9$.*

A distribution of actual test scores will not be *exactly* normal. However, the central limit theorem assures us that if the test is *sufficiently long*, the distribution of scores for a large group of students with simi-

---

[4]That is, $z(.2995) = 0.84$ in the notation of (8.4).

lar characteristics will be approximately normal. The normal curve will then be a good "model" for the resulting distribution.

The first step in obtaining the required cutoffs is to find the standard score values that separate the five grade ranges[5] in Figure 8.15. The grades A and F both require the most extreme 10% of the values in the distribution. That is, the proportion of the curve between the mean and the A cutoff is 0.40. The closest probability given in the body of the normal table is .3997, corresponding to a z value of 1.28. By symmetry, the z value separating the bottom 10% of the curve from the rest is -1.28; this is the cutoff between D and F.

Twenty percent of the normal distribution lies between the mean and the cutoff between B and C; this is the one-half of the C range of scores that is to the right of $z = 0$. The closest probability given in the normal table is .1985, corresponding to a z value of 0.52. By symmetry, the cutoff between C and D corresponds to a z value of -0.52. The only step remaining is to convert each of these z values to the original x scale, using the relationship above. For example, for $z = 1.28$, $x = 79 + 1.28 \times 9 = 90.52$. Students scoring 91 or above receive A's. Applying the same formula to each of the z values and rounding to the nearest whole number of items gives the results in Table 8.4.

TABLE 8.4

A Grading Policy

Grade	Percentage of students	Standard scores	Actual test score
A	10	Above 1.28	91 or above
B	20	0.52 to 1.28	84–90
C	40	−0.52 to 0.52	75–83
D	20	−1.28 to −0.52	68–74
F	10	Below −1.28	67 or below

Summary

When an individual is drawn at random from a population, a measurement on that individual is a *random variable*. The set of probabilities of the possible values of the random variable constitutes a *probability*

---

[5] That is, $z(0.10)$, $z(0.30)$, $z(0.70)$, and $z(0.90)$, respectively, in the notation of (8.4).

distribution. Random variables and probability distributions can also be defined mathematically.

The mean of a random variable or its *expected value* is the probability-weighted average of its values; it is represented by the symbol μ. The variance of a random variable is the expected value of the squared deviations from μ; it is represented by σ^2 and the standard deviation by σ.

One family of probability distributions is made up of the *uniform* distributions that describe, for example, the probabilities in choosing a number at random. One family of theoretical distributions is the family of *normal* distributions; these approximate the empirical distributions of many natural variables. The Central Limit Theorem explains why this is so often the case: any variable approaches normality if each value is made up of a large number of independent components, any of which could take on two or more values.

There is a different normal distribution for each pair of values of μ and σ. The *standard normal* distribution has mean 0 and standard deviation 1; it is the normal distribution of standard scores. The table of the standard normal distribution can be used to find probabilities in particular intervals, and to find values of the random variable that correspond to certain probability statements. About two-thirds (68.26%) of a normal population has values within one standard deviation of the mean. Most of a normal population (95.44%) has values within two standard deviations of the mean.

Exercises

8.1   The number of children under 18 in African-American families in the United States in 1990 is given in Table 8.5.

TABLE 8.5

Distribution of African-American Families by Number of Children Under 18 Years of Age (United States, 1990)

Number of children	0	1	2	3	4 or more
Proportion of families	0.42	0.24	0.19	0.10	0.05

source: *Statistical Abstract of the United States*: 1993, p. 60.

Use the proportions in the table as probabilities and the number of children as random variable x.

(a) Find $\Pr(x = 0)$.

(b) Find $\Pr(x > 0)$.

(c) Find $\Pr(x > 3)$.

(d) Find $\Pr(x < 4)$.

(e) Compare this distribution to the distribution for *all* families in Table 8.1. How would you characterize the differences between the numbers of children in African-American families and all American families in 1990?

8.2 Suppose that a random variable x has only two values, 0 and 1, and $\Pr(x = 0) = 0.5$. Find

(a) $\Pr(x = 1)$;

(b) the expected value $E(x)$;

(c) the variance of variable x;

(d) the standard deviation of variable x.

8.3 (continuation) What would the results of Exercise 8.2 be if $\Pr(x = 0) = 0.3$?

8.4 Suppose that a random variable y has only two values, 2 and 3, and that $\Pr(y = 2) = 0.5$.

(a) Find the expected value and standard deviation of y.

(b) Compare your results with those from Exercise 8.2.

8.5 Suppose that the number of daily newspapers published in cities of over 100,000 population has the probability distribution given in Table 8.6. Find the expected value and standard deviation of the number of newspapers published in these large cities.

TABLE 8.6
Hypothetical Probability Distribution of Number of Daily Newspapers Published in Large Cities

Number of newspapers	1	2	3	4	5	6
Proportion of cities	0.440	0.360	0.132	0.056	0.008	0.004

Hypothetical data.

8.6 (continuation) Assume that a new "publication tax" was imposed, and that every daily newspaper was required to pay its city government a fee of $1,000. Thus cities with one daily newspaper have a publication tax income of $1,000; cities with two daily newspapers have an income of $2,000; and so on. Construct a table like Table 8.6, but giving the list of city incomes instead of the number of newspapers published. Find the expected value and standard deviation of publication tax incomes. Compare these results with those obtained in Exercise 8.5.

8.7 If a random variable x includes the whole numbers from 1 to 20 and these are distributed uniformly, what is

(a) $Pr(x = 20)$?

(b) $Pr(6 \leq x \leq 15)$?

(c) $Pr(9 \leq x \leq 12)$?

(d) $Pr(10 \leq x \leq 11)$?

(e) the expected value $E(x)$?

8.8 Assume that variable z has a standard normal distribution in the population. Use the table in Appendix I to find the probability of a z value in each of the following intervals:

(a) 0 to 1.65;

(b) 1.95 to ∞;

(c) -1.30 to 0;

(d) $-\infty$ to 1.00.

8.9 Use the table in Appendix I to find the probability in each of the following intervals of the standard normal distribution:

(a) 0 to 1.25;

(b) 1.65 to ∞;

(c) -1.50 to 0;

(d) $-\infty$ to 2.00.

8.10 Suppose that variable z has a standard normal distribution. Find

(a) $Pr(z \geq 1.30)$,

(b) $Pr(z \geq -1.30)$,

(c) $Pr(z \leq -.75)$,

(d) $Pr(z \leq 1.58)$.

8.11 Suppose that variable z has a standard normal distribution. Find

(a) $Pr(1.10 \leq z \leq 2.10)$,

(b) $Pr(-3.00 \leq z \leq 3.00)$,

(c) $Pr(-1.25 \leq z \leq 2.12)$.

8.12 Suppose that variable z has a standard normal distribution. Use the table in Appendix I to find the z value (to the nearest $\frac{1}{100}$th) that has

(a) 60% of the distribution below it,

(b) 75% of the distribution below it,

(continued ...)

(c) 80% of the distribution below it,

(d) 35% of the distribution below it.

8.13   Write the $z(p)$ notation to represent each of the z values obtained in Exercise 8.12.

8.14   Suppose that variable z has a standard normal distribution. Use the table in Appendix I to find the particular z value (z_0) that satisfies

(a) $\Pr(z \geq z_0) = .0099$,

(b) $\Pr(z \leq z_0) = .8907$,

(c) $\Pr(z \leq z_0) = .0179$,

(d) $\Pr(z \geq z_0) = .6985$,

(e) $\Pr(-z_0 \leq z \leq z_0) = .5762$,

(f) $\Pr(z_0 \leq z \leq 1.00) = .7056$.

8.15   Write the $z(p)$ notation to denote

(a) each of the z values obtained in (a)–(d) of Exercise 8.14,

(b) the z values obtained in (e) and (f) of Exercise 8.14.

8.16   The College Entrance Examination scores are approximately normally distributed with a mean of 500 and a standard deviation of 100 for a particular subgroup of examinees (those who are high school students taking a college preparatory course). For purposes of this exercise, we shall treat this distribution as if it were continuous. Let y represent the score on the College Entrance Examination. Find

(a) the probability of a score less than 650, that is, $\Pr(y \leq 650)$;

(b) the probability of a score between 550 and 650, that is, $\Pr(550 \leq y \leq 650)$;

(c) the probability of a score between 400 and 550, that is, $\Pr(400 \leq y \leq 550)$;

(d) the probability of a score less than 304 or greater than 696, that is, $\Pr(y \leq 304) + \Pr(y \geq 696)$;

(e) the probability of a score greater than 665, that is, $\Pr(y \geq 665)$;

(f) the score such that the probability of exceeding it is .0668, that is, the value y_0 of y such that $\Pr(y \geq y_0) = .0668$.

8.17   Scores on a certain IQ test are approximately normally distributed with a mean of 100 and a standard deviation of 20 for a certain population of school children. For purposes of this exercise, treat this distribution as continuous. Find the proportion of the population having a score

(a) less than 130,

(b) between 110 and 130,

(c) between 80 and 110,

(continued...)

(d) less than 61 or greater than 139,

(e) greater than 133.

(f) Find the score which is the 90th percentile of this IQ distribution, that is, the score that has 90% of the population below it.

8.18   You are flying from New York to San Francisco and must change planes in Omaha. The New York to Omaha flight is scheduled to arrive at 9:00 PM. You hold a reservation for a flight which leaves promptly at 10:00 PM, and the only other flight out of Omaha that evening leaves promptly at 10:30. While on the flight to Omaha you read that your airline has been cited by the FAA because the arrival times of its New York to Omaha flight have a probability distribution which is normal with mean debarkation time 9:15 and standard deviation of 30 minutes. Assume that you have no baggage and that to change flights takes no time.

(a) What is the probability that you make the 10:00 connection?

(b) What is the probability that you miss the 10:00 connection, but make the 10:30 connection?

(c) What is the probability that you spend the night in Omaha?

8.19   The age at which students graduate from 4-year colleges is normally distributed with $\mu = 22.1$ years and $\sigma = 1.1$ years. Find

(a) The age range of the youngest 10% of college graduates,

(b) The age range of the "middle" 50% of college graduates, that is, 25% below μ and 25% above μ,

(c) The proportion of college graduates between 21.0 and 22.0 years of age.

8.20   The frequencies of heights of 1253 12-year-old boys is given in Table 8.7. Graph the histogram of this distribution.

8.21   (continuation) Compare this histogram with Figure 8.1, which is for 11-year-old boys.

8.22   Using the midpoints of the intervals in Table 8.7 as x values, find the mean and standard deviation of the height distribution. You may use SPSS or another computer program to do this if you like.

    Assuming that the mean and standard deviation are population values, use the table in Appendix I to compute the proportion of a perfect normal distribution that falls into each interval in Table 8.7.

    Compare the fit of the perfect normal "model" (from the previous step) to the actual data in Table 8.7.

8.23   Assuming that grades are based on a normal curve, find the range of standard scores corresponding to each letter grade according to grading policy I in Table 8.8.

TABLE 8.7

Heights of 12-Year-Old Boys

Height	Frequency	Relative frequency
45.5–46.5	1	0.0008
46.5–47.5	3	0.0023
47.5–48.5	7	0.0055
48.5–49.5	13	0.0103
49.5–50.5	31	0.0247
50.5–51.5	73	0.0582
51.5–52.5	111	0.0886
52.5–53.5	176	0.1404
53.5–54.5	189	0.1508
54.5–55.5	198	0.1580
55.5–56.5	162	0.1293
56.5–57.5	106	0.0846
57.5–58.5	77	0.0615
58.5–59.5	49	0.0391
59.5–60.5	31	0.0247
60.5–61.5	10	0.0080
61.5–62.5	9	0.0072
62.5–63.5	4	0.0032
63.5–64.5	1	0.0008
64.5–65.5	1	0.0008
65.5–66.5	1	0.0008
	1253	

source: Bowditch (1877).

TABLE 8.8

Grading Policies

Policy no.	Percentage of students receiving grade				
	A	B	C	D	F
I	15%	20%	30%	20%	15%
II	10	20	50	10	10
III	10	40	40	0	10
IV	30	30	30	0	10

8.24   Assuming that grades are based on a normal curve, find the range of standard scores corresponding to each letter grade according to grading policy II in Table 8.8.

8.25   Assuming that grades are based on a normal curve, find the range of standard scores corresponding to each letter grade according to grading policy III in Table 8.8.

8.26   Assuming that grades are based on a normal curve, find the range of standard scores corresponding to each letter grade according to grading policy IV in Table 8.8.

8.27   If grades are to be awarded based on a 100-point composite of tests and homework, and the composite has $\mu = 74$ and $\sigma = 9$, then what are the actual cutoff scores for

(a) Policy I in Table 8.8?

(b) Policy II in Table 8.8?

(c) Policy III in Table 8.8?

(d) Policy IV in Table 8.8?

9

Sampling Distributions

Introduction

Statistical inference is the process of drawing conclusions about a population of interest from a sample of data. In order to develop and evaluate methods for using sample information to obtain knowledge of the population, it is necessary to know how closely a descriptive quantity such as the mean or the median of a sample resembles the corresponding population quantity. In this chapter the ideas of probability will be used to study the sample-to-sample variability of these descriptive quantities. The ways in which one sample differs from another, and thus how they are both likely to differ from the corresponding population value, is the key theoretical concept underlying statistical inference.

This chapter introduces a number of statistical principles including some that are relatively mathematical in nature. For the reader interested primarily in the application of these principles to actual data, the most important concepts are those of random samples and the idea of a sampling distribution (Section 9.1), the elements that constitute the binomial distribution (Section 9.3), and especially the normal distribution of sample means (Section 9.5).

9.1   Sampling from a Population

Random Samples

In Section 7.8 we saw how a random sample could be drawn from a specified population of a finite number of individuals. A random sample (without replacement) of 13 cards from a deck of 52 cards is obtained by dealing 13 cards from the deck after shuffling; a random sample of 50 beads from a jar of 1000 beads is obtained by mixing the jar of beads and blindly taking out 50 beads; and a random sample of 50 students from a class of 1000 students is obtained by numbering the students and using 50 different random numbers between 1 and 1000. All possible sets of 13 cards are equally likely to be drawn, and all sets of 50 students are equally likely to appear as the random sample.

Several terms are used in describing these samples. The *size* of a sample is the number of individuals in it. In the examples above the size of the bridge hand is 13, and the size of the sample of beads is 50, as is the size of the sample of students.

A *random sample* of a given size is a group of members of a population obtained in such a way that all groups of that size are equally likely to appear in the sample. Here the word "random" means that the sample was produced by a *procedure* that was random, i.e., that gave equal chance of selection to every possible subset of that size.

Is a card a heart or is it not a heart? That question can be asked about each card in a bridge hand. Is an individual bead red or blue? What is the Scholastic Aptitude Test (SAT) score of an individual student? The variable that is assessed for each member of a sample chosen by a random process is a *random variable*, as discussed in Chapter 8.

In contrast, a characteristic of the sample is termed a *statistic*. Statistics that describe the samples above might include the *number* of cards in a bridge hand that are hearts, the *number* of red or blue beads in the sample, or the *mean* of the SAT scores for a randomly chosen group of individuals.

We know that the number of hearts varies from hand to hand. The number of beads that are red will vary in randomly drawn sets of 50 (if there are some red beads in the jar, but not all of them are red). The mean SAT score will vary in samples of students. It is important to study the sample-to-sample variation of a statistic to know how closely its value may be to the corresponding population characteristic.

Sampling Distributions

If we draw random samples of 5 individuals, "basketball teams," from among students at a college, the individuals composing these teams would vary, and the average heights of these teams would vary over a range of heights. It is unlikely that the mean height for any one of these teams will be equal to the mean of the height of all students at the college. Each basketball team selected is a *sample*. The mean height of a team is a *statistic* for that sample.

Different samples usually lead to different values of the statistic: most of the basketball teams will have different average heights. The *sampling distribution* of the statistic gives the probability of each possible value of the statistic.

We illustrate with an example that is simpler than that of choosing 5 persons from a college population. We shall choose samples of size 2 from a population of size 4. This population consists of the four students, Bill, Karen, Tracy, and Randy. Their heights are given in Table 9.1. We shall draw random samples of two (different) students, measure the heights of the two students, and compute the mean height of the sample. In this example we sample without replacement so that the same student does not appear as both members of a "pair."

If we use random digits to draw the sample, we number the students 1: Bill, 2: Karen, 3: Tracy, and 4: Randy. The first digit drawn between 1 and 4 determines the first student: the second digit drawn between 1 and 4 different from the first digit specifies the second student. (See Table 9.2.)

TABLE 9.1

Heights of Four Students

Name	Height
Bill	56″
Karen	60″
Tracy	68″
Randy	72″

Hypothetical data.

The probability of Bill being drawn as the first student is 1/4, and the conditional probability of Karen as the second student given Bill as the first is 1/3. The probability of the pair being Bill, Karen in that order is $1/4 \times 1/3 = 1/12$. Similarly the probability of Karen, Bill in that order is 1/12. The probability of each possible pair of students is 1/6.

TABLE 9.2

Samples of Two Students

Pairs of digits in order	Pairs of students in order	Pairs of students
1,2 2,1	Bill, Karen Karen, Bill	Bill, Karen
1,3 3,1	Bill, Tracy Tracy, Bill	Bill, Tracy
1,4 4,1	Bill, Randy Randy, Bill	Bill, Randy
2,3 3,2	Karen, Tracy Tracy, Karen	Karen, Tracy
2,4 4,2	Karen, Randy Randy, Karen	Karen, Randy
3,4 4,3	Tracy, Randy Randy, Tracy	Tracy, Randy

Viewed from another perspective, there are 12 possible ordered selections of two students listed in Table 9.2. Of these, two selections contain both Bill and Karen so that the probability of the pair being selected is 2/12 or 1/6.

In Table 9.3 we have given the 6 possible samples of students, the corresponding 6 possible pairs of heights, and the means of the pairs of heights. The example makes it clear that usually different samples lead to different values of the statistic. Here the 6 samples gave rise to 5 different values of the mean; the two samples (Bill, Randy) and (Karen, Tracy) both gave values of 64 for the sample mean.

TABLE 9.3

Values of the Sample Mean for Each Sample

Sample	Probability of this sample	Heights (inches)	Mean height
Bill, Karen	1/6	56, 60	58
Bill, Tracy	1/6	56, 68	62
Bill, Randy	1/6	56, 72	64
Karen, Tracy	1/6	60, 68	64
Karen, Randy	1/6	60, 72	66
Tracy, Randy	1/6	68, 72	70

TABLE 9.4

Sampling Distribution of the Sample Mean

Value of sample mean	Probability
58	1/6
62	1/6
64	2/6
66	1/6
70	1/6

In Table 9.4 we tabulate the 5 possible values of the sample mean and their probabilities. The probability of 64 is the sum of the probability of (Bill, Randy) and the probability of (Karen, Tracy). Table 9.4 gives the *sampling distribution* of the mean.

Independence of Random Variables

The concept of *independence* is essential to statistical reasoning about sampling distributions. The idea applies to coin tosses for which the outcome of one toss (head or tail) has no effect on the outcome of subsequent tosses. Two or more random variables are independent if the specific outcomes defined by the variables are independent.

Two events A and B were said (Section 7.7) to be independent if

$$\Pr(B \mid A) = \Pr(B).$$

When we are studying two random variables, the events may be defined by values of the two random variables. As an example, consider the two random variables—number of children in the family and number of magazines read regularly—for a person drawn randomly from the population described by Table 9.5. Here, event A may be defined as the occurrence of a particular value of the random variable x, number of children; and the event B may be defined as the occurrence of a particular value of the random variable y, the number of magazines read regularly. The table displays independence in terms of events. For example, let A be the event $x = 1$ and B the event $y = 2$. We compute

$$\Pr(B \mid A) = \Pr(y = 2 \mid x = 1) = 0.3,$$
$$\Pr(B) = \Pr(y = 2) = 0.3,$$

and thus see that A and B are independent. We verify that, for any integer a from 0 to 4 and integer b from 0 to 3,

$$\Pr(y = b \mid x = a) = \Pr(y = b).$$

We say that x and y are *independent random variables*.

TABLE 9.5

Independent Random Variables

		Number of magazines read regularly				
		0	*1*	*2*	*3*	*Total*
Number of children	*0*	40 (10%)	200 (50%)	120 (30%)	40 (10%)	400 (100%)
	1	20 (10%)	100 (50%)	60 (30%)	20 (10%)	200 (100%)
	2	20 (10%)	100 (50%)	60 (30%)	20 (10%)	200 (100%)
	3	10 (10%)	50 (50%)	30 (30%)	10 (10%)	100 (100%)
	4	10 (10%)	50 (50%)	30 (30%)	10 (10%)	100 (100%)
	Total	100 (10%)	500 (50%)	300 (30%)	100 (10%)	1000 (100%)

Hypothetical data.

Another way of stating independence of events A and B (Section 7.7) is

$$\Pr(A \text{ and } B) = \Pr(A) \times \Pr(B).$$

If the frequencies in Table 9.5 are divided by 1000 to obtain probabilities, then we see that

$$\Pr(x = a \text{ and } y = b) = \Pr(x = a) \times \Pr(x = b)$$

for all possible values of a and b.

The idea described in terms of Table 9.5 applies to any pair of random variables x, y for which probabilities $\Pr(x = a \text{ and } y = b)$ are defined.

DEFINITION

Two discrete random variables x and y are *independent* if
$$\Pr(x = a \text{ and } y = b) = \Pr(x = a) \times \Pr(y = b)$$
for all possible values of a and b.

A similar definition of independence applies to continuous random variables such as the normal if the outcomes are first defined as the random variable falling in intervals.

Sampling from a Probability Distribution

The definition of a random sample and the example of a sampling distribution in Section 9.1 involve sampling from a finite population, that is, a population with a finite number of members or elements. Samples are also drawn from "infinite populations," that is, populations defined by probability distributions and random variables. The number of possible tosses of a coin is unlimited. We describe this "infinite population" by the probability p of a head on a single toss of the coin and the probability $q = 1 - p$ of a tail. We consider the outcomes on two tosses to be *independent*. Thus, the probability of a head on each of two tosses is $p \times p$. A random sample of two consists of the outcomes on two independent tosses. A random sample of n consists of the outcomes on n independent tosses.

These ideas generalize to a random sample of any size drawn from a probability distribution:

DEFINITION   A random sample of size n of a random variable with a given probability distribution is a set of n independent random variables, each with this probability distribution.

Not all underlying random variables are as simple as the toss of a coin. A continuous random variable has more than any finite number of values. One continuous random variable is the normal random variable; the probability of a value in a particular interval is the area under the normal curve in this interval. A random sample of size n from a normal distribution is a set of n independent random variables, each of which is itself normal.

9.2   Sampling Distributions of a Sum and of a Mean

It is theoretically possible to determine the sampling distribution of the mean or any other statistic computed from a random sample from *any* given parent population. Sometimes this is difficult, sometimes easy.

Sampling Distribution of a Sum

We have shown how to obtain the sampling distribution of the mean of a random sample of size 2 from a population of size 4. As another example we consider the distribution of the *sum* of two observations from the uniform distribution over the integers $1, 2, \ldots, 6$. This sample can be drawn by rolling two balanced dice. The outcomes are 36 pairs $(1, 1), (1, 2), \ldots, (6, 6)$, each having probability $1/36$. Note that the outcome on one die does not affect the probability of any particular outcome on the other die.

Let the outcome on the first die be x_1 and on the second x_2. We want to find the sampling distribution of the sum $x_1 + x_2$. The values of the sum are graphed in Figure 9.1 at the ends of the diagonal line segments. The probability of a particular value of the sum is the sum of the probabilities of the points on that line. For instance,

$$\Pr(x_1 + x_2 = 3) = \Pr(x_1 = 1 \text{ and } x_2 = 2) + \Pr(x_1 = 2 \text{ and } x_2 = 1)$$

$$= \frac{1}{36} + \frac{1}{36} = \frac{2}{36}.$$

The probabilities of the sum are given in Table 9.6 and graphed in Figure 9.2. The sampling distribution is unimodal (with a mode of 7 for the sum) and is symmetric about the mode. In fact, the shape of the distribution is somewhat like the normal.

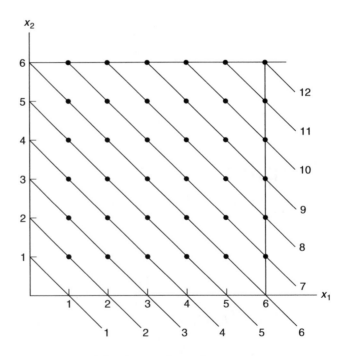

Figure 9.1   *Outcomes of the throw of two dice.*

TABLE 9.6
Probability Distribution for Sum of Outcomes on Two Dice

Sum	2	3	4	5	6	7	8	9	10	11	12
Mean	1	$1\frac{1}{2}$	2	$2\frac{1}{2}$	3	$3\frac{1}{2}$	4	$4\frac{1}{2}$	5	$5\frac{1}{2}$	6
Probability	$\frac{1}{36}$	$\frac{2}{36}$	$\frac{3}{36}$	$\frac{4}{36}$	$\frac{5}{36}$	$\frac{6}{36}$	$\frac{5}{36}$	$\frac{4}{36}$	$\frac{3}{36}$	$\frac{2}{36}$	$\frac{1}{36}$

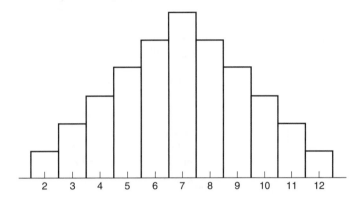

Figure 9.2   *Probabilities of the sum of values on two dice.*

The mean of the sampling distribution of the sum is 7. This is the *expected value*, obtained as described in Section 8.3; that is,

$$\sum p_j v_j = \left(\frac{1}{36} \times 2\right) + \left(\frac{2}{36} \times 3\right) + \cdots + \left(\frac{1}{36} \times 12\right) = 7.$$

Since each value v_j is the sum of values from two variables, x_1 and x_2 we designate this result as $E(x_1 + x_2)$. The mean of the sampling distribution of values on any *one* die would have been exactly one-half of this, $E(x_1) = E(x_2) = 3\frac{1}{2}$.

This illustrates an important property of sampling distributions generally, namely, *the mean of the sampling distribution of a sum of two variables is the sum of the means of the original two variables.* This is written

$$E(x_1 + x_2) = E(x_1) + E(x_2).$$

The same principle is true for any number of variables. For three variables

$$E(x_1 + x_2 + x_3) = E(x_1) + E(x_2) + E(x_3).$$

For *n* variables,

$$E(x_1 + x_2 + \cdots + x_n) = E(x_1) + E(x_2) + \cdots + E(x_n).$$

If $x_1, x_2, \ldots, x_n$ represent a random sample from a single parent population, as they do in the case of tosses of dice, the mean for any one variable (one die) may be represented as μ. Then the mean of the distribution of the sum of n variables (e.g., total spots on n dice) is

$$E(x_1 + x_2 + \cdots + x_n) = \mu + \mu + \cdots + \mu = n\mu.$$

The variance of the sampling distribution is obtained as described in Section 8.3. For the distribution of the sum $x_1 + x_2$, the expected value (computed above) is 7; the variance is

$$\sum p_j(v_j - 7)^2 = \frac{1}{36} \times (2-7)^2 + \frac{2}{36} \times (3-7)^2 + \cdots + \frac{1}{36} \times (12-7)^2 = 35/6.$$

The mean for *one* die having values 1, 2, 3, 4, 5, 6 each with probability $1/6$ is $\mu = 3.5$. The variance for one die is $\frac{1}{6} \times [(1 - 3.5)^2 + (2 - 3.5)^2 + \cdots + (6 - 3.5)^2] = 35/12$. The variance of the sum of 2 dice is exactly twice this value.

This illustrates a second general property of sampling distributions, namely, *the variance of a sum of two independent variables is the sum of their separate variances.* This is written

$$\sigma^2_{x_1 + x_2} = \sigma^2_{x_1} + \sigma^2_{x_2}.$$

The standard deviation is

$$\sigma_{x_1 + x_2} = \sqrt{\sigma^2_{x_1} + \sigma^2_{x_2}}.$$

The analogous rule holds for three, four, or any number of *independent* random variables:

$$\sigma^2_{x_1 + x_2 + \cdots + x_n} = \sigma^2_{x_1} + \sigma^2_{x_2} + \cdots + \sigma^2_{x_n}.$$

The standard deviation is

$$\sigma^2_{x_1 + x_2 + \cdots + x_n} = \sqrt{\sigma^2_{x_1} + \sigma^2_{x_2} + \cdots + \sigma^2_{x_n}}.$$

In particular, for a random sample $x_1, x_2, \ldots, x_n$ from a common parent population with variance σ^2, we have

$$\sigma^2_{x_1 + x_2 + \cdots + x_n} = n\sigma^2,$$

and

$$\sigma_{x_1 + x_2 + \cdots + x_n} = \sqrt{n}\sigma.$$

The example of the dice illustrates the general idea of finding the probability distribution of the sum of two random variables each of which takes on a finite number of values. We start with the probabilities of all possible pairs of values; the random variables need not be independent. To find the probability that the sum of the two variables is a given value we add the probabilities of all pairs that give rise to that value of the sum; geometrically, we add the probabilities of the points on a diagonal line as in Figure 9.1. If the random variables are continuous, the procedure is similar, but involves calculus.

To find the probability distribution of the sum of three independent random variables, we find the probability distribution of the sum of the first two of them and then find the probability distribution of the sum of that variable and the third. This procedure can be continued to find the sampling distribution of the sum of any number of independent random variables.

Sampling Distribution of the Sample Mean

The sampling distribution of the sample *mean* is closely related to the sampling distribution of the sum. In addition to the sums, Table 9.6 gives the values of the *mean* of two variables x_1 and x_2 (the outcomes on the two dice), namely, $(x_1 + x_2)/2$. The probabilities apply to the values of the sum and to the corresponding values of the mean as well.

The mean of the sampling distribution of means, $E[(x_1 + x_2)/2]$, is

$$\left(\frac{1}{36} \times 1 \right) + \left(\frac{2}{36} \times 1\frac{1}{2} \right) + \cdots + \left(\frac{1}{36} \times 6 \right) = 3\frac{1}{2}.$$

It is obvious that each value of the mean is the corresponding sum divided by 2; thus, the mean of the sampling distribution of means is equal to the mean of the sampling distribution of sums (7) divided by 2. The result, $3\frac{1}{2}$, is also the mean of the distribution of *each component variable* x_1 and x_2.

The variance of the means is also related to the variance of the sums and to the variance of x_1 and x_2. The deviations of values of the mean in Table 9.6 from their mean ($3\frac{1}{2}$) are only half as large as the deviations of the sums from their mean (7). For example, the two-dice *total* of 3 is exactly 4 dots below the average of 7; a two-dice *mean* of $1\frac{1}{2}$ is 2 dots below the average of $3\frac{1}{2}$. The *squared* deviations for the distribution of means are $(1/2)^2 = 1/4$ times the squared deviations for the distribution of sums. Since the variance is the average of squared deviations, the variance of the sampling distribution of the mean is $1/4$ times the variance of the sum. This can be demonstrated with actual figures. The variance of the means in Table 9.6 is

$$\frac{1}{36} \times \left(1 - 3\frac{1}{2} \right)^2 + \frac{2}{36} \times \left(1\frac{1}{2} - 3\frac{1}{2} \right)^2 + \cdots + \frac{1}{36} \times \left(6 - 3\frac{1}{2} \right)^2 = \frac{35}{24}.$$

This result is exactly $1/4$ times the variance of the sum (35/6).

The same reasoning discussed above for $n = 2$ applies to any sample of size n. The sampling distribution of the mean of *any* number of random variables from a single parent population has its center at μ, that is,

$$E(\bar{X}) = \mu.$$

The variance of the sampling distribution of the mean of n independent variables from a single parent population is $1/n^2$ times the variance of the sampling distribution of their sum. Since the variance of the sum of n independent random variables is $n\sigma^2$, the variance of the mean is $n\sigma^2/n^2 = \sigma^2/n$. This value is usually represented $\sigma_{\bar{x}}^2$, where the subscript $\bar{x}$ indicates that the variance is that of the distribution of sample means.

The discussion of the sampling distribution of a sum and a mean has been in terms of discrete random variables, but the same ideas also apply to continuous random variables such as normal random variables discussed in Chapter 8. The sampling distribution of the sum of two continuous random variables is calculated in a manner similar to that described above, but involves calculus and so is not given here. The expressions for the expectation and variance of sums of continuous random variables are the same as those given above for discrete random variables.

9.3   The Binomial Distribution

Sampling Distribution of the Number of Heads

In many situations we are interested in only two possible outcomes: a person is either male or female; a stamp collector or not; a person will either die or not die during the next year; a roll of two dice will either produce a seven or not.

We can consider each of these situations with only two possible outcomes as formally equivalent to the simple operation of tossing a coin with an arbitrary probability of heads. We associate one of the two possibilities with heads, denoted H; the other, with tails, denoted T. In another context the outcomes are termed "success" and "failure," respectively.

The probability of a head (or success) is labelled p, and the probability of a tail (or failure) is q. Because p and q are probabilities, we know p and q are each between 0 and 1. Since the two events H and T are mutually exclusive and exhaustive, $p + q = 1$. If the coin is fair or unbiased, $p = q = 1/2$. We sometimes call the occurrence of a head or tail with respective probabilities p and q a *Bernoulli trial.*[1]

We may toss the coin twice; the possible outcomes are two heads, a head on the first toss and a tail on the second, a tail on the first toss and a head on the second, and two tails. If we observe two 80-year-olds for a year, both may die, the first may die and the second not, the first may not die and the second may, and both may live. In each case we have two Bernoulli trials. We may consider the probability of a head (or success) on each trial to be p and we may consider the trials to be independent; this means

$$\text{Pr}(H \text{ on the first toss and } H \text{ on second toss})$$

$$= \text{Pr}(H \text{ on first toss}) \times \text{Pr}(H \text{ on second toss}),$$

which is $p \times p = p^2$.

We often want to consider an arbitrary number of Bernoulli trails, say n of them. We may toss the coin n times. If we observe 7000 80-year-olds, for a year, then $n = 7000$, and p is the probability that any one of them dies. In a group of 100 persons, each is either a stamp collector or not; here $n = 100$, and p is the probability that an individual is a stamp collector.

The formal definition of a set of n Bernoulli trials is as follows:

(i) There are n trials.
(ii) The same pair of outcomes are possible on all trials.
(iii) The probability of a specified outcome is the same on all trials.
(iv) The outcomes on the n trials are independent.

Many problems of statistics fit this model. Usually it is the total number of heads that is important; the number of heads in n trials is a *statistic.* We shall now derive the sampling distribution of this statistic.

Table 9.7 shows the possible outcomes when a coin is tossed once, twice, three times, or four times. We let n be the number of tosses, with $n = 1, 2, 3, 4$. The first row of the tables represents the two possible outcomes of a single toss; either we get a tail (T) or a head (H). Now we make a second toss. A tail on the first toss can be followed by a tail or by a head on the second toss, resulting in one of the two sequences

| First toss: | T | T |
| Second toss: | T | H |

These two sequences are listed in the left half of the two rows for $n = 2$, that is, under T in the row for $n = 1$, to emphasize that these two sequences can result from T on the first toss. A head on the first

---

[1] After James Bernoulli (1654–1705), who discussed such trials in his *Ars Conjectandi* (1713).

TABLE 9.7

Table of Possible Outcomes When Coin is Tossed n Times[a]

$n = 1$			T		H			

	T		TH		H	
$n = 2$	T		HT		H	

	T	TTH	THH	H
$n = 3$	T	THT	HTH	H
	T	HTT	HHT	H

	T	$TTTH$	$TTHTHH$	$THHH$	H
	T	$TTHT$	$THTHTH$	$HTHH$	H
$n = 4$	T	$THTT$	$HTTHHT$	$HHTH$	H
	T	$HTTT$	$HHHTTT$	$HHHT$	H

[a]Note: The sequence of tosses are indicated vertically. For example, for $n = 2$, the first toss is indicated by the first row " T TH H ," the second toss by the second row " T HT H ." Thus two tosses result in the sequence $\binom{T}{T}$, $\binom{T}{H}$, $\binom{H}{T}$, or $\binom{H}{H}$.

toss can also be followed by a tail or a head, resulting in one of the two sequences

First toss:	H	H
Second toss:	T	H

These two sequences are listed in the two rows for $n = 2$ at the right (under H in the row for $n = 1$). The two (vertical) sequences

$$TH$$
$$HT$$

are grouped together because each contains exactly one H.

For $n = 3$, there are 8 possible outcomes; each is obtained by appending an H or a T to one of the four outcomes for $n = 2$. These 8 possibilities are grouped to show that there is 1 possibility giving 0 heads, 3 possibilities giving exactly 1 head, 3 possibilities giving exactly 2 heads, and 1 possibility giving 3 heads. The 3 ways of getting exactly 2 heads,

First toss:	THH
Second toss:	HTH
Third toss:	HHT

arose from two sets of possible outcomes for $n = 2$, those which contain exactly 1 head and the one containing exactly 2 heads:

TH	H
HT	H

Thus the number of possibilities for $n = 3$ containing exactly 2 heads (3) is the number of possibilities for $n = 2$ containing exactly 1 head (2), plus the number of possibilities for $n = 2$ containing exactly 2 heads (1).

Pascal's Triangle. These numerical results are collected into the diagram given in Table 9.8, called *Pascal's triangle*. [Blaise Pascal (1623–1662) was a famous French philosopher and mathematician.]

TABLE 9.8
Pascal's Triangle of Binomial Coefficients to $n = 8$

				1	1			
			1	2	1			
		1	3	3	1			
	1	4	6	4	1			
1	5	10	10	5	1			
1	6	15	20	15	6	1		
1	7	21	35	35	21	7	1	
1	8	28	56	70	56	28	8	1

The entries in the first row are the numbers of different ways to obtain no head or one head, respectively, in one coin toss. The values in the second row are the numbers of different ways to obtain no head, one head, and two heads, respectively, in two coin tosses; and so on. There is a pattern to the entries that parallels the pattern in Table 9.7. For example, note that the 10 in the 5th row is $4 + 6$, the sum of the numbers above it. The 21 in the 7th row is $6 + 15$, the sum of the numbers above it.

The number of ways of obtaining exactly r heads in n tosses of a coin is denoted by the symbol C_r^n; it is the number of *combinations* of r objects out of n. An algebraic expression can be given for C_r^n involving the *factorial function*, defined as $n! = n \times (n - 1) \times \cdots \times 2 \times 1$, and $0! = 1$. Then

$$C_r^n = \frac{n!}{r!(n - r)!}$$

This expression can be used to calculate C_r^n for values of n that are not extremely large. For example, the number of ways of obtaining 3 heads in 4 tosses is

$$C_3^4 = \frac{4!}{3!(4 - 3)!} = \frac{4 \times 3 \times 2 \times 1}{(3 \times 2 \times 1) \times 1} = 4.$$

This is also the entry in the 4th row of Pascal's triangle corresponding to 3 heads (the 4th element).

Probabilities

As an example, consider the probability of obtaining 2 heads on 3 tosses of a coin. The number of combinations that yield 2 heads is

$$C_2^3 = \frac{3!}{2!(3-2)!} = \frac{3 \times 2 \times 1}{(2 \times 1) \times 1} = 3.$$

These are the sequences *HHT*, *HTH*, and *THH*, each containing two heads. (We now write the sequences horizontally for the sake of compactness.) Let us assume that the probability of a head is p, the probability of a tail is $q = 1 - p$, and the trials are independent.

Each of the three sequences has probability p^2q. To see this, first consider the sequence *HHT*. Using the independence of trials,

$$\Pr(HHT) = \Pr(H) \times \Pr(H) \times \Pr(T) = p \times p \times q = p^2q.$$

Similarly the probability of *any* sequence containing 2 heads and 1 tail in any order will be the product of two p's and one q, which is again p^2q.

What then is the probability of obtaining 2 heads in 3 tosses of a coin with heads probability p? There are 3 sequences containing exactly 2 H's, and each has probability p^2q, so the probability is $3p^2q$ (because this is the probability of an event which is composed of the 3 mutually exclusive events *HHT*, *HTH*, and *THH*).

This logic generalizes to the probability of any number of heads r out of any number of tosses n. Every sequence of coins with r heads will have the remaining $n - r$ coins showing tails. The probability of each such sequence is $p^r q^{n-r}$, as long as tosses are independent. For example, the probability of 2 heads out of 4 tosses is p^2q^2. The number of sequences of 4 coins of which 2 are heads is found in Pascal's triangle or obtains as $C_2^4 = 4!/(2! \times 2!) = 6$. Thus, the probability of 2 heads in 4 tosses is $6p^2q^2$.

In general, the probability of obtaining r heads and $n - r$ tails in a sequence of n tosses is

$$\Pr(\text{exactly } r \text{ heads in } n \text{ tosses}) = C_r^n p^r q^{n-r},$$

for $r = 0, 1, 2, \ldots, n$. This is called the *binomial probability distribution*. From a practical perspective, if n is large, it is difficult to compute C_r^n by the triangle or by this formula. Appendix II gives numerical values of $\Pr(x = r)$, $r = 0, 1, 2, \ldots, n$, for various values of n and p. The tabled p values only go as high as $1/2$; for larger values of p, use the rule $\Pr(r \text{ heads in } n \text{ tosses with probability } p) = \Pr(n - r \text{ heads in } n$ tosses of a coin with probability q).

The Case $p = 1/2$.   If the coin is fair ($p = 1/2$), all of the sequences of n H's and T's have the same probability. For example, when $n = 3$

there are 8 sequences and each has probability 1/8. The probabilities are then

$$\Pr(0 \text{ heads}) = \Pr(TTT) = 1/8,$$

$$\Pr(1 \text{ head}) = \Pr(TTH \text{ or } THT \text{ or } HTT)$$

$$= \Pr(TTH) + \Pr(THT) + \Pr(HTT) = 3/8,$$

$$\Pr(2 \text{ heads}) = \Pr(THH \text{ or } HTH \text{ or } HHT)$$

$$= \Pr(THH) + \Pr(HTH) + \Pr(HHT) = 3/8,$$

$$\Pr(3 \text{ heads}) = \Pr(HHH) = 1/8.$$

Table 9.9 gives the probabilities of different numbers of heads for $n = 1, 2, 3, 4, 5$ and $p = 1/2$. In Figure 9.3 the probabilities are graphed. Although the number of heads must be an integer, we have drawn the probability as an area centered at the number of heads.

TABLE 9.9
Binomial Distribution for $n = 1, 2, 3, 4, 5$ and $p = 1/2$

		\multicolumn{6}{Number of heads}					
		0	*1*	*2*	*3*	*4*	*5*
	1	$\frac{1}{2}$	$\frac{1}{2}$				
	2	$\frac{1}{4}$	$\frac{1}{2}$	$\frac{1}{4}$			
n	*3*	$\frac{1}{8}$	$\frac{3}{8}$	$\frac{3}{8}$	$\frac{1}{8}$		
	4	$\frac{1}{16}$	$\frac{4}{16}$	$\frac{6}{16}$	$\frac{4}{16}$	$\frac{1}{16}$	
	5	$\frac{1}{32}$	$\frac{5}{32}$	$\frac{10}{32}$	$\frac{10}{32}$	$\frac{5}{32}$	$\frac{1}{32}$

It will be observed that each histogram is centered at $\frac{1}{2}n$ and is symmetric about this point. The larger n is the more the probabilities are spread out.

The probability that a baby born is a boy is about 1/2. Table 9.9 can be interpreted as the probabilities of various numbers of boys in families of 1 to 5 children. If we observed many families of size 3, for example, we would expect no boy in about 1/8 of them and no girl in about 1/8 of them, and we would expect one boy in about 3/8 of them and one girl in about 3/8 of them.

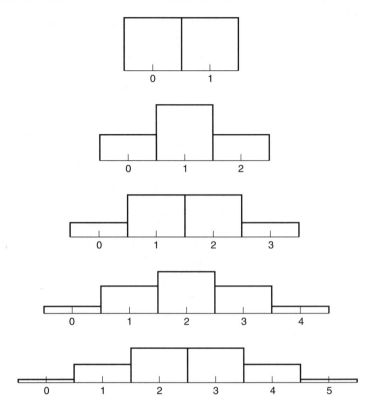

Figure 9.3   *Binomial distributions for n = 1, 2, 3, 4, 5, and p = 1/2.*

In many applications the *n* "coin tosses" correspond to observations of *n* members of a sample. In order that condition (iv) of the definition of Bernoulli trials hold, that is, in order that the *n* observations be strictly independent, the sample must be chosen with replacement; but if the population size is large, the observations are nearly independent even if the sampling is without replacement, and the binomial distribution provides a good approximation to the actual distribution, the hypergeometric distribution. [See, for example, Hoel, Port, and Stone (1972).] Suppose we have a jar of $n = 10,000$ beads, of which 5,000 are red, and take a sample of $n = 2$ beads. The probability of a red bead on the first draw is $5,000/10,000 = 1/2$ and the conditional probability of a red bead on the second draw given that the first bead was red is $4,999/9,999 = 0.49994999$, which is almost $1/2$. The probability of 2 red beads in drawing 2 beads (without replacement) is

$$\frac{1}{2} \times \frac{4,999}{9,999} = 0.24997,$$

which is almost $1/4$.

Proportion of Heads in Bernoulli Trials

Often we are interested in the *proportion* of heads in the tosses. In n tosses the proportion of heads can be $0/n = 0$, $1/n$, $2/n$, ..., or $n/n = 1$. The probability of a *proportion* being r/n is the same as the the probability of the *number* of heads being r. Thus Table 9.9 applies to proportions as well as numbers of heads.

When it comes to graphing these probabilities, however, there is a difference because the proportions must lie between 0 and 1. In Table 9.10 we have given the probabilities of proportions of heads in 1, 2, 3, and 4 tosses of an unbiased coin; these are graphed in Figure 9.4. Each probability is represented as an *area*. For instance, the probability of the proportion $1/2$ when $n = 4$ is $6/16$, or $3/8$. The bar is of width $1/4$ (from $3/8$ to $5/8$) and height $3/2$ to give area $1/4 \times 3/2 = 3/8$. Figures 9.4 and 9.5 show that the probability of a proportion being $1/2$ or near $1/2$ is greater for larger n.

TABLE 9.10

Probability Distributions of Proportions for $n = 1, 2, 3, 4, 5$ and $p = 1/2$

$n = 1$	Proportions	0	1			
	Probabilities	$\frac{1}{2}$	$\frac{1}{2}$			
$n = 2$	Proportions	0	$\frac{1}{2}$	1		
	Probabilities	$\frac{1}{4}$	$\frac{1}{2}$	$\frac{1}{4}$		
$n = 3$	Proportions	0	$\frac{1}{3}$	$\frac{2}{3}$	1	
	Probabilities	$\frac{1}{8}$	$\frac{3}{8}$	$\frac{3}{8}$	$\frac{1}{8}$	
$n = 4$	Proportions	0	$\frac{1}{4}$	$\frac{1}{2}$	$\frac{3}{4}$	1
	Probabilities	$\frac{1}{16}$	$\frac{4}{16}$	$\frac{6}{16}$	$\frac{4}{16}$	$\frac{1}{16}$
$n = 5$	Proportions	0	$\frac{1}{5}$	$\frac{2}{5}$	$\frac{3}{5}$	$\frac{4}{5}$	1
	Probabilities	$\frac{1}{32}$	$\frac{5}{32}$	$\frac{10}{32}$	$\frac{10}{32}$	$\frac{5}{32}$	$\frac{1}{32}$

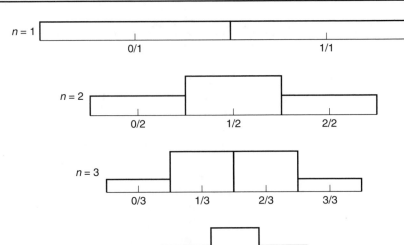

Figure 9.4   *Distribution of sample proportions for $n = 1, 2, 3, 4$, and $p = 1/2$.*

It is sometimes convenient to denote a head by a 1 and a tail by a 0. The sequence *HHT* is then written 110. In *n* trials we let x_i denote the number of heads on the *i*th trial, that is, $x_i = 1$ if there is a head on the *i*th trial and $x_i = 0$ if there is a tail on the *i*th trial. Then the sum of the x_i's,

$$\sum x_i,$$

is the number of of heads in the *n* trials. For example, in the sequence *HHT*, $x_1 = 1$, $x_2 = 1$, $x_3 = 0$, and

$$\sum x_i = x_1 + x_2 + x_3 = 1 + 1 + 0 = 2$$

is the number of heads in this sequence.

The proportion of heads in *n* trials is the number of heads divided by *n*, that is,

$$\frac{\sum x_i}{n},$$

which is the sample mean. (See Section 3.3.) Thus, any statements made about sample means apply to the proportion of heads in a number of trials. Conversely, the sampling distribution of the number of heads and proportion of heads are examples of the sampling distributions of sums and means.

The Mean and Variance of the Binomial Probability Distribution

The probability distribution representing a single coin toss is given in Table 9.11. The mean of this probability distribution is

$$\mu = p_1 v_1 + p_2 v_2 = p \times 1 + q \times 0 = p.$$

The variance of this probability distribution is

$$\sigma^2 = p_1(v_1 - \mu)^2 + p_2(v_2 - \mu)^2 = p(1 - p)^2 + q(0 - p)^2$$
$$= pq^2 + q(-p)^2 = pq^2 + p^2 q$$
$$= pq(q + p) = pq \times 1$$
$$= pq$$

The binomial probability distribution is the distribution of a sum of independent random variables $x_1, x_2, \ldots, x_n$, each having mean p and variance pq. Hence, using the rules for the mean of a sum of variables and for the variance of a sum of independent variables (Secton 9.2),

$$E(x_1 + x_2 + \cdots + x_n) = E(x_1) + E(x_2) + \cdots + E(x_n) = p + p + \cdots + p = np$$

and

$$\sigma^2_{x_1 + x_2 + \cdots + x_n} = \sigma^2_{x_1} + \sigma^2_{x_2} + \cdots + \sigma^2_{x_n} = pq + pq + \cdots + pq = npq.$$

Proportion.   The proportion of heads in one set of n independent Bernoulli trials shall be denoted by $\hat{p}$. It is the sample mean of $x_1, x_2, \ldots, x_n$, where x_i has the probability distribution of Table 9.11. Thus, for the sampling distribution of $\hat{p}$ over many sets,

$$E(\hat{p}) = p, \qquad \sigma^2_{\hat{p}} = \frac{pq}{n}.$$

These rules for the means and variances of sums and proportions can be verified numerically for Tables 9.9 and 9.10.

TABLE 9.11
Probability Distribution of a Single Coin Toss

Values	$v_1 = 1 \ (= \text{ head})$	$v_2 = 0 \ (= \text{ tail})$
Probabilities	$p_1 = p$	$p_2 = q$

9.4   The Law of Averages (Law of Large Numbers)*

The idea of the Law of Averages is that in the long run an event will happen about as frequently as it is supposed to, that is, according to its probability. This idea may seem like common sense to most of us. However, it also provides a justification for using random samples to learn about populations, that is, for statistical inference: if the sample is large enough, the sample information is likely to be very accurate.

To illustrate the Law of Averages in a familiar context, let us suppose that a restaurant has a list of 7 special entree's and chooses the entree to serve each day at random from among the 7. Thus the probability of a particular entree being served is 1/7 or about 0.14. The Law of Averages leads us to expect that, over years of operation, the restaurant would serve our favorite entree (say, veal piccata) on about 14% of the days.

We would not expect the same percentage over shorter time periods. For example, if the selection process is random, we would not be surprised if veal piccata was served twice in any particular week (that is, 2 out of 7 days) or, on rare occasion, even three times (3 out of 7). On the other hand, if veal piccata were to be served 4 times in a single week, this is so different from the probability 1/7 that we would suspect that entree's are not being selected at random. (Perhaps the restaurant is serving leftovers.)

The restaurant example illustrates the two main aspects of the Law of Averages. These are now discussed more formally using coin tosses as an example for which each toss results in a head or a tail being observed.

If one tosses an unbiased coin many times, one expects heads in about half the tosses. More precisely, if an event has probability p of occurring then, when a large number of independent trials are made, the probability is great that the proportion of trials on which the event occurs is approximately p. The larger the number of trials, the more likely it is that the observed proportion will be close to p, that is, within any prespecified small distance from p.

As an example, Table 9.12 gives the probabilities of actual coin tosses coming within a certain distance of 0.5, the probability of an unbiased coin. The first column ($n = 1$) is based on a single toss. If the coin shows a head, the proportion of tosses with a head is 1.00; if the coin shows a tail, the proportion of heads is 0. In either case, the observed proportion is within 0.5 of $p = 0.5$, that is, the probability of $\hat{p}$ being within $1/2$ of p is 1.00. In *neither* case is $\hat{p}$ within $1/4$ or $1/8$ of p; those probabilities are zero.

In two tosses, the proportion of heads can be 0, 0.5, or 1, with the outcome 0.5 being twice as likely as the other two values. All three outcomes are within $1/2$ of $p = 0.5$; the probability in the table is 1. Only the outcome 0.5 is within $1/4$ of p, and the proportion of 0.5's as an outcome is $1/2$. Similarly, 0.5 is within $1/8$ of p and the proportion of 0.5's as an outcome is $1/2$.

In 5 tosses of a coin the proportions of $2/5$ and $3/5$ are within $1/8$ of $1/2$ and each of these events has probability $10/32$; the probability that one of these two events occurs is $5/8$, as given in the last row and 5th column of Table 9.12. Also shown in Table 9.12 are the probabilities for 100 tosses, which is a large number of tosses; these are similarly computed from the binomial distribution. The probability that the proportion of heads in 100 tosses will fall between 0.375 and 0.625 is 0.98796; it is almost certain that the proportion will fall between 0.25 and 0.75. The calculations depend only on the probability p being $1/2$ and independence of events on the n tosses. This concentration of probability near $1/2$ can also be seen from Figure 9.4. If n is larger, the concentration around $1/2$ is greater.

TABLE 9.12

Probability That the Proportion of Heads Deviates from 1/2 by Various Amounts for Various Numbers of Tosses

	n					
	1	*2*	*3*	*4*	*5*	*100*
Deviation of at most $\frac{1}{2}$	1	1	1	1	1	1
Deviation of at most $\frac{1}{4}$	0	$\frac{1}{2}$	$\frac{3}{4}$	$\frac{7}{8}$	$\frac{5}{8}$	1.00000
Deviation of at most $\frac{1}{8}$	0	$\frac{1}{2}$	0	$\frac{3}{8}$	$\frac{5}{8}$	0.98796

In Bernoulli trials with a probability p of a head, the probability is high that the sample proportion of heads is close to the probability p if the number of tosses n is sufficiently large. The probability p is the mean of the Bernoulli distribution when a head is given the value 1 and a tail is given the value 0 and the proportion of heads is the sample mean.

The case of independent Bernoulli trials is an example of a more general form of the law of large numbers which states that the probability is high that the sample mean is close to the population mean if the sample size is large. This fact is easily made plausible when the random variable can take on a finite number of values, $v_1, \ldots, v_m$. Then the sample mean $\bar{x}$ can be written as

$$\bar{x} = \sum r_j v_j,$$

where r_j is the relative frequency of the value v_j in the sample. If the sample size is large, it is likely that each relative frequency r_j will be close to the corresponding population probability p_j. Thus $\sum r_j v_j$ will be close to $\sum p_j v_j$, that is, $\bar{x}$ will be close to μ. Thus the probability is large (near 1) that the sample mean $\bar{x}$ differs by only a small amount from the population mean

$$\mu = \sum p_j v_j.$$

The Law of Averages is also known as the Law of Large Numbers.

LAW OF LARGE NUMBERS
> If the sample size is large, the probability is high that the sample mean is close to the mean of the parent population.

It is stated in Section 9.2 that the mean of the sampling distribution of the sample mean $\bar{x}$ from a parent population with mean μ is that population mean μ; that is, the sampling distribution of the sample mean is located around the parent population mean. If the standard deviation of the parent population is σ, then the standard deviation of the sampling distribution of the mean of samples of size n is $\sigma/\sqrt{n}$.

Because n is in the denominator of this expression, it follows that the larger the sample size, the narrower or more concentrated the sampling distribution will be. More $\bar{x}$'s will be close to μ with larger samples; that is, the probability of $\bar{x}$ within any given distance of μ increases as n increases. This principle underlies the Central Limit Theorem as well, discussed in the next section.

Mathematical Note. We can make a rigorous definition of the Law of Large Numbers. Let "sample mean close to the population mean" be defined as

$$|\bar{x} - \mu| < a$$

for an arbitrarily small positive number a (for example, 0.1). Let "the probability is high" be defined as

$$\Pr(|\bar{x} - \mu| < a) > 1 - b,$$

for an arbitrarily small positive number b (for example, 0.05). The Law of Large Numbers states that for given a and b the above probability statement is true if n is at least as large as some suitable number (to be calculated on the basis of a and b).

9.5 The Normal Distribution of Sample Means

The Central Limit Theorem

The Central Limit Theorem provides a fundamental principle on which most of the inferential procedures discussed in this book are based. This theorem states that the sampling distribution of a mean is approximately a normal distribution. Because the properties of the normal distribution are well known, the Central Limit Theorem enables us to make probability statements about particular values of $\bar{x}$ *without actually observing the entire population* from which it was drawn.

In contrast to the Central Limit Theorem, the Law of Large Numbers is a rather crude statement; it states that the sampling distribution of a sample mean is "concentrated" around the mean of the parent population. The Central Limit Theorem is the more refined assertion that the sampling distribution has a particular form, that of a bell-shaped or normal curve.

In Chapter 8 we note that the distributions of many measurements such as height, weight, and test scores are approximately normal in shape. This is attributed to the fact that these variables are constituted from many separate components, each of which is itself a random variable. The same logic applies to the sampling distribution of the mean; a sample mean is comprised of many separate components—the individual observations—each of which has one of many possible values. It is a mathematical theorem as well as an empirical fact that the sampling distribution of the mean is approximately normal and becomes more accurately normal as the sample size increases.

This can be seen by examining the sampling distribution of the mean where the underlying variable is as simple as a 0–1 dichotomy. It is shown in Section 9.3 that the proportion of heads in n tosses of a fair coin is the *mean* of a variable scored "1" for a head and "0" for a tail. Figure 9.5 shows the sampling distribution of the proportion of heads ($\hat{p}$) for 2 tosses, 5 tosses, and 10 tosses,[2] respectively, when $p = 1/2$. Note that the distribution becomes increasingly normal in shape as n increases. For a larger n the resemblance would be still closer.

Figures 9.6 and 9.7 (truncated on the right) show the sampling distributions of $\bar{x}$ from two other parent populations, when n is 2, 6, and 25. These examples demonstrate that the sampling distribution of the mean approaches normality as n increases almost regardless of the nature of the parent population. The parent population may be symmetric

---

[2]Results for $n = 2$ and $n = 5$ are taken from Table 9.10.

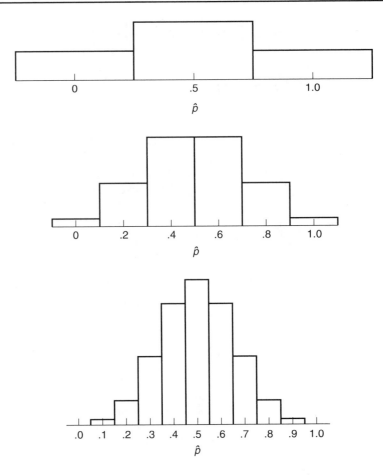

Figure 9.5   *Distributions of sample proportions for $p = 1/2$ and $n = 2, 5, 10$.*

(Figure 9.6) or skewed (Figure 9.7) and the underlying variable may be discrete or continuous. The sample size (n) that is required to have a "good" approximation to the normal, however, does depend on the shape of the parent distribution.

The Central Limit Theorem states this formally:

CENTRAL LIMIT THEOREM

If the sample size is large, the distribution of the sample mean of n independent observations is well approximated by a normal distribution.[3]

---

[3]A complete mathematical statement of the Central Limit Theorem may be found, for example, in Hoel, Port, and Stone (1972).

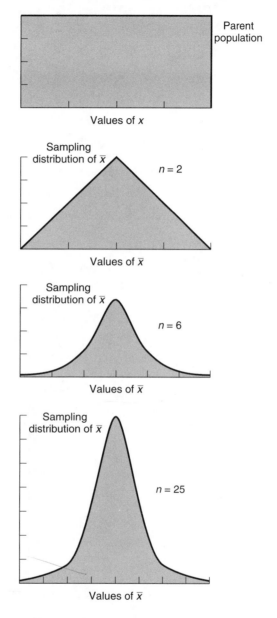

Figure 9.6   *Distributions of sample means from samples of various sizes taken from a uniform parent population.*

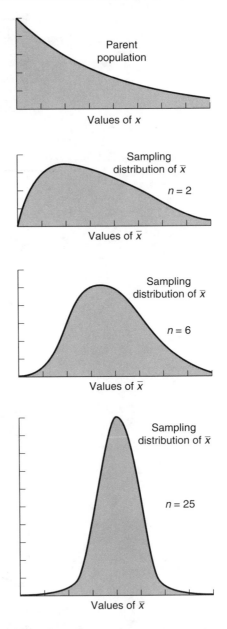

Figure 9.7   *Distributions of sample means from samples of various sizes taken from a skewed parent population.*

Note that a random sample of n observations (that is, $x_1, x_2, \ldots, x_n$) from a parent probability distribution is a set of n independent random variables. (See Section 9.1.) The sample mean is a linear combination of these random variables, namely,

$$\bar{x} = \frac{x_1 + x_2 + \cdots + x_n}{n} = \frac{1}{n}x_1 + \frac{1}{n}x_2 + \cdots + \frac{1}{n}x_n.$$

Thus, the Central Limit Theorem assures us mathematically that the sampling distribution of this linear combination of x_i's is almost normal as long as n is sufficiently large and the observations are independent. We shall see in later chapters that other statistics are also normally distributed as long as these two conditions hold.

An additional feature of the sampling distribution applies to variables with normally distributed parent populations. The sum (or mean) or difference of two or more independent random variables that have normal distributions themselves has a sampling distribution that is normal. Thus the sampling distribution of the mean is normal if the parent population is normal *regardless* of n. Since many variables have normal parent populations and since many samples are based on large n's, the normal distribution is frequently a good model of the sampling distribution of x to use in practice. This principle is illustrated in the chapters that follow.

Mean and Variance of the Sampling Distribution

The mean and variance are needed to complete the description of the sampling distribution of $\bar{x}$. It is intuitively obvious that the long-run average of all possible $\bar{x}$'s from a parent population whose center is at μ is also μ. This is demonstrated in Section 9.2 and can be stated as a general rule:

RULE 1

The *mean* of the sampling distribution of $\bar{x}$ is equal to the mean μ of the parent population:

$$E(\bar{x}) = \mu.$$

It is shown in Section 9.2 that deviations from the center of a sampling distribution of the mean are not as large as deviations of values from the center of the parent distribution. In other words, a distribution of possible values of $\bar{x}$ is not as spread out as the original population of x's. The relationship of the variance of the sampling distribution ($\sigma_{\bar{x}}^2$) to the variance of the parent population (σ^2) can be stated as a general rule:

RULE 2   The *variance* of the sampling distribution of $\bar{x}$ is the variance of the parent population divided by the sample size:

$$\sigma_{\bar{x}}^2 = \frac{\sigma^2}{n}.$$

Note that the variance of the sampling distribution is inversely proportional to n. Thus the larger the sample size, the less spread out the sampling distribution of $\bar{x}$ will be.

The standard deviation of the sampling distribution is just the square root of the variance:

RULE 3   The *standard deviation* of the sampling distribution of $\bar{x}$ is the standard deviation of the parent population divided by the square root of the sample size:

$$\sigma_{\bar{x}} = \frac{\sigma}{\sqrt{n}}.$$

The standard deviation $\sigma_{\bar{x}}$ is called the *standard error of the mean*.

Like the variance, the standard error indicates the spread of the sampling distribution of the mean. A larger value indicates that the distribution of sample means is relatively wide and a smaller value indicates that the distribution of means is relatively narrow, that is, the possible values of x are closer together. The size of the standard error also depends on n, with larger samples producing a smaller standard error. The standard error is an important component of inferential procedures that focus on population means.

In Rules 2 and 3 above it is assumed that the sampling is with replacement or from an infinite (hypothetical) population. This is not generally a serious consideration as long as the size of the population is large relative to the size of the sample.

The results above can be used to find the probabilities of sample means arising from a specified parent population. The example given here illustrates the principles of *statistical inference*, discussed more fully in the chapters that follow.

Suppose that the sampling distribution is constructed for the mean of $n = 25$ observations from a parent population in which $\mu = 500$ and $\sigma = 100$. (You might think of this as the distribution of $\bar{x}$'s on the Scholastic Assessment Tests for all possible high school classes of 25 students.) The sampling distribution of $\bar{x}$ has mean $\mu = 500$ and standard deviation of $\sigma_{\bar{x}} = 100/\sqrt{25} = 20$. Then what proportion of classes have mean SAT scores of 540 or above? That is, what is $\Pr(\bar{x} \geq 540)$?

The normal distribution that approximates a given sampling distribution is the one with the same population mean and standard deviation (e.g., 500 and 20). The probability of values 540 or above can be obtained by *standardizing* the value 540 and referring to the standard normal table in Appendix I. An $\bar{x}$ of 540 is exactly 2 standard deviations above the center of this normal distribution; specifically, $z = (540 - 500)/20 = 2$. From Appendix I $\Pr(0 \leq z \leq 2) = .4772$, and thus $\Pr(z \geq 2) = .5000 - .4772 = .0228$. The probability of obtaining a sample mean of 540 or greater from this parent population, based on an n of 25, is less than 0.03. It is a relatively rare event. If your class has an average SAT score of 542, for example, it must be one of the highest scoring classes in the country.

Note that an *individual score* of 540 is not that unusual. It is only 0.4 of a standard deviation above μ in the parent population, and many observations have values in this general range. A sample mean of 540 is less likely because sample means are less spread out than individual scores.

We see in this example that a sample mean may be expressed in standardized form relative to the mean and standard deviation of the sampling distribution. The value for z (2.0) was obtained by

$$\frac{\bar{x} - E(\bar{x})}{\sigma_{\bar{x}}} = \frac{\bar{x} - \mu}{\sigma/\sqrt{n}}.$$

Standardized means have a sampling distribution that is approximated by the standard normal distribution.

Normal Approximation to the Binomial Distribution

As noted in Section 9.3, the proportion of heads $\hat{p}$ in n trials is the sample mean of Bernoulli variables. Hence the Central Limit Theorem applies to $\hat{p}$ and implies that

$$\frac{\hat{p} - E(\hat{p})}{\sigma_{\hat{p}}} = \frac{\hat{p} - p}{\sqrt{pq/n}}$$

has a distribution which is approximated by that of the standard normal distribution. The approximation is good if n is large and p is not too small or large (too near 0 or 1). (As a rule of thumb we can say that the approximation may be used when $np > 5$ and $nq > 5$ if the "correction for continuity" explained in Appendix 9A is used.)

Let us use the normal distribution to approximate the probability that 100 tosses of a fair coin ($p = 0.5$) will yield 60% heads or fewer ($\hat{p} \le 0.6$). The value 0.6 can be *standardized* by

$$\frac{\hat{p} - p}{\sqrt{pq/n}} = \frac{0.6 - 0.5}{\sqrt{0.5 \times 0.5/100}} = 2.$$

From Appendix I, $\Pr(z < 2) = .5000 + .4772 = .9772$, the required probability.

If we let $x = n\hat{p}$ be the number of heads, then (multiplying both numerator and denominator by n)

$$\frac{\hat{p} - p}{\sqrt{pq/n}} = \frac{n(\hat{p} - p)}{n\sqrt{pq/n}} = \frac{x - np}{\sqrt{npq}} = \frac{x - \mu_x}{\sigma_x},$$

that is, the standardized version of the *number* of successes is equal to the standardized version of the *proportion* of successes, $\hat{p} = x/n$. This form is used to approximate the probability that the number of successes is less than or equal to 60 out of 100 tosses:

$$\frac{x - np}{\sqrt{npq}} = \frac{60 - 100 \times 0.5}{\sqrt{100 \times 0.5 \times 0.5}} = 2.$$

We obtain the same standardized value and the same probability as before because the proportion of successes is 0.6 if the number of successes in 100 tosses is 60.

As a second example, suppose the population is 24% Catholic (Figure 2.3). In a sample of 10,000 we expect about 2400 Catholics. What is the probability that the number of Catholics in a sample $n = 10,000$ will be less than or equal to 2440? Let x be the number of Catholics in the sample. Here $\mu = np = 10,000 \times 0.24 = 2400$, $\sigma = \sqrt{npq} = \sqrt{10,000 \times 0.24 \times 0.76} = \sqrt{10,000}\sqrt{0.24}\sqrt{0.76} = 100 \times 0.490 \times 0.872 = 42.7$. The value 2440 may be standardized to $(2440 - 2400)/42.7 = 0.94$, that is, 2440 is about 9/10 of a standard deviation above the mean of the parent population. From the standard normal table $\Pr(z < 0.94) = .5000 + .3264 = .8264$.

For a more accurate refinement of this use of the normal approximation see Appendix 9A.

Summary

A *random sample* of size n is a subset of n individuals obtained from the population in such a way that all groups of that size have the same chance of being the actual sample. In particular, this implies that all individuals have the same chance of being included in the actual sample.

A *statistic* is a characteristic of a sample, such as its mean, median, range, or standard deviation.

A *sampling distribution* is the probability distribution of a statistic. It gives the probability associated with each possible value of the statistic. In this context the distribution from which the sample is drawn is called the *parent distribution*.

The sampling distribution of the number of "heads" in n independent trials of the same two-outcome (heads/tails) experiment is the *binomial distribution*. It gives the probability of obtaining exactly r heads in such an experiment, $r = 0, 1, 2, \ldots, n$.

The proportion of heads in n independent coin-toss trials is an example of the mean of a sample. The sampling distribution of the proportion of heads is an example of the sampling distribution of a mean.

The *Law of Averages* (Law of Large Numbers) states that in large samples the mean of a random sample has a good chance of being close to the population mean.

The mean of the sampling distribution of the mean is the population mean. The variance of the sampling distribution of the mean is the variance of the parent population divided by the sample size. If the mean and variance of the parent population are μ and σ^2, respectively, then the mean and variance of the sampling distribution of $\bar{x}$ are μ and $\sigma_{\bar{x}}^2 = \sigma^2/n$, respectively. The standard deviation of the sampling distribution of the mean, $\sigma_{\bar{x}} = \sigma/\sqrt{n}$, is called the *standard error* of the mean.

Mathematically speaking, the mean of a random sample from a normal parent population has a normal distribution. More generally, almost regardless of the shape of the parent distribution, the mean of a random sample has a sampling distribution which is approximated by a normal distribution. This fact is known as the *Central Limit Theorem*.

In fact, the normal distribution plays a central role in probability and statistics, not only because many natural phenomena have distributions that are well approximated by normal distributions, but also because many statistics (not just the sample mean) have sampling distributions which are well approximated by normal distributions, almost regardless of the shape of the parent distribution.

Appendix 9A   The Correction for Continuity

The heights of 11-year-old boys are recorded in Table 2.12 to the nearest inch and are graphed in Figure 2.14 in a histogram in which the bars are of width one inch. In Figure 8.1 the histogram is repeated together with a continuous probability distribution which approximates it. This continuous distribution is a normal distribution that has the same mean and standard deviation as the histogram. In both cases area represents relative frequency or probability; the entire area in the histogram or under the curve is 1. The most marked difference between the histogram and the normal curve is that the former is discrete. This fact should be taken into account when the normal curve is used to approximate the relative frequencies or probabilities of the histogram.

In the histogram the relative frequency of 54 inches, for example, is .1446 (Table 2.12) and is represented by the area of the bar over the interval 53.5 to 54.5, the area of the entire histogram being 1. The relative frequency of heights up to and including 54 inches is the area to the left of 54.5, namely, .7393; it is the sum of the areas of the histograms to the left of 54.5. The area under the normal curve that approximates this area of the histogram is the area over the same part of the horizontal axis, namely to the left of 54.5. We can calculate this area under the normal curve by referring to Appendix I. The mean of the distribution is 52.9 and the standard deviation is 2.46. Therefore the value 54.5 is $(54.5 - 52.9)/2.46 = 1.6/2.46 = 0.65$ standard deviations from the mean. From Appendix I we find that the area to the left of this point is .7422 ($.5000 + .2422$) which is close to the observed relative frequency, .7393.

If, instead, we calculate the area under the normal curve to the left of 54, we obtain .6700, which is not as close to .7393. The reason for obtaining a poorer approximation is seen in Figure 8.1; the area under the normal curve to the left of 54 approximates the area of the histogram also to the left of 54, thus omitting half of the bar representing a height of 54 inches.

To compute the normal approximation to the probability of a height of 55 or 56 inches, we note that in a histogram these heights are represented by two bars whose bases extend from 54.5 to 56.5. Therefore we find the normal approximation to the probability of a height less than 56.5 and subtract from it the normal approximation to the probability of a height less than 54.5. The height 56.5 is $(56.5 - 52.9)/2.46 = 1.46$ standard deviations above the mean. From Appendix I we find that the probability corresponding to 1.46 is .4279. The height 54.5 is 0.65 standard deviations above the mean, and the corresponding problem is .2422. The difference is $.4279 - .2422 = .1857$. This is close to the actually observed relative frequency (from Table 2.12), which is $.1206 + .0742 = .1948$.

The general idea is that the area under the smooth normal curve should be calculated in the way best to approximate the relevant area of the histogram. The problem arises because of the fact that the histogram represents a discrete variable (in this case, height *to the nearest inch*); the above resolution of this problem is called *correction for continuity*.

This idea applies also to the use of the normal distribution to approximate theoretical probability distributions. In the approximation to the binomial distribution, the continuity correction is made by adding or subtracting 1/2 from the number of successes, as appropriate. For example, the probability that the number of heads is less than or equal to 60 is approximated by the area to the left of 60.5 under the normal curve with mean np and standard deviation $\sqrt{npq}$. If $n = 100$ and $p = 1/2$, we have

$$\Pr(x \le 60) = \Pr(x \le 60.5).$$

If 60.5 is converted to a standardized value, we have $(60.5-50)/5 = 2.1$. From Appendix I,

$$\Pr(z \le 2.1) = .9821,$$

as compared to the exact probability of .9824. The normal approximation without the continuity correction is .9772, an error of .0048 instead of .0003. Of course, we obtain the same approximation to the probability that the number of heads is less than 61, for we write

$$\Pr(x < 61) = \Pr(x \le 60.5)$$

and continue as before.

If $n = 10$ and $p = 1/2$ (so $np = nq = 5$), $\Pr(x \le 4) = .3770$ and the normal approximation with continuity correction is .3759; and $\Pr(x \le 2) = .0547$ and the normal approximation with the continuity correction is .0579. The normal approximation without the continuity correction gives .2635 and .0289, respectively; these are not satisfactory.

In general, when a continuous distribution is used to approximate the distribution of a discrete variable, the area of the continuous distribution that is closest to the area of the histogram is calculated. Usually this means calculating to a value halfway between the two adjacent discrete values.

Exercises for the Entire Class

9.1   Exercise 7.1 called for 25 tosses of two distinguishable coins by each student. Each student should report the number of heads obtained in each of the two sequences of 25 tosses. The results are to be compiled and recorded in a table like the one below. Compute the relative frequencies and the cumulative relative frequencies.

Number of heads observed in 25 tosses of a coin	Frequency	Relative frequency	Cumulative relative frequency	Expected relative frequency	Expected cumulative relative frequency
5				0.002	0.002
6				0.005	0.007
7				0.015	0.022
8				0.032	0.054
9				0.061	0.115
10				0.097	0.212
11				0.133	0.345
12				0.155	0.500
13				0.155	0.655
14				0.133	0.788
15				0.097	0.885
16				0.061	0.946
17				0.032	0.978
18				0.015	0.993
19				0.005	0.998
20				0.002	1.000
				1.000	

(a) How close are your empirically derived relative frequencies to the actual theoretically derived distribution?

(b) How could you modify your procedure to make the empirical results closer to the theoretically derived (exact) results?

9.2 (continuation) Compile the observed relative frequencies of the deviations from 1/2 as reported by members of the class.

Size of deviation from 1/2	Observed relative frequency	Expected relative frequency
at most 1/2		
at most 1/4		
at most 1/8		
at most 1/16		

9.3 (continuation) Let x be the number of heads occurring in 25 tosses, and consider the following events.

(a) 16 or fewer heads occur, that is, $x \leq 16$.

(b) 10 or fewer heads occur, that is, $x \leq 10$.

(c) 14 or more heads occur, that is, $x \geq 14$.

(d) No more than 16 and no fewer than 14 heads occur, that is, $14 \leq x \leq 16$.

For each of these events, give

(1) the observed relative frequency of occurrence (from Exercise 9.1);

(2) the normal approximation to the probability, without the correction for continuity;

(3) the normal approximation to the probability, with the correction for continuity (Appendix 9A).

Exercises for Each Student

9.4 Compute the sampling distribution of the mean for samples of size 3, drawn without replacement from the population of Table 9.3.

9.5 Compute the probability of getting 3 heads and 3 tails in 6 flips of a fair coin. Check your answer in Appendix II.

9.6 Compute the probability of getting 4 heads and 4 tails in 8 flips of a fair coin. Check your answer in Appendix II.

9.7 From Appendix II find the probability of getting exactly 2 fives in 6 rolls of a balanced die.

9.8 From Appendix II find the probability of getting no more than 2 fives in 6 rolls of a balanced die.

9.9 From Appendix II find the probability of getting exactly 2 fives in 5 rolls of a balanced die.

9.10 From Appendix II find the probability of getting no more than 2 fives in 5 rolls of a balanced die.

9.11 If in a multiple-choice test there are 8 questions with 4 answers to each question, what is the probability of getting 3 or more correct answers just by chance alone?

9.12 If in a multiple-choice test there are 10 questions with 4 answers to each question, what is the probability of getting 3 or more correct answers just by chance alone?

9.13 The probability of recovery from a certain disease is .75. Find the distribution of the number of recoveries among 3 patients and draw the histogram of this binomial distribution.

9.14   The probability of a patient recovering from a certain disease is .75. Find the distribution of the number of recoveries among 4 patients and draw the histogram of this binomial distribution.

9.15   Table 6.23 gives bivariate percentages for two random variables "offense" and "outcome." Assume that each percentage divided by 100 is a probability. Are the two variables independent?

9.16   (continuation) Keeping the row probabilities and column probabilities in the *margins* of Table 6.23, construct a hypothetical table of probabilities such that "offense" and "outcome" are independent according to the definition given in Section 9.1.

9.17   Find the probability distribution of the sum of the values on three dice by using the probability distribution of the sum of the values on two dice (given in Table 9.6) and the value on one die. Graph the probabilities. Does the graph look more normal than Figure 9.2? Verify that the mean is $3 \times 3\frac{1}{2}$ and the variance is $3 \times 35/12$.

9.18   From Exercise 9.13 find the probability that the proportion of recoveries differs from .75 by not more than (a) 1/2, (b) 1/4, (c) 5/12, and (d) 3/4.

9.19   From Exercise 9.14 find the probability that the proportion of recoveries differs from .75 by not more than (a) 0, (b) 1/4, (c) 1/2, and (d) 3/4.

9.20   Calculate the mean and standard deviation of the distribution of the number of recoveries in Exercise 9.13 by Section 8.3. Compare these results to those you would get using the formulas in Section 9.3.

9.21   Calculate the mean and standard deviation of the distribution of the number of recoveries in Exercise 9.14 by Section 8.3. Compare these results to those you would get using the formulas in Section 9.3.

9.22   Calculate the mean and standard deviation of the distribution of the proportion of recoveries out of 1000 patients if the probability of recovery is .75.

9.23   Calculate the mean and standard deviation of the distribution of the proportion of recoveries out of 10,000 patients if the probability of recovery is .75.

9.24   The variance of a distribution is a measure of uncertainty as to what will be the value of a random observation drawn from the distribution. In the case of a Bernoulli variable, and thus a binomial distribution, the degree of uncertainty is a function of p. Construct a graph that has values of p as the horizontal axis from 0 to 1.0 in intervals of $\frac{1}{10}$, and the variance on the vertical axis. At what p value is the variance the least? The greatest?

9.25   Suppose that random samples of 49 independent observations are drawn from a parent population in which $\mu = 19.4$ and $\sigma = 14.7$.

(a) What is the mean of the sampling distribution of $\bar{x}$?

(b) What is the standard deviation of the sampling distribution of $\bar{x}$?

(c) Use the normal curve to approximate $\Pr(\bar{x} \leq 19.4)$.

(d) Use the normal curve to approximate $\Pr(\bar{x} \geq 21.5)$.

(e) Use the normal curve to approximate $\Pr(\bar{x} \geq 13.1)$.

9.26   Suppose that random samples of 64 independent observations are drawn from a parent population in which $\mu = 92$ and $\sigma = 10$.

(a) What are the mean and standard deviation of the sampling distribution of $\bar{x}$?

(b) Evaluate the following probabilities, using a normal curve approximation:

(i) $\Pr(80.75 \leq \bar{x} \leq 93.25)$;

(ii) $\Pr(89.55 \leq \bar{x} \leq 94.45)$;

(iii) $\Pr(89 \leq \bar{x} \leq 95)$;

(iv) $\Pr(92 \leq \bar{x} \leq 94.06)$.

9.27   Random variable x has population mean μ and standard deviation $\sigma = 24$. Samples of $n = 144$ independent observations are drawn and the sampling distribution of $\bar{x}$ is constructed. The probability of an $\bar{x}$ greater than 10 is very close to .8665. What is μ?

9.28   The probability of being selected to serve in the United States armed forces, out of those taking the qualifying examination is $p = .45$. The examination is administered repeatedly to samples of 10,000 applicants. Use the normal distribution to approximate the proportion of applicant pools that will have 4,600 or more applicants selected to serve.

9.29   (continuation) What proportion of applicant pools will have 4,600 or more applicants selected if the selection probability is increased to

(a) $p = .50$?

(b) $p = .65$?

IV

Statistical Inference

10

Using a Sample to Estimate Characteristics of One Population

Introduction

In this section the theory developed in Part III is used to allow us to *infer* the characteristics of a population based on data from a sample. This is an essential part of statistical analysis because we often need to know about the parent population, but are not able to study every one of its members. For example, we might like to know what percentage of voters favor a particular political issue, but cannot survey all voters by phone; or we may need to know if a particular medication is effective, but cannot wait (or afford) to test it on every individual who contracts the disease before declaring it as effective or ineffective.

Statistical inference is usually discussed as two separate but related sets of procedures: *hypothesis testing* and *estimation*. The former allows us to make simple decisions about a population based on a sample of data, for example, "Does a majority of voters favor an increase in taxes?" or "Does this medication reduce the mortality rate from a particular disease?" Each of these is answered by "yes" or "no." Estimation of population values addresses the question of the *magnitude* or extent of the effect, for example, "What percentage of voters favor an increase in taxes?" or "What does the mortality rate become if the new medication is used as prescribed?" Both of these approaches are important parts of statistical inference, and both are included in most reports prepared from statistical data.

Chapter 10 discusses estimation of the characteristics of a single population. Two approaches are discussed. The first is obtaining a *point estimate* of a population value, that is, a single best guess. Since even a good point estimate is only an approximation, an *interval estimate* is useful as well. This is a range of plausible values for the population characteristic, or an interval in which we are *confident* the population value lies.

Chapter 11 discusses hypothesis testing for a single population, and Chapter 12 describes estimation and hypothesis testing for the comparison of two populations, for example, males and females, or a new experimental medication with that used previously. Hypothesis testing for a variance and the comparison of two variances are discussed together in Chapter 13 because they require probability distributions that are not discussed previously.

10.1   Estimation of a Mean by a Single Number

One of the main objectives of statistical inference is the *estimation* of population characteristics on the basis of the information in a sample. Responses to political polls are used to estimate the proportions of voters in favor of different candidates. The statistical evaluation of the polio vaccine trial involved the estimation of proportions of polio cases among those inoculated with vaccine and those inoculated with placebo. Thus, a sample proportion is an estimate of the population proportion; the arithmetic mean of a sample is an estimate of the population mean; and the sample variance is an estimate of the population variance. In this chapter we shall discuss how well these estimates, as well as estimates of medians and other percentiles, may correspond to the population quantities.

We introduce the idea of estimating a population characteristic by discussing how a patient's true cholesterol level might be estimated from several cholesterol determinations on a single blood sample. Because there is variability among cholesterol measurements due to the technique and equipment used, the true cholesterol level is considered to be the mean of a hypothetical population of repeated determinations of cholesterol concentration in a single blood sample. Three determinations are treated as a random sample from this population. Suppose the values (milligrams of cholesterol in 100 milliliters of blood) obtained are 215, 222, and 205 units. Then the true cholesterol level of the patient is estimated by the arithmetic sample mean,

$$\frac{215 + 222 + 205}{3} = 214 \text{ units.}$$

Suppose that from previous extensive testing of the procedure used here it is known that a population of many repeated determinations of blood cholesterol levels has a standard deviation of 12.6 units. Then from Chapter 9 we know that the standard deviation of the mean of three independent determinations is $12.6/\sqrt{3} = 12.6/1.73 = 7.28$. The sample mean is thus a more precise estimate of the true value than is a single observation, which has a standard deviation of 12.6.

Analogy Between Population Mean and Sample Mean

Now we discuss these ideas in a more general way. Consider a variable in which a finite population of N individuals has the values $y_1, y_2, \ldots, y_N$. The population mean μ of this variable is defined by the arithmetic operations expressed in the formula

$$\mu = \frac{\sum y_k}{N}.$$

Given a sample of n individuals with observations $x_1, x_2, \ldots, x_n$, it is natural to estimate the population mean μ by applying the same arithmetic operations to the sample; this produces the estimate

$$\bar{x} = \frac{\sum x_i}{n},$$

the arithmetic mean of the sample. Likewise, as shown in Section 8.3, the same operations that are applied to probabilities to obtain a population mean can be applied to sample relative frequencies to obtain the sample mean.

The sample mean $\bar{x}$ is used to estimate the population mean because $\bar{x}$ is constructed from the sample values in the same way that μ is constructed from population values. The value $\bar{x}$ is a *point estimate* of μ in the sense that it is a single "best guess;" in this context, $\bar{x}$ can also be represented as $\hat{\mu}$, where the "$\hat{}$" means "the estimate of" (the population value μ). Thus, $\bar{x}$ has two uses, as a descriptive measure of a sample of observations and as an inferential statistic to approximate the population mean.

Sampling Distribution of the Sample Mean

From sample to sample the arithmetic mean will vary, that is, it has a sampling distribution (Section 9.2). The arithmetic mean of a particular sample of n observations is one value from this distribution.

Two properties of the sampling distribution are especially important if we are to use the mean of one sample to estimate the mean of the population, μ. First, it is desirable that the sampling distribution be centered around the population mean. If this is the case, then a sample mean approximates μ without a systematic error. If the sampling distribution does not have its center at μ, samples drawn from the distribution will tend to systematically overestimate or systematically underestimate μ.

Second, it is desirable that the values in the sampling distribution be clustered closely together, that is, that the variance of the distribution is relatively small. If this is the case, then the possible values of $\bar{x}$ are close to each other and close to μ as well. If the variance of the sampling distribution is relatively large, then a particular $\bar{x}$ drawn from the distribution is more likely to be substantially below or substantially above μ.

It is shown in Section 9.2 that the mean of the sampling distribution of $\bar{x}$ is equal to μ, the mean of the parent population. Thus the sampling distribution of $\bar{x}$ is centered around the parameter being estimated, μ. This property, called *unbiasedness*, is discussed further in the section below.

The variance of the sampling distribution of $\bar{x}$ is $\sigma_{\bar{x}}^2$, which is related to the variance of the parent population by

$$\sigma_{\bar{x}}^2 = \frac{\sigma^2}{n}.$$

The standard deviation of the sampling distribution of the mean, also called the *standard error of the mean*, is

$$\sigma_{\bar{x}} = \frac{\sigma}{\sqrt{n}}.$$

The sampling distribution tends to be concentrated near the value of the parameter being estimated because the variability is reduced by n in the denominator. This property, called *precision*, is discussed further in the section below.

Besides the mean and variance of the sampling distribution, it is often possible to know its shape. The parent population of cholesterol determinations is known from previous testing to be approximately normal. Therefore the shape of the sampling distribution of the sample mean is close to normal. For other variables, the sampling distribution of the mean is approximately normal almost regardless of the shape of the parent population if the sample size is sufficiently large. (See Section 9.5.)

Unbiasedness In statistical notation, $\bar{x}$ is called an *unbiased* estimate of μ because $E(\bar{x}) = \mu$.

DEFINITION A statistic is an *unbiased estimate* of a parameter if the mean of the sampling distribution of the statistic is the parameter.

When an estimate is unbiased, its sampling distribution has the desirable property of being centered at the parameter. In this chapter we shall also discuss unbiased estimates of parameters other than a mean, for example, the population proportion and the variance. (In fact, it will be shown in Chapter 13 that the reason for dividing by $n - 1$ in computing a sample variance is to obtain an unbiased estimate of the population variance.)

Precision The standard deviation of any distribution is a measure of the extent to which values are clustered around the center. Thus the standard deviation of the sampling distribution of $\bar{x}$, namely $\sigma/\sqrt{n}$, is a measure of how closely $\bar{x}$'s are clustered around μ. If the entire distribution is highly concentrated, then it is more likely that the *particular* $\bar{x}$ we compute to estimate μ is close to μ as well.

The standard deviation of the sampling distribution is also called the standard error of $\bar{x}$. The larger the sample size, the smaller the standard error, and the more precisely we are likely to be estimating μ. Thus the standard error is a measure of precision of estimation. It is given without proof that in most cases the standard error of the mean is smaller than the standard error of either the median or the mode; thus the sample mean is the most precise estimator of location in the population of the three.

Summary To sum up, we use $\bar{x}$ to estimate μ because it is formed by applying the same operations to the sample values as are applied to the population values to form μ. Further, we can say that $\bar{x}$ is unbiased and has a standard deviation which is smaller than that of a single observation. These facts imply that $\bar{x}$ has a good chance of being close to μ.

10.2 Estimation of Variance and Standard Deviation

If the standard deviation σ is not known, it can also be estimated from the sample. If the weights of a random sample of three players from a large list of football players are $x_1 = 225$, $x_2 = 232$, and $x_3 = 215$ pounds, then the sample mean is $\bar{x} = (225 + 232 + 215)/3 = 224$, and the sample variance is

$$
\begin{aligned}
s^2 &= \frac{\sum (x_i - \bar{x})^2}{3 - 1} \\
&= \frac{(225 - 224)^2 + (232 - 224)^2 + (215 - 224)^2}{2} \\
&= 73.
\end{aligned}
$$

The sample standard deviation s is $\sqrt{73}$, or 8.54.

Analogy Between Population Variance and Sample Variance: Unbiasedness

Consider again a variable which in a finite population of N individuals has the values $y_1, y_2, \ldots, y_N$. The population variance σ^2 of this variable is defined by the arithmetic operations expressed in the formula

$$
\sigma^2 = \frac{\sum (y_k - \mu)^2}{N},
$$

where the population mean μ is as defined in the previous section. More generally, it is shown in Section 8.3 that the variance of a population can be obtained from a probability distribution. The result is still the population average or "expected value" of squared deviations from μ.

Given observations $x_1, x_2, \ldots, x_n$ arising from a random sample of n individuals, it is natural to estimate σ^2 by applying the same arithmetic operations to the sample. If we know that value of the population mean μ we can do this, producing the estimate

$$
\frac{\sum (x_i - \mu)^2}{n}.
$$

This is an unbiased estimate of the population variance. Usually, however, we cannot use this estimate because the population mean μ is unknown. Then we substitute the sample mean $\bar{x}$ for μ and use the estimate

$$s^2 = \frac{\sum(x_i - \bar{x})^2}{n - 1}.$$

An explanation for using $n - 1$ instead of n in the denominator of s^2 is given in Chapter 4. The value $n - 1$ is the number of *independent* pieces of information about dispersion in a sample of n observations; it is called the number of degrees of freedom. It is also the case the s^2 is an unbiased estimate of σ^2 only if $n - 1$ is in the denominator. Using $n - 1$, the mean of the sampling distribution of s^2 is σ^2, that is,

$$E(s^2) = \sigma^2.$$

This is true regardless of the nature of the parent population. If we were to use n in the denominator of the sample variance, then it would not be unbiased.[1]

The sample variance s^2 is a *point estimate* of the population variance and thus may be represented $\hat{\sigma}^2$. Its square root, the sample standard deviation, is used to estimate the population standard deviation and may be represented $\hat{\sigma}$, that is, the "estimate" of σ. The sampling distribution of s tends to be concentrated around the population standard deviation. The larger the sample size n the greater the concentration of s^2 around σ^2 and s around σ.

The Estimated Standard Error

The standard error of the sample mean, that is, the standard deviation of the sampling distribution of the sample mean, is $\sigma_{\bar{x}} = \sigma/\sqrt{n}$. If a value for σ is not available, it may be estimated from the sample. The estimated standard error is $s_{\bar{x}} = s/\sqrt{n}$ where $s_{\bar{x}}$ is used to indicate that it is a sample-based approximation to $\sigma_{\bar{x}}$.

10.3   An Interval of Plausible Values for a Mean

Confidence Intervals when the Standard Deviation is Known

If the statistician reports only a point estimate, that is, a single value, someone may ask how sure he or she is that the estimate is correct.

---

[1]If s^2 were computed with n in the denominator instead of $n - 1$, s^2 would consistently underestimate σ^2; it would estimate $\sigma^2(n - 1)/n$ instead.

Usually the point estimate will not be exactly correct. Since this is the case, the precision of the estimate must be indicated in some way. One way to do this is to give the standard deviation of the sampling distribution of the estimate (the standard error) along with the estimate itself, thus permitting others to gauge the concentration of the sampling distribution of the estimate. Another way is to specify an *interval*, which depends on both the point estimate and the standard error, and which contains a range of plausible values for the parameter.

The upper endpoint of the interval is obtained by adding to the sample mean a certain quantity, which is a multiple of the standard error of the mean. The lower endpoint of the interval is obtained by subtracting this quantity from the sample mean.

In the cholesterol example we supposed that the distribution of repeated determinations of cholesterol concentration for a single sample had standard deviation σ equal to 12.6 units, and that the distribution around the mean μ is approximately normal. Three determinations of the cholesterol concentration in a single sample are considered to be a random sample from such a normal distribution.

The mean of random samples of size 3 has a sampling distribution which has mean equal to μ, the true concentration, and standard error equal to $\sigma/\sqrt{3} = 12.6/1.73 = 7.28$. Furthermore, this sampling distribution is normal.[2]

From the table of the normal distribution we see that the probability is about 2/3 (actually .6826) that a sample mean $\bar{x}$ is within one standard error of μ. Similarly, the probability is .95 that $\bar{x}$ is within 1.96 standard errors of μ. This is because 2.5% of the area under the standard normal curve is to the right of the point $z = 1.96$, and 2.5% of the area is to the left of the point $z = -1.96$, and so 95% of the area is between -1.96 and $+1.96$. In cholesterol units, 1.96 standard errors is $1.96 \times 7.28 = 14.3$ units.

Thus the probability is .95 that $\bar{x}$ is within 14.3 units of μ, as shown in Figure 10.1. That is, the probability is .95 that the *distance between* μ and $\bar{x}$ is less than 14.3. In other words, the probability is .95 that we will obtain a value of $\bar{x}$ such that μ is less than $\bar{x} + 14.3$ and greater than $\bar{x} - 14.3$. In symbols,

$$\Pr(\bar{x} - 14.3 < \mu < \bar{x} + 14.3) = .95.$$

This probability statement can be used to obtain an interval estimate because it says that the probability is .95 of drawing a sample such that

---

[2]It is normal because the parent distribution is normal, and, if the sample size were larger, it would be nearly normal even if the parent population were not by the Central Limit Theorem.

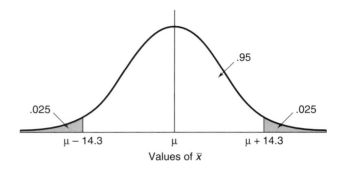

Figure 10.1 *Sampling distribution of $\bar{x}$.*

the interval from $\bar{x} - 14.3$ to $\bar{x} + 14.3$ contains the population mean μ. We shall adopt the notation that this interval is represented by "lower limit–comma–upper limit" in parentheses, for example $(\bar{x} - 14.3, \ \bar{x} + 14.3)$.

The procedure of interval estimation of μ consists of (a) computing the point estimate, the sample mean, and (b) subtracting from and adding to this a certain number of standard errors. The resulting interval is a range of plausible values for the population mean. In the example of Section 10.1, $\bar{x} = 214$. The upper endpoint of this interval is $\bar{x} + 14.3 = 214 + 14.3$, or about 228 units; the lower endpoint is $\bar{x} - 14.3 = 214 - 14.3$, or about 200 units. The results can be represented as the interval $(200, \ 228)$.

The more that is added to and subtracted from $\bar{x}$ to obtain the interval the more certain we are that μ is actually within the resulting interval. The probability .95 or 95% associated with this particular interval is called the *level of confidence* or *confidence coefficient*, and the interval estimate is called a *confidence interval*. The level of confidence is chosen in advance and this determines how many units are added and subtracted to obtain the resulting interval.

In this example, we have added 14.3 units to $\bar{x}$ and subtracted the same amount to obtain the interval. The number 14.3 is 1.96 times the standard error of the mean. If a sample of size n is drawn from a population with a standard deviation σ, a 95% confidence interval for the population mean μ is

$$\left(\bar{x} - 1.96\frac{\sigma}{\sqrt{n}}, \ \ \bar{x} + 1.96\frac{\sigma}{\sqrt{n}} \right).$$

This is the case because 1.96 is the 97.5 percentile of the standard normal distribution, this is, $z(.025)$. Also $z(.005) = 2.58$ is the 99.5

percentile of the standard normal distribution; thus,

$$\left(\bar{x} - 2.58\frac{\sigma}{\sqrt{n}}, \quad \bar{x} + 2.58\frac{\sigma}{\sqrt{n}} \right)$$

is a 99% confidence interval for μ. (See Figure 10.2.)

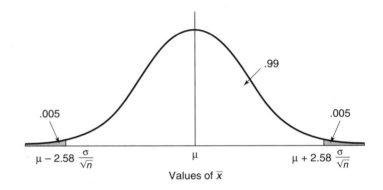

Figure 10.2   *Sampling distribution of* $\bar{x}$.

Confidence intervals with other confidence levels can be obtained by substituting other percentage points from the normal distribution for 1.96 or 2.58. For example, if we wished to obtain a 90% confidence interval for a mean, then we would compute $\bar{x}$ and add and subtract 1.64 standard errors; this value, $z(.05)$, is obtained from the table in Appendix I. It is obvious from the example above that if the level of confidence is increased, the interval becomes wider. In order to be more confident that μ is actually within the interval, a greater range of plausible values is necessary.

The range of plausible values for μ can be narrowed and greater precision attained by increasing the sample size. Since a certain number of standard errors are added to and subtracted from $\bar{x}$, the size of the standard error, $\sigma/\sqrt{n}$, influences the width of the interval as well. In a situation in which $\sigma = 10$ and $n = 25$, for example, the standard error is 2. A .95 confidence interval would require adding and subtracting $1.96 \times 2 = 3.92$ units, giving an interval whose width would be 7.84 units in all (3.92 units below $\bar{x}$ to 3.92 units above $\bar{x}$). If the sample size is increased to $n = 100$, then the standard error is 1 and the resulting interval width is 3.92 units, half of what it is with $n = 25$. This is an important principle because we are now just as confident that we know μ to within 2 units as we were previously that we knew μ to within 4 units; that is, μ is estimated more precisely.

In general, these procedures depend upon the fact that $\bar{x}$ has a sampling distribution which is approximated by a normal distribution.

By the Central Limit Theorem, we know this is true if the sample size is moderate to large. It is true even when n is fairly small if the parent distribution is unimodal and symmetric, except in unusual cases.

Interpretation   The probability statements leading to a confidence interval are based on the sampling distribution of $\bar{x}$. The probability refers to the entire procedure of drawing a sample, calculating the sample mean $\bar{x}$, and adding 14.3 to $\bar{x}$ and subtracting 14.3 from $\bar{x}$ to obtain an interval; the probability is .95 that an interval constructed in this way will include μ, the mean of the population sampled.

After the confidence interval statement has been made, we replace the word "probability" by "confidence" because the procedure has already been carried out and the conclusion that μ lies in the stated interval is now either true or false. As an analogy, suppose the instructor flips an unbiased coin but does not show the result to the class. You know that the probability of tossing a head is 1/2. Although the coin in this case has already been tossed and the face up has been determined, it is rational to bet in the same way as one would before the coin was tossed. We can call 1/2 our "confidence" that the coin is a head on this particular occasion, distinguishing this from the probability associated with the process of flipping.

To illustrate the idea that the confidence level is a probability that refers to a complete procedure, we have simulated drawing 10 random samples of size 3 from a normal population with a mean of 210 and a standard deviation of 12.6. The observations are shown in Table 10.1 and the 95% intervals in Figure 10.3. It happens that 8 of the 10 intervals cover the value $\mu = 210$, although if we repeated the procedure many

TABLE 10.1

95% Confidence Intervals from Ten Samples of Size 3 from a Normal Population with a Mean of 210 and a Standard Deviation of 12.6

Sample number	Sample	Mean	Confidence interval	Interval includes 210?
1	216, 227, 240	227.7	(213.4, 242.0)	No
2	206, 209, 214	209.7	(195.4, 224.0)	Yes
3	206, 226, 213	215.0	(200.7, 229.3)	Yes
4	208, 211, 178	199.0	(184.7, 213.3)	Yes
5	203, 308, 217	209.3	(195.0, 223.6)	Yes
6	190, 208, 195	197.7	(183.4, 212.0)	Yes
7	210, 217, 229	218.7	(204.4, 233.0)	Yes
8	206, 202, 219	209.0	(194.7, 223.3)	Yes
9	221, 227, 220	222.7	(208.4, 237.0)	Yes
10	198, 200, 187	195.0	(180.7, 209.3)	No

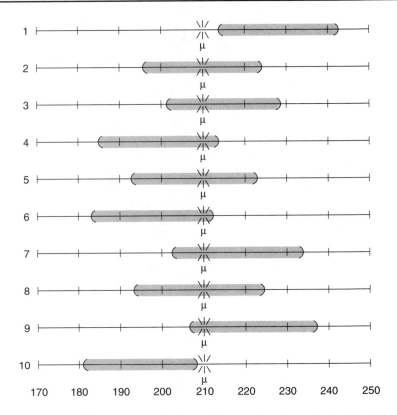

Figure 10.3 *Confidence intervals from 10 samples of size 3 (Table 10.1).*

times we would expect the proportion of intervals including μ to be close to .95. Note, however, that in practice only one of these intervals is computed; it may be one that includes μ or one that does not. Because there are so many more of the former, the odds are in our favor, and we can be *confident* that our single interval includes μ within its limits.

Number of Digits to Report. The reader may well wonder why we have not yet discussed the matter of how many digits to give when reporting the result of a calculation. This simple question is without a simple answer. A sensible answer to the question of how many digits to report involves consideration of the standard deviation of the result as well as the number of significant figures in the original data and the nature of the calculations involved. Usually one would not want to have to compute a standard deviation merely to decide how many digits to report. However, when we compute a confidence interval, we do compute a standard deviation and have at hand information relevant to how many digits it is sensible to report. There is no sense in reporting a mean as 167.243 when the associated 90% confidence interval is (147, 187). The ten's digit (the 6 in 167.243) is slightly uncertain; surely there is no point in reporting the 2, 4, and 3. To report "167" would suffice. Of course it would be good to give the confidence interval, too, or just say: "167 ± 20 (90% confidence interval)."

Confidence Intervals when the Standard Deviation is Estimated

If often happens that we would like to estimate a population mean but we do not know the population standard deviation, σ. In this case, the interval estimate will still have the sample mean as its center, but the width of the interval will depend on the sample standard deviation, s.

As an example, consider a random sample of $n = 10$ men from a population in which height is approximately normally distributed. Suppose the sample yields a mean height of $\bar{x} = 68.7$ inches and a standard deviation of 2.91 inches. We construct a 95% confidence interval by using the estimate of the standard error of the sample mean, which is

$$s_{\bar{x}} = \frac{s}{\sqrt{10}} = \frac{2.91}{\sqrt{10}} = 0.92.$$

We shall add to and subtract from the sample mean an appropriate multiple of 0.92. Because of the additional error introduced by having to estimate σ, this multiple will be *larger* than 1.96 used to achieve 95% confidence when σ is known.

Looking back on the development of the previous section, we can see that we know the probability that μ is in the interval

$$\left(\bar{x} - 1.96 \frac{\sigma}{\sqrt{n}}, \quad \bar{x} + 1.96 \frac{\sigma}{\sqrt{n}} \right)$$

is .95 because $\bar{x}$ has a sampling distribution which is at least approximately normal with mean μ and standard deviation $\sigma/\sqrt{n}$. A mathematically equivalent way of saying this is to say that if we had a sampling distribution of $\bar{x}$'s and "standardized" each $\bar{x}$ by computing

$$z = \frac{\bar{x} - \mu}{\sigma/\sqrt{n}},$$

then the resulting z values have a standard normal distribution. This fact allows us to find the value 1.96 by referring to the standard normal table in Appendix I.

If, however, we did not have the single correct value for σ but had to approximate it from each sample that gave us an $\bar{x}$, then we would have many different s's, one to accompany each $\bar{x}$. We can "standardize" the $\bar{x}$'s by computing

$$\frac{\bar{x} - \mu}{s/\sqrt{n}}$$

from each $\bar{x}$. Because s differs from one sample to the next, some being slightly larger than σ and some being smaller than σ, the resulting distribution will only approximate the standard normal curve. In particular, the distribution will be slightly wider than the standard normal since we have introduced an additional source of sampling variability. The statistic given above, with s replacing σ, is represented by the letter t rather than the letter z.

Student's t-distribution   The sampling distribution of the mean is the distribution of all possible $\bar{x}$'s, each based on n observations, from some parent population. If each $\bar{x}$ is standardized by computing

$$t = \frac{\bar{x} - \mu}{s/\sqrt{n}},$$

then the sampling distribution of this statistic is known as *Student's t-distribution*.[3] The conditions for t to have this distribution are the same as the conditions for the sampling distribution of $\bar{x}$ to be normal in the first place. That is, either the parent population must be normal or else n must be sufficiently large so that the Central Limit Theorem assures us that the sampling distribution of $\bar{x}$ is nearly normal. In either case, the sampling distribution of t will be very nearly Student's t-distribution as well.

Student's t-distribution resembles the standard normal distribution, but is more spread out and less peaked than the standard normal distribution, and consequently has a bigger variance. In fact, the variance of the t-distribution depends on the sample size n. If n is small, then the values of s^2 may vary quite a bit and some may be far from σ^2; the resulting t-distribution may be substantially wider than the standard normal.

More specifically, the distribution depends on the amount of information about σ^2 that is contained in s^2. This is the number of *degrees of freedom* in s^2, namely, the denominator $n - 1$.[4] We shall use the notation t_f to denote the distribution of the t-statistic based on f degrees

---

[3]W. S. Gosset, a chemist at Guiness Brewery, Dublin, in the early 1900's was the first to use this statistic. His company insisted that he publish only under a pen name, and he chose "Student."

[4]Because the sum of deviations from the mean is restricted to sum to zero, $\sum(x_i - \bar{x}) = 0$, there are only $n - 1$ separate or independent deviations that are summarized in the sample variance. The number of degrees of freedom of a statistic is, loosely speaking, the number of *unrestricted* variables associated with it.

of freedom. Figure 10.4 shows the t-distribution with 1 and 5 degrees of freedom, together with the normal distribution. The curve marked t_1 is much less peaked and has much bigger tails than the normal; t_5 is more like the normal. When the number of degrees of freedom is large, the corresponding t-distribution is very much like the standard normal.

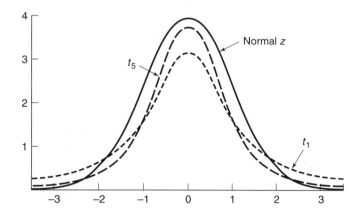

Figure 10.4   *Probability distributions of z, t_1, and t_5.*

Some percentage points of Student's t-distribution for various degrees of freedom are given in Appendix III. Note that the left-hand column lists the values of f, the number of degrees of freedom. Only five percentage points are given for each distribution. Each is a value that separates a portion in the *right-hand tail* of the curve from the rest of the particular t-distribution. The proportions of the curve in the right-hand tail are the column headings, .10, .05, .025, .01, and .005. Thus, in the t-distribution with 9 degrees of freedom (t_9), the t-value 1.833 separates 95% of the curve on the left from the most extreme 5% on the right. By symmetry, the t-value of -1.833 separates the 5% of the curve on the far left from the remaining 95% of the curve.

What values of various t-distributions encompass the central 95% of curve? For t_9 95% of the curve is contained between -2.262 and $+2.262$. These are the values that separate the most extreme 2.5% of the curve in both tails, leaving 95% in the middle; 2.262 is found in the column headed .025. For t_3, the values are -3.182 and $+3.182$. For t_{30} the values are -2.042 and $+2.042$.

As the degrees of freedom get larger, the t-distribution becomes more like the standard normal. If in theory the sample size approaches infinity, then the sample variance would become indistinguishable from the population variance and the t-distribution would "converge" to the normal distribution. Note that the values in the last row of Appendix III are exactly the percentage points of the standard normal distribution.

The central 95% of the curve lies between t (or z) values of -1.96 and $+1.96$.

Because the values in any one column of Appendix III get closer together as the degrees of freedom increase, only a few curves are given with more than 30 degrees of freedom. If percentage points are needed between those listed, linear interpolation will be sufficiently accurate for most purposes. For example, to obtain the t-value that separates the right-hand 5% of t_{35} from the rest of the curve, we may use a value half-way between 1.684 and 1.697, or just 1.69. Computer algorithms are readily available if more accuracy is required.

The notation t_f refers to an entire probability distribution, that is, Student's t-distribution with f degrees of freedom. The *specific numerical values* found in Appendix III will be designated $t_f(\alpha)$, where α is the proportion of area in the right-hand tail of the curve. Thus the value that separates the right-hand 5% of the t-distribution with 9 degrees of freedom from the rest of the curve is represented $t_9(.05) = 1.833$. The values in the t-distribution with 30 degrees of freedom that contain the central 95% of the curve are $-t_{30}(.025)$ and $t_{30}(.025)$ or -2.042 and $+2.042$.

Confidence Intervals Based on Student's t

For a 95% confidence interval based on a sample of 10 observations, we use percentage points of the t-distribution with 9 degrees of freedom. The appropriate percentile is $t_9(.025) = 2.262$. Thus, as shown in Figure 10.5, the probability is .95 that t will be between -2.262 and $+2.262$.

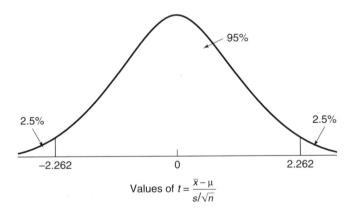

Figure 10.5 *Student's t-distribution for $n = 10$ (degrees of freedom $= 9$).*

In analogy to the case where σ is known, to form a 95% confidence interval for μ based on $n = 10$, $\bar{x} = 68.7$, and $s = 2.91$, we add to and subtract from 68.7 the multiple 2.262 of $s/\sqrt{n} = 0.92$; that is,

$2.262 \times 0.92 = 2.1$. The confidence interval has endpoints $68.7 - 2.1 = 66.6$ and $68.7 + 2.1 = 70.8$; it is $(66.6, \ 70.8)$.

In general, the 95% confidence interval based on a sample size $n = 10$ is

$$\left(\bar{x} - 2.262 \frac{s}{\sqrt{10}}, \quad \bar{x} + 2.262 \frac{s}{\sqrt{10}} \right).$$

This is because 2.262 is the upper 2.5% point (that is, 97.5 percentile) of the Student's t-distribution with $10 - 1 = 9$ degrees of freedom.

The 95% confidence interval for μ based on a sample size of n is

$$\left(\bar{x} - t_{n-1}(.025) \frac{2}{\sqrt{n}}, \quad \bar{x} + t_{n-1}(.025) \frac{s}{\sqrt{n}} \right),$$

where t_{n-1} (.025) denotes the upper 2.5% point of the student's t-distribution with $n - 1$ degrees of freedom (Figure 10.6). For other levels of confidence .025 is replaced by the appropriate number, for example, by .005 for a 99% confidence interval. If n is very large, percentage points from the standard normal table may be used in place of values from the t-distribution with very little loss of accuracy.

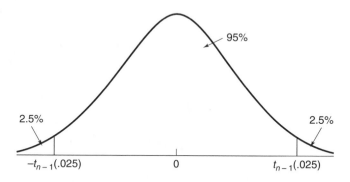

Figure 10.6 *Student's t-distribution with $n - 1$ degrees of freedom.*

Summary   A confidence interval for a population mean may be obtained if σ is known using percentage points from the standard normal distribution. If σ must be approximated by s, percentage points from the t-distribution with $n - 1$ degrees of freedom are used instead.

The use of the t-distribution rests on the assumption that the sampling distribution of the sample mean is normal. This is certain to be the case if the parent population from which x's are drawn is normal and will be approximately correct if the sample size is large, regardless of the parent distribution. There is one gap in our set of procedures, however. If n is small and the parent population is quite different from normal, other methods must be used; these are not discussed in this book.

10.4   Estimation of a Proportion

Point Estimation of a Proportion

Polling a random sample of 400 registered voters from a large population reveals that 224 intend to vote for Jones. The sample proportion $224/400 = 0.56$ estimates p, the proportion of the population intending to vote for Jones. We denote the estimate of p by $\hat{p}$ and write $\hat{p} = 0.56$. Again, the caret (^) or "hat" may be read "estimate of"; thus, $\hat{p}$ denotes an "estimate of p."

We can put this estimate in the context of the preceding section by letting x_i equal 1 or 0, according to whether or not the ith individual intends to vote for Jones. Polling the 400 voters leads to observations $x_1, x_2, \ldots, x_{400}$. The estimate $\hat{p}$ is simply the arithmetic mean of these dichotomous observations.

The variance of any of the dichotomous observations x is $\sigma^2 = pq$, where $q = 1 - p$. Since $\hat{p}$ is simply the arithmetic mean $\bar{x}$ of these observations, the variance of $\hat{p}$ is[5]

$$\sigma_{\hat{p}}^2 = \sigma_{\bar{x}}^2 = \frac{\sigma^2}{n} = \frac{pq}{n}.$$

The standard deviation of the distribution of $\hat{p}$, its *standard error*, is

$$\sigma_{\hat{p}} = \sqrt{\frac{pq}{n}}.$$

As with the sample mean, the standard error of a proportion can be estimated from a single sample. If $\hat{p}$ is the sample proportion and we let $\hat{q} = 1 - \hat{p}$, then the estimate of the standard error is

$$\sqrt{\frac{\hat{p}\hat{q}}{n}}.$$

---

[5]It is also shown in Section 9.3 that the variance of a proportion over repeated samples is pq/n.

Interval Estimation of a Proportion

A confidence interval for the proportion p of the population intending to vote for Jones is based on the idea that for sufficiently large sample size the sample mean is approximately normally distributed. To use a procedure parallel to that in Section 10.3, the standard error $\sigma/\sqrt{n}$ is replaced by the estimate $\sqrt{\hat{p}\hat{q}/n}$.

For the numerical example above the procedure is as follows. The estimate of the variance of $\hat{p}$ is $\hat{p}\hat{q}/n = 0.56 \times 0.44/400 = 0.2464/400$; the estimate of the standard error is the square root of this, $0.496/20$, or 0.0248. For a 95% confidence interval, this is multiplied by 1.96, the appropriate percentage point of the normal distribution. This gives $1.96 \times 0.0248 = 0.0486$, which is about 0.05. The upper and lower limits are, respectively,

$$0.56 + 0.05 = 0.61 \quad \text{and} \quad 0.56 - 0.05 = 0.51.$$

The justification for this procedure is that the standardized version of $\hat{p}$, which is

$$\frac{\hat{p} - E(\hat{p})}{\sigma_{\hat{p}}} = \frac{\hat{p} - p}{\sqrt{pq/n}}, \tag{10.1}$$

has approximately a standard normal distribution.[6] Since n is large here, we replace pq in (10.1) by $\hat{p}\hat{q}$; that is, we use the fact that

$$\frac{\hat{p} - p}{\sqrt{\hat{p}\hat{q}/n}}$$

is approximately the same as (10.1) and hence has approximately a standard normal distribution. A 95% confidence interval for a proportion in general is

$$\left(\hat{p} - 1.96\sqrt{\frac{\hat{p}\hat{q}}{n}}, \quad \hat{p} + 1.96\sqrt{\frac{\hat{p}\hat{q}}{n}} \right). \tag{10.2}$$

To obtain a confidence interval for any other level of confidence we replace 1.96 by the appropriate number from the normal tables. For example, for 99% we take 2.58 because the probability of a standard normal variable being between 0 and 2.58 is 0.495, corresponding to one-half of 99%.

---

[6]As mentioned in Chapter 9, the normal approximation is good if $np > 5$ and $nq > 5$ and the continuity correction is used; larger n is needed if the correction is not used. Because p and q are unknown, the researcher may check that using sample values, $n\hat{p} > 5$ and $n\hat{q} > 5$.

Short Cut for 95% Interval   The 95% confidence interval can be approximated in a simple way because 1.96 is about 2 and the maximum possible value of $\sqrt{pq}$ is $\sqrt{1/4} = 1/2$. Thus $1.96\sqrt{pq/n} < 1/\sqrt{n}$. The interval

$$\left(\hat{p} - \frac{1}{\sqrt{n}}, \quad \hat{p} + \frac{1}{\sqrt{n}} \right)$$

is slightly wider than confidence interval (10.2) but not much if $\hat{p}$ is between 1/4 and 3/4; it has confidence level greater than .95 in any case.

The normal approximation for p can be improved by use of a continuity correction. The importance of such an adjustment is explained in Appendix 9A, especially for small to moderate n. For a confidence interval, the amount $1/(2n)$ is subtracted from $\hat{p}$ to obtain the lower limit and added to $\hat{p}$ to obtain the upper limit. Thus, the approximate 95% a confidence interval for p, with continuity adjustment, is

$$\left(\hat{p} - \frac{1}{2n} - 1.96\sqrt{\frac{\hat{p}\hat{q}}{n}}, \quad \hat{p} + \frac{1}{2n} + 1.96\sqrt{\frac{\hat{p}\hat{q}}{n}} \right).$$

The effect of the correction is to increase the width of the confidence interval by an amount that becomes smaller as n increases.

Small Samples.   If the sample size is small, the confidence interval should be based on the exact binomial sampling distribution. Charts exist from which the confidence limits can be obtained given the sample size and the value of the sample proportion. Bever (1972) gives charts for obtaining 95% and 99% confidence limits for n = 5, 10, 15, 20, 30, 50, 100, 250, and 1000. Pearson and Hartley (1954) give a more extensive set of charts.

10.5   Estimation of a Median

Point Estimation of a Median

As an estimate of the population median it is natural to take the median of a random sample from that population. As an example, consider the lifetimes of a sample of 15 light bulbs which were (in hours)

$$499, \quad 26, \quad 614, \quad 231, \quad 719.$$
$$2063, \quad 2723, \quad 466, \quad 709, \quad 904,$$
$$2303, \quad 27, \quad 1374, \quad 981, \quad 654.$$

These numbers, arranged from smallest to largest, are

$$26, \quad 27, \quad 231, \quad 466, \quad 499,$$
$$614, \quad 654, \quad 709, \quad 719, \quad 904,$$
$$981, \quad 1374, \quad 2063, \quad 2303, \quad 2723.$$

The median of these 15 lifetimes is the 8th ranking value, or 709 hours. This is an estimate of the population median. For large random samples the sample median is likely to be close to the population median.

For certain types of variables and for certain distributions it is more appropriate to estimate the median than the mean. To obtain the median of a sample, the observations are ordered and the median is the value of the middle observation. To apply this procedure the observations only have to be made on an ordinal scale. Thus the median can be used more widely than the mean; the assumptions for its applicability are weaker. For example, an artist may rank the quality of five paintings. The median quality is the quality of the painting ranked third even though "quality" does not have a numerical scale.

The median is also preferable to the mean when a distribution is highly skewed; in this situation the mean and median can be quite different. (See Exercise 3.14.) From Figure 10.7 it can be seen that the distribution of light-bulb lifetimes in the sample is highly skewed to the right. Although 15 is only a small sample size, this is evidence that the population distribution is skewed. The median (709 hours) is a better measure of location than the mean (953 hours) because it is closer in value to more of the observations. The presence of "outliers" in a sample also tends to produce skewness. The median is a preferable

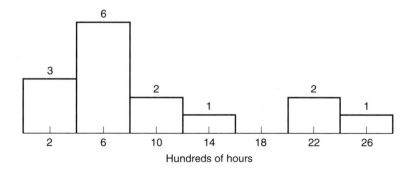

Figure 10.7   *Frequency distribution of lifetimes of light bulbs.*

index of location in this case because it is not sensitive to the value of these extreme observations.

For *symmetric* distributions the mean and the median coincide. To estimate the center of the distribution we can use either the median or the mean. However, for samples from *normal* distributions the sample mean is the better of these two estimates. In fact, for large random samples from normal distributions the variance of the sample median is about $\pi/2$ (about $1\frac{1}{2}$) times that of the sample mean, that is, the variance of the sample median is about $(\pi/2)(\sigma^2/n)$, or about $1.57\sigma^2/n$. This implies that the median of 157 observations is only as precise an estimate as the mean of 100 observations.

Interval Estimation of a Median*

We shall give a way of constructing a confidence interval for the median M of any population. That is, we shall give a *distribution-free* confidence interval for the median. By "distribution-free" we mean that the procedure achieves the desired confidence level regardless of the nature of the parent distribution.

Arrange the sample $x_1, x_2, \ldots, x_n$ in ascending order. Denote the *ordered sample* by $x_{(1)}, x_{(2)}, \ldots, x_{(n)}$. That is, $x_{(1)}$ is the value of the smallest observation and $x_{(n)}$ is the value of the largest.

Since exactly $1/2$ of the population is below the median and $1/2$ is above it, the probability of any particular value x_i being below M is $1/2$. If the sample size is 6, it follows from the binomial distribution with $p = 1/2$ that the probability that the *largest* value is less than M (that is, all 6 values less than M) is $1/64$. Likewise, the probability that all 6 values are greater than M is $1/64$. Hence,

$$\Pr(x_{(1)} < M < x_{(6)}) = 1 - \left(2 \times \frac{1}{64}\right) = \frac{62}{64} = 0.96875,$$

and the interval $(x_{(1)}, x_{(6)})$ is a confidence interval for M with confidence coefficient 0.96875. The interval $(x_{(2)}, x_{(5)})$ is a confidence interval for M with a smaller confidence coefficient.

For an arbitrary sample size n we shall use confidence intervals for M of the form $(x_{(r)}, x_{(n-r+1)})$. [For example, if $n = 6$ and $r = 2$, then $(x_{(2)}, x_{(5)})$ is the interval.] Given n, one can compute the value of r which gives an interval with approximate confidence coefficient .95. This value is

$$r = \frac{n+1}{2} - 1.96\frac{\sqrt{n}}{2} = \frac{n+1}{2} - 0.98\sqrt{n}.$$

This formula is derived using the normal approximation with continuity correction to the binomial distribution with $p = 1/2$. Since 0.98 is close to 1, the formula is very nearly

$$r = \frac{n+1}{2} - \sqrt{n}.$$

When $\frac{1}{2}(n+1) - \sqrt{n}$ is not an integer, it is rounded down to the largest integer less than it. For a 95% confidence interval for the median of the lifetimes of the population of light bulbs from which the 15 are a sample, we have

$$r = \frac{15+1}{2} - \sqrt{15} = 8.0 - 3.9 = 4.1,$$

which rounds down to 4. The confidence interval is $(x_{(4)}, x_{(12)})$ or (466, 1374).

For an interval with confidence coefficient .99, the formula for r is

$$r = \frac{n+1}{2} - 2.58\frac{\sqrt{n}}{2} = \frac{n+1}{2} - 1.2\sqrt{n}.$$

The rule of thumb, $np > 5$ and $nq > 5$ reduces in this case to $n > 10$. If $n \le 10$, the exact binomial distribution with $p = 1/2$ should be used.

10.6   Paired Measurements

Mean of a Population of Differences

There are many situations in which the members of a single sample are measured twice on the same scale. Such data arise in "before and after" or "pre and post" studies. For example, we might assess the severity of an illness before and after medication is administered, or the degree of adjustment of a patient before and after psychotherapy. In other situations we might have pairs of measurements to be compared, such as the strengths of left and right hands for a sample of right-handed people or the number of errors made by employees in the first half and second half of the work day.

In each case it is interesting to compare the means of the two measures in the sample and also in the population from which the sample is drawn. This can be accomplished statistically by reducing each pair of measures to a single *change* or *growth* or *difference score*. For example, Table 10.2 shows reading scores of 30 pupils before and after the second grade. Each pupil's score after second grade is compared with the score before second grade by computing the gain, after-minus-before. Of course, for the different scores to be meaningful, both the before and after tests should measure the same reading skills and be equally difficult. In general, pairs of measurements to be treated in this way must reflect the same phenomenon and be assessed on the same measurement scale.

Once the two measurements are reduced to a difference score, the statistical analysis can focus on just this single variable. The sample of observed differences can be viewed as random sample from a normally distributed population of differences. Point and interval estimates of the population mean difference μ_d are obtained as they are for a single mean in Sections 10.1 and 10.3.

The calculation of differences, mean of the differences, and standard deviation of the differences is summarized at the bottom of Table 10.2. Let x_i represent the "before" score for student i and y_i the "after" score for the same individual. Then the difference for this individual is $d_i = y_i - x_i$. In this example we subtract after-minus-before because this conforms to our idea of "growth" in a year of school; this is a good idea whenever the two measures are "before" and "after" measures.

TABLE 10.2

Reading Scores of 30 Pupils Before and After Second Grade

Pupil i	Scores Before x_i	After y_i	Difference $d_i = y_i - x_i$	Pupil i	Scores Before x_i	After y_i	Difference $d_i = y_i - x_i$
1	1.1	1.7	0.6	16	1.5	1.7	0.2
2	1.5	1.7	0.2	17	1.0	1.7	0.7
3	1.5	1.9	0.4	18	2.3	2.9	0.6
4	2.0	2.0	0.0	19	1.3	1.6	0.3
5	1.9	3.5	1.6	20	1.5	1.6	0.1
6	1.4	2.4	1.0	21	1.8	2.5	0.7
7	1.5	1.8	0.3	22	1.4	3.0	1.6
8	1.4	2.0	0.6	23	1.6	1.8	0.2
9	1.8	2.3	0.5	24	1.6	2.6	1.0
10	1.7	1.7	0.0	25	1.1	1.4	0.3
11	1.2	1.2	0.0	26	1.4	1.4	0.0
12	1.5	1.7	0.2	27	1.4	2.0	0.6
13	1.6	1.7	0.1	28	1.5	1.3	−0.2
14	1.7	3.1	1.4	29	1.7	3.1	1.4
15	1.2	1.8	0.6	30	1.6	1.9	0.3

Summary statistics: $n = 30$, $\sum d_i^2 = 14.81$,

$\sum d_i = 15.3$, $\sum (d_i - \bar{d})^2 = 15.61 - \dfrac{(15.3)^2}{30} = 7.01$,

$\bar{d} = 0.51$, $s_d^2 = 0.242$,

$s_d = 0.492$.

source: Records of a second grade class.

However, in general, the direction of subtraction makes *no difference whatsoever* except for the sign on the individual differences d_i and the sign on the mean difference. It is only important that the direction of subtraction is the same for all observations.

The sample mean difference is a point estimate of the population mean difference μ_d. For the 30 second-grade pupils, this is $\bar{d} = 0.51$, and the standard deviation of the difference scores is $s_d = 0.492$. A 95% confidence interval for μ_d is

$$\left(\bar{d} - t_{29}(0.25)\frac{s_d}{\sqrt{n}}, \quad \bar{d} + t_{29}(.025)\frac{s_d}{\sqrt{n}} \right)$$

From the table of the t-distribution with 29 degrees of freedom, $t_{29}(0.025) = 2.045$. The amount added and subtracted to obtain the interval is

$$t_{29}(.025)\frac{s_d}{\sqrt{n}} = 2.045 \times \frac{0.492}{\sqrt{30}} = 2.45 \times 0.090 = 0.22,$$

so that the confidence interval is

$$(0.51 - 0.22, \quad 0.51 + 0.22), \quad \text{or} \quad (0.29, \ 0.73).$$

In the population the mean improvement in reading is estimated to between 0.29 and 0.73 units during the second grade year.

It should be noted that the real objective in this procedure is to compare the "before" and "after" means, $\bar{x}$ and $\bar{y}$. The method of computing difference scores is a convenient way to do this because *the mean of the differences is equal to the difference of the means*; that is, $\bar{d} = \bar{y} - \bar{x}$. Similarly, the population mean difference μ_d is the difference of the means, $\mu_y - \mu_x$. It is obvious that this procedure is only possible when there are two comparable measurements taken from each observational unit. Thus if one group of youngsters was tested at the beginning of second grade and a different group at the end of second grade, difference scores could not be obtained for any individual, and another approach would be necessary to compare the before and after means. This is discussed in Chapter 12.

Matched Samples

The method of "matched samples" can be used to make a comparison between two treatments, such as two ways of teaching statistics, the effectiveness of two drugs, or for comparing the mean outcome between two members of a "dyad" (a matched pair of observations). In this approach, two measurements are taken, one from each member of the matched pair, and these are compared by the method of differences described above.

As an example, suppose two brands of food for puppies, Brand X and Brand Y, are to be compared. It is known that puppies within a litter are more similar than puppies from different litters. To avoid confusing differences between litters with differences between brands, an investigation is based on feeding Brand X to one puppy in a litter and Brand Y to another puppy from the same litter. Thus, from each of n litters two puppies are chosen randomly; one of each pair is assigned at random to be fed Brand X and the other Brand Y. The puppies are weighed at the end of ten weeks. Let

y_i = weight of the puppy from the ith litter fed Brand Y,

x_i = weight of the puppy from the ith litter fed Brand X,

$d_i = y_i - x_i$ = difference between weights of the two puppies
from the ith litter,

for $i = 1, \ldots, n$. Then $d_1, \ldots, d_n$ may be considered as a sample of n from a (hypothetical) population of differences of the effects of Brand Y and Brand X. The analysis described in the preceding section can be used to estimate μ_d, the population mean difference, and to give a confidence interval for μ_d.

The weights of the n puppies fed Brand X can be considered as a sample from a general population of puppies being fed Brand X; call the population mean μ_x. Similarly, the weights of the other n puppies constitute a sample from a general population, with mean μ_y, being fed Brand Y.

The pair of weights $(x_1, y_1), \ldots, (x_n, y_n)$ are considered as a sample from a population of pairs. The population mean of differences μ_d is the difference of the respective means $\mu_y - \mu_x$. Such a sample of pairs is often called a pair of "matched samples."

In this example taking a pair of puppies from a single litter "matches" the puppies in each pair. In a psychological study the investigator might identify pairs of individuals who have the same ability with regard to completing a particular task. One person from the pair can be assigned to attempt the task under highly speeded conditions and the other asked to complete the task at his or her own rate; the latter is considered as a "control." The differences between the members of these pairs can reveal the effect of speededness on any important outcome such as the quality of work performed, the person's memory for details of the task, or the stress experienced while the task was being performed as indicated by, say, heart rate or skin conductance.

The same procedure for studying the difference μ_d is useful when pairs of observations are matched by circumstances beyond the investigator's control. For example, suppose that married couples entering an auto agency are exposed to a high-pressure "hard-sell" approach. As they leave the agency, the two individuals are asked separately to judge the quality of the salesperson's presentation. To study the difference between husbands' and wives' reactions to the sales approach, a difference score may be obtained for each couple and examined as described in the preceding section.

10.7 Importance of Size of Population Relative to Sample Size*

In many instances the population of interest consists of a finite number of individuals or objects, such as all the light bulbs produced in a certain month, the blood cells in 0.10 milliliters of blood, the counties in the United States, or members of the national labor force. As was seen in Section 7.8, after one unit has been drawn from a finite population the remaining population is different from the original one because of the omission of that unit. After the second unit is drawn, the remaining population is different from the original because of omission of the two units. If the size of the original population is large relative to the size of the sample, this effect is not very large. For instance, consider a jar of 10,000 beads of which 5,000 are white and 5,000 are black. The original proportion of white beads is 1/2, which is the probability of drawing a white bead at random from the original population. After drawing a few beads this probability has not changed much. After 20 beads have been drawn, the probability that the 21st bead is white is at least $4980/9980 = 0.4991$ (which is the case if the first 20 beads drawn are white) and at most $5000/9980 = 0.5009$ (which is the case if the first 20 beads drawn are black).

If more white beads than black have been drawn, the probability of a white bead is less than 1/2; and if more black beads have been drawn, the probability is greater than 1/2; the next bead drawn is more likely to bring the ratio of white beads in the sample back towards 1/2 than away from it.

If the sampling is done with replacement, there is greater variability than when the sampling is done without replacement. In particular, the variability of the sampling distribution of the *sample mean* is greater when sampling with replacement than when sampling without replacement.

This fact manifests itself mathematically as follows. The formula given in Chapter 9 for the variance of the sampling distribution of the mean is

$$\sigma_{\bar{x}}^2 = \frac{\sigma^2}{n},$$

where σ^2 is the variance of the parent population. This formula is exactly true for sampling *with* replacement or for sampling from an infinite population. For sampling *without* replacement from a population of size N the variance of the sampling distribution of the mean is

$$\sigma_{\bar{x}}^2 = \frac{N-n}{N-1} \frac{\sigma^2}{n}.$$

The standard deviation of the sampling distribution of the mean is thus

$$\sigma_{\bar{x}} = \sqrt{\frac{N-n}{N-1}} \frac{\sigma}{\sqrt{n}}.$$

The factors $(N-n)/(N-1)$ and $\sqrt{(N-n)/(N-1)}$, which appear here, are called *finite population correction factors*. The values of the latter factor for various n and N are given in Table 10.3. Multiplying this factor by $\sigma/\sqrt{n}$ gives $\sigma_{\bar{x}}$. (The row labelled ∞ corresponds to sampling with replacement.) It will be seen that the factor is about 1 for all values of N from 10,000 to 1,000,000 except for $n = 3,600$ and $N = 10,000$, where the sample size is a relatively large fraction of the population size, that is, $n/N = 0.36$.

TABLE 10.3

Values of $\sqrt{(N-n)/(N-1)}$ for various N and n.

	Sample size n		
Population size N	*100*	*900*	*3600*
10,000	0.995	0.955	0.80
100,000	0.9995	0.9955	0.982
1,000,000	0.99995	0.99955	0.9982
∞	1.00000	1.00000	1.00000

For samples as large as $n = 100$ or larger, the sampling distribution of the sample mean is nearly normal. Figures 10.8 and 10.9 show the sampling distributions of sample means from a parent population with mean 150 and standard deviation 20 and for population sizes of $N = 10,000$ and $N = 1,000,000$. When the sample size is 900, as in Figure 10.8, there is little difference between the two sampling distributions. When the sample size is 3600 (four times as large), the sampling distribution for $N = 10,000$ is more concentrated than for $N = 1,000,000$. These examples emphasize that the variability depends primarily on the sample size and only slightly on what proportion of the population is sampled *unless* that proportion is fairly large.

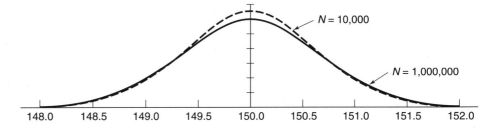

Figure 10.8   *Sampling distributions of the mean of samples of size n = 900 when μ = 150 and σ = 20.*

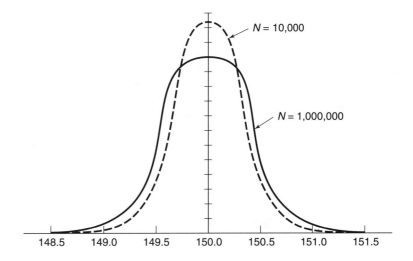

Figure 10.9   *Sampling distributions of the mean of samples of size n = 3600 when μ = 150 and σ = 20.*

The means of all the distributions in Figures 10.8 and 10.9 are the same (150.0). However, the size of $\sigma_{\bar{x}}$ is important because it affects the width of confidence intervals (as well as the power of tests of hypotheses, explained in Chapter 11). The values of $\sigma_{\bar{x}}$ corresponding to the four curves in Figures 10.8 and 10.9 are listed in Table 10.4.

Table 10.5 gives mathematical expressions for the variances and standard deviations of sums and means, for sampling with replacement and sampling without replacement.

If we draw a random sample without replacement from a large population the result is almost the same sampling *with* replacement. This is because, even if we replaced the individuals drawn, there is only a very slight chance that the same individual would be included in the sample more than once since the population is large. Consequently,

TABLE 10.4

Standard Deviation of the Sampling Distribution of the Mean of a Random Sample from a Parent Population with $\sigma = 20$ when $n = 900$, and $N = 10,000$, 1,000,000, and Infinity.

n	N	$\sigma_{\bar{x}}$
900	∞	0.667
900	1,000,000	0.666
900	10,000	0.637
3,600	∞	0.333
3,600	1,000,000	0.333
3,600	10,000	0.273

TABLE 10.5

Means, Variances, and Standard Deviation of Sums and Means

		Mean	Variance	Standard deviation
Sampling distribution of sample *sums* with replacement	In general	$n\mu$	$n\sigma^2$	$\sigma\sqrt{n}$
	For Bernoulli variables	np	npq	$\sqrt{npq}$
Sampling distribution of sample *sums* without replacement	In general	$n\mu$	$\dfrac{N-n}{N-1}n\sigma^2$	$\sigma\sqrt{n\dfrac{N-n}{N-1}}$
	For Bernoulli variables	np	$\dfrac{N-n}{N-1}npq$	$\sigma\sqrt{n\dfrac{N-n}{N-1}npq}$
Sampling distribution of sample *means* with replacement	In general	μ	$\dfrac{\sigma^2}{n}$	$\dfrac{\sigma}{\sqrt{n}}$
	For Bernoulli variables	p	$\dfrac{pq}{n}$	$\sqrt{\dfrac{pq}{n}}$
Sampling distribution of sample *means* without replacement	In general	μ	$\dfrac{N-n}{N-1}\dfrac{\sigma^2}{n}$	$\dfrac{\sigma}{\sqrt{n}}\sqrt{\dfrac{N-n}{N-1}}$
	For Bernoulli variables	p	$\dfrac{N-n}{N-1}\dfrac{pq}{n}$	$\sqrt{\dfrac{N-n}{N-1}\dfrac{pq}{n}}$

if the sample size is small relative to the population size, sampling without replacement will give about the same results as sampling with replacement, and we need make no adjustment.

As a rule of thumb, we can say that n/N is less than $1/10$, the finite population correction may be ignored. When the correction is ignored, the variability of the sample mean will be overestimated, and the resulting confidence interval will be conservative in the sense that it is widened and the confidence level is underestimated. When the sample is more than about $1/10$ of the population being studied the correction should be employed, as in the following example.

EXAMPLE

The officials of a town of population 5000 wish to conduct a survey to determine whether a majority of its citizens favor a certain proposition. Of a random sample of 1200 persons, 960 are in favor of the proposition; thus $\hat{p} = 0.80$. A 95% confidence interval is given by

$$\left(0.80 - 1.96\sqrt{\frac{N-n}{N-1}\frac{\hat{p}\hat{q}}{n}}, \quad 0.80 + 1.96\sqrt{\frac{N-n}{N-1}\frac{\hat{p}\hat{q}}{n}}\right).$$

The estimated standard deviation of the sample mean is

$$\sqrt{\frac{N-n}{N-1}\frac{\hat{p}\hat{q}}{n}} = \sqrt{\frac{5000-1200}{5000-1} \times \frac{0.80 \times 0.20}{1200}} = 0.01006.$$

(The finite population correction is 0.8718.) Since $1.96 \times 0.01006 = 0.020$, the interval is (0.780, 0.820). Thus we would be confident in saying that somewhere between 78% and 82% of the people are in favor of the proposition.

Summary

An important aspect of statistical inference is estimating population values (parameters) from samples of data. Two kinds of estimates are useful for this purpose, a *point estimate* or single "best guess" and an *interval estimate* that gives a range of plausible values for the parameter.

An estimate of a parameter is *unbiased* if the mean of its sampling distribution is equal to that parameter, no matter what the value of the parameter is. The sample mean is an unbiased estimate of the population mean. The sample variance is an unbiased estimate of the population variance.

A *confidence interval* with a specified *level of confidence* is an interval produced by a *procedure* that gives intervals that include the true parameter value with probability equal to the confidence level. The term "probability" applies to the procedure in general; "confidence" is used for a particular interval computed from a sample.

Student's t-distribution takes into account the extra variability introduced by having to estimate the population standard deviation.

A short-cut, approximate 95% interval for a proportion is obtained by adding to and subtracting from the sample proportion the quantity $1/\sqrt{n}$, where n is the sample size. A short-cut, approximate 95% interval for any parameter, the estimate of which is approximately normally distributed, is obtained by adding to and subtracting from the estimate twice its standard error, or estimated standard error.

The sample median is an estimate of the population median and is preferable to the mean for ordinal variables and for distributions that are highly skewed.

The sample size is much more important than the population size in determining the precision of the sample mean. This is how polls based on thousands of individuals can be relevant in predicting the behavior of a country of millions of people. When a population is finite, the *finite population correction* should be used when computing measures of variability if the sample contains about 1/10 of the population or more.

Appendix 10A   The Continuity Adjustment

The number of heads in n tosses of a coin can be any of the values $0, 1, \ldots, n$. In a histogram of the binomial distribution, the probability of exactly r heads is represented by the area of the rectangle whose base extends from $r - 1/2$ to $r + 1/2$. This probability is approximated by the area under the appropriate normal curve from $r - 1/2$ to $r + 1/2$. The adding and subtracting of $1/2$ is called the continuity adjustment, since we need to make this adjustment when approximating a discrete distribution by a continuous one.

As shown in Appendix 9A, if a and b are two integers, then

$$\Pr(a < \# \text{ heads} < b)$$

is approximated by standardizing the values $a + 1/2$ and $b - 1/2$ and using probabilities from the standard normal distribution. The approximation is

$$\Pr\left(z \le \frac{b - \frac{1}{2} - np}{\sqrt{npq}}\right) - \Pr\left(z \le \frac{a + \frac{1}{2} - np}{\sqrt{npq}}\right).$$

To make this probability approximately .95 we write

$$\frac{b - \frac{1}{2} - np}{\sqrt{npq}} = 1.96 \quad \text{and} \quad \frac{a + \frac{1}{2} - np}{\sqrt{npq}} = -1.96.$$

This gives

$$b = np + \frac{1}{2} + 1.96\sqrt{npq} \quad \text{and} \quad a = np - \frac{1}{2} - 1.96\sqrt{npq}.$$

Thus

$$.95 = \Pr\left(np - \frac{1}{2} - 1.96\sqrt{npq} < \# \text{ heads} < np + \frac{1}{2} + 1.96\sqrt{npq}\right),$$

or dividing through by n, since $\hat{p} = \# \text{ heads}/n$,

$$.95 = \Pr\left(p - \frac{1}{2n} - 1.96\frac{\sqrt{pq}}{\sqrt{n}} < \hat{p} < p + \frac{1}{2n} + 1.96\frac{\sqrt{pq}}{\sqrt{n}}\right),$$

which is equivalent to

$$.95 = \Pr\left[-\left(1.96\frac{\sqrt{pq}}{\sqrt{n}} + \frac{1}{2n}\right) < \hat{p} - p < \left(1.96\frac{\sqrt{pq}}{\sqrt{n}} + \frac{1}{2n}\right)\right].$$

Thus, with probability .95, $\hat{p}$ is within a distance

$$1.96\frac{\sqrt{pq}}{\sqrt{n}} + \frac{1}{2n}$$

of p; that is, p is within that distance of $\hat{p}$ with probability .95:

$$.95 = \Pr\left(\hat{p} - \frac{1}{2n} - 1.96\frac{\sqrt{pq}}{\sqrt{n}} < p < \hat{p} + \frac{1}{2n} + 1.96\frac{\sqrt{pq}}{\sqrt{n}}\right).$$

When we substitute $\hat{p}\hat{q}$ for pq we obtain an approximate 95% confidence interval of

$$\left(\hat{p} - \frac{1}{2n} - 1.96\frac{\sqrt{\hat{p}\hat{q}}}{\sqrt{n}}, \quad \hat{p} + \frac{1}{2n} + 1.96\frac{\sqrt{\hat{p}\hat{q}}}{\sqrt{n}}\right).$$

The effect of the continuity correction is to *increase* the width of the confidence interval.

Exercises

10.1 A population distribution of incomes has a standard deviation of $1,000. If the mean income for a random sample of 100 persons is $10,300, estimate the population mean by a point estimate and by a 95% confidence interval.

10.2   (continuation) Repeat Exercise 10.1, but make a 99% confidence interval.

10.3   (continuation) Repeat Exercise 10.1, but make a 90% confidence interval.

10.4   A population distribution of weights has a standard deviation of 20 pounds. If the mean weight for a random sample of 400 persons is 161 pounds, estimate the population mean by a 95% confidence interval.

10.5   (continuation) Repeat Exercise 10.4, but make a 99% interval.

10.6   (continuation) Repeat Exercise 10.4, but make a 90% interval.

10.7   The mean diastolic blood pressure for a random sample of 25 people was 90 millimeters of mercury. If the standard deviation of individual blood pressure readings is known to be 10 millimeters of mercury, estimate the population mean and construct a 95% confidence interval.

10.8   Suppose it is known that the standard deviation of the lifetimes of electric light bulbs is 480 hours, and we obtain a mean lifetime of 493 hours for a sample of 64 bulbs.

(a) Give a 95% confidence interval for the true mean lifetime.

(b) The electric company claims the true mean lifetime is 500 hours. Is their claim consistent with the data? [Hint: Does the confidence interval include the value of 500 hours?]

10.9   Use the table of Student's t-distribution (Appendix III) to find

(a) $t_{10}(.01)$,

(b) $t_{15}(.01)$,

(c) $t_{60}(.01)$,

(d) $t_{25}(.05)$,

(e) $t_{25}(.025)$,

(f) $t_{25}(.005)$.

10.10   Use the table of student's t-distribution (Appendix III) to find the probability of each of the following

(a) For t_8, what is $\Pr(t < 1.860)$?

(b) For t_{120}, what is $\Pr(t > 1.289)$?

(c) For t_{15}, what is $\Pr(t > -2.602)$?

(d) For t_{15}, what is $\Pr(-1.753 < t < 1.753)$?

(e) For t_{30}, what is $\Pr(-2.457 < t < 2.457)$?

(f) For t_2, what is $\Pr(2.920 < t < 6.964)$?

10.11 For the t-distribution with 60 degrees of freedom what value of t has 90% of the distribution above it? What value of z from the standard normal distribution has 90% of the distribution above it?

10.12 The average cholesterol level for a random sample of 25 people of age 30 is $\bar{x} = 214$ units (milligrams of cholesterol per 100 milliliters of blood). The sample variance is $s^2 = 456$.

(a) Give an 80% confidence interval for the population mean cholesterol level.

(b) What property of the parent distribution will justify using this procedure for the confidence interval?

10.13 Make a 90% confidence interval for the population mean years of teaching experience, using the data of Table 2.14 as a sample. The nationwide average is known to be 14.2 years. Are the data for teachers in rural schools in Tennessee consistent with this figure?

10.14 Use the sample mean blood pressure of Exercise 10.7 to construct a 95% confidence interval for the population mean, but assume that the standard deviation (10) was *obtained from the sample of 25 people*. How does this confidence interval compare to the one computed in Exercise 10.7?

10.15 Make a 90% confidence interval for the population mean height of 11-year-old boys, treating the 1293 boys represented in Table 2.12 as a sample.

10.16 (continuation) These data were published in 1877. Do you think 11-year-old boys today have the same height distribution? Document your opinion, if possible, by reference to a book or article or to an expert such as a pediatrician or teacher of physical education.

10.17 Use the data on percentage of minority students in Table 2.16 as a random sample to obtain an unbiased estimate of

(a) the mean percent minority enrolled in school districts with over 10,000 students;

(b) the variance of the percent minority enrolled in school districts with over 10,000 students;

(c) the proportion of school districts, out of those with over 10,000 students, that have one-quarter or more minority students.

10.18 In a study of dietary intake of 4- and 5-year-old children, Iannotti *et al.* (1994) computed the daily intake of a number of nutrients for each child. The means and standard deviations for several nutrients for a sample of 17 youngsters are given in Table 10.6.

(a) Use the data in Table 10.6 to construct a 90% confidence interval for the mean of the PFAT measure.

(continued...)

TABLE 10.6

Mean and Standard Deviation of Daily Nutrient Intakes for 17 Children

Nutrient	Mean	Standard deviation
Energy (calories)	1095.2	554.0
Sodium (milligrams)	1307.9	639.8
Percent of Energy Attributable to Fatty Acids (PFAT)	34.0	11.1
Percent of Energy Attributable to Saturated Fatty Acids (PSAT)	12.2	5.4

SOURCE: Adapted with permission of publisher from Iannotti *et al.* (1994), Comparison of dietary intake methods with young children. *Psychological Reports.* © Psychological Reports 1994, p. 886.

(b) Use the data in Table 10.6 to construct a 90% confidence interval for the mean of the PSAT measure.

(c) From the PFAT interval, is it plausible that, on average, 4- and 5-year-old children derive as much as 1/3 of their energy from fatty acids? As much as 40%? As much as 1/2?

(d) From the PSAT interval, is it plausible that, on average, 4- and 5-year-old children derive as little as 2% of their energy from saturated fatty acids? As little as 5%? As little as 10%?

10.19   (continuation) The standard deviation of caloric intake in Table 10.6 is approximately equal to one-half of the mean. The two-standard deviation rule of thumb would suggest that some individual children have daily intakes close to 0. Since this is certainly not the case, what do these values tell us about the distribution of this variable?

10.20   Three dozen people were invited to a party and asked to "RSVP." Table 10.7 is a cross classification according to RSVP and attendance.

TABLE 10.7

Cross-Tabulation of 36 Persons, by RSVP and Attendance

		Attended		
		No	Yes	Total
Result of RSVP	No response	7	5	12
	No	4	1	5
	Yes	1	18	19
	Total	12	24	36

Suppose you invite to a party 55 persons whose behavior could be expected to be similar to these three dozen, and 38 say they will attend,

5 say they will not, and the other 12 do not respond. How would you predict the attendance at your party?

10.21 In a random sample of 300 households in a large metropolitan community, it was found that a language other than English was spoken in 33 homes. Construct a 95% confidence interval for the proportion of households in the entire community in which a language other than English is spoken.

10.22 (continuation) Recompute the confidence interval of Exercise 10.21 supposing that the entire community has only 3900 households, using the finite population correction. How does this interval compare with that for a very large population as assumed in Exercise 10.21?

10.23 A random sample of 800 phone calls to an airlines reservation office results in 280 confirmed reservations. Obtain a point estimate and a 90% confidence interval for the proportion of all phone calls that result in reservations. How should the data on the 800 calls be obtained to assure that the estimate is unbiased?

10.24 The University Bookstore decides how many beginning chemistry books to order by looking at the class enrollment figures and ordering enough for a percentage of those students. Ordering too few books is a serious inconvenience to the students and ordering too many results in an unnecessary surplus. Last year's "typical" class of 340 students purchased 289 texts. The Bookstore wants to be 98% confident that it has not purchased too few or too many books for next year's class.

Use the proportion of students who purchased books last year as the point estimate of the proportion for next year's class. Construct a 98% confidence interval for the true proportion. From this interval, what is the minimum percentage of books that should be ordered to assure that there are enough, and what is the maximum percentage that should be ordered to avoid an unnecessary surplus?

10.25 On July 16, 1989, the *Mercury News* of San Jose (CA) reported the results of an opinion poll, concluding that 67% of Californians support a woman's right to obtain an abortion during the first trimester of pregnancy.[7] The article did not provide the sample size. Taking 67% as the proportion 0.67 and assuming that the sample was drawn at random, construct and compare two 90% confidence intervals for the population proportion as follows:

(a) Obtain the first confidence interval by assuming that the number of persons polled was 100.

(b) Obtain the second interval by assuming that the number of persons polled was 1,000.

(continued . . .)

---

[7]Copyright 1989 *San Jose Mercury News*. All rights reserved. Reproduced with permission.

(c) Compare the width of these two intervals.

(d) From these intervals, is it reasonable to conclude that *less than one-half* of the population supports a woman's right to an abortion? *More than three-quarters* of the population?

10.26   (continuation) Repeat the steps in Exercise 10.25 using the 99% level of confidence.

10.27   Suppose that in 1991 and 1992, a pollster reported the following data based on interviews with registered voters. Each respondent was asked "If Bill Clinton were the Democratic candidate for president and George Bush were the Republican candidate, which one would you like to see win?" The percentages are tabulated below. (The respondents who were undecided were not included in the calculation of the percentages.) A write-up of this survey would also include a statement that public opinion samplings of this size are subject to a margin of deviation of approximately 4% when the findings are at or very near the 50% mark.

	October 1991	December 1991	January 1992
Bush	52.3%	51.1%	50.2%
Clinton	47.7%	48.9%	49.8%

Hypothetical data.

(a) Compute 95% and 99% confidence intervals for the true proportion in favor of Bush at each of the three time periods. You may take $n = 1000$.

(b) Discuss what further information you would need in order to make probability statements about the differences between poll results obtained at different times. Are the proportions at the three times independent?

(c) What do you think was meant by a "margin of deviation of approximately 4% . . . ?"

10.28   Twins are of two types: identical and fraternal. We want to estimate the number of identical twins in the population.

Fraternal twins are as often of different sex as of the same sex. Identical twins are, of course, of the same sex. We know the total number of twins in the population is 1,000,000. We count the number of twins in pairs of different sex and find 300,000. Give an estimate of the number of identical twins in the population. Show work to indicate your reasoning.

10.29   The lifetimes of a random sample of 9 light bulbs were 1066, 1776, 1492, 753, 70, 353, 1984, 1945, and 1914 hours. Estimate the median lifetime in the population. Compute both a point estimate and an interval estimate with confidence approximately 0.90.

10.30 Use the data on percentage of minority students in Table 2.16 as a random sample to obtain a point and an interval estimate of the median percent minority enrolled in school districts with over 10,000 students. Select an appropriate confidence coefficient for your interval.

10.31 (continuation) Obtain a point and an interval estimate of the mean percent minority with approximately the same confidence level as in the preceding item. How do the point and interval estimates compare with those for the median? [Note: you may have computed the mean and variance for Exercise 10.17.]

10.32 A sample of 16 women go on a low-fat diet for 3 weeks (with very little cheating). Their initial and final weights are given in Table 10.8. What is the mean weight loss associated with this diet? Contruct an 80% confidence interval on the true mean weight loss that the promoters of the diet can use in advertising copy.

TABLE 10.8
Mean Before and After Weights of 16 Women (Pounds)

Dieter	Before	After	Dieter	Before	After
RA	106	102	MY	124	121
LF	132	118	NC	135	129
KD	121	116	NL	166	159
HP	154	139	DC	140	132
DJ	126	121	KV	122	123
AG	130	119	RP	148	141
DK	174	168	LB	139	136
LS	111	101	DD	110	107

Hypothetical data.

10.33 Compute a point estimate and a 95% interval estimate of the difference in effectiveness of the two sedatives for the 10 patients listed in Table 3.15.

10.34 In 1990 S. G. Ciancio and M. L. Mather reported a comparison of two electric toothbrushes with manual brushing on a sample of 30 patients with a history of high plaque formation. Each patient's teeth were cleaned professionally after which the individual brushed manually for one week and with an electric toothbrush for an additional week. Plaque formation was recorded after a week of manual brushing and again after a week of electric brushing. The measure recorded is the percentage of tooth surfaces that are plaque-free; higher values are preferable.

One-half of the patients ($n = 15$) used an Interplak electric toothbrush. At the end of the week of manual brushing the mean plaque index was 39; that is, on average 39% of participants' tooth surfaces had no noticeable plaque. At the end of the week of electric brushing the mean plaque score was 47. The standard deviation of *change*[8] in the plaque measure from the end of week 1 to the end of week 2 was approximately 14.8.

(a) Compute a 95% confidence interval for the change in mean plaque scores from the end of week 1 to the end of week 2.

(b) Is it plausible that, on average, brushing with the Interplak brush does not reduce the amount of plaque buildup that occurred during week 1? That is, does the confidence interval include the value 0?

10.35   (continuation) The other half of the patients ($n = 15$) used a Water Pik electric toothbrush. At the end of the week of manual brushing these individuals had an average plaque index of 35. At the end of the second week, after using the electric toothbrush, the average plaque index was 56. The standard deviation of *change*[8] in the plaque index from the end of week 1 to the end of week 2 was approximately 17.6.

(a) Compute a 95% confidence interval for the change in mean plaque scores from the end of week 1 to the end of week 2.

(b) Is it plausible that, on average, brushing with the Water Pik brush does not reduce the amount of plaque buildup that occurred during week 1? That is, does the confidence interval include the value 0?

10.36   In a national study of physicians the standard deviation of hours worked per week was 7 hours. A random sample of 49 physicians was taken from the staff of the Mayo Clinic, and a mean work-week of 50 hours was obtained. Give a 95% confidence interval for the mean work-week of all physicians in the Mayo Clinic. Assume there are 401 physicians altogether, and use a finite population correction.

10.37   According to the report *Academic Libraries: 1990* (Williams, 1992) there are 3,274 academic libraries in operation in the United States. Suppose that the size of the collection is obtained for a random sample of 200 libraries; the mean is found to be 172,400 volumes and the standard deviation is 35,360 volumes. Use a finite population correction to obtain a 90% confidence interval for the average size of the collections of all 3,274 academic libraries. Multiply the point estimate and both ends of the confidence interval by 3,274 to estimate the *total* number of volumes in all academic libraries.

---

[8]This value is not provided in the published report but was estimated from other information in the article.

10.38   (continuation) The actual total number of volumes reported in Williams (1992) is just over 717 million volumes. This information is obtained by asking each library administrator to complete a survey form listing the number of volumes in the school's collection. What sources of error might be introduced by this procedure that become incorporated into the reported total?

10.39   What statistic is being used as an estimate in the following military problem from Thucydides [Warner (1954), p. 172]?

> [The problem for the Athenians was] . . . to force their way over the enemy's surrounding wall . . . . Their methods were as follows: they constructed ladders to reach the top of the enemy's wall, and they did this by calculating the height of the wall from the number of layers of bricks at a point which was facing in their direction and had not been plastered. The layers were counted by a lot of people at the same time, and though some were likely to get the figure wrong, the majority would get it right, especially as they counted the layers frequently and were not so far away from the wall that they could not see it well enough for their purpose. Thus, guessing what the thickness of a single brick was, they calculated how long their ladders would have to be . . . ."

10.40   To estimate the mean length of logs floating down the river, we sit on the bank for 10 minutes and measure the lengths of the logs that pass during those 10 minutes and take the mean of those measurements.

(a) Is the resulting estimate unbiased? Why or why not?

(b) Suppose we choose n logs at random from among those in the river and measure their lengths. Is the resulting estimate unbiased?

11

Answering Questions about Population — Characteristics

Introduction

It is often the purpose of a statistical investigation to answer a yes-or-no question about some characteristic of a population. An election candidate, for example, may employ a pollster to determine whether the proportion of voters intending to vote for him does or does not exceed 1/2. The polio vaccine trial was designed so that medical researchers could decide whether the incidence rate of polio is or is not smaller in a population of persons inoculated with the vaccine than in a population of persons not inoculated with the vaccine. Industrial quality control involves determination as to whether the average strength, lifetime, or concentration of the product in each manufacturing batch does or does not fall within acceptable limits.

Chapter 10 deals with the use of samples to *estimate* population characteristics. (How many? How tall? What is the proportion in the population?) In this chapter we discuss methods for using these estimates to provide yes-or-no answers to questions about population characteristics.

The approach, called *hypothesis testing*, is a second form of statistical inference. (The first is estimation.) Again we will be using samples of data to draw conclusions about the parent population. It is neither practical nor desirable to poll all voters in a large community every few weeks to decide how many intend to vote for a candidate, or to test a vaccine on all individuals before we decide if it is effective. Yet we may need to know the answer to these questions for the entire group. Inferential statistics are useful for forming conclusions based on a sample that is only part of the population, and in some cases a small part.

A sample can, on occasion, lead to an incorrect conclusion about a population. That is, the conclusions drawn from statistical inference are subject to errors due to sampling variability. This chapter shows how to take account of sampling variability and how to evaluate and limit the probabilities of such errors.

The focus of this chapter is on questions concerning means and proportions. The methods introduced here are applied to questions about other parameters in the chapters that follow.

11.1   Testing a Hypothesis About a Mean

It is often desirable to know whether some characteristic of a population is larger than (or, in other cases, smaller than) a specified value. We may want to know, for instance, if the average body temperature of individuals with a particular disease exceeds the average for healthy individuals ($98.2°F$), whether the average reading speed of second-grade children for material of a certain level of difficulty is less than 120 words per minute, or whether the average concentration of carbon monoxide in downtown Chicago is or is not greater than some specified noxious concentration.

Since the value of the population characteristic is unknown, the information provided by a sample drawn from the population is used to answer the question of whether or not the population quantity is larger than the specified value. In the present discussion the sample is assumed to be *random*. We know that when we estimate the value of a parameter there is some sampling error; an estimate hardly ever coincides exactly with the parameter. Similarly, when a sample is used to answer a question about the population, sampling error may sometimes lead us astray and hence must be taken into account.

An Example and Terminology

A question that might be investigated by drawing a sample from a population is whether college students devote more hours to school work than the common 40-hour work week of other adults. To answer this question, suppose that a sample of 25 students was selected by assigning consecutive numbers to all registered students and choosing those 25 students whose numbers correspond to 25 different numbers from a table of random numbers. Each student in the sample was interviewed about current interests and activities. Among other things, the student reported how many hours were spent on studies during the week which corresponded to the week prior to the investigation. We ask the question, "In the population of *all* students at the university, is the average time spent on studies 40 hours per week or is it greater?" We use the information in the sample to arrive at the answer.

It seems reasonable that we should calculate the mean time spent on studies by the 25 students in the sample and compare that with the 40 hour baseline. If the sample mean is much larger than 40 hours, we will be led to believe that the population average this year is greater than 40. However, if the sample mean is less than 40 or only slightly more than 40, we would not conclude that the average study time in the population exceeds 40 hours. The investigator needs a way of determining whether the sample mean is large enough to warrant the conclusion that the population mean is large. Hypothesis testing is a statistical procedure for making such a decision.

The possibility that the population mean is 40 is called the *null hypothesis*, denoted H_0. The possibility that it is greater than 40 is called the *alternative hypothesis*, denoted H_1. If the null hypothesis is true, students invest no more time than workers in other occupations (that is, the additional effect on work hours of being a student is "null"); further investigation may be uninteresting. If the investigator decides the alternative is true, he or she will want to see how the demands of school differ from those of employment and how these demands affect students' attitudes and habits.

For random samples of 25 we know that the mean of the sampling distribution of the sample mean $E(\bar{x})$ is equal to the mean of the parent population, μ. The standard deviation of this sampling distribution (the standard error, $\sigma_{\bar{x}}$) is the standard deviation of the parent population σ divided by 5, the square root of the sample size. If the standard deviation of the parent population is $\sigma = 10$, then the standard error is $\sigma/5 = 10/5 = 2$.

According to the Central Limit Theorem, the sampling distribution of the mean is approximately normal. From the table of the normal distribution, we see that the probability is .05 that the sample mean exceeds the population mean by more than 1.645 standard errors. Suppose we use the rule that if the sample mean exceeds 40 hours by *more* than 1.645 standard errors we shall reject the null hypothesis, otherwise we shall accept the null hypothesis. The decision rule is diagrammed in Figure 11.1.

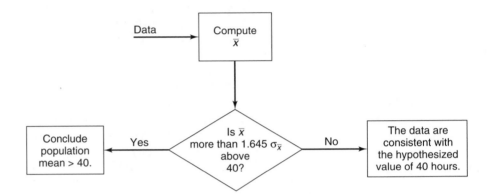

Figure 11.1   *Schematic diagram of hypothesis test concerning the average number of hours studied.*

The value .05 (or 5%) is called the *significance level* of the test. It defines how far above 40 the *sample* mean must be for us to conclude that the *population* mean is also above 40. It is the probability of making one kind of statistical error, namely, concluding that the null hypothesis is false when in reality it is true.[1] We choose the significance level to be small to reduce the risk of wasting time and money in follow up research based on a erroneous conclusion.

The value 1.645 is called the *significance point*. Values of $\bar{x}$ that are more than 1.645 standard errors above 40 are said to be *significantly different* from 40 at the 5% level.

Suppose that the mean for the sample of students is $\bar{x} = 46.2$ hours. This value is 6.2 hours above 40; if the standard error of the mean is 2, then $\bar{x}$ is 3.1 standard errors above 40. Our decision rule would lead us to reject the null hypothesis and conclude that the population mean is greater than 40. That is, we would conclude that students spend more time on the average on their studies than workers spend in a typical work week. We draw this conclusion because a sample mean this large or larger is very unlikely (probability less than .05) to be obtained if the null hypothesis is true. The number of standard errors between the sample mean and the hypothesized value of μ (that is, 3.1) is the value of the *test statistic*.

---

[1]This is discussed further in later sections of this chapter.

We could reduce the probability of an erroneous conclusion by taking a smaller probability as the significance level. If we used .01, then we would reject the null hypothesis if the sample mean is more than 2.33 standard errors above 40. The sample mean 46.2 would lead to reject the null hypothesis at this significance level, too. On the other hand, a sample mean of 44.2 is only 2.1 standard errors above 40. If this were the value of $\bar{x}$, it would lead us to reject the null hypothesis at the 5% level, but not at the 1% level.

Hypothesis testing is useful in many situations, and the probability of an erroneous conclusion must be weighed carefully in each application. In the study of group pressure described in Section 1.1, the null hypothesis was that the probability of making errors in the experimental situation, under group pressure, was the same as when not under pressure. The numbers of errors made by the two sets of subjects are so dramatically different that the null hypothesis must be rejected. The consequence of this was intensive study of various aspects of group pressure; if the experiment had not led to rejection of the null hypothesis, that experimental setup would not have been used in future experiments, and the concept of group pressure would have been studied in a different way.

In the case of testing of polio vaccine, the consequence of rejection of the null hypothesis was mass inoculation of children all over the country. Although the scientific question was whether the vaccine was or was not effective, the practical result was such that the experiment led to a decision whether to use the vaccine or not. Sometimes, as in this case, a test is considered a *decision procedure*; that is, some action will be taken or not taken according to the outcome of the test. The result of the test is not simply to reject or accept the null hypothesis; it is to take some action or not.

Types of Error.   Whenever a conclusion is drawn about a population based on a sample of observations, there is a possibility—no matter how small—that the conclusion is not correct. In the hours-of-study example the significance level is the probability of concluding that students spend more than 40 hours per week on their studies if, in fact, on the average they spend exactly 40 hours per week. By choosing the significance point appropriately, the test can be conducted so as to make this probability small.

There is another possible type of error as well, namely, concluding that students are *not* studying more than 40 hours if they actually *are*. These two types of possible errors—rejecting a null hypothesis that is true and accepting a null hypothesis that is false—are discussed in more detail in Section 11.2. The ways in which they may be controlled by the investigator, that is, kept unlikely, are described as well.

Hypothesis Testing Procedures

Hypotheses

Many research questions are addressed by testing statistical hypotheses about population values. For example, we may know that the average Scholastic Assessment Tests (SAT) score[2] of American students is 500 but suspect that it is higher among students of Asian descent. This question could be answered by administering the test to a random sample of Asian students and testing the null hypothesis that $\mu \leq 500$ against the alternative that $\mu > 500$. Likewise, physicians believe that the average birthweight of infants born to mothers who were chronic smokers while they were pregnant is lower than the nationwide average of full-term infants, 3300 grams. This belief could be tested by weighing a random sample of infants born to mothers who smoked and testing the null hypothesis that $\mu \geq 3300$ against the alternative that $\mu < 3300$.

In the examples above we are asking whether the population mean is *less than* some specific value or not, *or* we are asking whether the mean is *greater than* some specific value or not. In other situations we may want to know whether a population mean is equal to a pre-specified value or is different from it, either by being larger *or* smaller. For example, suppose that a company is producing Rabbit Grahams (cookies) in 8-ounce boxes. Periodically, the quality control supervisor draws a random sample of boxes from the production line and weighs them to see if the equipment is consistently putting the right weight of cookies in the boxes ($\mu = 8$) or is malfunctioning in either direction ($\mu \neq 8$). Of course if the conclusion is "malfunction," the direction of malfunction will need to be considered when deciding which way to adjust the machinery. Procedures for hypothesis testing when the alternative hypothesis is two-directional, as in the cookie example, differ in one important respect from testing one-directional alternatives. They are described separately in a later section of this chapter.

The null hypothesis represents a "statement of no effect;" for example, SAT scores are *not* higher among Asian students (H_0: $\mu \leq 500$) or birthweight is *not* reduced by cigarette smoking (H_0: $\mu \geq 3300$ grams). In contrast the alternative hypothesis represents the statement that there *is* an effect (H_1: $\mu > 500$ or H_1: $\mu < 3300$). We shall represent the specified constant (e.g., 500, 3300 grams, or 8 ounces) by the symbol μ_0. In order for a hypothesis test to be unambiguous, H_0 and H_1 should include all mathematical possibilities. Thus if the form of H_1 is $\mu > \mu_0$, then the form of H_0 is $\mu \leq \mu_0$; if the form of H_1 is $\mu < \mu_0$, then the form of H_0 is $\mu \geq \mu_0$. If the form of H_0 is $\mu = \mu_0$, then the form of H_1 is $\mu \neq \mu_0$.

---

[2]On either the verbal or quantitative scale.

Hypothesis testing is a procedure for using a sample of data to choose between the two descriptions of the population represented by H_0 and H_1. The data are examined to see if they are consistent with the null hypothesis or inconsistent with the null and conform more closely to the alternative hypothesis. The decision is stated in terms of H_0, that is, H_0 is either "accepted" or "rejected."

Steps in Testing Hypotheses

Once the null and alternative hypotheses have been formulated, three steps are needed. In Step 1 the sampling distribution of the appropriate statistic is described on the basis that the null hypothesis is true, and a significance level is selected. This yields the significance point. In Step 2 a sample of observations is drawn and the test statistic is computed. In Step 3 the test statistic is compared to the significance point to decide if the null hypothesis is accepted or rejected.

These steps are described in detail in the following paragraphs and summarized in Figure 11.3. We illustrate using the SAT example. The null and alternative hypotheses are H_0: $\mu \leq 500$ and H_1: $\mu > 500$, respectively; $\mu_0 = 500$ in this example. The research plan is to draw a random sample of 400 Asian students, record each student's SAT score, calculate the sample mean ($\bar{x}$), and decide if it is sufficiently above 500 to conclude that the population mean (μ) is also greater than 500.

Step 1: The sample value used to test a hypothesis about the population mean is the sample mean. The sampling distribution of the mean with its center at 500 answers the question, "what is the distribution of sample means that we obtain *if the null hypothesis is true?*"

According to the Central Limit Theorem the sampling distribution of the mean based on a sample of size 400 is highly normal. The standard error of the sampling distribution is $\sigma_{\bar{x}} = \sigma/\sqrt{n}$. To find this value we need to know the value of σ, the standard deviation of the parent population. The Scholastic Assessment Tests are scaled with $\sigma = 100$, so that the standard error[3] is $100/\sqrt{400} = 5$. A diagram of the normal distribution with mean 500 and standard deviation 5 is given in Figure 11.2.

Values of the mean are identified that are unlikely to be observed if $\mu = 500$ and more likely to be observed if the alternative hypothesis ($\mu > 500$) is correct. This is accomplished by selecting a *significance level,* for example, .05. This defines a portion of the distribution that is far above the center, the 5% of the distribution in the right-hand tail

---

[3]In general there are many situations in which we have access to a reasonable value for σ. We may have the results of prior population studies, the results of a production process monitored over a long period of time, or we may be studying a variable with a pre-set standard deviation such as the SAT. In other situations the standard deviation must be estimated from the sample. This issue is discussed in Section 11.3.

(the shaded area in Figure 11.2). This portion of the $\bar{x}$-axis is called the *region of rejection*; if the mean of the sample falls in this region, we will reject the null hypothesis. The significance level is represented by the symbol α; we say that we are testing our hypothesis at $\alpha = .05$, or at the 5% level of significance.

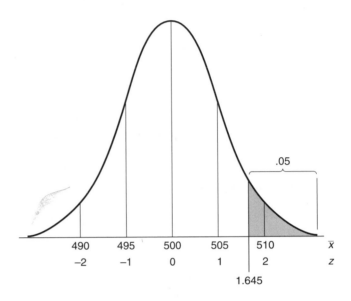

Figure 11.2 *Sampling distribution of the mean with $\mu = 500$, $\sigma_{\bar{x}} = 5$.*

Once a significance level is chosen, it is converted to a *significance point*, which is the corresponding z value from the table of the standard normal distribution (Appendix I). With $\alpha = .05$ the table is entered with probability .45, the area between the center of the curve and the 5% in the right-hand tail. The z value is between 1.64 or 1.65; we shall use 1.645. In words, the significance point is the number of standard errors above 500 that $\bar{x}$ must be to cause us to reject the null hypothesis and conclude that μ is greater than 500. It is represented statistically as $z(\alpha)$, for example, $z(.05) = 1.645$. If the region of rejection consists of the right-hand 1% of the distribution, the significance point is $z(.01) = 2.33$.

The choice of a significance level is obviously an important part of hypothesis testing procedure. In a sense, it "sets the standard" for determining whether a particular sample mean will be considered statistically significant or not. The significance level also determines the chance of being *wrong* if we conclude that H_0 is false when it is in fact true. Criteria for choosing α are presented with the discussion of statistical errors (Section 11.2).

Step 2: Data are collected on a random sample of observations, the sample mean ($\bar{x}$) is computed, and a standardized measure of the distance of $\bar{x}$ from μ_0 is obtained. For the SAT example, we could construct a list of all Asian students attending four-year universities in the United States, or perhaps with the cooperation of the test publisher obtain a list of all students who took the test and indicated that they were of Asian descent. A sample of 400 of these individuals could be chosen and their test scores recorded.

It is important that individuals are chosen from the population of Asian students at random. There is always some sampling error when a population value is estimated, and likewise when using a sample to test a hypothesis about a population. Random sampling will prevent the error from being systematic. For example, it would prevent us from oversampling students who had been in the United States for the longest period of time or those who were attending schools with the highest admission standards.

The sample mean is calculated from the data. Suppose that the mean SAT verbal score for the sample of 400 Asian students is $\bar{x} = 507.5$. The sample mean is converted to a *test statistic*. This is the distance between $\bar{x}$ and μ_0 expressed in standard errors. Since the standard error of the mean of 400 observations is $\sigma_{\bar{x}} = 5$, the test statistic is $(507.5 - 500)/5 = 1.5$. In symbols, the test statistic is

$$z = \frac{\bar{x} - \mu_0}{\sigma_{\bar{x}}}.$$

In the example the sample mean is higher than 500 by $1\frac{1}{2}$ standard errors.

Step 3: To decide whether to reject or accept the null hypothesis, the test statistic is compared with the significance point. H_0 is rejected if the test statistic lies in the region of rejection, that is, beyond the significance point; it is accepted otherwise.

In the SAT example, the test statistic is $z = 1.5$ and the significance point is 1.645. The test statistic does *not* lie in the region of rejection and H_0 is accepted. We have not obtained statistical significance at the 5% level (nor at the 1% level which has an even higher significance point). If some educational practice were based on the assumption of higher SAT scores for Asian students (e.g., different admissions policies) our finding would suggest that they are not warranted.

The distinction of sample and population is an essential concept in hypothesis testing. Sample means of 507.5 and 520, say, are both above 500. They both give a correct conclusion that *these 400 students in the sample* have an average SAT score above 500. But 507.5 is not large enough to convince us that the mean of the entire population is above 500. It lies near the center of the sampling distribution, that is, it is among typical values that might be obtained if a sample of 400 observations is drawn from a population in which $\mu = 500$. In contrast, an $\bar{x}$ of 520 is very unlikely to be obtained if the population mean is 500, and would cause us to reject the null hypothesis.

In the hours-of-study example discussed earlier, the test statistic (3.1) is larger than both the 5% significance value (1.645) and the 1% value (2.33). The null hypothesis is rejected. The sample mean is far enough above 40 to convince us that the mean study time of the entire population of students is greater than 40 hours per week.

Both the SAT and study-time examples give the same conclusion for $\alpha = .05$ as for $\alpha = .01$. It should be obvious, however, that this does not always have to be the case. An appropriate α level should be chosen before collecting the data in order for the conclusion to be unambiguous.

Summary and Another Example

The steps in testing a statistical hypothesis about a mean are summarized in Figure 11.3. Step 1 is based on statistical theory, in particular the concept of the sampling distribution. Step 2 involves the "field-work," actually collecting data and computing summary statistics. Step 3 is always a comparison of a test statistic with a significance point to reach a decision about H_0.

STEP 1: *Describe the sampling distribution of the mean with its center at μ_0. Choose the significance level α and convert it to a significance point, $z(\alpha)$, using the table of the standard normal distribution.*

STEP 2: *Draw a sample of observations, compute the sample mean, and convert it to a test statistic, $z = (\bar{x} - \mu_0)/\sigma_{\bar{x}}$.*

STEP 3: *Compare the test statistic with the significance point to see if the test statistic lies in the region of rejection. If so, H_0 is rejected. If not, H_0 is accepted.*

Figure 11.3   *Steps in testing a hypothesis about μ.*

These steps are illustrated in an example that uses the left tail of the normal distribution instead of the right.

EXAMPLE

The question of whether habitual smoking during pregnancy reduces birthweight can be answered by testing H_0: $\mu \geq 3300$ grams against the alternative H_1: $\mu < 3300$ grams. If we plan to sample 100 infants and the standard deviation of birthweights is known to be 516 grams, then the sampling distribution of the mean will be very nearly normal with standard deviation $\sigma_{\bar{x}} = 516/\sqrt{100} = 51.6$. A diagram of this distribution with center at 3300 grams is shown in Figure 11.4. We shall test the hypothesis at the 10% level of significance, that is, $\alpha = .10$.

Unlike the previous examples, the region of rejection is in the left tail of the normal distribution. (See Figure 11.4.) Only a sample mean far *below* 3300 grams will convince us that the population mean is also below 3300; any $\bar{x}$ below 3300 will give a negative test statistic. The value from the standard normal distribution that separates the most extreme 10% of the distribution on the right from the rest of the curve is $z(.10) = 1.28$. The significance point for a test using the left side of the distribution is its negative, $-z(.10) = -1.28$.

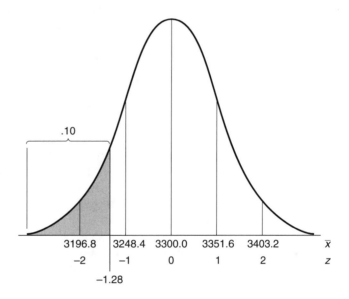

Figure 11.4 *Sampling distribution of the mean with $\mu = 3300$, $\sigma_{\bar{x}} = 51.6$.*

Suppose that a random sample of 100 births drawn from hospitals in several locations yields $\bar{x} = 3207$ grams. The test statistic is $z = (3207 - 3300)/51.6 = -1.80$, which lies well beyond the significance point in the region of rejection. The null hypothesis is rejected in favor of the conclusion that the mean birthweight of infants born to mothers who smoke is less than 3300 grams in the population, that is, $\mu < 3300$.

Had $\bar{x}$ been computed and found to be, say, 3248 grams, the test statistic would have been $z = -1.01$. This does not lie in the region of rejection and H_0 is accepted. We would conclude that mean birthweight is not reduced in the population even though the sample mean is somewhat below 3300.

Deciding Whether a Population Mean Differs from a Given Value

In many situations we want to know whether some characteristic of a population is equal to a given value or is different from it—either by being larger *or* smaller. We may want to know whether the mean weight of cookie packages is 8 ounces or know whether the packaging machinery is malfunctioning in either direction. The null hypothesis is that the mean package weight is 8 ounces ($\mu = 8$) and the alternative is that the mean weight is *different* from 8 ($\mu \neq 8$). Or we may ask whether the mean study time among college students is 40 hours or is *different* from 40 hours in either direction. Here the hypotheses are H_0: $\mu = 40$ and H_1: $\mu \neq 40$.

The procedure is to reject the null hypothesis if the observed sample mean is *either* very large or very small. To accomplish this, one change is made to the procedures discussed in the preceding section: the region of rejection is divided equally between the two tails of the normal distribution. For example, 5% of the standard normal distribution lies outside the interval from -1.96 and 1.96. (See Figure 11.5 and Appendix I.) The decision rule is to reject H_0 if the sample mean is more than 1.96 standard errors above μ_0 or more than 1.96 standard errors below μ_0, that is, if the absolute value of the test statistic is greater than 1.96. The test is called a *two-tailed* test.[4] The significance level α is the sum of the probabilities in the two tails of the normal distribution. The significance points are the values that separate the most extreme proportion $\alpha/2$ of the area in the standard normal distribution in each tail from the middle proportion $1 - \alpha$, that is, $-z(\alpha/2)$ and $z(\alpha/2)$.

The 1% significance level for a two-tailed test on μ requires the value from the standard normal distribution that has .005 of the distribution above it (in the right tail) and .005 of the distribution below its negative (in the left tail). From Appendix I this value is $z(.005) = 2.58$. The null hypothesis would be rejected if the test statistic is less than -2.58 or greater than 2.58.

We shall illustrate the procedure by testing the mean package weight of Rabbit Graham cookies for which the hypotheses are H_0: $\mu = 8$ and H_1: $\mu \neq 8$. Suppose that we plan to weigh a random sample of 100 packages from the line and test the hypothesis at the 5% level of significance. If the standard deviation of package weights is $\sigma = 0.2$ oz., then the sampling distribution of the mean is nearly normal with standard error $\sigma_{\bar{x}} = 0.2/\sqrt{100} = 0.02$ oz. A diagram of this distribution

---

[4]In contrast, the tests described in the preceding sections of this book are called *one-tailed* tests.

with its center at 8 ounces is presented in Figure 11.5. The region of rejection is indicated by the shaded areas consisting of $2\frac{1}{2}\%$ of the area in each tail. The significance points are $\pm z(.025)$, or -1.96 and 1.96.

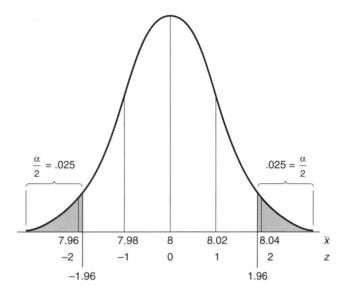

Figure 11.5   *Sampling distribution of the mean with $\mu = 8$, $\sigma_{\bar{x}} = 0.02$.*

If the 100 randomly chosen packages have an average weight of $\bar{x} = 8.05$ ounces, then the test statistic is $z = (8.05 - 8)/0.02 = 2.5$. This is greater than the upper-tail significance point and lies in the region of rejection. This leads us to conclude that the population mean is not 8. H_0 is rejected and the machinery can be adjusted to place fewer cookies in the boxes.

If the sample mean were $\bar{x} = 7.96$ ounces, the test statistic would be $z = -2.0$. This is less than the lower-tail significance point and again H_0 is rejected. The machinery is not working properly and should be adjusted to place *more* cookies in the boxes. In contrast sample means of 7.97 or 8.01 would give test statistics between -1.96 and 1.96 and the null hypothesis ($\mu = 8$) would be accepted.

EXAMPLE

As an additional example, suppose that quarter-pound burger patties are sold at a nationwide chain of hamburger stands. If the average weight of the patties is greater than 4 ounces, the company is suffering from reduced profits. If the average weight is less than 4 ounces, then customers are not receiving fair value for their money.[5] Periodically, the

---

[5]Note that this is the weight "before cooking." No promises are given regarding the after-cooking weight (or patty contents).

company weighs a random sample of 64 patties from the production line and tests H_0: $\mu = 4$ against H_1: $\mu \neq 4$. If they wish to test the hypothesis at the 90% level, then the significance points for the test are $\pm z(.05)$ or -1.645 and 1.645.

On this occasion the average weight of 64 patties is $\bar{x} = 4.03$ ounces. Assuming that the company knows that the long-run standard deviation of patty weights is $\sigma = 0.16$ ounces, then the standard error of the mean is $\sigma_{\bar{x}} = 0.16/\sqrt{64} = 0.02$. The test statistic is $z = (4.03 - 4.00)/0.02 = 1.5$. This value does not fall in either tail of the normal distribution; it is inside the interval delimited by -1.645 and 1.645. Thus H_0 is accepted and the company continues to produce patties as it has been up to now.

A two-tailed test allows us to detect a difference in either direction while a one-tailed test allows us only to decide if μ is greater than μ_0 (or alternatively if μ is less than μ_0). For example, suppose that customers had been complaining that their burger patties seemed too small. If the company tested H_0: $\mu \geq 4$ against H_1: $\mu < 4$, the region of rejection for the test would consist of only the lower tail of the normal distribution. The procedure would allow *no chance* for the company to discover that it might be making patties that are systematically too heavy.

It might seem that a two-tailed test is always preferable but this is not the case. The price that is paid for this advantage is an increased probability of a certain type of statistical error; this is considered as part of the discussion of errors (Section 11.2).

Relation of Two-Tailed Tests to Confidence Intervals

There is a close relationship between the results of a two-tailed hypothesis test and a confidence interval for the same parameter. Suppose that the null hypothesis is H_0: $\mu = \mu_0$, the alternative is H_1: $\mu \neq \mu_0$, and $\alpha = .05$. The test statistic is $z = (\bar{x} - \mu_0)/\sigma_{\bar{x}}$ and H_0 is accepted if z is between -1.96 and 1.96. In symbols, H_0 is accepted if

$$-1.96 < \frac{\bar{x} - \mu_0}{\sigma_{\bar{x}}} < 1.96.$$

Multiplying all parts of the inequality by $\sigma_{\bar{x}}$ and subtracting $\bar{x}$ gives

$$-\bar{x} - 1.96\sigma_{\bar{x}} < -\mu_0 < -\bar{x} + 1.96\sigma_{\bar{x}};$$

Multiplying all parts of the inequality by -1 and reordering the terms gives

$$\bar{x} - 1.96\sigma_{\bar{x}} < \mu_0 < \bar{x} + 1.96\sigma_{\bar{x}}.$$

TABLE 11.1
Length of Confidence Interval and Inclusion of Hypothetical Value

		Does confidence interval include μ_0?	
		Yes	*No*
Length of confidence interval	*Short*	(a)	(b)
	Long	(c)	(d)

In words, H_0 is accepted if μ_0 lies in the interval from 1.96 standard errors below $\bar{x}$ to 1.96 standard errors above $\bar{x}$. However, this range of values is exactly a 95% confidence interval on μ. Thus, H_0 *is accepted at level α if μ_0 is contained within the limits of the corresponding $1 - \alpha$ confidence interval, and rejected if the confidence interval does not contain the value μ_0.* Put another way, the confidence interval consists of all values of μ_0 for which the null hypothesis $H_0: \mu = \mu_0$ would be accepted.

The confidence interval provides the information that a hypothesis test does, but also more. We consider four examples to illustrate how the two may be used together. These are classified in Table 11.1 and are sketched in Figure 11.6.

In case (a) the interval is short and contains the hypothesized value. The null hypothesis is accepted with some conclusiveness since the investigator has confidence that the true value of the population mean is near μ_0; the interval may include only parameter values which have

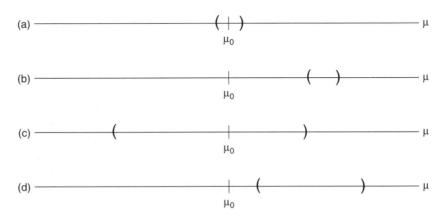

Figure 11.6 *Length of confidence interval and inclusion of hypothetical value (Table 11.1).*

the same meaning for the investigator as μ_0, that is, parameter values which are so close to μ_0 as to be equivalent to μ_0 in practical terms.

In case (b) the interval is short and does not contain the hypothesized value. The null hypothesis is rejected. From the confidence interval, the investigator is guided by some knowledge of what the true value of the parameter is; he or she may have confidence that it is quite different from the hypothetical value.

In case (c) the interval is long and contains the hypothetical mean. The null hypothesis is accepted, but other values of the population mean are acceptable; some of these are quite far from the hypothetical value.

In case (d) the interval is long and does not contain the hypothesized value. The null hypothesis is rejected, but the possible values of the population mean range widely; the investigator does not get much guidance from the interval.

Validity Conditions

The tests of hypotheses about μ described in the preceding sections rest on the assumptions that the sample mean is an appropriate measure of location and that the sampling distribution of $\bar{x}$ is nearly normal. According to the Central Limit Theorem the sample mean is approximately normally distributed almost regardless of the shape of the parent population. The normal approximation is good as long as the sample size is not small and observations are statistically independent. Hence, the procedures of this section apply even when the parent population is not normal, provided the sample size is sufficiently large.

Nevertheless, it is advisable to examine a histogram of the sample data for nonnormality. If the histogram is extremely skewed, a larger sample may be necessary in order to assure that the sampling distribution of the mean will be normal. If it is not possible to collect additional data, the sign test for the median may be preferable to the normal test for the mean. (See Section 11.6.)

In practice, steps must be taken to assure that observations are independent. In sampling units from a production line, for example, it is a good idea to avoid sampling items that are adjacent to one another if there is any chance that characteristics of one item affect the production of those next to it (for example, if the shortage of cookies in one package results in too many in the next). This may mean that randomization is not followed exactly and substitutions are made for some randomly selected observations. If the respondents are humans, they should not be aware of each others' answers in formulating their own. In general, the independence of observations is essential for the procedures described in this chapter to function correctly.

11.2   Errors and Power

Whenever an inference is made about a population based on a sample, there is a chance of being wrong. Nevertheless, statistical inference is used often because it allows us to draw conclusions from data when it is impractical or undesirable to observe all members of the relevant population. With careful planning, the probability of an error can be kept to an acceptably small level.

Types of Error

The null and alternative hypotheses in the SAT example are H_0: $\mu \leq 500$ and H_1: $\mu > 500$, respectively. The sampling distribution of the mean is drawn in Figure 11.2. This is the distribution of sample means that would be observed if H_0 is true and $\mu = 500$. It can be seen that is possible for a sample mean to exceed the significance point and lie in the region of rejection even though the null hypothesis is true. The hypothesis testing procedure leads us to reject H_0 if $\bar{x}$ lies in this region. Thus, it is possible to conclude that H_0 is false and to be wrong. This is called *Type I error.*

DEFINITION   A Type I error is *the error of rejecting the null hypothesis when it is true.*

Suppose, on the other hand, that H_0 is false and μ is 502. A sample drawn from such a population could have an $\bar{x}$ of, say, 506. If that were to happen, the hypothesis testing procedure would lead us to accept H_0 because 506 does not lie in the region rejection. We would have concluded incorrectly, making a *Type II error.*

DEFINITION   A Type II error *is the error of accepting the null hypothesis when it is false.*

All possible outcomes of a test of significance are shown in Table 11.2. If the null hypothesis is true, it is possible that we will reject

TABLE 11.2

Possible Errors in Drawing Statistical Conclusions

		Conclusion	
		Accept null hypothesis	*Reject null hypothesis*
True "State of Nature"	*Null hypothesis*	Correct conclusion	Type I error
	Alternative hypothesis	Type II error	Correct conclusion

H_0 based on a test of significance and thus make a Type I error. If the null hypothesis is false, that is, μ has some value other than μ_0, it is possible that we will accept H_0, making a Type II error. In any given situation the null hypothesis is either true or false. The true state of affairs is unknown to the investigator who must, therefore, protect against the possibilities of both types of error. It is important to keep the probabilities of both Type I and Type II errors small.

Similar errors can be made when individuals are given medical or psychological tests and classified into diagnostic categories. In such a situation the null hypothesis corresponds to the individual being healthy. If a healthy person is erroneously diagnosed as having a particular illness, the decision corresponds to a Type I error. If an individual is ill but is classified as healthy, this decision corresponds to a Type II error.

Probability of a Type I Error

The probability of a Type I error is the probability that the sample mean will fall in the region of rejection when the null hypothesis is true. From Figure 11.2 it can be seen that this probability is the significance level of the test, α. The *cost* of making such an error is an important consideration in choosing α.

For example, what is the cost if the study time problem results in a Type I error, leading to the erroneous conclusion that students study more than an average work week of 40 hours? If this is simply a debate between one professor and her colleagues, then there is no great loss except perhaps the colleagues' pride. They concede that the professor is right—that students study *a lot*—when this is not the case.

On the other hand, if this conclusion leads to a university-wide policy that class time and homework assignments are to be reduced, doing this based on an erroneous conclusion could cost the university its reputation and the quality of its graduates. In this case the cost of Type I error is high and we might protect against such an error by setting α at .01 or .005 or even .001.

Lowering α reduces the probability of making a Type I error, but a price is paid for this improvement. A consequence of choosing a small number for the significance level is a relatively extreme significance point. That is, we insist on a great deal of evidence from the sample before rejecting the null hypothesis. If this is carried to an extreme, it may become extraordinarily difficult to reject H_0 when it is false as well as when it is true. Also, there are situations in which a *higher* probability of a Type I error might be preferred. For example, we might want to warn pregnant women about the hazards of smoking if there is even modest evidence that birthweight is reduced. For the birthweight hypothesis, then, we might set α at .05 or even .10.

Probability of a Type II Error and Power

The probability of a Type II error, that is, the probability of accepting the null hypothesis when it is false, depends on "how false" it is. If the mean number of hours of study in the population is very great, say 55 hours, it is very unlikely that we would draw a random sample with a mean so small that we would accept the null hypothesis. On the other hand, if the population mean is only a little more than 40, that fact will be hard to discover; we are likely to accept the null hypothesis.

The probability of making a Type II error for a particular value of μ is represented as β. The probability of *not* making a Type II error for this value of μ is $1 - \beta$, which is called the *power* of the statistical test at the particular value of μ. High power is desirable since it is the probability of concluding correctly that H_0 is false, for example, of concluding that smoking reduces birthweight when the mean birthweight in the population is in fact reduced to 3200 grams.

We illustrate how power can be computed using the study time example in which the hypotheses are H_0: $\mu \leq 40$ and H_1: $\mu > 40$, and then discuss procedures for increasing statistical power and reducing the probability of a type II error.

Figure 11.7 shows three sampling distributions of the mean for a sample size of 25 drawn from normal distributions with a standard deviation of 10 and means of 40, 42, and 44, respectively. The top diagram is for $\mu = 40$ (H_0 is true). The null hypothesis will be rejected if the test statistic is 1.645 or greater, that is, if $\bar{x}$ is at least 1.645 standard errors above 40. Since the standard error is 2, this means that H_0 is rejected if $\bar{x}$ is $40 + (1.645 \times 2) = 43.29$ or greater. The probability that this will occur is .05, the significance level of the test.

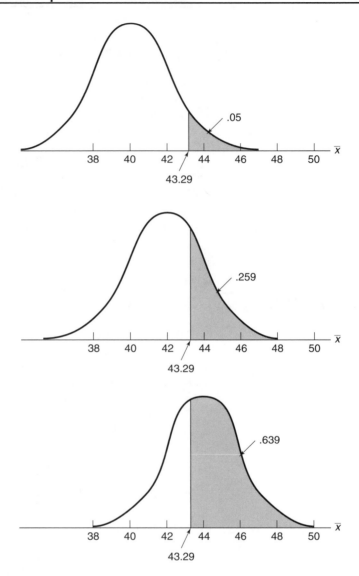

Figure 11.7   *Probability of $\bar{x} > 43.29$ for $\mu = 40, 42,$ and 44.*

But suppose that the null hypothesis is false and μ is actually 42, as in the middle diagram. What is the probability of observing an $\bar{x}$ of 43.29 or greater and (correctly) rejecting H_0? To determine this, the value 43.29 is standardized relative to 42, that is, $(43.29 - 42)/2 = 0.645$. The probability of a z value greater than this is 0.259 (found by interpolation in Appendix I). This value is the power of the test when the true population mean is 42. The probability of making a Type II error in this case is $1 - 0.259 = 0.741$.

By the same reasoning, if μ is actually 44, the power of the test is 0.639, as illustrated in the bottom diagram in Figure 11.7. The probability of making a Type II error is $1-0.639 = 0.361$. In general, the probability of rejecting the null hypothesis is high (and the probability of making a Type II error low) when the true value of μ is far from the hypothesized value μ_0.

Table 11.3 gives the probability that the sample mean exceeds 43.29 for various values of μ, and Figure 11.8 is a graph of these values. The probability of rejection as a function of μ, the mean of the parent population, is called the *power function* of the test. The power increases as μ increases, equals the level .05 when $\mu = 40$, and increases to 1 as μ gets larger. For values of μ less than 40, the probability of rejection is less than .05. Thus, the procedure distinguishes a large value of μ from values less than or equal to 40. Consequently, we write the null hypothesis as H_0: $\mu \leq 40$. If the data are such as to rule out a value of 40, they also rule out values smaller than 40.

TABLE 11.3

Power Function of 5%-level Test of H_0: $\mu \leq 40$ Against H_1: $\mu > 40$ when $\sigma = 10$ and $n = 25$

Value of μ	Power
38	0.004
39	0.016
40	0.050
41	0.126
42	0.259
43	0.442
44	0.639
45	0.804
46	0.912
47	0.968
48	0.991

For large values of μ, the investigator is almost certain to reject the null hypothesis, but for values of μ only a little larger than the hypothesized value of $\mu = 40$ the probability of rejection is only a little larger than the significance level of .05. In general, the further the actual population mean is above μ_0, the higher the probability of rejecting H_0; it is easier to detect big differences than small ones.

Effect of Sample Size. The power of a statistical test is directly related to the sample size. Suppose that the study-time hypothesis is tested with $n = 100$ instead of 25. Then the standard error of the distributions in

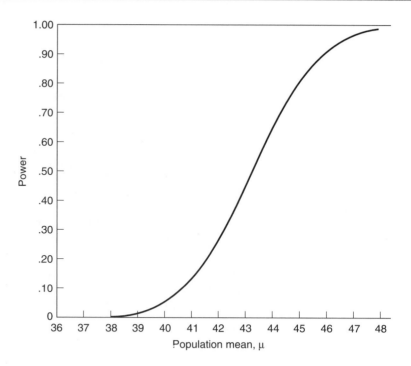

Figure 11.8   *Power function of 5%-level test of H_0: $\mu \leq 40$ against H_1: $\mu > 40$ when $\sigma = 10$ and $n = 25$ (Table 11.3).*

Figure 11.7 would be $10/\sqrt{100} = 1$ instead of 2, and the test would dictate that H_0 is to be rejected if $\bar{x}$ is greater than $40 + (1.645 \times 1) = 41.645$. If the population mean were truly 42, this would correspond to a standardized value of $(41.645 - 42)/1 = -0.355$. From the normal table, the probability of z exceeding -0.355 is 0.639.

The probability of rejecting H_0 when $\mu = 42$ has gone from 0.259 to 0.639 by increasing the sample size from 25 to 100 observations. The power function for $n = 100$, shown in Figure 11.9, is above the one for $n = 25$ for all values of μ above 40.

In general, increasing the sample size is the most straightforward way to increase the probability that statistical significance will be attained *when an effect is present in the population to be discovered*. It is possible to ask how large the sample has to be to give, say, 0.90 power in a given situation, once a particular value of μ_0 is specified. Several examples of this line of reasoning are given as chapter problems.

Effect of Significance Level.   The choice of significance level also has an effect on the power of the test. Suppose the study-time hypothesis H_0: $\mu \leq 40$ was tested at $\alpha = .01$ instead of .05. The significance

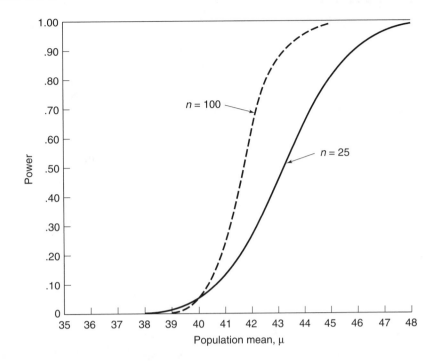

Figure 11.9   *Power functions of 5%-level test of* $\mu \leq 40$ *when* $\sigma = 10$ *for* $n = 25$ *and* $n = 100$.

point would be 2.33 instead of 1.645, which in turn corresponds to a mean of $40 + (2.33 \times 2) = 44.66$. If the true population mean is $\mu = 42$, the probability of obtaining an $\bar{x}$ greater than 44.66 based on 25 observations is only 0.092.

In other words, when the probability of a Type I error is reduced from .05 to .01, the power of the test is decreased and the probability of a Type II error is increased.

In choosing a significance level, it is important to balance the "costs" of a Type I and Type II error. An example was given of setting α at .05 and at a smaller value to test whether students study 40 hours per week or less. The risk we are willing to take of making a Type I error depends on how important the policy implications of the results will be. If university-wide changes are to be implemented if H_0 is rejected, then a smaller risk of a Type I error is preferable; we may be willing to risk a slightly higher probability of making a Type II error in this situation.

On the other hand, the potential harm of *not* warning pregnant women about the dangers of smoking is grave, even if birthweights are reduced only slightly. In order to increase the probability of detecting a small impact on mean birthweight, we might accept a somewhat higher risk of making a Type I error. Setting α at .10, for example, will result in a somewhat smaller value for the significance point. The probability of rejecting the null hypothesis is increased in general; in particular, the probability of rejecting the null hypothesis if it is false is increased as well.

In sum, in choosing the significance level it is necessary to balance the undesirabilities of the two types of error. We want to discover a difference from the null hypothesis if it exists, but we do not want to claim there is a difference if it does not exist. Although 5% and 1% are conventional levels of significance, there is no reason to restrict the choice to these. We can choose a significance level of 0.5% if we want to be careful in rejecting the null hypothesis, or choose 10% if we want a relatively small chance of overlooking a difference from the null hypothesis.[6]

Rational choice of a significance level involves simultaneous consideration of the power. If a 1%-level test gives reasonable power against reasonable alternatives, use it. Otherwise, use a 5%- or even a 10%-level test.

One-Tailed Tests Compared with Two-Tailed Tests.   The study-time example was conducted as a one-tailed test with H_0: $\mu \leq 40$ and H_1: $\mu > 40$. If we were interested in determining whether average study time is the same as a standard 40-hour work week or different from it in either direction, we could make a two-tailed test with H_0: $\mu = 40$ and H_1: $\mu \neq 40$. If $\alpha = .05$, the significance points for the two-tailed test are -1.96 and 1.96 instead of 1.645. With a sample of 25 observations, these significance points translate into $\bar{x}$ values of $40 - (1.96 \times 2) = 36.08$ and $40 + (1.96 \times 2) = 43.92$. The null hypothesis will be rejected if $\bar{x}$ is outside the range 36.08 to 43.92.

Suppose that the average study time in the population is actually $\mu = 42$ hours per week. The probability of a sample mean below 36.08, from the normal table, is .0015 and above 43.92 is .1685. Thus the power, or total probability of rejecting H_0 given that it is false when $\mu = 42$ is 0.170. This is smaller than the power of the one-tailed test (0.259) at the same significance level.

A two-tailed test allows us to detect a difference on either side of μ_0 while a one-tailed test allows us only to decide if μ is greater than μ_0 or if μ is less than μ_0, not both. It might seem that a two-tailed test is always preferable, but the example above shows that this is not the case.

---

[6]There is a technical reason for caution in using extremely small significance levels; the normal approximation to the sampling distribution of the mean is sometimes not very good for very extreme values.

In general, a one-tailed test has greater power on one side of μ_0 than a two-tailed test made under the same conditions, that is, with the same n and α. However, a one-tailed alternative hypothesis is more specific than a two-sided alternative in that it excludes certain possibilities. Thus, a one-tailed test has no power for detecting a difference in the the direction opposite that given by H_1.

If an investigator is interested only in detecting an effect in one direction, or if the only realistic alternative is that μ lies on one side of μ_0, then it is advantageous to make a one-tailed test. This will give a significance point with a smaller (absolute) value, and statistical significance is found more easily. Other situations call for a two-tailed test, for example, if there is no way of anticipating the direction in which μ may differ from μ_0, if the investigator needs to detect a difference in either direction (e.g., the malfunctioning cookie machinery), or if two well-informed sources differ in their prediction of the direction that μ will be affected. In any case, the decision between a one-tailed test and a two-tailed test should be considered carefully before conducting the test of significance.

11.3   Testing Hypotheses About a Mean when the Standard Deviation Is Unknown

When an investigator tests a hypothesis about a population mean, the standard deviation of the population may be unknown as well. In the examples in Section 11.1 it was assumed that σ was known either from previous studies or from scaling procedures that gave a prespecified value to σ. It is more often the case that information of this sort is not available and the test statistic as defined in Section 11.1 cannot be computed.

When this is the case, s, the standard deviation of the sample, can be calculated and used as an estimate of σ. Since the estimate and actual value of σ are different, this increases the sampling error in the procedure. Slightly larger significance points are used to compensate for this. These are drawn from the t-distribution with $n-1$ degrees of freedom (t_{n-1}) instead of the standard normal distribution.[7] The same conditions must be met for this test to be valid as for the normal-distribution test. (See "Validity Conditions.") Namely, the sampling distribution of the

---

[7]The substitution of the t-distribution for the normal when σ is unknown follows the same principles as discussed in Chapter 10 for a confidence interval on the mean.

mean should approximate a normal distribution and the n observations must be independent.

As an example, suppose that we wish to test the study-time hypothesis $H_0: \mu \leq 40$ against $H_1: \mu > 40$, but do not know the standard deviation of study times in the population of students. (This is in fact more realistic; who could be expected to know such an obscure piece of information?!) In this case we can draw our random sample of 25 students, ask them how much time they devoted to their studies last week, and calculate *both* $\bar{x}$ s and s from the data using the basic formulas given in Chapters 3 and 4. Instead of the test statistic

$$z = \frac{\bar{x} - \mu_0}{\sigma/\sqrt{n}},$$

we compute

$$t = \frac{\bar{x} - \mu_0}{s/\sqrt{n}}.$$

The latter can also be written $t = (\bar{x} - \mu_0)/s_{\bar{x}}$, using $s_{\bar{x}} = s/\sqrt{n}$ to represent the standard error calculated from the sample, that is, the estimate of $\sigma_{\bar{x}}$.

If σ were known exactly, the significance point from the standard normal distribution with $\alpha = .05$ would be 1.645. With σ estimated from a sample of 25 observations, the significance point is taken from t_{24}. From Appendix III the value that separates the right-hand 5% of this distribution from the rest is $t_{24}(.05) = 1.711$. It is somewhat larger than 1.645 to allow for the additional error in substituting s for σ. If we had been making a two-tailed test, the significance points would be $t_{24}(.025) = 2.064$ on the right and -2.064 on the left. The same convention about dividing α equally between the two tails of the distribution is applied whether we are using the standard normal distribution or t_{n-1}.

Note from the table of the t-distributions that the significance point gets larger as you read up the column and smaller as you read down. There is more room for error in estimating σ when the sample is small, and less error is likely with larger samples. If the sample is extremely large, the sample and population standard deviations are almost indistinguishable and the significance point from the t-distribution is the same as that from the normal.

The power of the t-test for a particular hypothesis (one-sided or two-sided) at a given significance level α and a given sample size n is slightly lower than the power of the corresponding test based on the normal distribution: the cost of not knowing σ is loss of power. It is advantageous to use σ and the normal distribution whenever a value for σ is available. Otherwise the only alternative is to substitute s and refer to percentage points of the appropriate t-distribution.

A Complete Example:
One-Tailed Test

The average reading speed of second-grade children when reading material that is "on grade level" is 120 words per minute. It is hypothesized that the average speed is reduced if word difficulty is increased even slightly. A random sample of 16 grade-two youngsters are presented with a second-grade reading passage that has a number of preselected "difficult" words interspersed at various points throughout the text. The average reading speeds of these youngsters, rounded to the nearest whole word per minute, are given in Table 11.4.

TABLE 11.4

Reading Speeds of 16 Grade-two children (words per minute)

120	114	113	134
110	123	142	118
100	112	106	113
131	101	104	115

Hypothetical data.

The null and alternative hypotheses for this investigation are H_0: $\mu \geq$ 120 and H_1: $\mu < 120$, respectively. The standard deviation of reading speed might be known by the test publisher for the population of second-grade children and may even be published in the test manual. Even so, it is highly unlikely that the standard deviation would be known under the unusual condition that more difficult words are interspersed in the passages. Thus, the sample standard deviation will be computed from the data in Table 11.4 instead, and the significance point drawn from t_{15}. If $\alpha = .05$, the t value that separates the most extreme 5% of the distribution from the rest is $t_{15}(.05) = 1.753$ from Appendix III. Since we are making a one-tailed test on the left side of the distribution, the significance point is -1.753.

From the data in Table 11.4, $\bar{x} = 116.0$, $s = 11.82$, and the test statistic is

$$t = \frac{116 - 120}{11.82/\sqrt{16}} = \frac{-4}{2.955} = -1.35.$$

Figure 11.10 is a diagram of the t-distribution with 15 degrees of freedom. The shaded area at the left side of the distribution is the region of rejection. It can be seen that the test statistic does not lie in the region of rejection, and thus we accept H_0. There is no compelling evidence

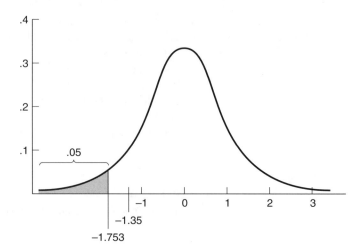

Figure 11.10   *The t-distribution with 15 degrees of freedom.*

that interspersing a few more difficult words in material read by second-grade pupils decreases average reading speed in the population.[8]

Discussion   In most studies both estimation and hypothesis testing are important. As discussed in Section 11.1, a confidence interval is closely related to the test of significance. If the confidence interval includes the hypothesized value (μ_0), the null hypothesis is accepted; if μ_0 is not in the interval, the hypothesis is rejected.

In general, tests of significance and estimates of population values go hand-in-hand. Every time a significant effect is found, the question remains "How big is the effect?" And most often a point or interval estimate leads to the question "Is it different from some recognized norm or standard?" A high-quality report that summarizes a statistical analysis should contain both types of information.

Even with both types of information, there are limitations to our statistics. For example, the *t*-test is based upon the sample mean and

---

[8]Of course we have not examined the impact of adding difficult words on the the youngsters' *comprehension* of the passage.

standard deviation. Other questions may be of interest such as whether *any* of the second graders were handicapped by the more complex reading material. Or, if a significant impact had been found, were *all* of the youngsters handicapped in reading speed? Inspection of the histogram of the scores could help to decide this.

Figures 11.11 and 11.12 might be histograms of reading speeds for two different groups of 25 students reading above-grade level material. The mean reading speed for the population when the material is *at* grade level is 120 words per minute. The two sample distributions have the same mean (116) and almost the same standard deviation (8.13 for

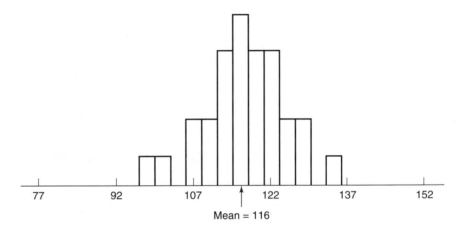

Figure 11.11 *Scores of 25 students: Mean = 116, standard deviation = 8.13.*

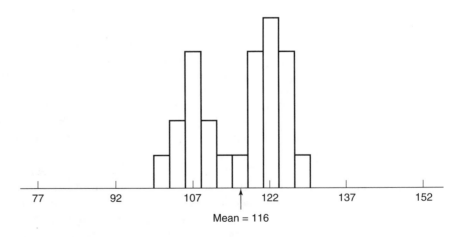

Figure 11.12 *A bimodal distribution with the same mean and standard deviation as Figure 11.11.*

Figure 11.11 and 8.17 for Figure 11.12). Thus, the value of Student's t is almost exactly the same in the two cases. However, the histogram in Figure 11.12 is bimodal, with one mode near 107 and another near 122, suggesting that there are two clusters of pupils. The lower cluster consists of people who were handicapped by the more difficult words and the upper cluster of those who where not. This figure would suggest that difficulty affects *some* individuals and poses further research questions: What distinguishes these two groups of individuals? Were pupils in the lower cluster slower readers initially? Are they primarily boys or girls? Are they individuals with language handicaps? Or are youngsters in the lower cluster reading more slowly but gaining a better understanding of the material they read?

EXAMPLE

Two-Tailed Test: As an additional example, suppose that a large retail firm with many branches reports that the average cost of items in its inventory is $36.00. For a host of reasons such as errors in recorded sales and purchases, theft, obsolescence, and the like, the average cost may be greater or less than this. Suppose that the company auditors draw a random sample of 121 items from the stock to test this possibility.

The null and alternative hypotheses are H_0: $\mu = \$36.00$ and H_1: $\mu \neq \$36.00$. If the average cost of the 121 items in the sample was $\bar{x} = \$33.60$ and $s = \$12.20$, then the test statistic is

$$t = \frac{\bar{x} - \mu_0}{s/\sqrt{n}}$$

$$= \frac{33.60 - 36.00}{12.20/\sqrt{121}} = -2.16.$$

Because the standard deviation of the sample was used in computing the test statistic, the probability distribution for finding significance points is t_{n-1} (that is, t_{120}) instead of the standard normal distribution. If the test is being conducted with $\alpha = .05$, the significance points from Appendix III are $t_{120}(.025) = 1.98$ on the right side of the distribution and -1.98 on the left. A diagram of this distribution is given in Figure 11.13. The shaded areas in the tails represent the region of rejection.

It can be seen that the test statistic is less than the significance point on the left side of the distribution and thus lies in the region of rejection. H_0 is rejected and the auditors conclude that the average cost of all items in the company's inventory has decreased to less than $36.00. The best single estimate of the current value is $\bar{x} = \hat{\mu} = \$33.60$; a 95% confidence interval would indicate that there is a range of possible values for μ, but $36.00 is not in that range.

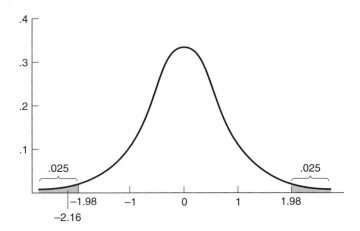

Figure 11.13   *The t-distribution with 120 degrees of freedom.*

In this application the population standard deviation of item costs may be known to a reasonable degree of accuracy from the previous year's complete inventory of the company's merchandise. The same factors that affect the mean, however, may also have changed the standard deviation. For example, if more expensive items are stolen, then there may be less dispersion at the upper end of the cost distribution. The use of the sample standard deviation allows us to circumvent this problem. In any case, since the sample size is reasonably large, the significance point from the *t*-distribution is close to the corresponding point from the standard normal distribution (1.96).

11.4   *P* Values: Another Way to Report Tests of Significance

In the discussion of tests of significance up to this point, the significance level α is chosen prior to conducting the test, leading to an unambiguous decision to reject or accept the null hypothesis. The probability of making a Type I error is no greater than α.

Another way that the results of a test of significance can be reported is to give a *P value* associated with the test statistic. The *P* value tells us the entire *range* of α levels for which H_0 would be rejected for this sample. The final decision to accept or reject H_0 is then left to the researcher or to individuals reading the research report after it is written. The *P* value is sometimes referred to as the "achieved significance level."

Technically, the P value for a test statistic is *the probability of obtaining a test statistic of that magnitude or greater if the null hypothesis is true*. As an example, consider the SAT problem given earlier for which the hypotheses are H_0: $\mu \le 500$ and H_1: $\mu > 500$. The mean of $n = 400$ observations in the sample is $\bar{x} = 507.5$; the standard deviation of the SAT test is $\sigma = 100$. From these values, the test statistic is $z = 1.5$. The P value corresponding to this test statistic is the probability of obtaining a test statistic of 1.5 or larger if the null hypothesis is true. Since z has a standard normal distribution (diagrammed in Figure 11.2) the probability may be found from the table in Appendix I. This is $\Pr(z > 1.5) = .5 - .4332 = .0668$.

This result is the *smallest* value of α for which H_0 can be rejected. In other words, H_0 would be rejected if α is set at any value greater than .0668, and H_0 would be accepted if α is set at .0668 or less.

The reasoning behind this is illustrated in Figure 11.14. The area under the normal curve to the right of the test statistic is .0668. Suppose that α is set at .05, for example, and the significance point is 1.645. The test statistic $z = 1.5$ has *more than 5% of the area under the curve to the right of it* and thus does not fall in the region of rejection; H_0 would be accepted. On the other hand, if α were set at .10, the significance point would be 1.28. The test statistic, having *less than 10% of the area to the right of it*, must lie in the region of rejection; H_0 would be rejected.

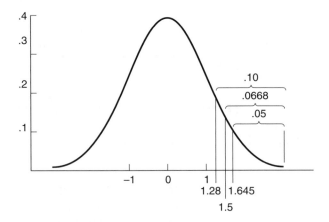

Figure 11.14   *Standard normal distribution.*

Using the P value approach, the researcher would report that P for the test of significance is .0668. He or she may also decide to accept or reject H_0 based on an α level decided in advance of the study or may leave that conclusion to readers of the research report.

Although it may seem that the researcher has not completed the statistical analysis unless an accept-or-reject decision is issued, this approach actually provides a different kind of information about the testing procedure. The *P* value not only specifies the range of α's at which H_0 would be rejected, but also tells the *extent* to which H_0 is refuted by the sample. Thus a small *P* value (for example, .002) arises from a relatively large test statistic. This occurs when the sample mean is quite far from μ_0; it tells us that H_0 can be rejected for a range of values of α including some that are quite small. A larger *P* value (for example, .130) arises from a smaller test statistic. This indicates that the sample mean is only slightly above μ_0 and that H_0 can only be rejected for values of α that are relatively high.

EXAMPLE

As a second illustration of the *P* value approach, consider the birthweight example for which the hypotheses are H_0: $\mu \geq 3300$ grams and H_1: $\mu < 3300$ grams. The sample mean of 100 observations is $\bar{x} = 3207$; the standard deviation is $\sigma = 516$. The test statistic is $z = (3207 - 3300)/(516/\sqrt{100}) = -1.80$.

The *P* value for a one-tailed test on the left side of the standard normal distribution is the probability of obtaining a *z* value of -1.80 or *less*, that is, $\Pr(z < -1.80)$, which is the same as the probability of a *z* value of $+1.80$ or greater. From Appendix I this is $.5 - .4641 = .0359$.

Thus, the null hypothesis would be rejected if α is set to any value of .0349 or greater and would be accepted if α is set at any value less than .0349. In Section 11.1 the α level was set at .10; if we compare *P* to this value of α we would conclude that H_0 is rejected. At the same time, had *P* been .011, say, H_0 would also be rejected, but the evidence provided by the data that H_0 is false would be stronger still.

One caution must be observed in using a *P* value to decide if H_0 is accepted or rejected. It is always tempting to look at a *P* value, especially when it is printed by a computer program, and use it *after the fact* to decide what the α level should be. For example, suppose we did not set α in advance of a study and obtained the *P* value of .0359. If we have a large personal investment in the outcome of the study, we might be tempted to set α at .05 in order to reject H_0, whereas we might have chosen .01 if α was set in advance. Likewise, seeing a *P* value of .0668 might tempt us to decide that α should be .10 for the SAT problem if we really "want" H_0 to be rejected.

To see the danger of choosing α after being given *P*, consider an investigator who is very anxious to get a "significant result", that is, reject the null hypothesis. The researcher notices that a significant result can always be obtained by choosing α to be a number slightly larger than the recorded P. By following this "system" they are sure to reject the null hypothesis (although it is not known in advance whether the resulting α will be impressive). The shortcoming of this "system" is that the probability of making a Type I error is 1; no matter whether the null hypothesis is true or not, it is rejected.

P Values for Two-Tailed Tests

When a two-tailed test is conducted, the P value must be computed using both tails of the standard normal distribution. For example, in Section 11.1 the hypothesis is tested that the mean package weight for Rabbit Graham cookies is 8 ounces. The hypotheses are H_0: $\mu = 8$ and H_1: $\mu \neq 8$. The test statistic based on the sample mean of $\bar{x} = 8.05$ is $z = 2.5$.

The P value for this test is the probability of obtaining a z value whose *absolute value* is greater than 2.5, that is, $\Pr(z > 2.5) + \Pr(z < -2.5)$. The sum of two probabilities corresponds to the fact that the region of rejection is divided between the two tails of the standard normal distribution. If we wish to compare P to α to decide whether to accept or reject H_0, we must compare the total area in the region of rejection to the total probability of obtaining a sample mean that far from μ_0 in either direction. For the Rabbit Graham example, the P value is $2 \times (.5 - .4938) = .0124$; this is illustrated in Figure 11.15. The null hypothesis would be rejected at any α level greater than .0124 (for example, at .05) and accepted for any α less than .0124 (for example, at .01).

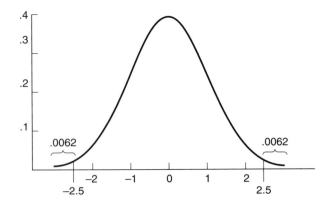

Figure 11.15   *Standard normal distribution.*

P Values when the Population Standard Deviation Is Unknown

Section 11.3 describes significance testing procedures when a value for the population standard deviation is σ is not available. The standard deviation of the sample s can be computed and substituted for σ and significance points are obtained from the t-distribution with $n - 1$ degrees of freedom (t_{n-1}) instead of the standard normal distribution.

Likewise, P values for a t-test are obtained from the table of the t-distribution in Appendix III. As an example, consider the reading speed problem for which data are given in Table 11.4. The hypotheses are H_0: $\mu \geq 120$ and H_1: $\mu < 120$. The sample of $n = 16$ observations yields $\bar{x} = 116$ and the test statistic is $t = -1.35$. The P value for this test is the probability of obtaining a value of -1.35 or less from the t-distribution with 15 degrees of freedom. This is also equal to the probability of obtaining a value of $+1.35$ or greater from the same distribution.

Appendix III lists just five percentile points for each t-distribution. However, we can use the table to find a P value *interval* that is usually sufficient for our purposes. From Appendix III, we see that the probability of obtaining a t value of 1.341 or greater from the distribution t_{15} is .10 and the probability of a value 1.753 or greater is .05. (See Figure 11.16.) Since 1.35 is between 1.341 and 1.753, we can report that the P value is between .05 and .10, that is, $.05 < P < .10$. The null hypothesis would be rejected at an α level of .10 or greater and accepted at any α value of .05 or less.

Since an α value is rarely chosen to be between .05 and .10, the interval for P is sufficient for most accept-or-reject decisions. The interval

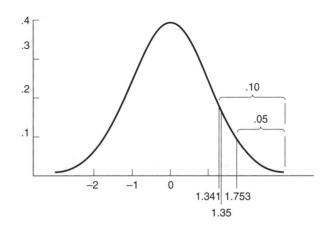

Figure 11.16   *The t-distribution with 15 degrees of freedom.*

also tells us that the data refute H_0 more strongly than a test statistic with a P value of .10 or greater, but not as strongly as a test statistic with a P between, say, .01 and .005.

Very small and very large test statistics fall outside the range of values given in Appendix III. For example suppose we were making a one-tailed test of H_0: $\mu \leq 10$ against H_1: $\mu > 10$ and a sample of 21 observations gave us the result $t = 5.80$. The largest value tabled for t_{20} is 2.845, the point with .005 of the distribution to the right of it. Using this table, we can only conclude that $P < .005$. Of course this is a very small P value and indicates that the null hypothesis would be rejected even if α were set as low as .01 or .005. In contrast, if the sample of 21 observations resulted in the test statistic $t = 0.90$, this is smaller than 1.325, the smallest value tabled for t_{20}. We could only conclude that $P > .10$ and that H_0 would be accepted at any usual α level such as .01, .05, or .10.

When making a two-tailed t-test, the P value is the probability of obtaining a value from t_{n-1} greater than the *absolute value* of the test statistic. Section 11.3 presents an example in which the average value of items in a company's inventory is hypothesized to be \$36.00. The hypotheses are H_0: $\mu = \$36.00$ and H_1: $\mu \neq \$36.00$. A sample of 121 items has a mean of \$33.60, resulting in the test statistic $t = -2.16$. To obtain a P value for this test, we refer to the table of the t-distribution with 120 degrees of freedom. The probability of a t value of 2.16 or larger is between .01 and .025. Likewise, the probability of a t value of -2.16 or smaller is between .01 and .025. Thus P for the two-tailed test is $.02 < P < .05$. The null hypothesis would be rejected at $\alpha = .05$ but not at $\alpha = .01$.

Fortunately, the interpretation of P values is made more precise when data are analyzed by computer. In place of a P interval, most computer programs give a single P value for every test statistic which is accurate to four or more decimal places. The data analyst only needs to take note of whether a one-tailed or two-tailed test is to be conducted.

11.5   Testing Hypotheses About a Proportion

Testing Hypotheses About the Probability of a Success

Information about the probability of an event is given by the proportion of trials in which the event occurs. The sample proportion is used to answer questions about the probability, in particular, the question of whether the probability is equal to some specified number. If the sample proportion is very different from the hypothesized probability, the null hypothesis will be rejected. As in testing a hypothesis about μ, the decision depends on whether the alternative hypothesis is that the parameter is greater than, less than, or *either* greater or less than the hypothesized value.

The same general steps in testing a hypothesis about a mean apply to tests on other parameters, including a probability. Once the null and alternative hypotheses are stated (1) the sampling distribution of the appropriate statistic is described and a region of rejection is identified by specifying a significance level, (2) a sample of observations is drawn and a test statistic is calculated, and (3) a decision is made to reject or accept H_0 by determining whether the test statistic lies within the region of rejection or not, respectively.

Hypotheses

Suppose psychotherapists have claimed success in treating a certain mental disorder. Their claim is to be studied by treating 200 patients having the disorder and then observing the recovery rate. Previous studies have shown that 2/3 of all patients with this disorder recover "spontaneously," that is, without treatment.

The psychotherapists' claim is that the *probability* that a patient can be treated effectively exceeds the spontaneous recovery rate of 2/3. Representing the probability by p, the null and alternative hypotheses are H_0: $p \leq 2/3$ and H_1: $p > 2/3$. Here as in the earlier examples, the null hypothesis is the "statement of no effect." These hypotheses, of course, require a one-tailed test.

Sampling Distribution and Region of Rejection

The procedure for testing hypotheses about the mean of a normal distribution can be applied to testing hypotheses concerning the probability of a success because the normal distribution provides a good approximation to the binomial distribution if n is large. More precisely, let x be the number of successes in n Bernoulli trials with probability p of a success. The proportion $\hat{p} = x/n$ is approximately normally distributed[9] with mean $E(\hat{p}) = p$ and variance $\sigma_{\hat{p}}^2 = pq/n$, where $q = 1 - p$.

Applying these principles to the psychotherapy study, the sampling distribution of the proportion is approximately normal. If the null hypothesis is true and $p = 2/3$, then the center of the distribution is at 2/3 and the variance is

$$\sigma_{\hat{p}}^2 = \frac{(2/3) \times (1/3)}{200} = \frac{1}{900}.$$

Its square root, the standard error, is $\sigma_{\hat{p}} = 1/30$. A diagram of the sampling distribution is given in Figure 11.17. The scale labeled $\hat{p}$

[9]See Section 9.3.

represents different possible values of the proportion that might, in theory, be observed. The z scale re-expresses each point as a standard normal deviate, that is, the number of standard errors between each possible value of $\hat{p}$ and the center of the distribution (2/3).

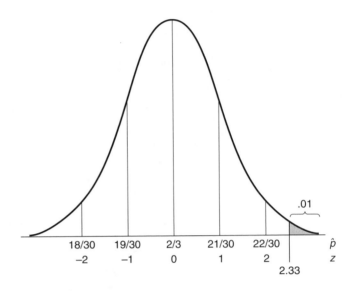

Figure 11.17   *Sampling distribution of $\hat{p}$ with $p = 2/3$, $\sigma_0 = 1/30$.*

Since the public is somewhat skeptical about the efficacy of psychotherapy, we might test these hypotheses with a relatively small α, say .01. This keeps the probability of Type I error small, and we can have high confidence that the treatment really works if the null hypothesis is rejected. From the table of the standard normal distribution, the z value that separates the right hand 1% from the rest of the distribution is 2.33; this is the significance point.

Data, Test Statistic, and Decision

Suppose that in the sample, 150 of the 200 patients recover, or a sample proportion of 150/200 = 0.75. The sample proportion exceeds 2/3, but does it exceed 2/3 sufficiently to convince us that the psychotherapy works for the population generally?

To answer this, the sample proportion is converted to a test statistic, that is, the number of standard errors that .75 lies above (or below) 2/3. For these data the test statistic is

$$\frac{0.750 - 0.667}{0.033} = 2.5.$$

Since 2.5 is greater than the significance point, H_0 is rejected at the .01 level of significance.[10]

In symbols, the test statistic is

$$z = \frac{\hat{p} - p_0}{\sigma_{\hat{p}}},$$

where p_0 is the hypothesized value of the probability (2/3) and $\sigma_{\hat{p}} = \sqrt{p_0 \times (1 - p_0)/n}$. The reader may note that, unlike the test on the mean, the value p_0 is all that is needed to compute both the hypothetical mean and the standard deviation of the parent population.

Example of a Two-Tailed Test

In a study of racial awareness, Clark and Clark (1958) investigated the question of whether African-American children distinguish the "race of dolls, under the assumption that the behavior of distinguishing indicates racial awareness and . . . (preference)." In this study, 253 African-American children (ages 3 to 7) were presented four dolls, two white and two nonwhite. Each child was told: "Give me the doll that you would like to play with." The results were as follows:

	83	children chose a nonwhite doll
	169	children chose a white doll
	1	child chose no doll
Total	253	children

We shall consider our sample to consist of the 252 children who chose one of the dolls, that is, $n = 252$.

In this study randomness arises from two sources: (a) selection of children—the 253 children are a sample from a large population (actually, 134 from Arkansas and 119 from Massachusetts) and (b) choice of doll—the phenomenon under observation is nondeterministic (random) in the sense that a given child who has a preference for one type of doll may not *always* choose that color of doll. Each child's behavior was observed on only one occasion; on another occasion the child may have behaved differently.

Here the success probability p represents the probability of selecting a child who will choose a nonwhite doll, and $q = 1 - p$ then represents the probability of selecting a child who will choose a white doll. Our null hypothesis is H_0: $p = 1/2$, that is, that a child selected is just as

---

[10]Greater accuracy could be obtained by using additional decimal places and the continuity correction described in Appendix 11A.

likely to choose one type of doll as the other. What kind of result would cause us to decide that a preference exists? Even if the null hypothesis is true, we know that we cannot expect *exactly* half of the children to choose white dolls and half to pick nonwhite dolls. The sample proportion will not be exactly 1/2. There will be some sampling variability. But if the sample proportion is *quite different* from 1/2 we will find the null hypothesis untenable.

On which side of 1/2 would we expect the proportion to be if there is racial awareness or preference? Toward white dolls (p less than 1/2)? Toward nonwhite dolls (p greater than 1/2)? Or either way (p not equal to 1/2)? In the absence of strong prior information we take our alternative hypothesis to be H_1: $p \neq 1/2$. In general, we are making fewer prior assumptions when we use a two-sided alternative hypothesis.

The sampling distribution of the proportion is approximately normal for large samples; this assertion certainly applies to an n of 252. If the hypothesis is tested at $\alpha = .05$, the significance points from the standard normal distribution are -1.96 and 1.96. If the null hypothesis is true and $p = 1/2$, the standard deviation of the sampling distribution (the standard error) is

$$\sigma_{\hat{p}} = \sqrt{\frac{1}{2} \times \frac{1}{2}} / \sqrt{252} = 0.0315.$$

Of the 252 respondents, 83 chose a nonwhite doll, so that the proportion was $\hat{p} = 83/252 = 0.329$. The test statistic is

$$z = \frac{0.329 - 0.500}{0.315} = -5.4.$$

Since $-5.4 < -1.96$, the test statistic lies in the region of rejection at the left side of the normal distribution. H_0 is rejected in favor of the alternative that $p \neq 0.5$. We take this evidence that the hypothesis of no racial awareness or preference is untenable. In particular, African-American children in this age category show preference for the white play object.

If we had taken a P value approach in this study, we would evaluate the probability that $|z| > 5.4$. This value is less than $2 \times \Pr(|z| > 5)$. From Appendix I (footnote), P is less than $2 \times .0000003$, that is, $P < .0000006$. The study provides very convincing evidence of racial preference; H_0 would be rejected even for extremely small values of α.

Other more detailed questions about racial attitudes, for example, the effect of age and state of residence upon the proportion choosing nonwhite dolls, were investigated in the study. The children were given other instructions too; for example, "Give me the doll that is a nice doll." Altogether, the results were a rather striking demonstration of racial awareness in young children.

11.6   Testing Hypotheses About a Median: The Sign Test*

The median is a measure of location defined even for an ordinal scale. It is not highly dependent on extreme values and thus is more appropriate than the mean for severely skewed distributions. The *sign test* for a hypothesized value of the median similarly does not depend on extreme values and is suitable for data on an ordinal scale (as well as on an interval or ratio scale). The sign test is also very easy to carry out; at the 5% significance level with a large sample the two-tailed test can be done without using pencil and paper.

The median divides the population in half in the sense that when drawing at random the probability is 1/2 of obtaining a value less than the median and 1/2 of obtaining a value greater than the median. (We assume that the probability of drawing an observation equal to the median is 0.) To test the hypothesis that a specified number, say M_0, is the median of a population, one can test the equivalent hypothesis that the probability of drawing an observation less than M_0 is 1/2. If the proportion of observations in a sample less than M_0 is sufficiently different from 1/2, the hypothesis is rejected.

Let us now put this in formal terms. Let M be the median[11] of the population sampled, and suppose that the probability of M is 0, that is,

$$\Pr(x < M) = \frac{1}{2} = \Pr(x > M),$$

where x is the variable. The null hypothesis for a two-tailed test is

$$H_0: M = M_0,$$

where M_0 is a specified number. The alternative hypothesis is $H_1: M \neq M_0$. An equivalent null hypothesis is $\Pr(x < M_0) = 1/2$, and an equivalent alternative hypothesis is $\Pr(x < M_0) \neq 1/2$. Suppose the test is to be made at the 5% level of significance.

In the sample of size n count the number of observations less than M_0 and denote this number by y. With random sampling y has a binomial distribution with a success probability of $\Pr(x < M_0)$. If the null hypothesis is true, this probability is 1/2. The standard deviation

---

[11]We assume that the median is uniquely defined.

of the proportion y/n is then[12] $\sqrt{(1/2) \times (1/2)/n} = (1/2) \times (1/\sqrt{n})$. The test statistic compares the sample ratio y/n to $1/2$ in standard deviation units, that is,

$$z = \frac{\frac{y}{n} - \frac{1}{2}}{\frac{1}{2} \times \frac{1}{\sqrt{n}}}.$$

In large random samples the distribution of this statistic is well approximated by the standard normal distribution. Thus, for example, H_0 is rejected at the 5% level if $z < -1.96$ or if $z > 1.96$.

There is an easy way to perform this test if we replace 1.96 by 2, a close approximation. In this case, H_0 is rejected if

$$\frac{\left|\frac{y}{n} - \frac{1}{2}\right|}{\frac{1}{2} \times \frac{1}{\sqrt{n}}} > 2,$$

which is equivalent to

$$\left|\frac{y}{n} - \frac{1}{2}\right| > \frac{1}{\sqrt{n}}.$$

In words, the proportion of the sample below M_0 is compared with $1/2$. H_0 is rejected if the absolute difference is bigger than $1/\sqrt{n}$, a term that involves only the sample size.

Like other tests of significance, the sign test on the median can be one-tailed or two-tailed. If H_1 is one-sided, for example, $M < M_0$, an equivalent alternative hypothesis is $\Pr(x < M_0) > 1/2$. The only procedural change is to use a region of rejection that is entirely in one tail of the standard normal distribution.[13] If the one-tailed test described here is conducted at $\alpha = .05$, H_0 is rejected at the 5% level if $z > 1.645$.

For other significance levels z is compared with other values from the standard normal table. The continuity correction may be useful to improve accuracy, especially if n is not very large. To accomplish this, the numerator of z is modified by $1/(2n)$ to make it numerically smaller.

EXAMPLE   Jones claims the median lifetime of light bulbs is 500 hours. This claim is incorrect if the median lifetime is either less than or greater than 500 hours. In a sample of 100 light bulbs 54 burned out before 500 hours had passed. Then the test statistic is

$$z = \frac{\frac{54}{100} - \frac{1}{2}}{\frac{1}{2} \times \frac{1}{\sqrt{100}}} = \frac{0.54 - 0.5}{0.5 \times 0.1} = \frac{0.04}{0.05} = 0.8.$$

---

[12]See Section 9.3.

[13]The "easy way" to perform the test cannot be used to make a one-tailed test, however.

From the table of the standard normal distribution, the probability of an (absolute) z value greater than this is $2 \times (.5 - .2881) = .4238$; this is the P value. The null hypothesis would not be rejected at any reasonable significance level. The data substantiate Jones' claim.

This procedure is called the "sign test" because it is based on counting the number of minus signs among the differences between the observations and M_0. If the parent distribution is symmetric the mean and median are identical and the sign test is appropriate to test a hypothesized value of the mean.

11.7 Paired Measurements

Testing Hypotheses About the Mean of a Population of Differences

Pairs Paired measurements arise when two measurements are made on one unit of observation. Several examples are given in Chapter 10. These include the assessment of the severity of an illness before and after medication, the degree of adjustment of a patient before and after psychotherapy, or the number of errors made by employees in the first four hours and the last four hours of the work day.

In other situations, the *method of matched samples* is used to choose pairs of individuals who have similar abilities, one of which is assigned to one experimental condition and the other to a second condition. For example, suppose that a manufacturer of monitors for computers wishes to know whether monochrome (MO) or multi-color (MC) screens produce greater eye strain. An experiment is designed in which clerical workers are to be assigned to use one type of screen or the other for 3 hours, following which the degree of eye strain is assessed on a 20-point scale. Since there is great variability among individuals in visual acuity and because poor acuity can also lead to eye strain, a test of visual acuity is given before the experiment begins. Pairs of individuals are identified who have similar acuity scores by listing the workers in order by acuity and selecting the top two, the next two, and so on. One member of each pair is randomly assigned to work at an MO screen and the other at an MC screen. A table of random numbers can be used to assign a digit to the first member of each pair, who is assigned to MO if the number is odd and MC if it is even. (See Table 11.5.) Then the comparison of the two types of screens cannot

TABLE 11.5

Use of Random Numbers to Assign Pair Members to MO or MC Screens

Employee	Visual acuity	Random digits for first member of pair	Type of screen
Karen	131	9	MO
Bill	128		MC
John	114	6	MC
Albert	109		MO
Ashley	103	7	MO
Heather	102		MC
Randy	99	3	MO
Bart	97		MC
etc.			

Hypothetical data.

be biased by systematically assigning the individuals with better acuity to MO.[14]

In each situation it may be of interest to examine the difference between the two measures and the mean of the differences. For example, we could compare the eye strain of the MO member of each pair of clerical workers with the eye strain of the MC member to assess how much additional strain was caused by one screen or the other. To understand the effect generally, we could examine the mean of the MO − MC differences. Of course, in order for such comparison to be meaningful, eye strain must be assessed on the same scale for MO and MC workers.

When the *difference* of two measurements is the variable of interest, a test of the hypothesis that the mean difference is 0 in the population can be obtained from the differences of pairs of measurements in the sample. This is a particularly useful application because a mean *difference* of 0 signifies that the mean of one measure is identical to the mean of the other measure, while a nonzero difference signifies that one measure has a higher mean than the other; the direction of the mean difference reveals which is higher. The procedure for testing the mean

---

[14]An alternative method of carrying out this experiment is to select randomly one-half of the employees to have MO displays and assign MC displays to the remaining half. This experimental procedure would also be unbiased, but it could be expected to be less efficient and to have lower statistical power. Stress depends on visual acuity as well as on the type of display so that creating homogeneous pairs with respect to acuity can be expected to reduce the variation in the stress measure.

of the differences is called the "matched t-test" or "t-test for correlated samples."

Example and Procedure

Education was one factor under consideration in Whelpton and Kiser's 1950 report *Social and Psychological Factors Affecting Fertility*. The sample consisted of $n = 153$ married couples. Each husband was asked to state the "highest school grade" completed by his wife. Each wife was also asked the highest grade she had completed. The information obtained from each husband is compared to that obtained from his wife to see whether husbands tend to exaggerate their wives' educational accomplishments. The difference between the husband's and wife's reports is the variable of interest. We want to test the null hypothesis that the population mean difference is 0 against the alternative that it is positive; the alternative reflects the idea that a respondent may tend to exaggerate his wife's education.

The basic data layout (Table 11.6) is the same as discussed in Section 10.6 of this book. Let x_i be the wife's statement of the highest grade completed and y_i be the husband's statement. The difference between their two statements is $d_i = y_i - x_i$. We are interested in the difference between the mean years as stated by the husband and the actual mean years completed by the wife, that is, $\bar{y} - \bar{x}$. It is convenient that this difference is identical to the mean of the individual differences, $\bar{d}$. Likewise, the difference between the population means $\mu_y - \mu_x$ is equal to the population mean of the differences, μ_d. Thus the test of significance for the *difference of two means* can be conducted simply by testing the *mean of the difference* variable.[15]

TABLE 11.6

Data Layout for Test of Mean Differences

Couple i	Wife's years of education x_i	Husband's statement y_i	Difference $d_i = y_i - x_i$
1	x_1	y_1	$d_1 = y_1 - x_1$
2	x_2	y_2	$d_2 = y_2 - x_2$
$\vdots$	$\vdots$	$\vdots$	$\vdots$
n	x_n	y_n	$d_n = y_n - x_n$
Sample mean	$\bar{x}$	$\bar{y}$	$\bar{d} = \bar{y} - \bar{x}$
Population mean	μ_x	μ_y	$\mu_d = \mu_y - \mu_x$
Sample standard deviation	s_x	s_y	s_d

[15]This is only possible when observations are paired, that is, when two measurements on the same scale are obtained from a single individual or matched pair. Procedures for comparing the means of two separate (independent) groups of observations are discussed in Chapter 12.

Once the difference variable is defined it can be used to state H_0 and H_1 and to complete the test of significance. The hypotheses are based on the fact that $\mu_d = 0$ if and only if $\mu_x = \mu_y$. If $\mu_x > \mu_y$, then $\mu_d < 0$, and if $\mu_x < \mu_y$, then $\mu_d > 0$. A one-tailed test is used here because the researcher's hypothesis is that husbands exaggerate their wives' education. This is represented as H_1: $\mu_x < \mu_y$, or equivalently H_1: $\mu_d > 0$. The null hypothesis is H_0: $\mu_d \leq 0$. (If all the subtractions had been done in the direction $x - y$, all of the differences would be in the opposite direction but this would have no effect whatsoever on the conclusions.)

The test procedure for a t-test on a single mean is followed with $\mu_0 = 0$. The Whelpton and Kiser report gives the mean and standard deviation of the differences as $\bar{d} = 0.32$ years and $s_{\bar{d}} = 1.07$ years. The standard error of the mean of the differences is $s_{\bar{d}} = 1.07/\sqrt{153} = 0.0865$, and the test statistic is

$$t = \frac{0.32}{0.0865} = 3.70.$$

This is the same test statistic for μ as defined in Section 11.3 except that the difference variable is substituted for x, that is,

$$t = \frac{\bar{d} - 0}{s_d/\sqrt{n}}.$$

The statistic is referred to percentage points of t_{n-1}.

Student's t-distribution with 152 degrees of freedom is almost the same as the standard normal distribution. For a one-tailed test at the 1% level, the null hypothesis is rejected if $t > 2.326$. Since $3.70 > 2.326$, the null hypothesis is rejected. The data have ruled out chance alone as a plausible explanation of the observed mean difference. The observed difference may be due to the fact that husbands tend to exaggerate their wives' education while wives tend to be accurate; or it may be due to the fact that the wives exaggerate, but the husbands exaggerate more; or it may even be due to some extraneous factor.

EXAMPLE

As a second example of the matched t-test, Table 11.7 lists data for 21 junior high school students drawn at random from a large survey of students in Washington, DC public schools (Iannotti and Bush, 1992). The survey asked about students' use of abusable substances generally. One question in the survey asked respondents to indicate the number of friends (up to a maximum of four) who were drinking alcohol, the number who were smoking cigarettes, and the number who were using marijuana. The resulting total ranges from 0 to 12. The same students were surveyed when they were in seventh grade and again when they were in eighth grade. For this example, we shall assume that the group of 21 students is a random sample from the Washington school district and ask whether, on average, friends' use of abusable substances increased, decreased, or remained the same between grades 7 and 8.

TABLE 11.7

Friends' Use of Abusable Substance in Grades 7 and 8

Respondent	Grade 7 (x)	Grade 8 (y)	Change (d)	Respondent	Grade 7 (x)	Grade 8 (y)	Change (d)
1	3	5	2	12	1	1	0
2	1	0	-1	13	7	10	3
3	0	0	0	14	0	3	3
4	2	2	0	15	2	4	2
5	0	5	5	16	0	4	4
6	0	0	0	17	0	0	0
7	5	9	4	18	2	4	2
8	2	2	0	19	4	10	6
9	8	2	-6	20	0	2	2
10	0	0	0	21	1	4	3
11	7	4	-3				
	$\bar{x} = 2.143$				$\bar{y} = 3.381$		
	$\bar{d} = \bar{y} - \bar{x} = 1.238$				$s_d = 2.737$		

source: Iannotti and Bush (1992).

Since we are interested in determining whether the use of abusable substances increased or decreased during this one-year period, we shall make a two-tailed test. The null and alternative hypotheses are H_0: $\mu_d = 0$ and H_1: $\mu_d \neq 0$, respectively. The test statistic is

$$t = \frac{\bar{d} - 0}{s_d/\sqrt{n}} = \frac{1.238}{2.737/\sqrt{21}} = 2.073.$$

The significance points from the distribution t_{20} for a two-tailed test at $\alpha = .05$ are $\pm t_{20}(.025) = \pm 2.086$. Since 2.073 is less than 2.086, the test statistic does not lie in the region of rejection. H_0 is accepted; we have no compelling evidence that friends' average use of abusable substances changes from grade 7 to grade 8.

Testing the Hypothesis of Equality of Proportions

The measurements in a pair can also be dichotomous. The patients who undergo psychotherapy can be asked before and after treatment "would you recommend psychotherapy to a friend who was seriously depressed?" The proportions answering "yes" before and after psychotherapy can be compared statistically. Or the husband and wife in a couple may each vote "pro" or "con" on a particular political issue and we can ask whether the proportion of married men who favor the issue is greater, the same, or less than the proportion of married women voting yes.

Turnover Tables

In Section 6.1 it is noted that one of the ways a 2×2 table arises is when we ask the same persons the same question at two different times. The resulting data are termed "change-in-time," and the resulting 2×2 table a "turnover table."

Jones, a political candidate, had polls taken in August and October; she wished to compare the proportions of voters favoring her at the two different times. If the same persons were polled both times, the results would appear as in Table 11.8. The results can be summarized in a "turnover table" like Table 11.9. The number of persons changing to Jones was 224, more than the 176 who changed away from her, causing the proportion of the sample in favor of Jones to grow from 37% in August to 40% in October. Does this reflect a real change in the population from which these 1600 persons are a sample?

The population is represented in Table 11.10. The proportions of the population in favor of Jones in August and October are $p_1 = (A + B)/N$ and $p_2 = (A + C)/N$, respectively. We wish to test the hypothesis of equality of these two proportions. Note that p_2 has C where p_1 has B; they are equal if and only if $B = C$, that is, if and only if the number changing to Jones is the same a the number changing away from Jones. Moreover, the equality $B = C$ is equivalent to $C/(B + C) = 1/2$. Note

TABLE 11.8

Table of Results of Poll

Identification of individual	Did individual favor Jones?	
	August	*October*
1	Yes	Yes
2	Yes	No
⋮	⋮	⋮
n	No	Yes

TABLE 11.9
Turnover Table for Poll

		October		
		For Jones	*For other candidates*	*Total*
August	*For Jones*	416	176	592 (37%)
	For other candidates	224	784	1008 (63%)
	Total	640 (40%)	960 (60%)	1600 (100%)

Hypothetical data.

TABLE 11.10
Turnover Table for the Population

		October		
		Jones	*Others*	*Total*
August	*Jones*	A	B	$A + B$
	Others	C	D	$C + D$
	Total	$A + C$	$B + D$	$N = A + B + C + D$

that $B + C$ is the total number of persons who changed their minds, and C is the number who switched to Jones. The number $p = C/(B + C)$ is the proportion, among those who changed, who changed to Jones. We wish to test the null hypothesis H_0: $p = 1/2$ against the alternative H_1: $p \neq 1/2$.

The persons in the sample who changed their minds were the 224 who switched to Jones and the 176 who switched away from Jones. These $224 + 176 = 400$ persons are considered a sample from the population of $C + B$ persons who changed their minds.

The observed sample proportion who changed to Jones is

$$\hat{p} = \frac{224}{400} = 0.56.$$

If the null hypothesis is true, the population proportion is $1/2$ and the standard deviation is $\sigma_{\hat{p}} = \sqrt{(1/2)(1/2)(1/400)} = 0.025$. The test statistic is

$$z = \frac{0.56 - 0.50}{0.025} = 2.4.$$

This exceeds 1.96, the significance point for a two-tailed 5%-level test from the standard normal distribution. The null hypothesis is rejected.

General Procedure.   We can describe a general problem of this type as follows. Each of n individuals is asked the same question at two different times. The sample joint frequency distribution takes the form of Table 11.11. The estimate of p, the proportion of changers in the population who change from No to Yes, is the sample proportion

$$\hat{p} = \frac{c}{b + c}.$$

The test statistic for testing $c/(b + c) = 1/2$ is[16]

$$z = \frac{\hat{p} - \frac{1}{2}}{\sqrt{\frac{1}{2} \times \frac{1}{2}/(b + c)}}.$$

The value of this statistic is compared with a value from the standard normal distribution to decide whether the sample result is significant. A large sample is important in order for the normal approximation to be appropriate for a test on p; in particular, the number of changers $(b + c)$ should be large, that is, 25 or more.

TABLE 11.11
2 × 2 Table: Change-in-Time Data

		Time 2		
		Yes	*No*	*Total*
Time 1	Yes	a	b	$a + b$
	No	c	d	$c + d$
	Total	$a + c$	$b + d$	n

Matched Samples.   The comparison of numbers of patients recommending psychotherapy before and after treatment can be made using a turnover table as described in the preceding discussion. Sometimes the proportions to be compared are not obtained from the same observations but from "matched pairs" such as the husband and wife of a married couple, an employer and his or her employee, or two observations that have been judged by an investigator as being similar on a number of background characteristics. When the basic response is dichotomous, the same procedures for testing homogeneity can be applied.

---

[16]This test is referred to as "McNemar's test," developed originally by Quinn McNemar (1947).

As an example, suppose that convicts due to be released in one year's time were grouped into matched pairs according to age, type of crime, and amount of prison time served. One member of each pair was selected at random for a special rehabilitation program. For each pair it was noted whether the convict in the program returned to prison within one year of release and whether the convict not in the program returned. The frequency table for pairs is Table 11.12. The numbers which pertain to whether the program reduced returns are c and b.

TABLE 11.12
Return of Convicts

| | | Convict not in program | | |
		Returned	*Not returned*	*Total*
Convict in program	*Returned*	a	b	$a + b$
	Not returned	c	d	$c + d$
	Total	$a + c$	$b + d$	n

Summary

Hypothesis testing is a procedure for reaching a yes/no decision about a population value based on a sample. It enables an investigator to decide whether a *null hypothesis* is a correct description of the population or whether the null hypothesis is to be rejected in favor of an *alternative*. The null hypothesis is a "statement of no effect" and includes the assertion that the parameter is equal to the specified value. The alternative hypothesis may be one-sided (for example, $H_1: \mu > \mu_0$ or $H_1: \mu < \mu_0$) or two-sided (for example, $H_1: \mu \neq \mu_0$). These require a *one-tailed test* or *two-tailed test*, respectively.

Hypotheses may be tested about various parameters including the population mean, median, proportion, and others described in later chapters. In each application three steps are involved: (1) The sampling distribution of the appropriate statistic is described, a significance level is chosen, and a *region of rejection* is identified. The size of the region of rejection is called the *significance level*, and is converted to a *significance point* by reference to the appropriate probability distribution; (2) a random sample is drawn, the relevant statistic computed and re-expressed in the form of a *test statistic*; and (3) the test statistic is compared with the significance point to determine if the null hypothesis is accepted or rejected.

The test statistic has a specific form and is referred to a specific probability distribution depending on the parameter being tested. For a test of a population mean, the test statistic can be computed using the population standard deviation (σ) or the standard deviation computed from the sample (s). The former should be used if it is available, and the resulting test statistic is compared with significance point(s) from the standard normal distribution. If no value for σ is available, s may be substituted. The resulting test statistic is compared with significance points from the *t*-distribution with $n - 1$ degrees of freedom, on the condition that the parent population is normal. For a test of a proportion, the test statistic is compared with significance point(s) from the standard normal distribution; this test, however, requires a reasonably large sample.

A *Type I error* consists of rejecting the null hypothesis when it is true. We control the probability of such an error to be equal to some small prescribed probability; this prescribed probability is the *significance level* and is represented by the symbol α.

A *Type II error* consists of accepting the null hypothesis when it is false. The probability of rejecting the null hypothesis when it is false is called the *power*. This probability depends upon the true value of the parameter, the sample size, and the significance level. A rule of thumb for the choice of significance level is to use 1% if it gives reasonable power against reasonable alternative parameter values, otherwise, use 5% or even 10%.

When the test statistic is approximately normally distributed, a two-tailed hypothesis test may be carried out by forming the appropriate confidence interval and noting whether or not the hypothesized parameter value falls in the interval.

As an alternative to (or in addition to) reporting that a null hypothesis is accepted or rejected, the results of a test of significance may be reported in terms of a *P value*. The *P* value defines the entire range of α levels for which the null hypothesis would be rejected and also conveys the degree to which the null hypothesis is refuted by the data.

Appendix 11A   The Continuity Correction

Consider a *discrete* random variable which can take on successive integer values. Then successive possible values of the mean of a sample of size 25, for example, differ by 1/25, or 0.04. The amount of the continuity correction is one-half of this, or 0.02. For a sample of size n, the continuity correction is $1/(2n)$. The continuity adjustment consists of adding or subtracting this constant as appropriate. For example, in testing $H_0: p \leq p_0$ against the alternative $H_1: p > p_0$ we would subtract $1/(2n)$ from the sample proportion ($\hat{p}$) before computing the test statistic. The same significance point is used from the standard normal distribution, for example, 1.645 if $\alpha = .05$.

In making a two-tailed test $1/(2n)$ is subtracted from $\hat{p}$ when comparing to the upper-tail significance point, and $1/(2n)$ is added before computing a test statistic to compare to the lower-tail significance point. (A way to remember whether to add or subtract the $1/(2n)$ is to recall that its effect is to make it "harder" to reject the null hypothesis.) If n is large, the continuity correction is small and may be ignored.

Binomial variables and ranks are, of course, discrete. The continuity correction for the number of successes is $\pm 1/2$, and for the proportion of successes is $\pm 1/(2n)$.

Exercises

11.1 The Graduate Record Examination (GRE), like the Scholastic Assessment Tests, is scaled to have a mean for the population of 500 and standard deviation of 100. A sample of 225 college seniors who are applying to graduate schools in education obtains an average GRE score of 512. Test the hypothesis that the mean GRE score of the population of applicants to education graduate schools is less than or equal to 500 against the alternative that it is greater than 500. Use an α level of .05. In your answer, give

(a) The null and alternative hypotheses in statistical form,

(b) The probability distribution to be used in testing the hypothesis,

(c) The .05 significance point from the distribution indicated in (b),

(d) The value of the test statistic,

(e) A decision to accept or reject the null hypothesis.

11.2 (continuation) Use the table of the standard normal distribution to approximate the P value for the test statistic in Exercise 11.1. From the P value state whether the null hypothesis would be rejected at

(a) $\alpha = .10$,

(b) $\alpha = .02$,

(c) $\alpha = .01$,

(d) $\alpha = .001$.

11.3 The Graduate Record Examination (GRE), like the Scholastic Assessment Tests, is scaled to have a mean for the population of 500 and a standard deviation of 100. A sample of 144 college seniors who are applying to graduate schools in engineering obtains an average GRE score of 492. Test the hypothesis that the mean GRE score of the population of applicants to engineering graduate schools is greater than or equal to 500 against the alternative that it is less than 500. Use an α level of .01. In your answer, give

(a) The null and alternative hypotheses in statistical form,

(b) The probability distribution to be used in testing the hypothesis,

(c) The .01 significance point from the distribution indicated in (b),

(d) The value of the test statistic,

(e) A decision to accept or reject the null hypothesis.

11.4   (continuation) Use the table of the standard normal distribution to approximate the P value for the test statistic in Exercise 11.3. From the P value state whether the null hypothesis would be rejected at

(a) $\alpha = .15$,

(b) $\alpha = .10$,

(c) $\alpha = .05$,

(d) $\alpha = .001$.

11.5   A random sample of 25 observations drawn from a population that is approximately normally distributed has mean $\bar{x} = 44.5$ and $s = 3.6$. Use these data to test the null hypothesis that the population mean is equal to 43 against the alternative that the population mean does not equal 43. Use an α level of .01. In your answer, give

(a) The null and alternative hypotheses in statistical form,

(b) The probability distribution to be used in testing the hypothesis,

(c) The .01 significance point from the distribution indicated in (b),

(d) The value of the test statistic,

(e) A decision to accept or reject the null hypothesis.

11.6   (continuation) Using the table of the t-distribution, determine a P value interval for the test statistic in the preceding exercise. *From the interval*, state the entire range of α values for which the null hypothesis would be rejected. Does this range include

(a) .10,

(b) .05,

(c) .01?

11.7   The average body temperature of healthy human adults is 98.2°F.[17] Close friends speculate that medical students have an average body

[17]This figure was released to the public in an article in the *Washington Post* on September 23, 1992.

temperature lower than this. A random sample of 30 medical students yields a mean body temperature of $\bar{x} = 97.9°F$ and standard deviation $s = 0.16°F$. Test the null hypothesis at the 5% level that the mean temperature in the population of medical students is greater than or equal to 98.2 degrees. In your answer, give

(a) The null and alternative hypotheses in statistical form,

(b) The probability distribution to be used in testing the hypothesis,

(c) the 5% significance point from the distribution indicated in (b),

(d) The value of the test statistic,

(e) A decision to accept or reject the null hypothesis.

11.8 (continuation) Using the table of the t-distribution, determine a P value interval for the test statistic in the preceding exercise. *From the interval,* state the entire range of α values for which the null hypothesis would be rejected. Does this range include

(a) .10,

(b) .05,

(c) .01?

11.9 ✓ Table 11.13 gives the best times of 10 sprinters in Great Britain who ran 200 meters in under 21.20 seconds in 1988 and who also recorded a time for 100 meters.

TABLE 11.13

Best Times of British Sprinters (in Seconds)

Athlete	200 meters best	100 meters best
L. Christie	20.09	9.97
J. Regis	20.32	10.31
M. Rosswess	20.51	10.40
A. Carrott	20.76	10.56
T. Bennett	20.90	10.92
A. Mafe	20.94	10.64
D. Reid	21.00	10.54
P. Snoddy	21.14	10.85
L. Stapleton	21.17	10.71
C. Jackson	21.19	10.56

source: British Amateur Athletic Board, Amateur Athletic Association, and National Union of Track Statisticians (1989).

90% confidence level

Assuming normality, test at the 10% level the hypothesis that the mean time for the population of the fastest 200-meter sprinters *running 100 meters* is no greater than 10 seconds. In your answer give

(a) The null and alternative hypotheses in statistical form,

(b) The probability distribution to be used in testing the hypothesis and the significance point from that distribution,

(c) The value of the test statistic,

(d) A decision to accept or reject the null hypothesis,

(e) A verbal statement about whether the mean time in the *sample* is less than, equal to, or greater than 10 seconds and your conclusion about the *population*.

11.10   Table 10.6 (Exercise 10.18) gives average daily nutrient intakes for a sample of 17 four- and five-year-old children. Use these data to test two hypotheses, (a) and (b) below, about the population of children from which the sample was drawn. For each test, state the null and alternative hypotheses in statistical form, paying close attention to whether a one-sided alternative or two-sided alternative is called for; name the probability distribution used in testing the hypotheses and the significance point from that distribution; give the value of the test statistic; and state your decision about whether the null hypothesis is accepted or rejected.

(a) Do the data indicate that the mean daily sodium intake of four- and five-year-old children is greater than 1000 milligrams? (Hint: Is this the null or alternative hypothesis?) Use the 5% significance level.

(b) Test the hypothesis that the proportion of energy attributable to fatty acids is equal to 1/3, that is, 33.33%. Use the 10% significance level.

11.11   (continuation) Using the table of the *t*-distribution, determine a *P* value interval for each test statistic in the preceding exercise.

11.12 ✓ The acceleration times (0–40 miles per hour) of a sample of 25 sports cars manufactured by a particular company are given in Table 11.14. Are these data consistent with a population mean of 3 seconds, as advertised? Make a two-tailed test at the 5% level of significance. In your answer give

TABLE 11.14

Acceleration Times (0–40 miles per hour) in Seconds for 25 Sports Cars

3.00	3.02	2.93	3.05	3.03
2.95	2.95	2.95	2.96	2.94
3.00	2.93	2.95	2.94	2.93
2.94	2.96	2.95	2.95	2.96
3.00	3.06	3.05	3.07	3.05

Hypothetical data.

(a) The null and alternative hypotheses in statistical form,

(b) The probability distribution to be used in testing the hypothesis,

(c) The 5% significance points from the distribution indicated in (b),

(d) The value of the test statistic,

(e) A decision to accept or reject the null hypothesis.

11.13 (continuation) Determine a P value interval for the test statistic in the preceding exercise. *From the interval* state the entire range of α values for which the null hypothesis would be rejected.

11.14 Hyman, Wright, and Hopkins (1962) conducted a study of 96 participants in a summer Encampment for Citizenship. Each participant was given a Civil Liberties Test at the beginning and at the end of the summer. The score at the beginning was subtracted from the score at the end to give a difference score for each participant. A *lower score* (negative difference) indicates improvement in attitude in the direction of civil liberties. The data are shown in Table 11.15.

TABLE 11.15

Campers' Scores on Civil Liberties Test at Beginning and End of the Summer

		Score at end of summer										Total
		0	*1*	*2*	*3*	*4*	*5*	*6*	*7*	*8*	*9*	*10*	
	0	10	2	1	1	2							16
	1	5	2	1									8
	2	3	2	13	1	1							20
Score at	*3*	2		2	2								6
beginning	*4*	5	4	4	1	6		1					21
of summer	*5*		1		3	1	0		1				6
	6	1		1		4	2	2	2				12
	7			1		1	1		0				3
	8	1								0			1
	9			1						1	0		2
	10									1		0	1
	Total	27	11	23	9	15	3	3	3	2	0	0	96

source: Hyman, Wright, and Hopkins (1962), Table D-6, p. 392. Copyright 1962 by the Regents of the University of California; reprinted by permission of the University of California.

Test at the 5% level the hypothesis that the population mean difference is zero against the one-sided alternative that the population mean difference is negative (indicating improvement). The sample variance of the differences is 4.14. In your answer give

(a) The null and alternative hypotheses in statistical form,

(b) The probability distribution to be used in testing the hypothesis, and the significance point from that distribution,

(c) The value of the test statistic,

(d) A decision to accept or reject the null hypothesis,

(e) A conclusion regarding the effectiveness or ineffectiveness of the program in improving civil liberties scores.

11.15   Use the data of Table 3.15 to test the null hypothesis that the average increase in hours of sleep due to the two drugs tested are the same. Use a two-tailed test at the 5% significance level.

11.16   Exercise 10.34 describes a study by Ciancio and Mather (1990) in which 15 dental patients brushed their teeth manually for one week and then with an Interplak electric toothbrush for one week. The mean on the plaque measure at the end of the first week was 39 and at the end of the second week was 47. The standard deviation of the *changes* from one week to the next was estimated to be 14.8.

Use these results to test the hypothesis that in the population there is no change in mean calculus buildup using the Interplak brush for one week. Use the 10% level of significance. In your answer give

(a) The null and alternate hypotheses, being careful to label the first-week and second-week means clearly,

(b) The probability distribution used to test the hypothesis and the significance point(s) from that distribution,

(c) The computed value of the test statistic,

(d) The decision to accept or reject the null hypothesis and a verbal statement of whether there is an increase, decrease, or no change in calculus buildup with one week's use of the Interplak brush.

(e) If you completed Exercise 10.34, compare your results with the conclusion you reached from the confidence interval. That is, do you find no significant change if 0 is contained in the 95% confidence interval and significant change if 0 is not in the interval?

11.17 ✓   A psychologist hypothesized that an individual's anxiety level may be increased by viewing a movie that contains violent scenes. To test this hypothesis she showed a half-hour movie containing a great deal of violence to a random sample of 9 college freshmen. The 50-item Hayes Anxiety Measure (HAM) was administered to the students before and after viewing the movie. Scores are given in Table 11.16. Use these data to test the psychologist's hypothesis at the 1% level of significance. State the null and alternative hypotheses, give the significance point, the value of the test statistic, and state the conclusion that the psychologist should draw from the results.

TABLE 11.16

Anxiety Scores Before and After Viewing Violence-Filled Movie

Student (i)	Anxiety before movie	Anxiety after movie
1	21	29
2	28	30
3	17	21
4	24	25
5	27	33
6	18	22
7	20	19
8	23	29
9	28	26

Hypothetical data.

11.18 An insurance company made a change in the rules for its health insurance plan. Effective January 1, 1988, the insurance company would no longer pay 100% of hospital bills; the patient would have to pay 25% of them. The joint distribution of number of hospital stays in 1987 and 1988 for a random sample of individuals is given in Table 11.17. Was there a significant decrease in usage between 1987 and 1988? (Choose an α level and test the significance of the sample mean difference.)

TABLE 11.17

Distribution of Hospital Stays in 1987 and 1988

		Number of hospital stays in 1988			
		0	1	2	Total
Number of hospital stays in 1987	0	2219	40	8	2267
	1	65	105	51	221
	2	54	8	50	112
	Total	2338	153	109	2600

Hypothetical data.

11.19 In a random sample of 1600 of the state's registered voters, 850 favor Double Talk for governor. Let p represent the proportion of all the state's registered voters in favor of Double Talk. Test H_0: $p \leq 1/2$ against the alternative H_1: $p > 1/2$ at the 5% level.

11.20   In a random sample of 3600 of the country's registered voters, 1850 favor Mr. Big for president. Let p represent the proportion of all the country's voters in favor of Mr. Big. Test H_0: $p \leq 1/2$ against the alternative H_1: $p > 1/2$ at the 5% level.

11.21   The Ad Manager says 25% of the cars sold with five-speed manual transmissions are four-door sedans. Super Salesperson, believing that four-door sedans constitute a smaller percentage of total sales, examines the data of a random sample of 400 cars from among the thousands sold with five-speed manual transmissions, finds only 40 four-door cars among them, and tells the Ad Manager that the 25% is wrong. Do you think the Ad Manager's figure of 25% is wrong?

11.22   Exercise 10.25 cites data from the *San Jose Mercury News* indicating that 67% of respondents to an opinion poll supported a woman's right to obtain an abortion during the first trimester of pregnancy. Assuming that the sample was comprised of 100 individuals drawn at random from a very large population of adults, test the hypothesis that no more than one-half of the population supports this proposition. Use the 5% level of significance. In your answer give

(a) The null and alternative hypotheses in statistical form,

(b) The probability distribution to be used in testing the hypothesis and the significance point from that distribution,

(c) The value of the test statistic,

(d) The decision to accept or reject the null hypothesis and a verbal statement of whether more than half of the population does or does not support the proposition.

11.23   Jerome Carlin (1962) presents information about the activities of solo legal practitioners in Chicago, and the social system within which these lawyers work. Many of the lawyers are limited by the availability of cases to those which demand little legal competence. If they are to remain solvent their work must at times be done in ways which are not condoned by the Bar Association's Canon of Ethics. The Chicago Bar Association might provide an arena in which the ethical dilemmas of the solo lawyers could be worked out. However, this probably would not be done if solo lawyers were not active in that association. Carlin presents a table to compare the frequency of solo practitioners holding office in the Chicago Bar Association with the proportion of solo lawyers in practice in Chicago. (See Table 11.18.)

Is the leadership of the Chicago Bar Association representative of the solo practitioners (or is it biased in favor of other lawyers)? Test the null hypothesis that the 95 committee chairpersons and vice chairpersons have, in effect, been drawn randomly from a population of lawyers in which the proportion of solo practitioners is greater than or equal

TABLE 11.18

Chicago Lawyers Holding Official Positions in the Chicago Bar Association, by Status in Practice, 1956

Status in practice	Percentage of all Chicago lawyers (1956)	Officers and members of board of mgrs. (1955–1957)	Committee chairs and vice chairpersons (1956)
Individual practitioners	54%	10	18
Lawyers in firms of 2 to 9	23%	23	29
Lawyers in firms of 10 or more	7%	28	38
Not in private practice	16%	2	10
Total	100%	63	95
Number of lawyers: 12,000			

source: *Lawyers on Their Own: A Study of Individual Practitioners in Chicago*, by Jerome E. Carlin, copyright 1962 © by Rutgers, The State University, p. 203. Reprinted by permission of the Rutgers University.

to 0.54. Use significance level 1%. The alternative hypothesis is that the proportion in the population drawn from is less than 0.54.

11.24   A doctor reported that the normal death rate for patients with extensive burns (more than 30% of skin area) has been cut by 25% through the use of silver nitrate compresses. Of 46 burned adults treated, 21 could have been expected to die. The number who died was 16. The compress treatment seemed especially beneficial to children. Of 27 children under 13, 2 died, while the expected mortality rate was 6 or 7.

The report implies that the expected mortality rate for adult burn patients before the new treatment was about $21/46 = 0.46$ and that the rate for children was about $6/27 = 0.22$.

(a) Test the hypothesis that the mortality rate for treated adults is 0.46 against the alternative that it is less than 0.46 at the 5% level.

(b) Test the hypothesis that the mortality rate for treated children is 0.22 against the alternative that it is less than 0.22 at the 5% level. (The sample size for the children is a bit small for use of the normal approximation, but use it for this exercise.)

11.25   In a study of attitudes in Elmira, New York, in 1948, Paul Lazarsfeld and associates found the following changes in attitudes toward the likelihood of war in the next ten years among the same persons interviewed in June and again in October. Treat these 597 people (Table 11.19) as a random sample from a large population, and test the hypothesis that the proportion of the population expecting war was the same at the two times.

TABLE 11.19
2 × 2 Table: Change-In-Time Data

If null was true the numbers would be approx the same

		October response		
		Expects war	*Does not expect war*	*Total*
June response	*Expects war*	194	45	239
	Does not expect war	147	211	358
	Total	341	256	597

source: Lazarsfeld, Berelson, and Gaudet (1968).

Marg. hypothesis: there has been no change in response from June to October. [turnover table]

11.26 Table 2.16 lists the percentage of minority students in a sample of the nation's largest school districts (over 10,000 students). The percentage of students in the entire country who are members of identified minority groups is 24.7%. Test that the median percent minority in large school districts is 25 against the alternative hypothesis that it is either less than or greater than 25. (Use the sign test.) Approximate the P value for the test and state the range of α values for which the null hypothesis would be rejected.

11.27 Jones claims the median daily precipitation during spring is $1/2$ inch per day. On 49 of the 90 spring days the rainfall exceeded $1/2$ inch. Using the sign test, test at the 5% level the hypothesis that the true median daily rainfall is one-half inch per day.

11.28 (continuation) The sign test is based on the assumption that the observations are independent. Do you think this assumption is justified? Is it reasonable to assume, for example, that the conditional probability that rainfall exceeds $1/2$ inch tomorrow, given that it exceeded $1/2$ inch today, is equal to the probability that rainfall exceeds $1/2$ inch tomorrow, given that it did not exceed $1/2$ inch today?

11.29 Give an approximate P value or P value interval for each of the following tests of significance: (the t or z value is the computed test statistic.)
(a) H_0: $\mu \leq 1200$, H_1: $\mu > 1200$, $n = 31$, $z = 1.49$;
(b) H_0: $\mu \geq 4.6$, H_1: $\mu < 4.6$, $n = 60$, $z = -2.66$;
(c) H_0: $\mu = 0$, H_1: $\mu \neq 0$, $n = 1,000$, $z = -2.12$;
(d) H_0: $\mu \leq 3.52$, H_1: $\mu > 3.52$, $n = 21$, $t = 1.84$;
(continued ...)

(e) H_0: $\mu \geq 33{,}400$, H_1: $\mu < 33{,}400$, $n = 18$, $t = -1.45$;

(f) H_0: $\mu = 28$, H_1: $\mu \neq 28$, $n = 612$, $t = -3.81$.

11.30 Give an approximate P value or P value interval for each of the following tests of significance: (The t or z value is the computed test statistic.)

(a) H_0: $p \leq .333$, H_1: $p > .333$, $n = 40$, $z = 1.87$;

(b) H_0: $\mu_d \geq 0$, H_1: $\mu_d < 0$, $n = 22$, $t = -1.12$;

(c) H_0: $p = .90$, H_1: $p \neq .90$, $n = 1{,}550$, $z = -2.81$;

(d) H_0: $\mu_d = 0$, H_1: $\mu_d \neq 0$, $n = 41$, $t = 3.18$;

11.31 Section 11.7 presents an example of the matched t-test for comparing the use of abusable substances among students in grade 7 and grade 8. Suppose that the researcher speculated that substance use *increased* from grade 7 to grade 8; thus the test would be a one-tailed test instead of two-tailed as presented. Using the data presented in Table 11.7, obtain the P value interval for the one-tailed test. From this interval, state the entire range of α values for which the null hypothesis would be rejected.

11.32 In each of the situations below a hypothesis test is required. In each case state the null hypothesis and the alternative hypothesis. If you use letters like p or μ, state what they mean. Do not do any calculations.

(a) It is well known that the lengths of the tails of newborn Louisiana swamp rats have a mean of 4.0 centimeters and a standard deviation of 0.8 centimeters. An experimenter is wondering whether or not adverse conditions on the rat mother shorten the tails of the young. In his lab the rat mothers were kept on a rotating turntable. The 131 young rats born from these mothers had a mean tail length of 3.8 centimeters.

(b) The U. S. Government issues \$1.00 coins to commemorate that the federal budget has at last been balanced. Coins are guaranteed to contain *at least* 1/10 ounce of gold. An investor purchases 2,000 of these coins and has them assayed to determine the gold content. The mean gold content was 0.098 ounce and the standard deviation was 0.002 ounce. He wishes to test whether the government was fulfilling its promise in general.

(c) A roulette wheel has 38 slots; there are 18 red slots, 18 black slots, and 2 green slots. Since the two green slots (0 and 00) are favorable to the casino, a wheel weighted by the casino owners would probably have a higher probability for the green slots. You suspect that the wheel at the Sodom and Gomorrah Casino has been weighted. Along with some friends you watch the wheel for 1000 successive spins. During these 1000 turns the green slots came up 27 times.

11.33 In each of the situations below a hypothesis test is required. In each case give

(i) the null hypothesis,
(ii) the alternative hypothesis, and
(iii) the test statistic that should be used to decide between H_0 and H_1.

If you use letters like p or μ, state what they mean. Do not do any calculations.

(a) The federal government provides funds for school children classified as disabled, up to a maximum of 12% of a district's enrollment. If a district classifies more than 12% of its youngsters as disabled, it is obliged to spend its own funds for special services for disabled students in excess of the 12% base. If it classifies fewer than 12% of its youngsters as disabled, it is sacrificing the additional money to which it may be entitled. A large school district sampled files of 1000 youngsters and discovered that 180 of these pupils were classified as disabled. Do these data indicate that the district is losing money (for either reason)?

(b) One-quart containers of soda pop are supposed to contain an average of 16 fluid ounces of liquid when filled. Any more than this results in a loss to the manufacturer. Any less than this creates the risk of customer complaints. A random sample of 100 bottles is drawn from the production line and the contents are found to average 16.1 fluid ounces. The standard deviation of bottled contents is known from years of monitoring to be 0.15 fluid ounce.

(c) Court cases involving juveniles are required by law to be adjudicated within 21 days on average. A sample of 60 cases processed by the Harper Valley County Court yields a mean time-to-adjudication of 18.2 days with a standard deviation of 2.2 days. Are these data consistent with the legal time requirement?

11.34   Suppose that you are making a two-tailed test on a mean and that σ is known.

(a) Is it possible for the null hypothesis to be rejected if $\alpha = .05$ but not if $\alpha = .01$? Explain, giving numerical facts that are relevant to your explanation.

(b) Is it possible for the null hypothesis to be rejected if $\alpha = .01$ but not if $\alpha = .05$? Explain.

(c) Does the lack of statistical significance at the 5% level imply that the results will be nonsignificant at the 1% level? Explain.

(d) Does the lack of statistical significance at 1% level imply that the results will be nonsignificant at the 5% level? Explain.

11.35   Suppose that you are making a test on a proportion at the 1% level of significance.

(a) Is it possible that the same data will be statistically significant making a one-tailed test but not if you make a two-tailed test? Explain, giving numerical values that are relevant to your answer.

(b) Is it possible that the same results will be statistically significant making a two-tailed test but not if you make a one-tailed test? Explain.

(c) Does the lack of statistical significance for a one-tailed test imply that the results will be nonsignificant if you make a two-tailed test using the same data? Explain.

(d) Does the lack of statistical significance for a two-tailed test imply that the results will be nonsignificant if you make a one-tailed test using the same data? Explain.

11.36   Suppose that you are testing $H_0: \mu \leq \mu_0$ at $\alpha = .05$. Compare the z- and t-tests by constructing a table in which the left-hand column contains the sample size from $n = 2$ to $n = 30$ in intervals of 2, and the right-hand column contains the difference between the significance point from the appropriate t-distribution and that from the standard normal distribution. Construct a graph of these values and summarize the pattern that is seen.

11.37   (continuation) Repeat Exercise 11.36 with $\alpha = .01$. What are the similarities and differences between these findings and those for $\alpha = .05$?

11.38   Each of the problems below gives two types of information: (i) a null hypothesis to be tested at a particular α, and (ii) a $1 - \alpha$ confidence interval that was constructed from the data collected to test the hypothesis. In each problem, use the confidence interval to decide whether the hypothesis is accepted or rejected, or whether there is insufficient information given to make such a decision.

(a) Hypothesis $H_0: \mu = 17.4$,   Interval: $13.2 \leq \mu \leq 17.0$;

(b) Hypothesis $H_0: p = 0.75$,   Interval: $0.63 \leq p \leq 0.81$;

(c) Hypothesis $H_0: \mu_d = 0$,   Interval: $0.6 \leq \mu_d \leq 2.2$;

(d) Hypothesis $H_0: \mu_d = 0$,   Interval: $-4.1 \leq \mu_d \leq 12.0$;

(e) Hypothesis $H_0: p \leq 0.5$,   Interval: $0.31 \leq p \leq 0.48$;

(f) Hypothesis $H_0: \mu \geq 100$,   Interval: $90 \leq \mu \leq 96$.

11.39   Suppose that a test is made of $H_0: \mu \leq 500$ against the alternative $H_1: \mu > 500$ using an α value of .05. A sample of n observations is drawn from a population in which $\sigma = 25$. What is the power of the test if

(a) the actual value of μ is 503 and $n = 64$?

(b) the actual value of μ is 503 and $n = 100$?

(continued...)

(c) the actual value of μ is 508 and $n = 64$?

(d) the actual value of μ is 510 and $n = 100$?

(e) the actual value of μ is 500 and $n = 64$?

State three principles that are illustrated by comparing these results.

11.40   Repeat the computations for Exercise 11.39 using an α value of .01. How do these results compare with those for $\alpha = .05$?

11.41   Assume that a test is made of H_0: $\mu = 98.6$ against the alternative that $\mu \neq 98.6$ at the 1% level of significance. A sample 25 observations is drawn from the population in which $\sigma = 0.15$. What is the power of the test if the actual value of μ is

(a) 98.645?

(b) 98.555?

(c) 98.660?

(d) 98.600?

11.42   A researcher wishes to test H_0: $\mu \geq 12$ against H_1: $\mu < 12$ at the 5% significance level. If the actual value of μ is 11.1 and $\sigma = 1.8$,

(a) How large a sample is needed to have an 80% chance of rejecting H_0 (0.80 power)?

(b) How large a sample is needed to have a 90% chance of rejecting H_0?

(c) How large a sample is needed to have an 80% chance of rejecting H_0 if $\alpha = .01$ instead of .05?

12
Differences Between Populations

Introduction

Frequently an investigator wishes to compare or contrast two populations—sets of individuals or objects. This may be done on the basis of a sample from each of the two populations, as when average incomes in two groups, average driving skills of males and females, or average attendance rates in two school districts are compared. The polio vaccine trial compared the incidence rate of polio in the hypothetical population of children who might be inoculated with the vaccine and the rate in the hypothetical population of those who might not be inoculated; the two groups of children observed were considered as samples from these respective (hypothetical) populations. This example illustrates an experiment in which a *group receiving an experimental treatment is compared with a "control" group*. Ideally, the control group is similar to the experimental group in every way except that its members are not given the treatment.

Other experiments involve *the comparison of two methods of accomplishing an objective*. A teacher may compare the reading achievement

455

of children taught in two different ways. A metallurgist may compare two methods of baking metal plates to improve their tensile strengths. A physician may compare the effectiveness of two dosages of a medicine used to reduce high blood pressure.

The steps involved in the experimental comparison of two treatments or methods are diagrammed in Figure 12.1. A random sample is drawn

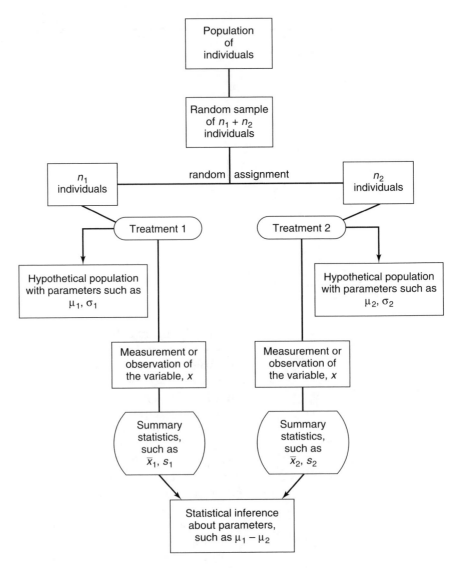

Figure 12.1   *Outline of steps in an experiment for comparing two treatments.*

from the basic population of all individuals. Members of this sample are randomly assigned to Treatment 1 or Treatment 2. There are thus two hypothetical populations of the values of the variable x—one which would result if all individuals in the basic population received Treatment 1 and one which would result if they all received Treatment 2.[1] The inference refers to parameters of these two hypothetical populations. The individuals given a particular treatment are considered to be a sample from the hypothetical population of individuals given that treatment.

In this chapter we discuss methods for making inferences about the difference between the means of two populations on the basis of two independent random samples, one from each population.[2] This is followed by a discussion of procedures for comparing the proportions of two populations that have a particular feature of interest.

Like the test of a single mean discussed in Chapter 11, different approaches are required for comparing two means depending on whether the population variance is known or must be estimated from the samples; these approaches are described separately in Section 12.1 and then in Sections 12.2 and 12.3. When two means are compared, we focus on the *differences* $\mu_1 - \mu_2$ and $\bar{x}_1 - \bar{x}_2$ rather than on the value of a single μ or $\bar{x}$. But the steps in testing hypotheses about the difference between two populations are generally the same as those presented for one mean. Once the null and alternative hypotheses are specified, (1) the sampling distribution of the *difference* between two sample means is constructed, a significance level is chosen and expressed as a significance point. (2) Samples are drawn from the two populations, and the difference observed is expressed in standardized form, that is, a test statistic. (3) The test statistic is compared with the significance point to decide if the observed difference is consistent with or incompatible with the null hypothesis, and H_0 is accepted or rejected, respectively. The details needed to apply these steps are described in Sections 12.1 through 12.3 for comparing two means and in Section 12.4 for comparing two proportions.

Is is also important to be able to estimate the *size* or *magnitude* of the difference between two population means or two proportions. Methods for obtaining point and interval estimates of these differences are described together with the tests of significance.

---

[1]Note that the word "population" is used to refer not only to a set of *individuals* but also to the set of *values* of x corresponding to those individuals.

[2]Note that this differs from the t-test for paired measurements (Section 11.6) which is appropriate only comparing two means from a *single population* of observations or of paired observations.

12.1   Comparison of Two Independent Sample Means When the Population Standard Deviations Are Known

One-Tailed Tests

In this section, the question is whether the means of two populations are equal or whether the mean of the first population is greater than the mean of the second. In formal terms, the null hypothesis is that the first population mean is equal to (or less than) the second, and the alternative hypothesis is that the first mean is greater than the other.

As an example, suppose that an educator wants to see if youngsters who live in two-parent families perform better on a particular standardized test than youngsters from one-parent families. In his class of 29 second grade pupils, 22 comes from two-parent families and 7 from one-parent families. These are considered as samples from the populations of second grade pupils from two-parent and one-parent families. The null hypothesis that the average performance of children from two-parent families is less than or equal to the average performance of children from one-parent families is tested against the alternative that the average for children from two-parent families is greater. The data are given in Table 12.1 and plotted in Figure 12.2.

The average test score of children from two-parent families is higher than the average for those from one-parent families, but there is con-

TABLE 12.1

Scores on a Standardized Test.

Two-parent families			One-parent families
97	83	81	89
95	85	90	104
86	110	121	107
102	119	105	85
119	117	104	70
99	108	96	91
87	74		96
101	93		
	$n_1 = 22$		$n_2 = 7$
	$\bar{x}_1 = 98.7$		$\bar{x}_2 = 91.7$

source: A second-grade teacher.

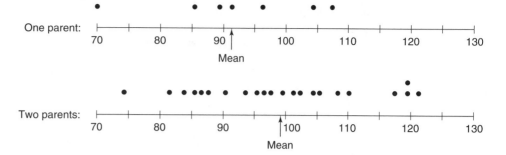

Figure 12.2   *Scores of two groups of children (Table 12.1).*

siderable overlap between the two sets of measurements. We ask if the difference between the two means, $98.7 - 91.7 = 7.0$, is large enough to cause us to believe there is a difference in population means, or whether a difference of this size can plausibly be expected to arise if the null hypothesis is true and there is no difference between population means. We test the hypothesis at the 10% level of significance rather than 5% or 1% to increase power for detecting a population difference that may be small.

To reach a decision, the difference of 7.0 points is re-expressed as a test statistic, that is, in standard errors.[3] The test statistic for comparing the means of two independent samples is

$$z = \frac{\bar{x}_1 - \bar{x}_2}{\sqrt{\dfrac{\sigma^2}{n_1} + \dfrac{\sigma^2}{n_2}}}.$$

The denominator in the expression is the *standard of error of the difference*. If the standard deviation of scores on the test is know to be 12 points, then the variance is $\sigma^2 = 12^2 = 144$ and the value of the test statistic is

$$z = \frac{98.7 - 91.7}{\sqrt{\dfrac{144}{22} + \dfrac{144}{7}}} = 1.34.$$

This value is compared with percentage points of the standard normal distribution. For a one-tailed test at the 10% level of significance, the appropriate significance point is 1.28. The test statistic, 1.34, is greater than this and the null hypothesis is rejected. We conclude that, in the population, the average performance of students from two-parent families is greater than that of children from one-parent families on this particular measure.

---

[3]The complete test procedure is derived in the next section.

General Test Procedure

The means of two populations to be compared may be represented as μ_1 and μ_2 and their variances as σ_1^2 and σ_2^2, respectively. Using this notation, the null hypothesis is

$$H_0: \mu_1 \leq \mu_2$$

or, equivalently,

$$H_0: \mu_1 - \mu_2 \leq 0.$$

This null hypothesis states that the difference between population means is zero or less than zero. The alternative hypothesis is $H_1: \mu_1 > \mu_2$, or, equivalently, $H_1: \mu_1 - \mu_2 > 0$. The null hypothesis and alternative reflect the idea of determining whether or not μ_1 is greater than μ_2.

A random sample

$$x_{11}, \; x_{12}, \ldots, x_{1n_1}$$

is drawn from the first population and another

$$x_{21}, \; x_{22}, \ldots, x_{2n_2}$$

from the second. The symbol x_{1i} denotes the value for the ith individual in the first sample, and x_{2j} denotes that of the jth individual in the second sample. The sample means are

$$\bar{x}_1 = \frac{\sum x_{1i}}{n_1} \quad \text{and} \quad \bar{x}_2 = \frac{\sum x_{2j}}{n_2}.$$

The sample mean difference is $\bar{x}_1 - \bar{x}_2$. To conduct the test of significance, the observed mean difference is compared with the distribution of differences that might be observed if the null hypothesis is true, that is, the *sampling distribution of the mean difference*. We can use the theory discussed in Chapters 8 and 9 to derive the mean, standard deviation and most likely the shape of this sampling distribution.

If a random sample of n_1 observations is drawn from the first population, then according to the Central Limit Theorem the sample mean will be approximately normally distributed with mean $E(\bar{x}_1) = \mu_1$ and variance σ_1^2/n_1; the sampling distribution of the mean for the second sample is approximately normal with mean $E(\bar{x}_2) = \mu_2$ and variance σ_2^2/n_2, and the two sample means are statistically independent. The sampling distribution of the *difference* of two sample means, $\bar{x}_1 - \bar{x}_2$, can be derived from these basic facts; in particular, the sampling distribution of the difference is approximately normal with mean $\mu_1 - \mu_2$ and variance $\sigma_1^2/n_1 + \sigma_2^2/n_2$. Each of these properties is discussed in turn.

The normality of the sampling distribution of $\bar{x}_1 - \bar{x}_2$ derives from the fact that the difference (or sum) of variables with normally distributed parent populations is also normally distributed. (See Section 9.5.) Variables $\bar{x}_1$ and $\bar{x}_2$ are normally distributed if n_1 and n_2 are reasonably large, thus $\bar{x}_1 - \bar{x}_2$, the difference between them, is also approximately normal. In fact, since $\bar{x}_1 - \bar{x}_2$ is obtained indirectly from all $n_1 + n_2$ observations, the approximation to the normal is good if even the *total* number of observations is moderate to large. It is not as good if the total sample size is small, however.

It is shown in Section 9.2 that the expected value of the sum of two random variables is the sum of their respective expected values, or $E(x + y) = E(x) + E(y)$. If we let the first variable be $x = \bar{x}_1$ and the second variable be $y = -\bar{x}_2$ so that $x + y = \bar{x}_1 + (-\bar{x}_2) = \bar{x}_1 - \bar{x}_2$, then $E(\bar{x}_1 - \bar{x}_2) = E(\bar{x}_1) + E(-\bar{x}_2) = \mu_1 - \mu_2$. Likewise, it is shown in Section 9.2 that the variance of a sum of two independent random variables is the sum of their separate variances, or $\sigma^2_{x+y} = \sigma^2_x + \sigma^2_y$. That variance of $\bar{x}_1$ is σ^2_1/n_1 and the variance of $\bar{x}_2$ is σ^2_2/n_2. The variance of $-\bar{x}_2$ is also σ^2_2/n_2 since $-\bar{x}_2 = (-1) \times \bar{x}_2$, and the effect of multiplying a variable by a constant is to multiply its variance by the square of the constant; since $(-1)^2 = 1$, the variance does not change. Thus the variance of the difference $\bar{x}_1 - \bar{x}_2$ is

$$\sigma^2_{\bar{x}_1 - \bar{x}_2} = \frac{\sigma^2_1}{n_1} + \frac{\sigma^2_2}{n_2}. \tag{12.1}$$

The standard error is the square root

$$\sigma_{\bar{x}_1 - \bar{x}_2} = \sqrt{\frac{\sigma^2_1}{n_1} + \frac{\sigma^2_2}{n_2}}.$$

Thus hypotheses about the differences are tested by standardizing the observed mean difference; the standardized form is

$$z = \frac{(\bar{x}_1 - \bar{x}_2) - (\mu_1 - \mu_2)}{\sigma_{\bar{x}_1 - \bar{x}_2}}.$$

If the null hypothesis is true and $\mu_1 = \mu_2$ then $\mu_1 - \mu_2 = 0$. Thus the test statistic in simpler form is the observed mean difference in standard errors,

$$z = \frac{\bar{x}_1 - \bar{x}_2}{\sigma_{\bar{x}_1 - \bar{x}_2}}. \tag{12.2}$$

The null hypothesis is rejected at 5% level if the z exceeds the corresponding significance point from the standard normal distribution (1.645) and at the 1% level if z is greater than 2.33. (Note that this is a one-tailed test.)

EXAMPLE

As an additional sample, consider a medication that is hypothesized to increase the life expectancy of patients with a particular terminal disease. Suppose that of a sample of 55 patients $n_1 = 30$ are randomly selected to receive the medication and the remaining $n_2 = 25$ patients receive a placebo ("sugar pill"). If μ_1 is the mean number of years remaining for patients who receive the new medication and μ_2 is the mean number of years for patients who receive the placebo, the null and alternative hypotheses are H_0: $\mu_1 \leq \mu_2$ and H_1: $\mu_1 > \mu_2$.

Suppose that in the samples the mean years of survival are $\bar{x}_1 = 9.4$ and $\bar{x}_2 = 6.6$. From many years of monitoring the survival histories of patients with this disease, the standard deviation of survival times is known to be $\sigma = 3.4$ years. This is the standard deviation if H_0 is true and we shall assume that it applies to the treated group as well. Do these data provide evidence at the 1% level that the new medication increases life expectancies generally?

The test statistic is

$$z = \frac{\bar{x}_1 - \bar{x}_2}{\sqrt{\dfrac{\sigma_1^2}{n_1} + \dfrac{\sigma_2^2}{n_2}}} = \frac{9.4 - 6.6}{\sqrt{\dfrac{3.4^2}{30} + \dfrac{3.4^2}{25}}}$$

$$= \frac{2.8}{0.921} = 3.04.$$

The statistic z is referred to the table of the standard normal distribution. The probability of obtaining a z of 3.04 or greater if the null hypothesis is true is approximately $.5 - .4988 = .0012$. The P value is very small indicating a relatively strong effect of the experimental medication; H_0 is rejected at $\alpha = .01$ and we conclude that the medication does increase life expectancies for the population of patients with this disease. The increase in the sample, compared to a placebo, was an average of 2.8 years.

Hypothesis in the Other Direction.   The preceding discussion gives procedures for testing H_0: $\mu_1 \leq \mu_2$ against the alternative H_1: $\mu_1 > \mu_2$. In practice the direction of the hypotheses might well be the reverse, that is, H_0: $\mu_1 \geq \mu_2$ and H_1: $\mu_1 < \mu_2$. (These may also be stated as H_0: $\mu_1 - \mu_2 \geq 0$ and H_1: $\mu_1 - \mu_2 < 0$.) For example, we might ask whether a tartar-control toothpaste reduces the average amount of dental calculus formed in a three-month period in comparison to the same toothpaste without tartar-control additives. If we consider the first population to be individuals brushing with the tartar-control product and the second population to be individuals using the toothpaste without the additives, then the hypotheses would be in this form.

The only change in the hypothesis testing procedure is to use significance points from the *left-hand tail* of the standard normal distribution. That is, we would take a sample of individuals using the tartar-control formula and measure the amount of calculus that is formed to obtain $\bar{x}_1$; a sample of individuals not using the additives gives us $\bar{x}_2$. The test statistic (12.2) is calculated with the difference $\bar{x}_1 - \bar{x}_2$ in the numerator. H_0 is rejected at the 5% level of significance if z is less than -1.645, and at the 1% significance level if z is less than -2.33. We would conclude that the tartar-control additives reduce the average amount of calculus that is formed over a three-month period. If $\bar{x}_1$ is not far enough below $\bar{x}_2$ to yield a negative test statistic of this magnitude, H_0 is accepted; we would conclude that, on average, the tartar-control additives do not reduce the formation of dental calculus.

Two-Tailed Tests

In many situations a researcher is interested in detecting a difference between two population means in either direction. For example, anti-tartar agents may be added either to toothpaste or the mouthwash, whichever is more effective in reducing dental calculus. The first population consists of individuals who use the mouthwash regularly and the second population consists of individual who brush with the toothpaste regularly. Samples are taken from each population and the average amount of calculus formed is obtained for both samples at the end of a three-month period.

Likewise, we may wish to determine whether either males or females have more automobile accidents, on the average, for every 1,000 miles driven, or whether juvenile delinquents who appear before Judge A are given more severe or less severe punishments, on average, then delinquents who appear before Judge F. In the latter instance the populations consist of all delinquency cases adjudicated by Judges A and F, respectively. In each of these examples, the investigator is asking whether *either* population mean is higher than the other.

As a specific example, suppose that a recruiter for a particular type of job wants to know if the average aptitude to perform the required work is higher among high-school students in urban or rural areas. The recruiter will focus hiring efforts in either location depending on the outcome of the test of significance. An aptitude test for this job has been used for many years and is known to have a standard deviation of 12 points, and thus a variance of 144. Random samples of 100 urban and 25 rural students are tested and the resulting data are summarized in Table 12.2.

TABLE 12.2

Scores for Two Groups of Students on a Job Aptitude Test Having a Standard Deviation of 12.

	Group 1 (urban)	Group 2 (rural)
Number of students	$n_1 = 100$	$n_2 = 25$
Mean	$\bar{x}_1 = 103.92$	$\bar{x}_2 = 101.50$
Variance	$\sigma_1^2 = 144$	$\sigma_2^2 = 144$
Variance of mean	$\sigma_{\bar{x}_1}^2 = \dfrac{144}{100} = 1.44$	$\sigma_{\bar{x}_2}^2 = \dfrac{144}{25} = 5.76$

Hypothetical data.

The variance of the difference in means is

$$\sigma_{\bar{x}_1 - \bar{x}_2}^2 = \sigma_{\bar{x}_1}^2 + \sigma_{\bar{x}_2}^2 = 1.44 + 5.76 = 7.20,$$

so that the standard error is

$$\sigma_{\bar{x}_1 - \bar{x}_2} = \sqrt{7.20} = 2.68.$$

The difference in means is

$$103.92 - 101.50 = 2.42,$$

and the test statistic is

$$z = \frac{2.42}{2.68} = 0.90.$$

That is, the difference between the two sample means is 9/10 of a standard error in favor of urban youngsters.

The hypothesis H_0: $\mu_1 = \mu_2$ is to be tested against the alternative H_1: $\mu_1 \neq \mu_2$. At the 1% level, one rejects H_0 if the observed difference exceeds 2.58 in either direction, positive or negative. At the 5% level, one rejects H_0 if the observed difference is larger than 1.96 standard errors in either direction, positive or negative. Here the null hypothesis is not rejected. The recruiter will choose to focus in urban or rural areas based on criteria other than the aptitudes of the applicant pools.

Confidence Intervals

Like a confidence interval on a mean, a confidence interval can be obtained that contains the range of plausible values of the *difference between two means*. The test of significance answers the yes-no question "Is there a difference between μ_1 and μ_2?" while the confidence interval addresses the question "How big is the difference?"

Suppose, for example, that a job recruiter wanted to know how much difference there was between the tested aptitudes of individuals rated by their supervisors as most successful in performing their jobs and those rated as least successful. A sample of 25 "most successful" employees is chosen and another sample of 40 "least successful" employees. Their aptitude scores are transcribed from their employment records. (The test was given before the employees were hired.) The aptitude measure is known to have a standard deviation of 12 points and variance of 144. The data are summarized in Table 12.3.

TABLE 12.3

Mean Scores for Two Groups of Employees on an Aptitude Test Having a Standard Deviation of 12.

	Group 1 Most successful	Group 2 Least successful
Number of employees	$n_1 = 25$	$n_2 = 40$
Mean	$\bar{x}_1 = 113.71$	$\bar{x}_2 = 102.50$
Variance of mean	$\sigma_{\bar{x}_1}^2 = \dfrac{144}{25} = 5.76$	$\sigma_{\bar{x}_2}^2 = \dfrac{144}{40} = 3.60$

Hypothetical data.

A confidence interval for the difference of two means is centered at the sample difference (which is the *point estimate* of $\mu_1 - \mu_2$). The center in the example is $\bar{x}_1 - \bar{x}_2 = 113.71 - 102.50 = 11.21$. The mean difference may also be expressed relative to the standard deviation of the scale, that is, $(\bar{x}_1 - \bar{x}_2)/\sigma = 11.21/12 = 0.93$. This index, called an "effect size," is a convenient way of expressing a difference between means when the variable is not measured in familiar units. The difference between group means of 11.21 points is equivalent to a difference of about 9/10 of a standard deviation.

The confidence interval is obtained by adding to and subtracting from the point estimate a certain number of standard errors. It encompasses a range of values from $(\bar{x}_1 - \bar{x}_2) - Q\sigma_{\bar{x}_1 - \bar{x}_2}$ to $(\bar{x}_1 - \bar{x}_2) + Q\sigma_{\bar{x}_1 - \bar{x}_2}$. "How many" standard errors (Q) depends on the degree of confidence desired that the interval contains the true difference between population means.

For example, for 95% confidence, 1.96 standard errors are added to and subtracted from $\bar{x}_1 - \bar{x}_2$. This is the value from the standard normal distribution that separates the middle 95% of the distribution from the most extreme $2\frac{1}{2}\%$ in each tail, that is, $z(.025)$. The variance of the difference between means for the example is

$$\sigma_{\bar{x}_1 - \bar{x}_2}^2 = \sigma_{\bar{x}_1}^2 + \sigma_{\bar{x}_2}^2 = 5.76 + 3.60 = 9.36,$$

and the standard error is

$$\sigma_{\bar{x}_1 - \bar{x}_2} = \sqrt{9.36} = 3.06.$$

Thus the 95% confidence interval ranges from $11.21 - (1.96 \times 3.06)$ to $11.21 + (1.96 \times 3.60)$, or $(5.21, 17.21)$. Because the interval contains only positive values, we are confident that μ_1 is greater than μ_2. The difference between the two means is estimated to be at least 5.21 points but no more than 17.21 points. Expressed as a range of effect sizes, the difference is between 0.43 and 1.43 standard deviations.

If both endpoints of a confidence interval on $\mu_1 - \mu_2$ are negative, then all plausible values for the difference of the means are negative and we can be confident that μ_2 is greater than μ_1. It is possible for a confidence interval on a difference of two means to contain both negative and positive values, for example $(-4.97, 7.21)$. In this case we cannot be confident which of the two means is greater or, in fact, if μ_1 and μ_2 differ at all since 0 is also contained in the interval. A narrow interval that contains 0, for example $(-1.2, 2.3)$, makes it plausible that μ_1 and μ_2 are nearly equal.

Like a confidence interval for a single mean, these confidence intervals contain all "acceptable" values of the difference between two means, that is, all values for which the null hypothesis is accepted if a two-tailed test is made at the same α level. Thus, we would accept the null hypothesis H_0: $\mu_1 = \mu_2$ (or $\mu_1 - \mu_2 = 0$) if the confidence interval contains 0 and would reject this H_0 if the interval contains only positive or only negative values.

Increasing either sample size, n_1 or n_2, reduces the width of the interval just as increasing n does for a confidence interval on a single mean. Increasing the confidence level increases the interval width. For example, if we want to construct a 99% confidence interval for the data of Table 12.3, then 2.58 standard errors are added to and subtracted from the mean difference. The resulting interval is $(3.32, 19.10)$.

The statistical basis for this confidence-interval procedure is that with random sampling

$$\frac{(\bar{x}_1 - \bar{x}_2) - (\mu_1 - \mu_2)}{\sigma_{\bar{x}_1 - \bar{x}_2}}. \tag{12.3}$$

has a standard normal distribution or (by the Central Limit Theorem) approximately a standard normal distribution. Thus the probability is .95 that (12.3) will be between -1.96 and 1.96. Substitution for the standard error in (12.3) yields

$$\frac{(\bar{x}_1 - \bar{x}_2) - (\mu_1 - \mu_2)}{\sqrt{\sigma_1^2/n_1 + \sigma_2^2/n_2}}, \tag{12.4}$$

and the probability is .95 that (12.4) will be between -1.96 and 1.96. In other words, the probability is .95 that the sample mean difference is within 1.96 standard errors of the population mean difference. Thus, if we sample at random from two populations and obtain $\bar{x}_1 - \bar{x}_2$, we can be 95% confident that the difference $\mu_1 - \mu_2$ is within 1.96 standard errors of this value.[4] The 95% confidence interval on the difference of two means is

$$\left(\bar{x}_1 - \bar{x}_2 - 1.96\sqrt{\frac{\sigma_1^2}{n_1} + \frac{\sigma_2^2}{n_2}}, \quad \bar{x}_1 - \bar{x}_2 + 1.96\sqrt{\frac{\sigma_1^2}{n_1} + \frac{\sigma_2^2}{n_2}} \right).$$

For other confidence levels, corresponding values are substituted from the standard normal table. For example, the 90% confidence interval is

$$\left(\bar{x}_1 - \bar{x}_2 - 1.645\sqrt{\frac{\sigma_1^2}{n_1} + \frac{\sigma_2^2}{n_2}}, \quad \bar{x}_1 - \bar{x}_2 + 1.645\sqrt{\frac{\sigma_1^2}{n_1} + \frac{\sigma_2^2}{n_2}} \right).$$

Validity Conditions

Like the test and confidence interval for a single mean, statistical tests and estimates of the difference of two means rest on a set of assumptions that should be considered each time the procedures are applied. For the comparison to be informative the mean should be a suitable measure of location. This means that the measured outcome variable should have nearly equal intervals and that its distribution should be (roughly) symmetric. Likewise, the variance (or standard deviation) should be a suitable measure of variability.

Three statistical conditions are also essential to the validity of the procedures described in the preceding discussion. First and foremost, the $n_1 + n_2$ observations must be independent. Sampling at random is the first step to assure that this condition will be met. In addition, certain procedural safeguards may be taken such as making certain that adjacent observations are not included in a sample or that respondents do not have the opportunity to hear or copy each other's responses. The independence of measurements in a sample is essential to the validity of all inferential procedures discussed in this book.

Second, both the significance tests and confidence intervals require that the sampling distribution of the difference of two means is nearly normal. According to the Central Limit Theorem, this condition will be met if the total sample size is large, almost regardless of the shape of the distribution in the parent population. If the parent populations are not normal and the sample sizes are small, then an alternative method for comparing locations should be considered; one such method is the sign test described in Section 12.5.

The procedures described in the preceding sections assume that we know the value of the variances, σ_1^2 and σ_2^2. In practice, it is much

---

[4]The same logic is applied to obtain the confidence interval for one mean in Section 10.3.

more common that the population variances are not known and must be estimated from the sample. The most frequently used alternative is the *t*-test described in Section 12.2, in which the population variance is replaced by the sample variance. This test rests on the assumption that the variance is the same in both populations being compared. If it is not clear whether this condition is met, the variances may be tested for equality using the methods presented in Chapter 13. A test that does not require equal population variances is presented in Section 12.3, but is useful only in certain limited situations.

12.2   Comparison of Two Independent Sample Means When the Population Standard Deviations are Unknown but Treated as Equal

If the standard deviations of the two parent populations are unknown, the standard error of the difference of the sample means has to be estimated from the data. How this estimation is done depends on whether or not the standard deviations of the parent populations can be considered to be equal. The test procedure when the two standard deviations can be considered equal is described in this section; in practice, this procedure is applied more commonly than any of the other tests for comparing two means. The procedure for situations in which the two standard deviations cannot be assumed to be equal is described in Section 12.3.

Like the *z*-test described in Section 12.1, both procedures require that the sampling distribution of the difference of two sample means is approximately normal. This is assured if the underlying parent populations have normal distributions and is usually the case, no matter what the shape of the underlying distribution, if the samples sizes are large.

We begin with an example. In a statistics course taken by both undergraduate and graduate students, the undergraduates claim they should be graded separately from the graduate students on the grounds that graduate students tend to do better. The graduate students claim they should be graded separately from the undergraduates on the grounds that undergraduates tend to do better! We ask the question whether or not, on average, undergraduates do about as well as graduate students in statistics courses of this type.

Scores of the 39 undergraduates and the 29 graduates who took the course one year are given in Table 12.4. We shall treat these two groups as samples from the (hypothetical) population of graduate students taking such courses and the (hypothetical) population of undergraduate students taking such courses. The null hypothesis is that the population means are equal. Since we are interested in both positive and negative differences in means, we shall make a two-tailed test.

TABLE 12.4
Final Grades in a Statistics Course

Undergraduates: $n_1 = 39$						Graduates: $n_2 = 29$			
Student	*Score*	*Student*	*Score*	*Student*	*Score*	*Student*	*Score*	*Student*	*Score*
1	153	14	101	27	113	40	145	55	139
2	109	15	109	28	130	41	117	56	107
3	157	16	159	29	139	42	110	57	137
4	145	17	99	30	129	43	160	58	138
5	131	18	143	31	152	44	109	59	132
6	161	19	158	32	161	45	135	60	136
7	124	20	143	33	108	46	141	61	160
8	158	21	153	34	126	47	132	62	94
9	131	22	149	35	147	48	124	63	144
10	120	23	99	36	164	49	153	64	144
11	153	24	162	37	119	50	126	65	85
12	113	25	122	38	137	51	133	66	109
13	134	26	165	39	145	52	157	67	105
						53	93	68	135
						54	83		

$$\bar{x}_1 = 136.44$$
$$\sum(x_{1i} - \bar{x}_1)^2 = 15{,}551$$
$$s_1^2 = 409.2$$
$$s_1 = 20.2$$

$$\bar{x}_2 = 127.00$$
$$\sum(x_{2i} - \bar{x}_2)^2 = 13{,}388$$
$$s_2^2 = 478.1$$
$$s_2 = 21.9$$

source: A professor who requests anonymity.

Because we are not dealing with scores on a standardized test, the standard deviations in the two populations are unknown and must be estimated. We suppose that the standard deviations in the two populations are the same.[5]

The common value of the standard deviation or of the variance is estimated on the basis of the deviations of the values in each sample from their respective sample means. We average the squared deviations over

---

[5]The ratio of sample variances is $s_2^2/s_1^2 = 1.17$. That is, s_2^2 is only about 20% larger than s_1^2. It can be shown that for samples of this size this difference is well within the range of variability expected when $\sigma_1^2 = \sigma_2^2$. A formal test for the difference of two variances is described in Chapter 13.

the two samples to estimate the variance by adding together ("pooling") the sums of squared deviations in the two groups,

$$15{,}551 + 13{,}388 = 28{,}939,$$

and dividing this total by the appropriate number of degrees of freedom. The number of degrees of freedom associated with 15,551 is $n_1 - 1 = 39 - 1 = 38$; the number associated with 13,388 is $n_2 - 1 = 29 - 1 = 28$. The number associated with their sum is $38 + 28 = 66$. The estimate of variance is the pooled sum of squared deviations divided by the number of degrees of freedom, that is,

$$s_p^2 = \frac{28{,}939}{66} = 438.5.$$

The estimated variance of the sampling distribution of the difference between means (namely, the estimate of $\sigma_1^2/n_1 + \sigma_2^2/n_2$) is

$$\frac{438.5}{39} + \frac{438.5}{29} = 26.36.$$

The corresponding standard error is $\sqrt{26.36} = 5.13$. The difference between means measured in standard errors is

$$t = \frac{136.44 - 127.00}{5.13} = \frac{9.44}{5.13} = 1.83.$$

This test statistic is compared with percentage points of Student's t-distribution with $(n_1 - 1) + (n_2 - 1) = 66$ degrees of freedom. For a two-tailed test at the 5% level, the significance point for 60 degrees of freedom is 2.00. The test statistic is not this large and the null hypothesis is accepted; we have no evidence that the population of undergraduate students or of graduate students performs better in the statistics course.

General Test Procedure   The null and alternative hypotheses for a two-tailed test are

$$H_0: \mu_1 = \mu_2$$

and

$$H_1: \mu_1 \neq \mu_2,$$

respectively. For a one-tailed test, the null hypothesis may be either $H_0: \mu_1 \leq \mu_2$ or $H_0: \mu_1 \geq \mu_2$; these hypotheses are discussed in Section 12.1.

The variances of the parent populations are denoted by σ_1^2 and σ_2^2. In many problems it is reasonable to consider these variances equal. It is convenient to call this common variance σ^2. Then the variance of the difference of sample means,

$$\sigma_{\bar{x}_1 - \bar{x}_2}^2 = \frac{\sigma_1^2}{n_1} + \frac{\sigma_2^2}{n_2},$$

can be written

$$\sigma_{\bar{x}_1 - \bar{x}_2}^2 = \frac{\sigma^2}{n_1} + \frac{\sigma^2}{n_2}$$

or, factoring out σ^2,

$$\sigma_{\bar{x}_1 - \bar{x}_2}^2 = \sigma^2 \left(\frac{1}{n_1} + \frac{1}{n_2} \right).$$

An estimate of $\sigma_{\bar{x}_1 - \bar{x}_2}^2$ is

$$s_p^2 \left(\frac{1}{n_1} + \frac{1}{n_2} \right),$$

where s_p^2 is the "pooled" estimate of the common variance σ^2, namely,

$$s_p^2 = \frac{\sum (x_{1i} - \bar{x}_1)^2 + \sum (x_{2j} - \bar{x}_2)^2}{n_1 + n_2 - 2}. \tag{12.5}$$

The divisor of s_p^2 is the number of degrees of freedom associated with the two sums of squared deviations together, $(n_1 - 1) + (n_2 - 1) = n_1 + n_2 - 2$. The statistic s_p^2 is an unbiased estimate of σ^2.

The pooled variance is just a "weighted average" of the separate variances, s_1^2 and s_2^2. The sum of squared deviations from the mean in the first sample, $\sum (x_{1i} - \bar{x}_1)^2$, is $(n_1 - 1)$ times the variance in the first sample, s_1^2. (Remember that the variance is just the sum of squared deviations divided by $n_1 - 1$.) Likewise, $\sum (x_{2i} - \bar{x}_2) = (n_2 - 1) \times s_2^2$. Thus the pooled variance is

$$s_p^2 = \frac{(n_1 - 1) \times s_1^2 + (n_2 - 1) \times s_2^2}{n_1 + n_2 - 2} \tag{12.6}$$

$$= \frac{n_1 - 1}{n_1 + n_2 - 2} s_1^2 + \frac{n_2 - 1}{n_1 + n_2 - 2} s_2^2.$$

The value of s_p^2 will always be between that of s_1^2 and s_2^2 but closer to the variance of the larger sample. Expression (12.5) is easier to use for s_p^2 when beginning with raw data; expression (12.6) is simpler if standard deviations have already been computed for the two samples.

Once the common population variance is estimated by s_p^2, the test statistic is computed using this value; this is

$$t = \frac{\bar{x}_1 - \bar{x}_2}{\sqrt{s_p^2\left(\dfrac{1}{n_1} + \dfrac{1}{n_2}\right)}} = \frac{\bar{x}_1 - \bar{x}_2}{s_p\sqrt{\dfrac{1}{n_1} + \dfrac{1}{n_2}}}.$$

The sampling distribution of this statistic is Student's t-distribution with $n_1 + n_2 - 2$ degrees of freedom. (This statistic has the same form as (12.2) but does not require that the population variance is known exactly.)

If we are making a one-tailed test of H_0: $\mu_1 \leq \mu_2$ against the alternative H_1: $\mu_1 > \mu_2$ at a given α level, then H_0 is rejected if

$$t > t_{n_1+n_2-2}(\alpha);$$

we would conclude that the first population mean is larger than the second. If the hypotheses are in the reverse direction, that is, H_0: $\mu_1 \geq \mu_2$ and H_1: $\mu_1 < \mu_2$, then H_0 is rejected if the test statistic falls in the region of rejection on the left side of the t-distribution, that is, if $t < -t_{n_1+n_2-2}(\alpha)$. If we are making a two-tailed test of H_0: $\mu_1 = \mu_2$ against the alternative H_1: $\mu_1 \neq \mu_2$ at a given α level, then H_0 is rejected if

$$t > t_{n_1+n_2-2}(\alpha/2) \qquad \text{or} \qquad t < -t_{n_1+n_2-2}(\alpha/2).$$

EXAMPLE   Table 2.24 lists the numbers of words recalled by $n_1 = 27$ children classified as fast learners and $n_2 = 26$ children classified as slow learners. The researchers hypothesized that, having twice learned the material to the same degree of accuracy, fast and slow learners would not differ in the average number of words they could recall. If μ_1 and μ_2 are the mean numbers of words recalled for fast and slower learners in the population, then the null and alternative hypotheses are H_0: $\mu_1 = \mu_2$ and H_1: $\mu_1 \neq \mu_2$, respectively; we shall test these hypotheses at the 1% level.

Because the poem constituted a new stimulus, no information was available from which to obtain a population standard deviation σ. Thus we shall estimate the standard deviation from the sample and use the t-procedure.

The mean and standard deviation in the sample of fast learners are $\bar{x}_1 = 48.44$ words and $s_1 = 6.14$ words; the mean and standard deviation in the sample of slow learners are $\bar{x}_2 = 41.31$ words and $s_2 = 8.97$ words. The pooled variance obtained from (12.6) is

$$s_p^2 = \frac{(n_1 - 1) \times s_1^2 + (n_2 - 1) \times s_2^2}{n_1 + n_2 - 2}$$

$$= \frac{(26 \times 6.14^2) + (25 \times 8.97^2)}{27 + 26 - 2} = 58.66.$$

The test statistic is

$$t = \frac{\bar{x}_1 - \bar{x}_2}{\sqrt{s_p^2 \left(\dfrac{1}{n_1} + \dfrac{1}{n_2} \right)}} = \frac{48.44 - 41.31}{\sqrt{58.66 \times \left(\dfrac{1}{27} + \dfrac{1}{26} \right)}}$$

$$= \frac{7.13}{2.10} = 3.40.$$

The test statistic is compared to percentage points of the t-distribution with $n_1 + n_2 - 2 = 27 + 26 - 2 = 51$ degrees of freedom. Although this distribution is not given in Appendix III, the more conservative percentage points from t_{40} may be used instead. The .01 significance points for the two-tailed test are $\pm t_{40}(.005) = \pm 2.704$. The t value of 3.40 exceeds 2.704 and H_0 is rejected; in the population, fast learners are able to recall more words than slow learners even after both groups have attained equal accuracy at an earlier time.

The P value interval for the two-tailed test is the probability if H_0 is true of obtaining a test statistic whose *absolute value* is 3.40 or greater. From Appendix III we can see only that this is $P < .01$.

Note for One-Tailed Test.   Suppose that the researchers had hypothesized at the outset that fast learners would recall more words than slower learners. Then the null and alternative hypotheses would have been $H_0\colon \mu_1 \leq \mu_2$ and $H_1\colon \mu_1 > \mu_2$. The test statistic would still be $t = 3.40$ and in this instance the conclusion would remain the same as well. However, the 1% significance point from t_{40} is $t_{40}(.01) = 2.423$ and the P interval for the statistic $t = 3.40$ is $P < .005$.

Confidence Intervals

The confidence-interval procedures for the difference $\mu_1 - \mu_2$ as presented in Section 12.1 can be appropriately modified; the standard error $\sigma_{\bar{x}_1 - \bar{x}_2}$ is replaced by its estimate $s_p \sqrt{1/n_1 + 1/n_2}$ and the value from the normal table is replaced by a value from the t table. For example, a 95% confidence interval for $\mu_1 - \mu_2$ is

$$\bar{x}_1 - \bar{x}_2 - t_{n_1 + n_2 - 2}(.025) s_p \sqrt{\frac{1}{n_1} + \frac{1}{n_2}} < \mu_1 - \mu_2$$

$$< \bar{x}_1 - \bar{x}_2 + t_{n_1 + n_2 - 2}(.025) s_p \sqrt{\frac{1}{n_1} + \frac{1}{n_2}}.$$

EXAMPLE   In the preceding *t*-test example we concluded that fast learners recall more words on average than slow learners even if both groups have memorized a passage to the same mastery level. With this finding in mind, we would also want to know *how many* more words fast learners recall, that is, the magnitude of the effect. The sample means are $\bar{x}_1 = 48.44$ and $\bar{x}_2 = 41.31$. The point estimate of the population mean difference $\mu_1 - \mu_2$ is $\bar{x}_1 - \bar{x}_2 = 48.44 - 41.31 = 7.13$ words. In other words, in a poem containing 67 words, it is estimated that fast learners recall about 7 words (about 10.6%) more than slow learners.

To compute an interval estimate, a multiple of the standard error of the difference is subtracted from and added to the point estimate. The standard error, computed previously, is $\sqrt{s_p^2(1/n_1 + 1/n_2)} = 2.10$. The confidence interval uses percentage points from the *t*-distribution with $n_1 + n_2 - 2 = 51$ degrees of freedom. For 95% confidence the multiplier is $t_{51}(.025)$.

Appendix III does not list this distribution but gives the upper 2.5% points from t_{40} and t_{60}; these are 2.021 and 2.000, respectively. It is sufficiently accurate for most applications to interpolate between these two values. That is, since 51 is 11/20 of the distance between 40 and 60, we can estimate $t_{51}(.025)$ by computing a value 11/20 of the way from 2.021 to 2.000. The difference is $2.021 - 2.000 = 0.021$, and 11/20 of this is 0.012. Subtracting, we estimate $t_{51}(.025) = 2.021 - 0.012 = 2.009$.

The 95% confidence interval is from $7.13 - 2.009 \times 2.10 = 2.91$ to $7.13 + 2.009 \times 2.10 = 11.35$; we represent this interval by $(2.91, 11.35)$. It is estimated that fast learners recall at least 2.91 words (4.3%) more than slow learners, on average, and perhaps as many as 11.35 words (16.9%) more than slow learners.

12.3   Comparison of Two Independent Sample Means When the Population Standard Deviations Are Unknown and Not Treated as Equal*

As we have seen, the standard deviation of the sampling distribution of the difference of two sample means is

$$\sigma_{\bar{x}_1 - \bar{x}_2} = \sqrt{\frac{\sigma_1^2}{n_1} + \frac{\sigma_2^2}{n_2}}.$$

To test the null hypothesis H_0: $\mu_1 = \mu_2$ when the population variances are known, the test statistic is

$$z = \frac{\bar{x}_1 - \bar{x}_2}{\sqrt{\dfrac{\sigma_1^2}{n_1} + \dfrac{\sigma_2^2}{n_2}}}. \qquad (12.7)$$

This is compared with percentage points of the standard normal distribution as described in Section 12.1, *whether or not* the two variances (σ_1^2 and σ_2^2) are equal.

When population variances are unknown and there is reason to believe that the two populations have different variances, then separate sample variances may be used in the test statistic instead of the pooled value s_p^2. That is, s_1^2 and s_2^2 are taken to be separate estimates of σ_1^2 and σ_2^2, respectively, and the standard error of the difference of two sample means is estimated by

$$\sqrt{\frac{s_1^2}{n_1} + \frac{s_2^2}{n_2}}.$$

The test statistic is

$$\frac{\bar{x}_1 - \bar{x}_2}{\sqrt{\dfrac{s_1^2}{n_1} + \dfrac{s_2^2}{n_2}}} \qquad (12.8)$$

instead of (12.7).

The probability distribution to use in deciding if there is a difference between population means when variances are unequal depends on the n's. If n_1 and n_2 are both reasonably large (each greater than 30 as a rule of thumb), then (12.8) can be referred to the normal distribution since the Central Limit Theorem insures that the sampling distribution of the differences in sample means is at least approximately a normal distribution and there is little sampling variability in the estimate of $\sigma_{\bar{x}_1 - \bar{x}_2}$.

For samples of moderate size the statistic (12.8) can be referred to the table of the t-distribution with $n_1 + n_2 - 2$ degrees of freedom. This procedure gives an acceptable, although inexact, test of the equality of two means when the population variances are not very different and if the ratio n_1/n_2 is not very different from 1.

If the population variances are believed to be very different from one another, if the ratio n_1/n_2 is far below or above 1, *or* if the sample sizes are small, then an alternative approach to comparing the means is necessary. One possibility is the sign test described in Section 12.5; another is the analysis of ranks described in Section 15.3.

The considerations in choosing a test statistic and probability distribution for comparing two population means are summarized in Figure 12.3. All of these procedures (except those termed "alternative") are based on the assumption that the sampling distribution of the difference between two sample means is approximately normal. If it is clear that this condition is not met in a particular analysis, then the alternative approaches to comparing means should be considered in any case.

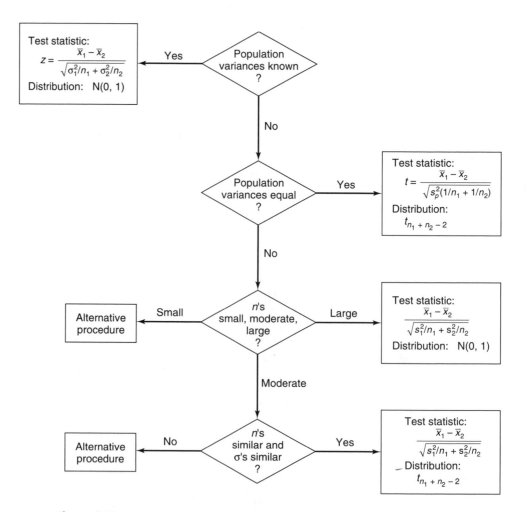

Figure 12.3   *Summary of decisions in choosing a test procedure for comparing two population means.*

12.4   Comparison of Two Independent Sample Proportions

Hypothesis Tests

A frequently asked question is whether the proportions of individuals or objects in two populations having some characteristic are equal. Is the proportion of urban voters intending to vote for the Republican candidate equal to the proportion of rural voters? Is the probability of contracting polio the same in the population of children inoculated with vaccine and the population of children not inoculated? To answer such a question a sample is taken from each population and the relevant proportions in the two samples are compared.

As an example, we return to the Asch experiment on the effect of group pressure on perception (Section 1.1) in which the individuals in the test group were under pressure to give wrong responses. Instead of counting the number of errors made by each subject, account will be taken only of whether a subject made at least one error or made no error. This condensation of Table 1.7 results in Table 12.5.

TABLE 12.5
Summary of Results of One of Asch's Early Experiments

	Test group	Control group	Both groups
One or more errors	37 (74.0%)	2 (5.4%)	39 (44.8%)
No errors	13 (26.0%)	35 (94.6%)	48 (55.2%)
Total	$n_1 = 50$ (100%)	$n_2 = 37$ (100%)	87 (100%)

source: Table 1.7.

In $n_1 = 50$ cases of individuals subjected to group pressure are considered to constitute a sample from the (hypothetical) population of individuals who might be subjected to pressure. In this population the probability of one or more errors is p_1. Similarly the $n_2 = 37$ cases in the "control group" are treated as a sample from a population not under pressure, with probability p_2 of at least one error. The researchers anticipate that the probability of an error is increased by the social pressure. The null hypothesis is $H_0: p_1 \leq p_2$ and the alternative is $H_1: p_1 > p_2$. The null hypothesis is rejected if the sample proportion $\hat{p}_1$ is much greater than the sample proportion $\hat{p}_2$, that is, if the difference

$\hat{p}_1 - \hat{p}_2$ is large. In this case $\hat{p}_1 = 0.740$ and $\hat{p}_2 = 0.054$. Is the difference $\hat{p}_1 - \hat{p}_2 = 0.740 - 0.054 = 0.686$ large enough?

Because the sampling distribution of the sample proportion is approximately normal, the distribution of the *difference* of two proportions is approximately normal as well. If the null hypothesis is true and $p_1 = p_2$, then the standard deviations of the two populations are also equal. The standard deviations[6] are $\sqrt{p_1 \times (1 - p_1)}$ and $\sqrt{p_2 \times (1 - p_2)}$.

Let p represent the common value of the probability $p_1 = p_2$. The estimate of p is the proportion of subjects in both groups making one or more errors, $\hat{p} = 0.448$. The estimate of the standard deviation of the sampling distribution of the difference between the two independent proportions is

$$\sqrt{\hat{p}(1 - \hat{p})\left(\frac{1}{n_1} + \frac{1}{n_2}\right)} = \sqrt{0.448(1 - 0.448)\left(\frac{1}{50} + \frac{1}{37}\right)}$$

$$= \sqrt{0.247(0.0200 + 0.0270)}$$

$$= \sqrt{0.247 \times 0.0470}$$

$$= 0.108.$$

This is the "standard error of the difference" between the proportions. The difference in proportions divided by the standard error is

$$\frac{0.740 - 0.054}{0.108} = \frac{0.686}{0.108} = 6.36.$$

If this result is compared to upper percentage points of the standard normal distribution, we see that the 2×2 table gives overwhelming evidence that group pressure did affect perception.

General Test Procedure

Now we explain this procedure in general terms. The observations are Bernoulli variables, taking the values 0 or 1. Each observation in the first sample is from a population with mean $\mu_1 = p_1$ and variance $\sigma_1^2 = p_1(1 - p_1) = p_1 q_1$; similarly, each observation in the second sample is from a population with mean $\mu_2 = p_2$ and variance $\sigma_2^2 = p_2(1 - p_2) = p_2 q_2$. The null hypothesis is

$$H_0: p_1 = p_2$$

(which may be extended to $H_0: p_1 \leq p_2$ if the alternative is $H_1: p_1 > p_2$ and may be extended to $H_0: p_1 \geq p_2$ if the alternative is $H_1: p_1 < p_2$).

Here the means $\bar{x}_1$ and $\bar{x}_2$ are the sample proportions, $\hat{p}_1$ and $\hat{p}_2$. The difference $\bar{x}_1 - \bar{x}_2$ is $\hat{p}_1 - \hat{p}_2$. The variance of the sampling distribution

---

[6]See Section 9.3.

of the difference in sample proportions is

$$\sigma^2_{\hat{p}_1-\hat{p}_2} = \frac{p_1 q_1}{n_1} + \frac{p_2 q_2}{n_2}.$$

If H_0 is true and $p_1 = p_2 \, (= p)$,

$$\sigma^2_{\hat{p}_1-\hat{p}_2} = \frac{p_1 q_1}{n_1} + \frac{p_2 q_2}{n_2} = pq \left(\frac{1}{n_1} + \frac{1}{n_2} \right).$$

The standard error is the square root

$$\sigma_{\hat{p}_1-\hat{p}_2} = \sqrt{pq(1/n_1 + 1/n_2)},$$

and the standardized version of $\hat{p}_1 - \hat{p}_2$ is

$$\frac{\hat{p}_1 - \hat{p}_2}{\sigma_{\hat{p}_1-\hat{p}_2}} = \frac{(\hat{p}_1 - \hat{p}_2)}{\sqrt{pq(1/n_1 + 1/n_2)}}.$$

The frequencies of individuals having and not having a particular characteristic, as in Table 12.5, are represented in symbols in Table 12.6. The estimate of p_1 is $\hat{p}_1 = a/n_1$ and of p_2 is $\hat{p}_2 = b/n_2$. The estimate of p, the proportion in both samples combined, is $\hat{p} = (a + b)/n$.

TABLE 12.6

Classification of Two Samples According to Presence or Absence of Character-istic A

	Sample 1	Sample 2	Both samples
A present	a	b	$a + b$
A absent	c	d	$c + d$
	$n_1 = a + c$	$n_2 = b + d$	$n = n_1 + n_2 = a + b + c + d$

The test statistic is

$$z = \frac{\hat{p}_1 - \hat{p}_2}{\sqrt{\hat{p}\hat{q}(1/n_1 + 1/n_2)}},$$

where $\hat{q} = 1 - \hat{p}$. The distribution of this statistic approximates a standard normal distribution for moderate to large sample sizes n_1 and n_2. (For greater accuracy the continuity correction given in Appendix 12A should be used.)

EXAMPLE   As a second example, consider a hypothetical manufacturing company that employs large numbers of males and females. Some of the female employees claim that promotions to managerial positions are given disproportionately to males. To test this claim, a review is conducted of the records of $n = 80$ randomly-chosen employees hired between 5 and 8 years previously who are still working for this company. (We ignore the fact that there may also be disproportionate layoffs or resignations for males and females.) The null and alternative hypotheses are $H_0: p_1 \leq p_2$ and $H_1: p_1 > p_2$, respectively. We shall use the 10% significance level.

The sample results are as follows: Of the $n_1 = 46$ males in the sample, $a = 9$ currently hold managerial positions; thus $\hat{p}_1 = 9/46 = 0.196$. Of the $n_2 = 34$ females in the sample, $b = 4$ currently hold managerial positions, thus $\hat{p}_2 = 5/34 = 0.118$. The estimate of the total proportion of employees promoted in this period is $\hat{p} = (9 + 4)/80 = 0.162$, and thus $\hat{q} = 1 - 0.162 = 0.838$. The test statistic is

$$z = \frac{\hat{p}_1 - \hat{p}_2}{\sqrt{\hat{p}\hat{q}(1/n_1 + 1/n_2)}} = \frac{0.196 - 0.118}{\sqrt{.0162 \times 0.838 \times (1/46 + 1/34)}}$$

$$= \frac{0.078}{0.083} = 0.94.$$

If $z = 0.94$ is compared to upper percentage points of the standard normal distribution, the approximate P value is $.5 - .3264 = .1736$. The null hypothesis is not rejected if α is set at .10. That is, the test statistic 0.94 does not exceed the 10% significance point of 1.28. The data do *not* provide compelling evidence that a disproportionate number of males is promoted by this firm.

Confidence Intervals

Confidence intervals for $p_1 - p_2$ are obtained by using the fact that

$$\frac{(\hat{p}_1 - \hat{p}_2) - (p_1 - p_2)}{\sqrt{\hat{p}_1\hat{q}_1/n_1 + \hat{p}_2\hat{q}_2/n_2}}$$

has a sampling distribution which is approximately standard normal. A 95% confidence interval for $p_1 - p_2$ is

$$\hat{p}_1 - \hat{p}_2 - 1.96\sqrt{\frac{\hat{p}_1\hat{q}_1}{n_1} + \frac{\hat{p}_2\hat{q}_2}{n_2}} < p_1 - p_2 < \hat{p}_1 - \hat{p}_2 + 1.96\sqrt{\frac{\hat{p}_1\hat{q}_1}{n_1} + \frac{\hat{p}_2\hat{q}_2}{n_2}}.$$

From the data of Table 12.5, the proportion of the test group who made one or more errors is 0.740, the proportion of the control group who made one or more errors is 0.054, and the difference is $\hat{p}_1 - \hat{p}_2 =$

$-.686$. The standard error used in the confidence interval is

$$\sqrt{\frac{\hat{p}_1\hat{q}_1}{n_1} + \frac{\hat{p}_2\hat{q}_2}{n_2}} = \sqrt{\frac{0.740 \times 0.260}{50} + \frac{0.054 \times 0.946}{37}}$$
$$= 0.072.$$

The confidence interval is from $0.686 - (1.96 \times 0.072)$ to $0.686 + (1.96 \times 0.072)$ or $(0.545, 0.827)$. The effect of social pressure is to increase the likelihood of making one or more errors by at least 54% and perhaps by as much as 80% or more.

Note that this confidence interval does not correspond to the test of the hypothesis H_0: $p_1 = p_2$. In the hypothesis test we use $\sqrt{\hat{p}\hat{q}(1/n_1 + 1/n_2)}$ as an estimate of the standard error of $\hat{p}_1 - \hat{p}_2$ because the significance level is a probability based on the assumption that the null hypothesis is true and $p_1 = p_2 = p$. In forming a confidence interval we do not assume that the population proportions are equal and thus the standard error is $\sqrt{\hat{p}_1\hat{q}_1/n_1 + \hat{p}_2\hat{q}_2/n_2}$.

12.5   The Sign Test for a Difference in Locations*

In this section, as in Section 11.6, we discuss a test that is based on the medians of two distributions.[7] Because the median is not severely affected by extreme values and because it is a good measure of location for ordinal scales, the sign test is appropriate for comparing locations of skewed distributions or distributions based on ordinal data. Sometimes, because it is simple, the sign test is used even when a t-test would be valid.

The so-called "sign test" of the null hypothesis of no difference in locations of two populations consists in comparing the observations in the two samples with the median of the entire set of observations. The null hypothesis is that the distribution of x_1 is the same as the distribution of x_2. The alternative is that the distributions differ with respect to location, that is, that observations from one distribution tend to be larger than those from the other. Two samples of sizes $n_1 = 7$ and $n_2 = 9$ are shown in Figure 12.4. The median M of the combined sample is $\frac{1}{2}(18 + 20) = 19$. There is only one value in the first sample which exceeds the median of the combined sample, while there are 7 such values in the second sample. The numbers of observations above

---
[7]This test is also referred to as the "median test."

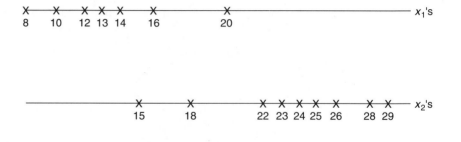

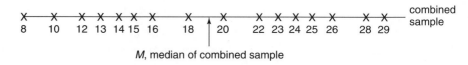

Figure 12.4   *Data plot for two-sample sign test.*

and below the median in the two samples are recorded in Table 12.7. The fact that most of the observations in the first sample are less than the median and most in the second greater suggests that the null hypothesis may be implausible.

If the null hypothesis is true and the two distributions are the same, all of the observations may be considered as a single sample from the common distribution, and the probability of any one observation being greater than the median is $1/2$. If the two distributions are different in such a way that the x_2's tend to be larger than the x_1's, the probability that an x_2 observation is greater than the median of the sample exceeds $1/2$. The null hypothesis corresponds to the probability of an x_1 observation exceeding the common median and the probability of an x_2 observation exceeding the common median both being $1/2$.

TABLE 12.7
Summary Data for Two-Sample Sign Test

	First sample	*Second sample*	*Combined samples*
Greater than common median	1 (14.3%)	7 (77.8%)	8 (50.0%)
Less than common median	6 (85.7%)	2 (22.2%)	8 (50.0%)
Total	7 (100.0%)	9 (100.0%)	16 (100.0%)

source: Figure 12.4.

The procedure of Section 12.4 for testing equality of proportions can be used. This procedure requires a moderate to large sample size, that is, a total n in the range of 30 or more. Although the sample in Figure 12.4 and Table 12.7 is smaller than this, we shall complete the example for purposes of illustration and follow with another example that has a more appropriate n.

Let $\hat{p}_1 = 1/7$ be the proportion greater than the common median in the first sample and $\hat{p}_2 = 7/9$ in the second. In the pooled sample $\hat{p}$ must be $1/2$ by definition of the common median M. The test statistic is the standardized version of $\hat{p}_1 - \hat{p}_2$, which is

$$z = \frac{\hat{p}_1 - \hat{p}_2}{\sqrt{\hat{p}\hat{q}\left(\dfrac{1}{n_1} + \dfrac{1}{n_2}\right)}} = \frac{\hat{p}_1 - \hat{p}_2}{\sqrt{\dfrac{1}{2} \times \dfrac{1}{2} \times \left(\dfrac{1}{n_1} + \dfrac{1}{n_2}\right)}} = 2\sqrt{\frac{n_1 n_2}{n_1 + n_2}}(\hat{p}_1 - \hat{p}_2).$$

For the data of Figure 12.4,

$$z = 2 \times \sqrt{\frac{7 \times 9}{7 + 9}}(1/7 - 7/9) = -2.52.$$

This value is less than the significance point -1.96 from the standard normal distribution for a two-tailed test at the 5% level. The null hypothesis is rejected.

The procedure is called the "sign test" because the frequencies in Table 12.7 are obtained by counting the number of positive and negative $(x_{1i} - M)$'s and $(x_{2j} - M)$'s. In practice it is necessary to compute only one of the frequencies in the 2×2 table because all the marginal totals are known.

EXAMPLE

As another example, suppose that we were concerned about salary inequities between males and females performing the same half-time clerical job. The data are given in Table 12.8. The median of the combined sample is 94 hundred dollars. When $n_1 + n_2$ is odd, the median is equal to one of the observations. Since this observation cannot be

TABLE 12.8

Salaries (100's of Dollars) of 18 Males and 15 Females for a Half-Time Clerical Job

Males	22, 36, 41, 53, 64, 67, 84, 95, 100, 101, 115, 116, 127, 131, 138, 140, 146, 172
Females	9, 11, 18, 43, 47, 54, 61, 69, 78, 94, 96, 96, 106, 110, 112

Hypothetical data.

TABLE 12.9

Summary Information for Sign Test of Salaries of Males and Females

		Gender		Total
		Male	*Female*	
Income	*Greater than 94*	11 (61.1%)	5 (35.7%)	16 (50.0%)
	Less than 94	7 (38.9%)	9 (64.3%)	16 (50.0%)
	Total	18 (100.0%)	14 (100.0%)	32 (100.0%)

source: Table 12.8.

classified as above or below the median, we do not include it in counting observations for the sign test. The frequencies of males and females above and below the median are given in Table 12.9.

The sample proportions of males and females with salaries above the common median are $\hat{p}_1 = 11/18$ and $\hat{p}_2 = 5/14$, respectively. The test statistic is

$$z = 2\sqrt{\frac{n_1 n_2}{n_1 + n_2}}(\hat{p}_1 - \hat{p}_2) = 2 \times \sqrt{\frac{18 \times 14}{18 + 14}}(11/18 - 5/14)$$

$$= 2 \times 2.806 \times (0.611 - 0.357) = 1.43.$$

The test statistic does not exceed 1.645, the 5% significance point from the standard normal distribution making a one-tailed test. The null hypothesis that males and females have the same median salary for this clerical job is accepted. The P value for the test statistic $z = 1.43$ is $.5 - .4236 = .0764$; H_0 would be accepted at any α value less than this.

Summary

Many statistical studies involve the comparison of two populations, whether they are distinguished by experimental interventions or by naturally occurring circumstances. In this chapter, techniques are presented for comparing the means of two populations on a numerical scale, and the proportions of two populations that have some characteristic of interest. In both cases, samples are drawn from the two populations and the *difference* between sample means (or proportions) becomes the focus of the analysis.

Tests of significance are used to determine if the difference in the sample is large enough to conclude that there is a difference between the two population values as well. The tests are based on the sampling distribution of the difference. The standard deviation of the sampling distribution (the standard error of the difference) is used to standardize the sample difference, that is, to re-express it in the form of a test statistic. Several test procedures are discussed for comparing means: a z-statistic that is compared with percentage points of the standard normal distribution if the population variance is known, and a t-statistic that is compared with percentage points of Student's t-distribution if the variance of the measure must be estimated from the sample. The t-test is based on the assumption that the variances in the two populations are the same; a variation is discussed for situations in which this cannot be assumed. The test statistic for the equality of two proportions is referred to percentage points of the standard normal distribution; equality of the two proportions implies that the corresponding variances are equal. This test also requires moderate to large sample sizes.

Confidence intervals for two populations means or proportions give the range of plausible values for the true population difference. These have the same general form as confidence intervals on a single mean or proportion. A multiple of the standard error is added to and subtracted from the sample difference. "How many" depends on the level of confidence that is required. Once the confidence level is decided, the appropriate multiple is obtained from the table of the standard normal distribution or from the t-distribution; the same distribution is used for constructing a confidence interval as for testing a hypothesis about a difference between population values.

Section 11.6 of this book also discusses a set of procedures for comparing two means. It is important to distinguish the test of a difference between "paired measurements" in that section from the tests discussed in this chapter. Here we describe procedures for comparing two *independent* samples; the observed values in the first sample are in no way related to those of the second. The concept of a correlation existing between the observations in the two samples is not meaningful because they are not "paired" or "matched" in any sense; in fact, the two samples may have different numbers of observations. In contrast, a basic concept in Section 11.6 is that there are meaningful pairs of values, whether they arise from the same individual tested on two occasions or from two individuals who are matched on some background characteristic(s). There are the same number of observed values in both columns of data and a difference score is a meaningful representation of change or growth or differential effect of some independent variable on the outcome measure. Whenever two means are to be compared, it is important to identify whether they are based on paired measurements (Chapter 11) or on independent samples (Chapter 12) in order to apply the appropriate statistical strategy.

Appendix 12A   The Continuity Adjustment

The continuity adjustment for a difference of two proportions consists in adding to or subtracting from $\hat{p}_1 - \hat{p}_2$ the continuity correction

$$\frac{n_1 + n_2}{2n_1 n_2}.$$

The effect of the adjustment is always to make test statistics less significant and confidence intervals wider. For example, suppose a confidence interval is being computed for $p_1 - p_2$, and $\hat{p}_1$ is larger than $\hat{p}_2$. Then, in forming the upper limit, $\hat{p}_1 - \hat{p}_2$ is replaced by

$$\hat{p}_1 - \hat{p}_2 + \frac{n_1 + n_2}{2n_1 n_2}$$

and, in forming the lower limit, by

$$\hat{p}_1 - \hat{p}_2 - \frac{n_1 + n_2}{2n_1 n_2}.$$

The quantity $(n_1 + n_2)/2n_1 n_2$ is also the continuity correction for the two-sample sign test.

Exercises

12.1   In the hope of being hired by a certain company, some individuals take a course to prepare themselves for the Job Aptitude Test administered to all applicants. The mean test score of 25 job applicants who completed the course is 104.2, and of 40 applicants who did not complete the course is 99.6. The standard deviation of scores on the test is known to be $\sigma = 12$ points. Assume that these individuals are random samples from the population of all individuals who might apply to the company. At the 5% level of significance, do these data support the hypothesis that taking the course increases the population mean on the test?

(a) State H_0 and H_1. (Is a one-tailed or two-tailed test required?)

(continued ...)

(b) Compute the difference of the two sample means and the standard error of the difference.

(c) Compute the test statistic and find the significance point from the table of the standard normal distribution.

(d) Conclude whether the null hypothesis is accepted or rejected.

12.2 (continuation) Do you think it is reasonable to assume that those who took the course and those who did not are random samples from a single population of applicants? If so, justify your conclusion. If not, state how the results might be affected if the groups represent two distinct populations.

12.3 (continuation) Approximate the P value for the test statistic in the preceding exercise. From the P value state whether the null hypothesis would be rejected at

(a) $\alpha = .10$,

(b) $\alpha = .01$,

(c) $\alpha = .005$.

12.4 (continuation) Construct a 95% confidence interval for the difference between the means of individuals who take the course and those who do not. Does the interval lead to an unambiguous conclusion that the mean of those who take the course is higher than for those who do not? From the interval, what is the *maximum* plausible benefit of taking the course in terms of points on the test? In standard deviation units (the effect size)?

12.5 The Graduate Record Examination (GRE) is scaled to have a national mean of 500 and standard deviation of $\sigma = 100$ points. A random sample of 30 graduate students at university H has a mean GRE score of $\bar{x}_1 = 621$, and a random sample of 35 graduate students at university S has a mean GRE score of $\bar{x}_2 = 616$. Use these data to test the null hypothesis that the mean of all graduate students at university H is the same as the mean of all graduate students at university S against the alternative hypothesis that the two means are not the same. Use the 5% level of significance.

(a) State H_0 and H_1.

(b) Compute the difference of the two sample means and the standard error of the difference.

(c) Compute the test statistic.

(d) Approximate the P value for the test statistic and state the range of α values for which H_0 could be rejected.

(e) Is the null hypothesis rejected? Would it have been rejected if α was set at .01?

12.6   (continuation) Construct a 95% confidence interval for the difference between the mean GRE scores of all graduate students at university H and graduate students at university S.

(a) Is 0 in this interval? (That is, is a conclusion of no difference between the two schools on mean GRE scores plausible?)

(b) Re-express the lower and upper limits of the interval in standard deviation units, that is, as effect sizes.

12.7   A 1989 issue of the *Journal of the American Medical Association* (Blair, *et al.*) describes a "prospective" study of men and women who received a complete preventive medical examination at a clinic between 1970 and 1981 and who then were followed up an average of 8 years after their original examinations. Only individuals who were healthy at the original examination and who had no history of severe medical problems were included in the study. The mean age of participants in the study was 41.5 years and the standard deviation of ages was about $9\frac{1}{2}$ years. Table 12.10 summarizes some of the baseline characteristics of individuals who survived until the time of the followup and those who were deceased by that time.

TABLE 12.10

Baseline Characteristics of Surviving and Deceased Male and Female Patients

	Men				Women			
	Surviving (*n* = 9984)		*Deceased* (*n* = 240)		*Surviving* (*n* = 3077)		*Deceased* (*n* = 43)	
	Mean	*s.d.*	*Mean*	*s.d.*	*Mean*	*s.d.*	*Mean*	*s.d.*
Weight (kilograms)	81.9	12.1	83.2	13.3	59.9	10.2	60.7	9.4
Height (centimeters)	178.8	6.3	180.1	22.1[c]	164.3	5.6	164.1	5.6
Body mass index[a]	25.6	3.3	25.8	3.4	22.2	3.5	23.2	3.7
Systolic blood pressure[b]	120.4	13.1	126.1	17.1	112.2	14.0	124.1	19.5
Diastolic blood pressure[b]	79.7	9.1	82.4	11.2	74.5	9.1	79.2	9.4
Current smoker, %	28.5	—	47.5	—	21.2	—	32.6	—

Adapted from Table 1 of Blair *et al.* (1989). Physical fitness and all-cause mortality. *Journal of the American Medical Association, 262,* 2395–2401. © 1989, American Medical Association.
[a]The body mass index is computed as W/H^2 where W is the individual's weight in kilograms and H is height in meters. Very high values are associated with a variety of health risks.
[b]Millimeters of mercury.
[c]A standard deviation of 22.1 centimeters = 8.7 inches is absurd; this figure was obviously published in error.

(a) Decide which of the five measured characteristics of men are associated with mortality by comparing the mean of surviving men with the mean of deceased men on each of weight, height, body mass index, and two blood pressure measures. For each comparison, make a two-tailed test at the 10% level of significance, using the known values of σ given in the table.

(b) For each statistically significant variable found in (a), express the magnitude of the difference as an "effect size," that is, $(\bar{x}_1 - \bar{x}_2)/\sigma$ where σ is the standard deviation for survivors. Which variables have the largest survivor-deceased difference?

(c) Based on your knowledge of factors important to health and longevity, and on recent reports in the news media, is there sufficient information about any of these variables to justify making a one-tailed test instead of a two-tailed test? Explain.

12.8   (continuation) Repeat the analysis of Exercise 12.7 using the data for women. What differences are apparent between the mortality risk factors for men and for women?

12.9   (continuation) Describe four health-related factors that may explain why some individuals are deceased while others are surviving after 8 years, other than those reported in Table 12.10. Explain how each factor may have affected the two groups differently.

12.10   For the data in Exercise 2.14 (Table 2.21) compare the mean final grades of students who had taken at least one course on probability and statistics with the mean final grades of students who had no prior course on probability and statistics. Make a one-tailed test in which the null hypothesis is that the mean is the same or lower for students with no prior course and the alternative is that the mean is increased by taking prior coursework. Use the 5% level of significance and assume that the population variances are equal.

12.11   A researcher wishes to test H_0: $\mu_1 \leq \mu_2$ against the alternative H_1: $\mu_1 > \mu_2$ at the 10% level of significance. She draws a random sample of 21 observations from population 1 and computes $\bar{x}_1 = 22.7$ and $s_1 = 4.6$. A random sample of 16 observations from population 2 gives $\bar{x}_2 = 19.5$ and $s_2 = 4.8$.

(a) Find the common variance s_p^2 for these data.

(b) Compute the standard error of the difference between the sample means.

(c) Compute the test statistic for comparing μ_1 and μ_2.

(d) Give the significance point for the test of the researcher's hypothesis.

(e) State whether H_0 is accepted or rejected.

12.12   A researcher wishes to test H_0: $\mu_1 \geq \mu_2$ against the alternative H_1: $\mu_1 < \mu_2$ at the 5% level of significance. She draws a random sample of 36 observations from population 1 and computes $\bar{x}_1 = 7.28$ and $s_1 = 0.44$, and a random sample of 26 observations from population 2 gives $\bar{x}_2 = 6.31$ and $s_2 = 0.41$.

Use these data to complete steps (a) through (e) listed for Exercise 12.11.

12.13 A researcher wishes to test H_0: $\mu_1 = \mu_2$ against the alternative H_1: $\mu_1 \neq \mu_2$. She draws a random sample of 22 observations from population 1 and computes $\bar{x}_1 = 98.4$. A random sample of 20 observations from population 2 gives $\bar{x}_2 = 107.7$. The standard error of the difference is $s_{\bar{x}_1 - \bar{x}_2} = 3.88$.

(a) Give the value of the test statistic for testing the researcher's hypothesis.

(b) Give the significance point(s) for testing the researcher's hypothesis at the 5% level of significance.

(c) State whether H_0 is accepted or rejected.

12.14 (continuation) Give the significance point(s) and conclusion about accepting or rejecting H_0 in Exercise 12.13 if (a) the 2% level of significance is chosen instead of the 5% level, and (b) the 1% level of significance is chosen instead of the 5% level.

12.15 (continuation) Determine a P value interval for the test statistic obtained in Exercise 12.13. *From the interval*, state the entire range of α values for which the null hypothesis would be rejected.

12.16 R. W. O'Brien and R. J. Iannotti (1994) were concerned with the ways in which mothers' "Type A" behavior (i.e., impatience, aggression, competitiveness) influence their relationships with their preschool-aged children. A sample of 82 mothers with high levels of Type A behavior and 76 mothers with low levels of Type A behavior participated. Each mother was asked to rate the child's level of Type A behavior on a scale that ranges from 17 to 65. In addition, each child's preschool teacher or day care supervisor rated the child on the same measure. The results are summarized in Table 12.11.

TABLE 12.11

Mothers' and Teachers' Ratings of Children's Type A Behavior

		Mothers		Teachers	
	n	*Mean*	*s.d.*	*Mean*	*s.d.*
High type A mothers	82	54.78	10.04	42.56	10.05
Low type A mothers	76	50.59	9.23	44.09	9.64

source: O'Brien, R. W., & Iannotti, R. J. (1994). *Behavioral Medicine, 19,* 162–168. Reprinted with permission of the Helen Dwight Reid Educational Foundation. Published by Heldref Publications, 1319 Eighteenth St., N.W., Washington, D.C. 20036-1802. © 1994.

Use the *mothers'* ratings to test the hypothesis that the mean Type A behavior for children with high Type A mothers is equal to the mean Type A behavior for children with low Type A mothers; use the 5% significance level. Assuming equal population variances the pooled variance estimate is $s_p^2 = 93.30$, and the standard error of the difference is $s_{\bar{x}_1 - \bar{x}_2} = 1.54$. In your answer give

(a) The null and alternative hypotheses in statistical form,

(b) The value of the test statistic computed from the data,

(c) The name of the probability distribution used to test the hypothesis and the significance points from that distribution,

(d) A decision to accept or reject the null hypothesis.

12.17 (continuation) Repeat all of the steps in Exercise 12.16 using the *teachers'* ratings of the children's Type A behavior. For these ratings the pooled variance estimate is $s_p^2 = 97.12$ and the standard error of the difference is $s_{\bar{x}_1 - \bar{x}_2} = 1.57$.

12.18 (continuation) Compare the conclusions you drew in Exercises 12.16 and 12.17. If the two conclusions are different, how would you explain these differences in terms of the mothers' or children's behavior?

12.19 S. Schneer and F. Reitman (1993) studied career-related outcomes of 269 women who had received their M.B.A. degrees from two large northeastern universities between 1975 and 1980. Of this sample, 121 women reported that they were married with no children; the mean and standard deviation of their annual incomes were 61.11 and 24.51, respectively. (Note: The published report does not describe the scale of the income measure; for purposes of this exercise assume that the values are thousands of dollars.) Another 62 women reported that they were married with children; the mean and standard deviation of their annual incomes were 56.03 and 24.04, respectively. Assuming that these are representative samples of women with M.B.A. degrees, test the researchers' contention that having children is associated with *lower* mean income among married women with M.B.A.'s. In your answer give

(a) The null and alternative hypotheses in statistical form,

(b) The value of the pooled variance estimate (assuming equal population variances), the standard error of the mean difference, and the test statistic,

(c) The name of the probability distribution used to test the hypothesis,

(d) A P value interval for the test,

(e) Your conclusion about whether or not the researchers' hypothesis is supported by the data.

12.20   (continuation) Obtain a 90% confidence interval for the difference between the mean income of married women with M.B.A. degrees who have children and who do not have children. Use the data of Exercise 12.19. From this interval, what is the largest plausible financial advantage associated with having no children (on average)?

12.21   Table 3.16 (Exercise 3.15) gives the numbers of books borrowed from two university libraries by random samples of students at those schools. Use these data to test the hypothesis of no difference between the mean number of books withdrawn at the two libraries. Make a two-tailed test assuming equal population variances. Use the sample sizes and sample means computed for Exercise 3.15.

(a) Compute the variance and standard deviation of numbers of books borrowed at each university, and the pooled estimate of the variance.

(b) Give the null and alternative hypotheses in symbols.

(c) Compute the test statistic.

(d) Obtain a P value interval for the test statistic.

(e) From the P value interval, would you reject or accept H_0? Why, or why not?

12.22   Table 2.22 (Exercise 2.15) gives test scores for a sample of firefighter applicants in a large city in the northeastern United States. Assume that these individuals constitute a random sample from a population of applicants. Use SPSS or another computer program to compare the average performance of white and minority applicants on the Scaled Agility Score and also (separately) on the Written Test Score. For each comparison, make a two-tailed test at the 5% level of significance, assuming equal population variances.
    Make a table that contains the following for each measure:

(a) The mean and standard deviation for white applicants and for minorities,

(b) the mean difference between whites and minorities,

(c) the mean difference in standard deviation units (that is, an "effect size") where the standard deviation is the square root of the pooled variance s_p^2,

(d) the standard error of the mean difference,

(e) the test statistic,

(f) a P value interval for the test statistic.

Summarize your findings in a written paragraph. Be sure to include a statement about whether each difference is or is not statistically significant at the 5% level and, if it is, the direction and magnitude of the difference.

12.23   (continuation) Repeat the process outlined in Exercise 12.22, but compare the performance of male and female applicants instead of whites and minorities. If you were going to devise an employment plan that is "fair" to white and minority applicants and to male and female applicants, how would you take the findings of these two analyses into consideration?

12.24   A number of students in a ninth-grade General Mathematics course were divided into two groups. One group (the experimental group) received instruction in a new probability-and-statistics unit; the other group (the control group) did not. Such instructional changes are not instituted without the possibility of consequent decline in the learning of traditional skills; accordingly, all the students were given a test of computational skill before and after the semester. The score analyzed was the *gain* in skill. The results are summarized in Table 12.12.

TABLE 12.12
Summary of Students' Scores on Test of Computational Skill

Experimental group[a]	Control group[b]
$n_1 = 281$	$n_2 = 311$
$\bar{x}_1 = -0.11$	$\bar{x}_2 = 3.54$
$s_1^2 = 74.0$	$s_2^2 = 80.0$
$s_1 = 8.60$	$s_2 = 8.95$

source: Shulte (1970). Reprinted from *The Mathematics Teacher*, January 1970, Vol. 63, pp. 56–64 (© 1970 by the National Council of Teachers of Mathematics). Used by permission.
[a]Experimental group: Course included a unit on probability and statistics.
[b]Control group: Course did not include instruction in probability and statistics.

(a) The pooled sum of squared deviations is 45,583.96. Test the null hypothesis of no difference between the mean gains in the two populations. Treat the population standard deviations as equal and use the 5% level of significance.

(b) Test the null hypothesis that the mean gain in the experimental population is zero. Make a two-tailed test at the 5% level of significance.

(c) Test the null hypothesis that the mean gain in the control population is zero, using the 5% level of significance.

(d) Write a paragraph interpreting the results of (a), (b), and (c). (A sentence for each will suffice.)

12.25   Weiss, Whitten, and Leddy (1972) reported the figures in Table 12.13, which provide a comparison between the lead content of human hair

TABLE 12.13

Lead in Hair (micrograms of lead per gram of hair)

		Populations	
		Antique	*Contemporary*
Age group	*Children*	$\bar{x} = 164.24$ $s = 124$ $n = 36$ $s^2/n = 428$	$\bar{x} = 16.23$ $s = 10.6$ $n = 119$ $s^2/n = 0.941$
	Adults	$\bar{x} = 93.36$ $s = 72.9$ $n = 20$ $s^2/n = 266$	$\bar{x} = 6.55$ $s = 6.19$ $n = 28$ $s^2/n = 1.37$

source: Adapted with permission from Weiss, D., Whitten, B., & Leddy, D. (1972). Lead content of human hair (1871–1971). *Science, 178*, 69–70. © 1972, American Association for the Advancement of Science.

removed from persons between 1871 and 1923 (the "antique" population) with that from present day populations. (A microgram is a millionth of a gram.) Make the following comparisons by doing two-tailed *t*-tests at the 1% level, using the separate variance estimates:

(a) Is there a significant difference between the antique and contemporary groups (i) among children? (ii) among adults?

(b) Is there a significant difference between the children and the adults (i) in the antique group? (ii) in the contemporary group?

12.26   Of the students for whom data are given in Table 12.4, the students numbered 28, 39, 40, 42, 44, 48, 49, 52, 53, 54, 56, 59, 62, 66, and 67 had a native language other than English. It is anticipated that such students might suffer a disadvantage when enrolled in a course taught in English. However, it is frequently observed that such students invest extra effort in the courses and often perform better than those who have English as a native language.

(a) Make a two-tailed test of the difference in means at the 5% level of significance assuming that the population variances are equal.

(b) Construct a 95% confidence interval for the difference between the means. What does the interval tell you about the possible values of the difference between these populations of students?

12.27   (continuation) Arrange the scores in Table 12.4 into a two-way table like Table 12.14.

(a) Compute the sum of squared deviations for each of the four cells of the table.

(continued . . .)

TABLE 12.14

Scores of Students in a Statistics Course, by Student Status and Native Language

		Student status	
		Undergraduate	*Graduate*
Native language	*English*		
	Other		

(b) Add these four sums of squared deviations to obtain a pooled sum of squared deviations.

(c) What is the number of degrees of freedom associated with this pooled sum of squared deviations? [Hint: Each cell contributes one less than the sample size for that cell.]

(d) Compute a pooled estimate of variance by dividing the number of degrees of freedom into the pooled sum of squared deviations. Use this in the denominator of *t*-statistics in the rest of the exercise.

(e) Test the hypothesis of no difference between undergraduate and graduate means (i) for those whose native language is English and (ii) for the others. Compare the result with that obtained in Section 12.2; interpret any discrepancy.

(f) Test the hypothesis of no difference between means of students whose native language is English and the other students (i) for the undergraduates and (ii) for the graduates. Compare the result with that obtained in Exercise 12.26; interpret any discrepancy.

(g) Does the average effect of language depend upon student status? [Hint: Compute the difference between language groups for undergraduates and for graduates. Then compute the difference between these differences. The estimated variance of this combination of four means has the form

$$s_p^2 \left(\frac{1}{n_1} + \frac{1}{n_2} + \frac{1}{n_3} + \frac{1}{n_4} \right)$$

where s_p^2 is the pooled estimate of variance.]

12.28 (continuation) Write a paragraph summarizing this analysis of the data in Table 12.14.

12.29 Use the data in Table 12.10 to test the null hypothesis that the mortality rate for men is the same as for women. Make a two-tailed test at the 1% level of significance. In your answer give

(a) the null and alternative hypotheses in statistical form,

(b) the values of $\hat{p}_1$ and $\hat{p}_2$ and the test statistic,

(continued ...)

(c) the name of the probability distribution used to test the hypothesis and the significance points from the distribution,

(d) your decision to accept or reject the null hypothesis,

(e) a verbal statement that the mortality rate is found to be higher for men, the same for men and women, or higher for women.

12.30   Two types of syringes, A and B, were used for injection of medical patients with a certain medication. The patients were allocated at random to two groups, one to receive the injection from syringe A, the other to receive the injection from syringe B. Table 12.15 shows the number of patients showing reactions to the injection. Test at the 1% level the null hypothesis of no difference between the true proportions of patients giving reactions to syringes A and B. State the alternative hypothesis clearly.

TABLE 12.15

Results of Injections with Two Types of Syringes

| | | Number of patients | | |
		With reactions	*Without reactions*	*Total*
Type of	*A*	44	6	50
syringe	*B*	20	30	50
	Total	64	36	100

Hypothetical data.

12.31   Suppose that in the context of the previous exercise the figures had been as in Table 12.16. At the 1% level, would you accept the hypothesis of no difference between proportions?

TABLE 12.16

Results of Injections with Two Types of Syringes

| | | Number of patients | | |
		With reactions	*Without reactions*	Total
Type of	*A*	22	3	25
syringe	*B*	10	15	25
	Total	32	18	50

Hypothetical data.

12.32   Stewart, Marcus, Christenson, and Lin (1994) administered a questionnaire to a group of patients at a local dental clinic. Based on their responses the patients were classified as "high dental anxiety" (HDA) or "low dental anxiety" (LDA); 21 patients were classified into each group. At the end of 2 years, 9 of the HDA patients and 13 of the LDA patients, together with the dentist, had outlined a plan for complete dental treatment. Assume that these patients represent random samples of the population of HDA and LDA dental patients. Test at the 5% significance level the null hypothesis that the proportion of HDA patients preparing a treatment plan was greater than or equal to the proportion of LDA patients preparing a plan. In your answer give

(a) the null and alternative hypotheses in statistical form, being careful to label which proportion is for the HDA group and which is for the LDA group,

(b) the values of $\hat{p}_1$ and $\hat{p}_2$, the standard error of the difference, and the test statistic,

(c) the name of the probability distribution used to test the hypothesis and the significance point from the distribution,

(d) your decision to accept or reject the null hypothesis,

(e) a verbal statement that a greater proportion of LDA patients than HDA patients prepare a treatment plan *or* that no greater proportion of LDA patients prepare a treatment plan.

12.33   (continuation) The report by Stewart et al. (1994) cited in the previous exercise also noted that, in 2 years, one of the 21 HDA patients and 7 of the 21 LDA patients actually completed the prescribed course of treatment. Test the null hypothesis that the proportion of HDA patients completing the treatment was greater than or equal to the proportion of LDA patients who completed their treatment. In your answer give

(a) the null and alternative hypotheses in statistical form, being careful to label which proportion is for the HDA and the LDA group,

(b) the values of $\hat{p}_1$ and $\hat{p}_2$, the standard error of the difference, and the test statistic,

(c) the name of the probability distribution used to test the hypothesis,

(d) a P value for the test,

(e) your conclusion about whether the null hypothesis should be rejected; that is, do a greater percentage of LDA patients actually complete their recommended treatment? Are the data convincing?

12.34   Use the data of Table 12.10 (Exercise 12.7) to test whether, at the time of the interviews, there were more cigarette smokers among participants who are now deceased than surviving. Make two one-tailed tests at the 1% level of significance, one for males and one for females. For each test state H_0 and H_1, give the test statistic, the significance point, and your conclusion regarding whether smoking is associated with an increased risk of mortality.

12.35   (continuation) Use the cigarette smoking data from Table 12.10 to obtain a 90% confidence interval for the difference between the proportion of deceased and surviving males who smoked cigarettes. Repeat the process with the data for females. Summarize the conclusions that may be drawn from these intervals.

12.36   For the data from the 1954 polio vaccine trial (Section 1.1, Table 1.4), test the hypothesis of no difference between incidence rates in the control and experimental populations against the one-sided alternative that the rate in the experimental population is *less*. Use $\alpha = .01$.

12.37   Table 12.17 has been adapted from a study by M. C. Ware and M. F. Stuck (1985) of the ways in which men and women were portrayed in illustrations in three mass market computer magazines. Their sample included a total of 465 illustrations of men and 196 illustrations of women. Table 12.17 gives the number and percentage of male and female illustrations in each of the four most common categories and all others combined.

TABLE 12.17

Illustrations of Males and Females in Various Roles in Computer Magazines

		Men	Women	Both
	Seller	124 (26.7%)	81 (41.3%)	205 (31.0%)
	Manager	101 (21.7%)	13 (6.6%)	114 (17.2%)
Role	Clerical	40 (8.6%)	47 (24.0%)	87 (13.2%)
	Computer expert	51 (11.0%)	2 (1.0%)	53 (8.0%)
	All others	149 (32.0%)	53 (27.0%)	202 (30.6%)
	Total	465 (100.0%)	196 (100.0%)	661 (100.0%)

source: Adapted from Ware and Stuck (1985), p. 209.

Use these data to test at the 5% significance level the null hypothesis that equal proportions of men and women are portrayed in computer magazines as (i) Sellers, (ii) Managers of businesses, (iii) Clerical workers, and (iv) Computer experts, as follows:

(a) State the null and alternative hypotheses in statistical form.

(b) Name the probability distribution used to test the hypothesis and the significance points from the distribution.

(continued . . .)

(c) Construct a table that has only the four "roles" as rows (eliminate "others" and recompute the totals). Make a column for each of the following:

- the proportion of men in that role,
- the proportion of women in that role,
- the value of the test statistic for comparing the two proportions,
- the decision to accept or reject the null hypothesis,
- a statement about which group is represented more commonly in the particular role, that is, "men" if there is a significantly higher proportion of men in the role, "women" if there is a significantly higher proportion of women in the role, or "neither" if the difference is not significant.

12.38   (continuation) Approximate a P value for each of the test statistics obtained in the preceding exercise. From these P values state which differences would have been significant if the 1% significance level had been used instead of the 5% level.

12.39   (continuation) Write a brief report summarizing the findings of Exercise 12.37 and give your explanation about *why* the differences that you found, or did not find, characterized popular computer magazines in 1983? Do you think that the same differences would be found today?

12.40   (continuation) Using the data of Table 12.17 construct a 95% confidence interval for the difference between the proportion of males and the proportion of females who are portrayed in computer magazines as Managers. Repeat the process for the proportions of males and females portrayed as Clerical workers. Write a brief summary of your findings, comparing these two confidence intervals.

12.41   Carry out the sign test on the two samples shown in Table 12.1 and Figure 12.2. Use a two-tailed test at the 5% level of significance.

12.42   For each of the following research problems, give

(1) H_0 and H_1,
(2) the formula for the test statistic that should be used to test the hypothesis,
(3) the probability distribution to which the test statistic should be referred to see if H_0 is accepted or rejected.

(a) An English composition written by a college freshman was sent to a random sample of 50 freshman English instructors who were asked to rate the quality of the composition on a 100-point scale. Half of the instructors were told that the student was admitted to college through an "open admissions" policy and half were told that the student was admitted on an academic scholarship. Does this information affect the ratings that are given to the composition?

(b) One hundred obese adults undertake a particular one-month diet. Of the 50 dieters whose weight loss was greatest, 3 admitted eating ice cream at least once during the month. Of the 50 whose weight loss was not as great, 12 admitted eating ice cream during the month. Is ice cream consumption associated with *less* weight loss?

(c) Two popular tennis magazines are being compared to decide which should become the official magazine of the Quality International Tennis Society (QUITS). A random sample of 25 tennis pros are asked to rate magazine A on a scale from 1 to 10, with 10 denoting the very highest opinion of the magazine, and also to rate magazine B on the same scale. In the population of tennis pros, is there a difference in the mean ratings of the two magazines?

(d) IQ scores on the Wechsler Intelligence Scale for Children are scaled so that the mean in the population of normal-intelligence youngsters is 100 and the standard deviation is 16. Children with learning problems in school are sometimes classified by school psychologists as "mildly mentally retarded" and placed in special education classes where they receive more individualized attention. Other youngsters are classified as "learning disabled" by a completely different set of criteria. A sample of 30 mildly retarded youngsters and 30 learning disabled youngsters is chosen from a large school district and tested on the Wechsler scale. Do the data indicate that mildly retarded youngsters actually have *lower* IQs, on average, than learning disabled youngsters?

13

Variability in One Population and in Two Populations

Introduction

The expression, "A chain is only as strong as its weakest link," may be construed as an admonition to consider the *variability* of the links as well as their average strength. In comparing distributions, averages alone are not always adequate. Figures 3.5 and 3.6 show two telephone waiting-time distributions with equal means (1.1 seconds) but very different shapes. In Section 4.3 we pointed out that their standard deviations, 0.41 second and 0.69 second, were quite different. If the pupils in each of two school classes have mean IQs of 100, but Class A has a standard deviation of 10, while Class B has a standard deviation of 20, teaching the relatively homogeneous Class A may be very different from teaching the relatively heterogeneous Class B.

In this chapter we consider methods of answering questions about population variances. We begin with a review of what we learned in Chapters 4 and 10 about the variance and its sampling distribution. This discussion is extended by consideration of a family of probability distributions called the chi-square distributions. We then consider interval estimates and hypothesis tests for a population variance. Finally, we consider statistical inference concerning two population variances. This necessitates the introduction of another family of probability distributions, the *F*-distributions.

13.1   Variability in One Population

The Sampling Distribution of the Sum of Squared Deviations

In Section 4.3 we defined the variance of a sample. In Section 10.2 we noted that the sample variance is an unbiased estimate of the population variance; that is, the mean of the sampling distribution of the sample variance is the population variance, if the population is infinite.

Table 13.1 shows ten random samples of size 10 from a normal distribution with a mean of 69 and a standard deviation of 3, that is, a

TABLE 13.1

Ten Samples of Size 10 from a Normal Population with a Mean of 69 and a Variance of 9

Sample number	Sample	Mean	Sum of squared deviations	Sample variance	80% confidence interval	Interval includes 9?
1	72.1 67.6 72.8 79.6 70.7 63.4 69.6 72.6 67.5 68.2	70.41	168.9	18.8	(11.5, 40.5)	No
2	73.2 67.3 69.1 70.0 77.8 74.9 68.2 70.2 70.3 68.9	70.99	98.0	10.9	(6.7, 23.5)	Yes
3	71.7 67.5 67.4 70.8 71.6 66.2 73.7 69.5 63.3 70.1	69.18	85.7	9.5	(5.8, 20.5)	Yes
4	72.5 65.8 69.0 71.3 71.9 71.1 72.3 67.1 68.2 66.9	69.63	56.4	6.3	(3.8, 13.5)	Yes
5	64.5 67.5 68.5 68.6 72.1 69.9 70.3 71.2 67.7 67.7	68.77	42.3	4.7	(2.9, 10.1)	Yes
6	66.9 71.3 64.1 68.0 67.5 62.8 67.6 68.3 71.6 67.6	67.57	66.1	7.3	(4.5, 15.9)	Yes
7	73.1 69.7 70.1 71.3 69.5 66.8 71.9 64.4 68.2 69.4	69.44	57.3	6.4	(3.9, 13.7)	Yes
8	67.6 74.0 68.8 65.3 67.5 71.6 67.5 63.1 60.5 68.3	67.42	134.5	14.9	(9.2, 32.3)	No
9	73.2 65.5 66.3 72.7 68.4 68.3 72.7 61.3 67.3 69.2	68.49	125.0	13.9	(8.5, 30.0)	Yes
10	63.6 68.2 72.7 72.1 67.5 64.1 68.6 67.8 67.1 70.7	68.24	81.7	9.1	(5.6, 19.6)	Yes

variance of 9. The values of the sample variance, given in Table 13.1, vary from sample to sample. The mean of these ten sample variances is 10.18, fairly close to the true value of 9. The standard deviation of the sample variances is 5.27; the values range from 4.7 to 18.8. These facts indicate considerable variability of s^2 in samples of size $n = 10$.

For random samples of $n = 10$ observations from a normal distribution with variance σ^2, there is a theoretical sampling distribution of s^2. (The principle of deducing such a distribution was described in Chapter 9.) The relative frequencies are tabulated in Table 13.2 and graphed in Figure 13.1. Roughly speaking, a number in Table 13.2 indicates the probability that s^2 for $n = 10$ is within $\frac{1}{2}$ of 7 is about 0.104. The entire area under the curve in Figure 13.1 is 1. The probability that s^2 is greater than 14, for instance, is the area below the curve and to the right of 14. A location parameter of the distribution, the mean, is $\sigma^2 = 9$.

TABLE 13.2

Sampling Distributions of s^2 for $\sigma^2 = 9$ and $n = 10$ and 28.

Value of s^2	Height of probability curve	
	$n = 10$	$n = 28$
0	0.000	0.000
1	0.002	0.000
2	0.016	0.000
3	0.040	0.001
4	0.066	0.012
5	0.087	0.042
6	0.100	0.092
7	0.104	0.141
8	0.101	0.166
9	0.092	0.162
10	0.081	0.135
11	0.069	0.099
12	0.056	0.066
13	0.045	0.040
14	0.036	0.022
15	0.027	0.012
16	0.021	0.006
17	0.016	0.003
18	0.012	0.001
19	0.009	0.001
20	0.006	0.000
21	0.004	0.000
22	0.003	0.000

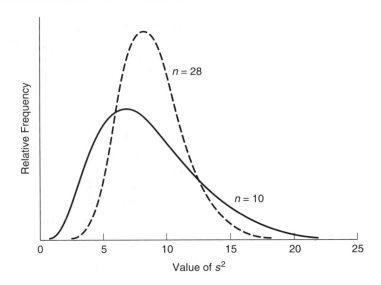

Figure 13.1 *Sampling distributions of s^2 for $\sigma^2 = 9$ and $n = 10$ and 28.*

Table 13.2 also gives the relative frequencies of s^2 for a sample of size $n = 28$. The curve of relative frequencies, also graphed in Figure 13.1, shows that the probability for this larger sample size is more concentrated about the value $\sigma^2 = 9$. Comparison of the two curves suggests that the greater the sample size n the greater the concentration around the population variance and the less the variability of s^2.

As explained in Section 4.3, the sample variance, $s^2 = \sum(x_i - \bar{x})^2/(n-1)$, has $n-1$ degrees of freedom. If $x_1, x_2, \ldots, x_n$ are a random sample from a normal distribution with variance σ^2, the distribution of $(n-1)s^2/\sigma^2$ is known as the *chi-square distribution with $n-1$ degrees of freedom.* (Chi is the Greek letter χ.) If each value of s^2 in Table 13.2 were multiplied by $n-1 = 9$ and divided by $\sigma^2 = 9$, then the resulting probability distribution would be the chi-square distribution with 9 degrees of freedom; of course, the values of $(n-1)s^2/\sigma^2$ would be identical to s^2 in this instance so that the solid line in Figure 13.1 is actually the chi-square distribution itself. If each value of s^2 in Table 13.2 were multiplied by $n-1 = 27$ and divided by $\sigma^2 = 9$, the resulting probability distribution would be the chi-square distribution with 27 degrees of freedom.

The curve of relative frequencies is graphed in Figure 13.2 for 6 degrees of freedom. The curves for other numbers of degrees of freedom are similar. If f, the number of degrees of freedom, is greater than 6, the curve is to the right of the curve in Figure 13.2. The mean of the distribution is f, the standard deviation is $\sqrt{2f}$, and the mode (the value of χ^2 at which the maximum of the curve occurs) is $f - 2$. Thus

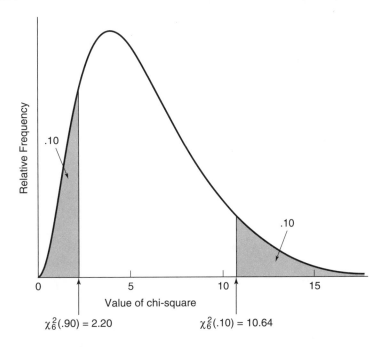

Figure 13.2 *Relative frequency curve of the chi-square distribution for 6 degrees of freedom with some percentiles indicated.*

the distribution of χ_f^2 is centered at f with a spread proportional to $\sqrt{f}$.

We shall denote a particular point on the chi-square distribution with f degrees of freedom as $\chi_f^2(p)$, where the subscript f indicates the number of degrees of freedom and the value in parentheses indicates the proportion of the distribution to the right of this particular value of χ^2. For example, Figure 13.2 indicates $\chi_6^2(.10)$. This is the value of chi-square with 6 degrees of freedom that is exceeded with probability .10; that is, the 90th percentile of the distribution. A table of values of $\chi_f^2(p)$ is given in Appendix IV. The number of degrees of freedom associated with s^2 is $f = n - 1$. For example, if $n = 10$ ($n - 1 = 9$), we find from the table that $\chi_9^2(.975) = 2.700$ and $\chi_9^2(.025) = 19.022$. Thus $\Pr(2.700 < \chi_9^2 < 19.022) = .95$. For $n = 20$ ($n - 1 = 19$), we find $\chi_{19}^2(.05) = 30.143$; thus, $\Pr(\chi_{19}^2 < 30.143) = .95$.

The family of chi-square distributions is useful for statistical inference for the sample variance because of the close relationship of the distribution of $(n - 1)s^2/\sigma^2$ to that of s^2 itself.[1] If we rearrange the expression for χ^2, we see that s^2 is distributed as $\sigma^2\chi_{n-1}^2/(n - 1)$. The

---

[1]Note that the numerator, $(n - 1)s^2$, is just the sum of squared deviations from the mean, $\sum(x_i - \bar{x})^2$, that is, all but the last step in computing the sample variance.

curve of relative frequency of s^2 is similar to that of χ^2_{n-1}. Its mean is $\sigma^2(n-1)/(n-1) = \sigma^2$, and its standard deviation is $\sigma^2\sqrt{2(n-1)}/(n-1) = \sigma^2\sqrt{2}/\sqrt{n-1}$. The larger n is, the smaller the standard deviation of the sampling distribution of the sample variance.

Testing the Hypothesis that the Variance Equals a Given Number

A One-Tailed Test

All gold ingots produced by a certain company are supposed to have very nearly the same weight; it is important that the weight of the ingots not vary much in either direction from the mean. The standard deviation is an average distance from the mean. (See Section 4.3.) Suppose that a standard deviation of 2 grams or less is tolerable. A sample of 20 ingots yields a standard deviation of 2.1 grams. Is this consistent with population standard deviation of 2 grams or less? In formal terms, the null hypothesis is H_0: $\sigma \leq 2$ grams, and the alternative hypothesis is H_1: $\sigma > 2$ grams. Suppose that the 5% level is used.

The 20 observations are treated as a sample from a normal distribution. With random sampling the variable $\chi^2 = (n-1)s^2/\sigma^2$ has a chi-square distribution with 19 degrees of freedom. If H_0 is true, then $\sigma^2 = 4$. To test H_0, the variable $\chi^2 = 19s^2/4 = 4.75s^2$ is to be referred to the chi-square distribution with 19 degrees of freedom. From the table in Appendix IV we find $\chi^2_{19}(.05) = 30.143$; we shall reject H_0 if the observed χ^2 exceeds 30.143. The sample variance of $s^2 = 2.1^2 = 4.41$ gives $\chi^2 = 4.75 \times 4.41 = 20.9$, which does not exceed 30.143. We therefore accept H_0; the conclusion is that weights of the gold ingots do not vary too much.

The general procedure for making a one-tailed test of a variance is as follows. The null and alternative hypotheses are of the form H_0: $\sigma \leq \sigma_0$ and H_1: $\sigma > \sigma_0$, where σ_0 is a prespecified constant value (2 grams in the example). A random sample of n observations is drawn from the population and the sample variance s^2 is computed. The test statistic, $\chi^2 = (n-1)s^2/\sigma^2_0$, is compared to percentage points from the chi-square distribution with $n-1$ degrees of freedom. H_0 is rejected at significance level α if $\chi^2 > \chi^2_{n-1}(\alpha)$.

The P value approach may be used as well. For example, a test statistic of $\chi^2 = 50.3$ based on 36 observations would be referred to the χ^2-distribution with 35 degrees of freedom. We would conclude that the P interval for the test is $.025 < P < .05$ because this value is between the .05 and .025 significance points, 49.802 and 53.203, respectively.

A Two-Tailed Test

A job application examination has a standard deviation of 4 units in the relevant national population. A sample of 10 applicants who have taken a special preparatory course is tested. The question is raised whether the variability in the population of individuals who take the course is the same as in the national population.

Let σ be the standard deviation test scores in this population. We want to test the hypothesis H_0: $\sigma = 4$ against the alternative H_1: $\sigma \neq 4$. We shall reject H_0 if the observed sample variance is too small or too large. In random samples of 10 from the normal distribution with variance 16, the variable $\chi^2 = (n-1)s^2/\sigma^2 = 9s^2/16$ has a chi-square distribution with $n-1 = 9$ degrees of freedom. From Appendix IV, $\chi_9^2(.975) = 2.700$ and $\chi_9^2(.025) = 19.022$; these are the significance points for a two-tailed test at the 5% level. If the observed value of χ^2 is less than 2.700 or greater than 19.022, then H_0 is rejected; if χ^2 falls between these two values, then H_0 will be accepted.

Suppose that the sample variance computed from the 10 observations is $s^2 = 7.96$. Then the value of the test statistic is $\chi^2 = 9 \times 7.96/16 = 4.48$. In other words, s^2 is greater than 4 but not by enough to cause the test statistic to be larger than the upper significance point. The null hypothesis is accepted and we conclude that the preparatory course does not change variation in scores on the job application test.

Confidence Intervals for the Variance*

We continue with the same example to illustrate the construction of a confidence interval for the population variance. Whatever the value of σ^2, the random variable $\chi^2 = (n-1)s^2/\sigma^2$ is distributed according to the chi-square distribution with 9 degrees of freedom. Hence,

$$.95 = \Pr\left[\chi_9^2(.975) \leq \chi^2 \leq \chi_9^2(.025)\right]$$

$$= \Pr(2.700 \leq \chi^2 \leq 19.022)$$

$$= \Pr(2.700 \leq 9s^2/\sigma^2 \leq 19.022)$$

$$= \Pr(1/2.700 \geq \sigma^2/(9s^2) \geq 1/19.022)$$

$$= \Pr(9s^2/2.700 \geq \sigma^2 \geq 9s^2/19.022)$$

$$= \Pr(3.33s^2 \geq \sigma^2 \geq .473s^2).$$

(The fourth pair of inequalities follows from the fact that if $a > 0$ and $b > 0$, then $a \leq b$ implies $1/a \geq 1/b$.) Hence, a 95% confidence interval for the population variance σ^2 is $(.473s^2, 3.33s^2)$, where s^2 is the observed sample variance. If the observed variance is 7.96, for example, the interval is $(.473 \times 7.96, 3.33 \times 7.96)$, which is $(3.77, 26.5)$. Since the population standard deviation is simply the square root of the population variance, the corresponding interval for the population standard deviation σ is $(\sqrt{3.77}, \sqrt{26.5})$, which is $(1.94, 5.15)$. If we observe that $s^2 = 7.96$, we assert that $1.94 \leq \sigma \leq 5.15$ with confidence .95.

In general, an interval for the population variance σ^2 with confidence level $1 - \alpha$ is

$$\left[(n-1)s^2/\chi^2_{n-1}(\alpha/2), \quad (n-1)s^2/\chi^2_{n-1}(1-\alpha/2)\right].$$

The square roots of the limits are confidence limits for the population standard deviation σ.

The confidence interval consists of those values of σ^2 (or σ) that are "acceptable" as hypothesized values in hypothesis tests. In the example with $s^2 = 7.96$, at the 5% level one would accept the hypothesis that the variance is 16, but would reject the hypothesis that it is 35.

The tests and confidence intervals for σ^2, based on chi-square distributions, depend on the sampled population being normal or approximately normal. This feature is to be contrasted with the inference for μ, based on the sample mean (Chapters 10 and 11). The Central Limit Theorem implies that in random samples $\bar{x}$ is approximately normally distributed with variance σ^2/n if n is reasonably large, even if the sampled population is different from normal. If the observations are not drawn from a normal population, however, $(n-1)s^2/\sigma^2$ may have a sampling distribution quite different from chi-square with $n-1$ degrees of freedom. Hence, the test and confidence interval procedures just described may be quite inaccurate if the sampled population is different from normal; that is, significance levels and confidence levels may not be the asserted levels. The technique should be used only when there is reason to believe that the normal distribution provides a good approximation to the distribution sampled.

A Sampling Experiment.   The 80% confidence intervals for each of ten samples are given in Table 13.1. These intervals were formed by the method just presented, which consists in this case of dividing the sample sum of squared deviations by the 10th and 90th percentiles of the chi-square distributions with 9 degrees of freedom to obtain the upper and lower limits. These percentiles are 4.168 and 14.684. This method of construction of the interval guarantees that 80% of all samples will give rise to a confidence interval that includes the true value. It happened that in this sampling of only ten samples exactly eight included the true value.

13.2   Variability in Two Populations

The Sampling Distribution of the Ratio of Two Sample Variances

When two populations are being studied, it is often of interest to compare their variabilities. (In Chapter 12 the difference between location parameters, particularly the means, was studied.) Since we usually measure variability by the standard deviation or variance, comparison of variability will be made on the basis of the standard deviations σ_1 and σ_2 of the two populations or on the basis of the variances σ_1^2 and σ_2^2. The ratio σ_1/σ_2 or the ratio σ_1^2/σ_2^2 is a desirable measure of the relationship of variabilities because such a ratio does not depend on the units of measurement. For example, if $\sigma_1 = 2$ feet and $\sigma_2 = 1$ foot, the ratio is $\sigma_1/\sigma_2 = 2$; if the unit of measurement were inches, the ratio would be $24/12 = 2$, the same as in feet. (Note that if $\sigma_1 - \sigma_2$ were used, the comparison would depend on the units.) We would say that the first population is twice as spread as the second. If the variabilities of the two populations are the same, $\sigma_1/\sigma_2 = 1$ and $\sigma_1^2/\sigma_2^2 = 1$.

When we have a sample from each of two populations, we use the sample variances s_1^2 and s_2^2 to estimate the population variances σ_1^2 and σ_2^2. If we want to use the sample variances to make inferences about the comparison of population variances, it is appropriate to use the ratio of variances s_1^2/s_2^2. It turns out that the ratio of variances is more useful than the ratio of standard deviations because the distribution of a variance ratio is simpler and it has direct application in other statistical analysis. (See Chapter 16.)

When making inferences about a single mean or the difference of two means, it was necessary to consider the sampling distribution of the mean or of the difference of two means, respectively. Likewise, to test a hypothesis about the ratio of two variances, it is necessary to describe the distribution of ratios that could be obtained if the null hypothesis is true ($\sigma_1^2 = \sigma_2^2$). We begin by assuming that n_1 observations are drawn at random from the first normally distributed population and s_1^2 is computed, and n_2 observations are drawn from a second, independent, normally distributed population and s_2^2 is computed. If the two population variances are equal, then the sampling distribution of s_1^2/s_2^2 is known as the F-distribution with $m = n_1 - 1$ and $n = n_2 - 1$ degrees of freedom. There are two separate parameters for the F-distribution: m, the number of degrees of freedom of the variance in the numerator of the F-ratio, and n, the number of degrees of freedom of the variance in the denominator.

Since both s_1^2 and s_2^2 are positive, $F = s_1^2/s_2^2$ is positive. The sampling distribution of F for 10 and 20 degrees of freedom is tabulated in Table 13.3 and graphed in Figure 13.3. Since the distribution of either s_j^2 is rather concentrated around σ_j^2, the distribution of F is rather concentrated around 1 when $\sigma_1^2 = \sigma_2^2$. The value exceeded by a probability p is denoted by $f_{m,n}(p)$. These values are tabulated in Appendix V for various values of m, n, and p. The significance point $F_{10,20}(.05)$, illustrated in Figure 13.3, is found in Appendix V in the row labeled $m = 10$, $n = 20$, and in the column labeled $p = .05$; the F-value is

TABLE 13.3
Probability curve of $F = s_1^2/s_2^2$ when $m = n_1 - 1 = 10$, $n = n_2 - 1 = 20$, and $\sigma_1^2/\sigma_2^2 = 1$

Value of F	Height of probability curve
0.0	0.000
0.1	0.015
0.2	0.120
0.3	0.311
0.4	0.520
0.5	0.688
0.6	0.792
0.7	0.833
0.8	0.824
0.9	0.779
1.0	0.714
1.25	0.525
1.5	0.358
2.0	0.153
2.5	0.064
3.0	0.027
3.5	0.012

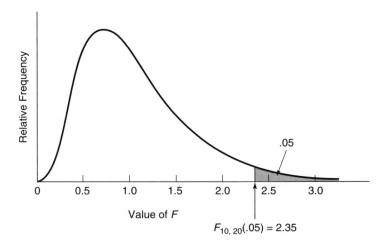

Figure 13.3   *Probability curve of $F = s_1^2/s_2^2$ when $m = n_1 - 1 = 10$, $n = n_2 - 1 = 20$, and $\sigma_1^2/\sigma_2^2 = 1$.*

2.3479. This is the value that would be used to make a one-tailed test at the 5% level if $n_1 = 11$ and $n_2 = 21$.

Note that even though the F-distributions are not symmetric, significance points are given only for small values of p, that is, points in the upper tail. This is because significance points on the left side of the distribution can be found easily from upper-tail values. For any p,

$$p = \Pr\left\{\frac{s_1^2}{s_2^2} \geq F_{m,n}(p)\right\}$$

$$= \Pr\left\{\frac{s_2^2}{s_1^2} \leq \frac{1}{F_{m,n}(p)}\right\}$$

$$= 1 - \Pr\left\{\frac{s_2^2}{s_1^2} \geq \frac{1}{F_{m,n}(p)}\right\}.$$

When we compare this equation with the definition of $F_{n,m}(1-p)$

$$\Pr\left\{\frac{s_2^2}{s_1^2} \geq F_{n,m}(1-p)\right\} = 1 - p,$$

we see that

$$F_{m,n}(p) = \frac{1}{F_{n,m}(1-p)}. \tag{13.1}$$

That is, to find a significance point on the left side of a given F-distribution, we interchange m and n, look up the significance point in the right-hand tail of the F-distribution with n degrees of freedom in the numerator and m in the denominator, and take the reciprocal of that value. For example, suppose that we require the value from the F-distribution with 10 and 20 degrees of freedom that has 5% of the distribution to the left of it (and 95% to the right). This is obtained by finding the value from the F-distribution with 20 and 10 degrees of freedom that has 5% of the distribution to the right of it (2.7740) and taking its reciprocal. That is,

$$F_{10,20}(.95) = \frac{1}{F_{20,10}(.05)} = \frac{1}{2.7740} = 0.36.$$

As might be expected, the F-distributions are closely related to the chi-square distributions. In particular, $(n_1 - 1)s_1^2/\sigma_1^2$ has a chi-square distribution with $n_1 - 1$ degrees of freedom and $(n_2 - 1)s_2^2/\sigma_2^2$ has a chi-square distribution with $n_2 - 1$ degrees of freedom. The ratio of these two variables is

$$\frac{(n_1 - 1)s_1^2/\sigma_1^2}{(n_2 - 1)s_2^2/\sigma_2^2}.$$

In general, the F-distribution is the distribution of the ratio of two chi-square variables each divided by its number of degrees of freedom. Thus the F-distribution with $n_1 - 1$ and $n_2 - 1$ degrees of freedom is the sampling distribution of $(s_1^2/\sigma_1^2)/(s_2^2/\sigma_2^2)$. If the null hypothesis is true and $\sigma_1^2 = \sigma_2^2$, this statistic is exactly the F-ratio s_1^2/s_2^2.

The F-distribution was named for Sir Ronald A. Fisher, who in 1924 introduced the concepts and methods underlying the comparisons of variances. (Actually Fisher used not F itself but another variable directly related to F. Tables for the sampling distribution of F were first constructed by George W. Snedecor.) Fisher was a great contributor not only to statistical methodology but to the development of scientific methods of inquiry in general. Two books that are among Fisher's most important and influential works are listed in the references (Fisher, 1960, 1970).

Testing the Hypothesis of Equality of Two Variances

The F-distribution is used to test whether the variances of two separate populations are the same. As an example, suppose that a class of 32 students was randomly divided into a group of $n_1 = 11$ that received instruction in intermediate algebra through an "individually programmed instruction" (IPI) approach and $n_2 = 21$ that were taught in the usual lecture-discussion format. The IPI approach allows students to proceed at their own paces with access to a wide variety of resources including teaching machines, mathematics tutors, workbooks at all levels of difficulty, and more. Evaluations of IPI in different subject areas have shown little if any *average* improvement in learning compared to traditional teaching approaches. At the same time, IPI is hypothesized to increase the *dispersion* in learning outcomes because the most highly motivated students can study more material in great depth; less motivated students may not take much initiative and will learn less than they would if a teacher was monitoring their work every day. To test this proposition for algebra, all 32 students were given a test at the end of the semester. The data were examined to determine if the standard deviation in the population of students learning by IPI, σ_1, is greater than the standard deviation in the population who learn in a traditional format, σ_2.

The sample standard deviations in this experiment were $s_1 = 3.05$ and $s_2 = 1.73$. The problem is to test H_0: $\sigma_1 \leq \sigma_2$ against the alternative H_1: $\sigma_1 > \sigma_2$. Note that H_0 and H_1 can be stated in terms of the ratio of population variances as H_0: $\sigma_1^2/\sigma_2^2 \leq 1$ and H_1: $\sigma_1^2/\sigma_2^2 > 1$. The ratio of sample variances is an estimate of the ratio of population variances, and one rejects H_0 if the ratio of sample variances is sufficiently greater than 1. Here the test statistic is $F = s_1^2/s_2^2 = (3.05)^2/(1.73)^2 = 9.30/2.99 =$

3.11. This is the estimate of σ_1^2/σ_2^2. Is it sufficiently greater than 1 to warrant rejection of H_0? Since $3.11 > F_{10,20}(.05) = 2.3479$, H_0 is rejected at the .05 level. We conclude that IPI results in a greater variability in intermediate algebra scores.

Two-Tailed Test   The scores of $n_1 = 21$ males and $n_2 = 31$ females in a driver education course gave respective sample variances of $s_1^2 = 478$ and $s_2^2 = 372$. The ratio of variances is $s_1^2/s_2^2 = 1.28$. Is this within the range of variability expected when the population variances are equal? That is, is the hypothesis of equality of population variances consistent with the data?

The null hypothesis is H_0: $\sigma_1^2/\sigma_2^2 = 1$ and the alternative is H_1: $\sigma_1^2/\sigma_2^2 \neq 1$. If the observed ratio $F = s_1^2/s_2^2$ is close to 1, H_0 is accepted. If the observed ratio falls in the extreme left-hand portion or in the upper tail of the F-distribution, H_0 will be rejected, that is, if $F < F_{m,n}(1 - \alpha/2)$ or if $F > F_{m,n}(\alpha/2)$. In the example $m = n_1 - 1 = 20$ and $n = n_2 - 1 = 30$.

Suppose that the test is made at the 5% level of significance, that is, $\alpha = .05$. The region of rejection is indicated by the two shaded portions in Figure 13.4. The .025 significance point in the upper tail of the F-distribution is found directly from Appendix V; this is $F_{20,30}(.025) = 2.1952$. The significance point on the left $F_{20,30}(.975)$ is obtained by applying (13.1). First use the table to find the .025 upper-tail point from the F-distribution with 30 and 20 degrees of freedom; this is

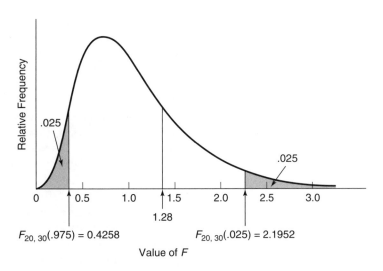

Figure 13.4 *Probability curve of* $F = s_1^2/s_2^2$ *when* $m = n_1 - 1 = 20$ *and* $n = n_2 - 1 = 30$ *and* $\sigma_1^2/\sigma_2^2 = 1.$

$F_{30,20}(.025) = 2.3486$. Next take its reciprocal: $F_{20,30}(.975) = 1/2.3486 = 0.4258$. The null hypothesis will be rejected if the test statistic is less than 0.4258 or greater than 2.1952, and accepted otherwise.

In this example the observed $F = 1.28$ is between these values and H_0 is accepted. While males and females were not equally heterogeneous in the sample, the difference was not large enough to convince us that there is a difference in dispersions between the two populations.

EXAMPLE   A measure of lead content in hair removed from persons in two samples, "antique" and "contemporary," is summarized in Table 12.13. In Exercise 12.25 the mean lead content is compared between antique and contemporary populations and between children and adults. In addition, we might ask whether the four populations differ in the *variability* of their lead content. For this illustration we shall use the contemporary data only and ask whether lead readings are more variable in the population of children, in the population of adults, or whether the variabilities of lead readings for these two age groups are indistinguishable.

Since we have no prior reason to believe that one group is more or less variable than the other, we shall perform a two-tailed test. The null and alternative hypotheses are $H_0: \sigma_1 = \sigma_2$ and $H_1: \sigma_1 \neq \sigma_2$, respectively. The sample standard deviations are $s_1 = 10.6$ and $s_2 = 6.19$. The test statistic is $F = 10.6^2/6.19^2 = 2.93$; this F-ratio has $m = n_1 - 1 = 118$ degrees of freedom in the numerator and $n = n_2 - 1 = 27$ degrees of freedom in the denominator.

Although the F-distribution with 118 and 27 degrees of freedom is not listed in Appendix V, we shall use the more conservative significance points from the F-distribution with 100 and 25 degrees of freedom. At the 5% level of significance, the upper-tail significance point is $F_{100,25}(.025) = 1.9955$. The lower-tail significance point is obtained using (13.1). First, the upper-tail significance point is obtained from the F-distribution with the degrees of freedom reversed, that is, with 25 and 100 degrees of freedom. From Appendix V this is $F_{25,100}(.025) = 1.7705$. Second, the reciprocal is taken to obtain the required lower-side value: $F_{100,25}(.975) = 1/1.7705 = 0.5648$. The null hypothesis will be rejected if the test statistic is less than 0.5648 or greater than 1.9955.

The computed test statistic $F = 2.93$ is greater than 1.9955 and H_0 is rejected. The variance of lead measurements for children in the sample was almost 3 times as large as the variance of measurements for adults and thus we conclude that $\sigma_1 > \sigma_2$ in the populations as well.

Had we taken a P value approach to this test, we would note that 2.93 also exceeds 2.5191, the .005 upper-tail point of F with 100 and 25 degrees of freedom. Since we are conducting a two-tailed test, the P interval for the test would be $P < .01$.

Note on One-Tailed Test.   If we had prior reason to suspect that one population is more variable than the other, the variances could be compared through a one-tailed test. If our reasoning or prior observation suggested that children would be more variable than adults, the null and alternative hypotheses would be $H_0: \sigma_1 \leq \sigma_2$ and $H_1: \sigma_1 > \sigma_2$, respectively. The 5% significance point is $F_{100,25}(.05) = 1.7794$.

Medical researchers usually find that humans become *more* variable as they age, however. On this basis the most reasonable one-sided hypothesis would be $H_0: \sigma_1 \geq \sigma_2$ and $H_1: \sigma_1 < \sigma_2$. The significance point would be on the left side of the F-distribution, that is, less than 1. In the sample $s_1 > s_2$, however, and the test statistic would be greater than 1. Thus we know without conducting the test that H_0 would be accepted.

Comparison of variances can be of interest in itself. Also, variances are sometimes compared before making a comparison of means where assumptions are made about equality of variances. However, the F test for equality of variances is sensitive to departures from normality; one should not make an F-test of equality of variances unless the sample distributions seem fairly close to normal.

Confidence Intervals for the Ratio of Two Variances*

Regardless of whether H_0 is true, the random variable $F^* = (s_1^2/\sigma_1^2) \div (s_2^2/\sigma_2^2)$ has an F-distribution when the samples are from normal distributions [because $(n_1 - 1)s_1^2/\sigma_1^2$ and $(n_2 - 1)s_2^2/\sigma_2^2$ have chi-square distributions with $n_1 - 1$ and $n_2 - 1$ degrees of freedom, respectively]. This fact can be used to construct confidence intervals for the ratio of population variances σ_1^2/σ_2^2. For example,

$$.90 = \Pr\left[F_{m,n}(.95) \leq F^* \leq F_{m,n}(.05)\right]$$

$$= \Pr\left[F_{m,n}(.95) \leq \frac{s_1^2/\sigma_1^2}{s_2^2/\sigma_2^2} \leq F_{m,n}(.05)\right]$$

$$= \Pr\left[\frac{1}{F_{m,n}(.95)} \geq \frac{s_2^2}{s_1^2} \times \frac{\sigma_1^2}{\sigma_2^2} \geq \frac{1}{F_{m,n}(.05)}\right]$$

$$= \Pr\left[\frac{1}{F_{m,n}(.95)} \times \frac{s_1^2}{s_2^2} \geq \frac{\sigma_1^2}{\sigma_2^2} \geq \frac{1}{F_{m,n}(.05)} \times \frac{s_1^2}{s_2^2}\right].$$

Rewriting the last expression with the signs reversed, we have

$$\Pr\left[\frac{1}{F_{m,n}(.05)} \times \frac{s_1^2}{s_2^2} \leq \frac{\sigma_1^2}{\sigma_2^2} \leq \frac{1}{F_{m,n}(.95)} \times \frac{s_1^2}{s_2^2}\right].$$

This expression is the basis for using s_1^2 and s_2^2 to construct a 90% confidence interval for σ_1^2/σ_2^2. The interval is

$$\left[\frac{1}{F_{m,n}(.05)} \times \frac{s_1^2}{s_2^2}, \quad \frac{1}{F_{m,n}(.95)} \times \frac{s_1^2}{s_2^2} \right].$$

When the observed s_1^2/s_2^2 is 1.28, and $n_1 = 21$ and $n_2 = 31$ as in the previous example, the interval is (1.28/1.9317, 1.28/0.4904), which is (0.663, 2.61). The confidence interval consists of those values of the ratio of population variances that are "acceptable" in a two-tailed test. For instance, the hypothesis $\sigma_1^2/\sigma_2^2 = 1$ would be accepted, but the hypothesis $\sigma_1^2/\sigma_2^2 = 3.1$ would be rejected. For intervals with other confidence levels, other significance points are substituted for $F_{m,n}(.05)$ and $F_{m,n}(.95)$ in the expression above.

Summary

The sum of squared deviations from the mean of a random sample of size n from a normal distribution with variance σ^2, divided by σ^2, has a chi-square distribution with $n - 1$ degrees of freedom. Hypothesis tests and confidence intervals for the population variance are based on this fact.

If s_1^2 is the variance of a random sample of n_1 observations from a normal distribution with variance σ_1^2 and s_2^2 is the variance of an independent random sample of n_2 observations from a normal distribution with variance σ_2^2, then the variable $(s_1^2/\sigma_1^2) \div (s_2^2/\sigma_2^2)$ is distributed according to the F-distribution with degrees-of-freedom parameters $m = n_1 - 1$ and $n = n_2 - 1$. Inferences about the ratio of population variances are based on this fact.

Exercises

13.1   From the table in Appendix IV, find the following percentage points.

(a) $\chi_2^2(.05)$,

(b) $\chi_2^2(.95)$,

(c) $\chi_4^2(.05)$,

(d) $\chi_4^2(.95)$.

13.2   From the table in Appendix IV, find the following percentage points.

(a) $\chi_2^2(.10)$,

(b) $\chi_2^2(.90)$,

(c) $\chi_{20}^2(.10)$,

(d) $\chi_{20}^2(.90)$.

13.3   A geneticist has some data from 1870 giving heights of human males who were then 30 years old. The standard deviation of these heights is 3 inches, that is, the variance is 9 squared inches. One of the implications of the geneticist's theory is that the variance σ^2 of the distribution of heights of human males 30 years old in 1970 is 10% smaller than this. From a sample of size $n = 41$, the geneticist finds $s^2 = 8.29$ squared inches. (Assume the distribution of heights is normal.)

(a) Give a 99% confidence interval for the variance of the distribution of heights in 1970.

(b) Do you reject at the 1% level the hypothesis that $\sigma^2 = 8.1$ squared inches? Use a two-sided alternative.

13.4   In the context of Exercise 13.3, do the following:

(a) Give an 80% confidence interval for the variance of the distribution of heights in 1970.

(b) Do you reject at the 20% level the hypothesis that $\sigma^2 = 8.1$ squared inches? Use a two-sided alternative.

13.5   Suppose that $s^2 = 7.96$ and $n = 10$, as in the example in Section 13.1. Construct an 80% confidence interval for the population variance. Compare the interval with the 95% interval given in the text.

13.6   Suppose that $s^2 = 7.96$, as in Section 13.1, but $n = 20$ instead of 10. Construct a 95% confidence interval for the population variance. Compare the interval with the interval given in the text, which was constructed for $n = 10$.

13.7   The standard deviation of a random sample of 20 observations from a normal distribution is 2.1. Construct an 80% confidence interval for the population standard deviation.

13.8   The standard deviation of a random sample of 20 observations from a normal distribution is 2.1. Construct a 95% confidence interval for the population standard deviation.

13.9   The Scholastic Assessment Tests (SAT) are scored so that the mean and standard deviation of respondents in the population are $\mu = 500$ and $\sigma = 100$, respectively. A prestigious university on the west coast claims that its student body has aptitudes that are homogeneously higher than the nationwide norm, that is, the mean of its students is greater than

500 and the standard deviation is less than 100. A random sample of 41 students was selected and their SAT scores were recorded. The mean and standard deviation were $\bar{x} = 534.2$ and $s = 77.31$.

(a) Test the null hypothesis that the standard deviation of SAT scores in the institution is greater than or equal to the nationwide values against the alternative that scores in this institution are more homogeneous. Use the 5% level of significance. What is the significance point for the test? What is the value of the test statistic?

(b) Test the null hypothesis that the mean SAT score in the institution is less than or equal to the nationwide mean against the alternative that the mean is higher among students in this school. Use the 5% level of significance. What is the significance point? What is the value of the test statistic? (Should this hypothesis be tested using the standard normal distribution or the t-distribution?)

(c) Summarize your findings regarding SAT scores of the population of students who attend this school.

13.10   From the table in Appendix V, find the following percentage points:

(a) $F_{10,20}(.05)$,

(b) $F_{10,20}(.10)$,

(c) $F_{2,10}(.01)$,

(d) $F_{2,30}(.01)$.

13.11   From the table in Appendix V, find the following percentage points:

(a) $F_{10,10}(.025)$,

(b) $F_{4,10}(.05)$,

(c) $F_{4,10}(.10)$,

(d) $F_{3,60}(.10)$.

13.12   From the table in Appendix V, compute the following percentage points:

(a) $F_{10,20}(.95)$,

(b) $F_{6,10}(.90)$,

(c) $F_{10,10}(.975)$.

13.13   In Sample 1, $n_1 = 11$ and $s_1^2 = 18$. In Sample 2, $n_2 = 21$ and $s_2^2 = 36$. Assuming normality, test the hypothesis of equality of population variances against a two-sided alternative at the 5% level.

13.14   In Sample 1, $n_1 = 21$ and $s_1^2 = 30$. In Sample 2, $n_2 = 31$ and $s_2^2 = 10$. Assuming normality, test the null hypothesis that $\sigma_1^2 \leq \sigma_2^2$ against the alternative that $\sigma_1^2 > \sigma_2^2$ at the 10% level of significance.

13.15   In Section 12.2 we considered the scores of $n_1 = 39$ undergraduates and $n_2 = 29$ graduate students in a statistics course. The sample variances were 409.2 and 478.1, respectively. We stated that the ratio of variances, 1.17, was well within the range of variability expected when the population variances are equal. Explain this by conducting the appropriate two-tailed test at $\alpha = .10$. (Since the F-distribution with 38 and 28 degrees of freedom is not tabled in Appendix V, what significance points should be used instead?)

13.16   In Exercise 12.25 we recommended that separate variance estimates be used in making t tests. To support this recommendation, test the hypothesis of equality of variances in the adult antique population and the adult contemporary population against the alternative that the variances are different. Use the 1% level of significance. Give your answer in the form of a P value interval and state the range of α values for which H_0 would be rejected.

13.17   Use the tables of the F-distribution (Appendix V) to obtain a P value interval for each of the following test statistics:

(a) $F = 4.51$ with 12 and 18 degrees of freedom (1-tailed test, upper tail);

(b) $F = 2.92$ with 20 and 15 degrees of freedom (1-tailed test, upper tail);

(c) $F = 1.58$ with 30 and 30 degrees of freedom (1-tailed test, upper tail);

(d) $F = 2.92$ with 20 and 15 degrees of freedom (2-tailed test);

(e) $F = 6.39$ with 8 and 8 degrees of freedom (2-tailed test);

(f) $F = 0.27$ with 9 and 20 degrees of freedom (2-tailed test).

13.18   Table 4.1 lists waiting times for 10 patients who were seen by Doctor A and 10 who were seen by Doctor F. Use these data to test the hypothesis of equal population variances for the two doctors against the two-sided alternative that the variances are different. Use the 10% level of significance. (Note that the standard deviations are computed for Exercise 4.3.)

(a) State the null and alternative hypotheses.

(b) State the significance points for the test.

(c) Give the test statistic.

(d) State your conclusion regarding H_0.

(e) Make a recommendation to potential patients based on your findings.

13.19   The study by Ciancio and Mather (1990) comparing plaque formation after using an electric toothbrush (Exercise 10.34) reports plaque buildup after one week of brushing manually and one week with the electric brush. Lower plaque scores are preferable. For 15 patients using the Interplak toothbrush, the mean was 47% and the standard deviation

was 8.92%. For 15 patients using the Water Pik toothbrush the mean
was 56% and the standard deviation was 15.40%.

Use these results to test the hypothesis that the two electric toothbrushes are equally *variable* with respect to plaque buildup. Use the
1% significance level. In your answer give

(a) The null and alternative hypotheses in statistical form,

(b) The value of the test statistic,

(c) The name of the specific probability distribution used to test the
hypothesis, and the significance points from that distribution,

(d) The decision to accept or reject the null hypothesis,

(e) A verbal statement about your conclusion, that is, does Interplak
result in more variability, Water Pik, or do both electric toothbrushes
result in equal variability?

13.20   (continuation) Considering the plaque scores of the preceding exercise, what combination of mean and standard deviation would be the
most desirable; that is, would you recommend that your friends use a
toothbrush with a high mean and large standard deviation, high mean
and small standard deviation, low mean and large standard deviation,
or low mean and small standard deviation? Give the rationale for your
answer.

13.21   The *sample coefficient of variation* is $s/\bar{x}$; the *population coefficient of
variation* is σ/μ. How do you estimate this parameter? Estimate it for
the waiting times for Doctor A and for Doctor F given in Table 4.1. (The
standard deviations are computed in Exercise 4.3.)

13.22   Compute the coefficients of variation for each variable in each population in Table 13.4. Compare the results. What does the coefficient of
variation reveal about differences among the populations?

TABLE 13.4
Values of (μ, σ)

		Population	
Variable	*A*	*B*	*C*
x	$(2, 4)$	$(3, 9)$	$(4, 16)$
y	$(3, 6)$	$(4, 12)$	$(5, 20)$
z	$(4, 8)$	$(5, 15)$	$(6, 24)$
w	$(5, 10)$	$(6, 18)$	$(7, 28)$

13.23   Calculate $s/\bar{x}$ for each sample summarized in Table 12.13 and compare
the four values. What reason(s) can you suggest for the differences you
obtained?

V

Statistical Methods for Other Problems

14

Inference on Categorical Data

Introduction

In this chapter we present some methods for treatment of categorical data. The methods involve the comparison of a set of observed frequencies with frequencies specified by some hypothesis to be tested. In Section 14.1 the hypothesis is that one categorical variable has a specific distribution. A test of such a hypothesis is called a test of *goodness of fit*.

The notion of independence of two categorical variables was developed in Chapter 6. In Section 14.2 we show how to test the hypothesis that two categorical variables are independent. The test statistics discussed have sampling distributions that are approximated by chi-square distributions. For this reason the tests are often called *chi-square tests*.

A measure of association for 2×2 tables was given in Chapter 6. In Section 14.3 we shall give several measures of association for two-way tables that are valid for larger tables.

14.1   Tests of Goodness of Fit

When statistical data consist of one categorical variable it is often informative to ask whether the proportions of responses in the categories conform to a particular pattern. The test of significance is called a *goodness of fit* test. It is based on comparison of an observed frequency distribution with a hypothesized distribution. In the simplest situation there are only two categories such as whether a particular characteristic is present or absent in the members of the sample. We could test whether the probability of the characteristic being present is 1/2; here the hypothesized distribution has values "present" with probability 1/2, and "absent," also with probability 1/2.

If, on the other hand, our business had three branch stores, we might test whether the proportions of customers who shop at the branches are equal. If not, we might shift salespeople or other resources to the branch that has more customers than the others. Here the hypothesized distribution has values "Branch 1" with probability 1/3, "Branch 2" with probability 1/3, and "Branch 3" with probability 1/3. The goodness-of-fit test enables us to compare the observed distribution based on a random sample of customers with the hypothesized distribution.[1]

In the following sections we describe the goodness-of-fit test for a two-category variable and then for a variable with more than two categories. In the section below we show that the goodness-of-fit test for two categories gives exactly the same conclusion as the test of one proportion discussed in Chapter 11. The goodness-of-fit approach generalizes readily to more than two categories as well as to tests for the independence of two categorical variables.

Two Categories—Dichotomous Data

We begin our discussion with the case of a variable defined by just two categories, which one may think of as Yes and No. Let p be the probability of Yes. Then $q = 1 - p$ is the probability of No, and the distribution of the variable is specified when the value of p is specified. We shall first review the procedure described in Chapter 11 for testing a hypothesis about p. Next we illustrate how the test statistic and significance point are transformed to the more generalizable goodness-

---

[1]The hypothetical distributions in the preceding examples are based on dividing the total population equally among the categories. This need not be the case, however, and other hypothetical distributions can be derived from logic or prior research.

of-fit forms. Third, we show how goodness-of-fit statistics are obtained directly from a sample of data.

In Section 11.5 we considered a test of whether African-American children were just as likely to choose a white as a nonwhite doll, that is, a test of the hypothesis H_0 that the proportion who would choose a nonwhite doll is $p_0 = 1/2$. In a sample of 252 children, 83 chose a nonwhite doll. We developed a test based on consideration of the observed sample proportion $\hat{p} = 83/252 = 0.329$. If the null hypothesis is true the standard deviation of a proportion in a random sample of 252 is $\sqrt{p_0 q_0}/\sqrt{n} = \sqrt{0.5 \times 0.5}/\sqrt{252} = 0.0315$, where $q_0 = 1 - p_0$.

The test statistic is

$$z = \frac{\hat{p} - p_0}{\sqrt{p_0 q_0/n}}$$

$$= \frac{0.329 - 0.5}{0.0315} = -5.43.$$

The value of z is referred to the standard normal distribution. If a two-tailed test is being made at the 5% level, the significance points are -1.96 and 1.96 and H_0 is rejected.

An alternative method of carrying out this procedure is to use the fact that the square of a standard normal variable has a chi-square distribution with 1 degree of freedom. In the above example the observed value, $(-5.43)^2 = 29.5$, is to be compared with $1.96^2 = 3.84$ for a test at the 5% level. (In Appendix IV, $\chi_1^2(.05) = 3.843$.) The conclusion—reject H_0—is obviously the same. The χ^2 test is equivalent to a two-tailed test using z.

It is instructive to examine this alternative method algebraically. To do so, let x represent the number of Yes responses out of a sample of n individuals, so that the sample proportion is $\hat{p} = x/n$ and also $x = n\hat{p}$. The sample proportion of No's is $\hat{q} = 1 - \hat{p}$.

Let p_0 represent the hypothesized value of the probability of Yes and q_0 represent the hypothesized probability of No, with $q_0 = 1 - p_0$. Note that np_0 can be interpreted as the hypothesized or *expected number* of Yes responses in the sample of n individuals, and nq_0 the hypothesized or *expected number* of No responses. The null hypothesis can be stated as $H_0: p = p_0$ or equivalently as $H_0: 1 - p = 1 - p_0$ or equivalently as $H_0: q = q_0$.

The test statistic in the alternative form is

$$z^2 = \left(\frac{\hat{p} - p_0}{\sqrt{p_0 q_0}/\sqrt{n}}\right)^2 = \frac{(\hat{p} - p_0)^2}{p_0 q_0/n} = n\frac{(\hat{p} - p_0)^2}{p_0 q_0}$$

$$= \frac{[n(\hat{p} - p_0)]^2}{np_0 q_0} = \frac{(x - np_0)^2}{np_0 q_0}. \qquad (14.1)$$

In terms of the No response, the test statistic is

$$\left(\frac{\hat{q} - q_0}{\sqrt{q_0 p_0}/\sqrt{n}}\right)^2 = \frac{(n - x - nq_0)^2}{np_0 q_0}. \tag{14.2}$$

Since $n - x - nq_0 = n - x - n(1 - p_0) = -(x - np_0)$, criterion (14.2) is identical to criterion (14.1).

In general, if two values are equal, then their average or any weighted average of them is also equal to those values. (Try it.) The sum of the two terms—expression (14.1) multiplied by $q_0 = q_0/(p_0 + q_0)$ plus (14.2) multiplied by $p_0 = p_0/(p_0 + q_0)$—is a weighted average of (14.1) and (14.2) because $p_0 + q_0 = 1$. Thus the test statistic is equivalent to

$$\frac{(x - np_0)^2}{np_0} + \frac{(n - x - nq_0)^2}{nq_0}. \tag{14.3}$$

The goodness-of-fit test is to compare the observed value of (14.3) with a value from the table of chi-square with 1 degree of freedom.

In the example, $p_0 = q_0 = 1/2$, $n = 252$, $np_0 = nq_0 = 126$, $x = 83$, and $n - x = 169$. The test statistic is

$$\frac{(83 - 126)^2}{126} + \frac{(169 - 126)^2}{126} = \frac{(-43)^2}{126} + \frac{(43)^2}{126}$$

$$= 14.7 + 14.7$$

$$= 29.4,$$

which is within rounding error of $29.5 = (-5.43)^2$.

The purpose of developing (14.3) is to treat Yes and No (or nonwhite and white) symmetrically because this expression can be generalized to more than two categories. In (14.3) np_0 is the *expected* number of Yes responses under the null hypothesis, and nq_0 is the expected number of No responses. Each fraction in (14.3) is of the form

$$\frac{(\text{Observed Number} - \text{Expected Number})^2}{\text{Expected Number}}. \tag{14.4}$$

The test statistic is the sum of these terms.

Any Number of Categories

Suppose that the outcomes on a variable can fall into one of k categories. Let $p_1, p_2, \ldots, p_k$ denote the respective probabilities of these outcomes. Consider the null hypothesis

$$H_0: p_1 = p_{10}, \quad p_2 = p_{20}, \ldots, \quad p_k = p_{k0},$$

where $p_{10}, p_{20}, \ldots, p_{k0}$ are specified positive numbers adding to 1. Suppose that n individuals are observed. The *expected number* of outcomes in the ith category is $e_i = np_{i0}$; the actually observed number is denoted by n_i. The test statistic is

$$X^2 = \sum \frac{(n_i - e_i)^2}{e_i}. \tag{14.5}$$

This statistic is similar to (14.3) but extended to k categories ($k \geq 2$). It is the sum of k terms, and has a sampling distribution that is approximately chi-square with $k - 1$ degrees of freedom. We reject H_0 at level α if the observed value of the test statistic exceeds $\chi^2_{k-1}(\alpha)$.

EXAMPLE   One might think that how long a person survives beyond a birthday is a random event such that each day or week or month of death (within a year) beyond the birthday is equally likely. In particular, in some specified population, it is expected that the probability of the death of an individual occurring in the first month after a birthday should be 1/12, etc. However, a person in a situation where death is not unlikely who is looking forward to celebrating the next birthday might stave off death, that is, reduce the probability of death shortly before the birthday.

Phillips (1972) investigated the tendency of famous people to survive until their next birthdays. Using a number of anthologies of famous people, he noted that such people are least likely to die in the month before their birth month; there is a dip in the death rates corresponding to the month before the birth month. Correspondingly there is an apparent rise in the death rate during the months immediately following the birth month. Some data are given in Table 14.1. Are the observed frequencies 24, 31, etc., consistent with a probability of 1/12 for each month?

If the probability is in fact 1/12 for each month, then out of 348 people one would expect 348/12 = 29 to die during each month. The expected value e_i of 29 is compared with the observed value n_i for each month. The differences $n_i - e_i$ are shown in the table. To construct a test statistic, each deviation $n_i - e_i$ is squared and divided by e_i. The resulting values $(n_i - e_i)^2/e_i$ are summed to obtain

$$\frac{(-5)^2}{29.0} + \frac{(+2)^2}{29.0} + \frac{(-9)^2}{29.0} + \frac{(-6)^2}{29.0} + \frac{(+5)^2}{29.0} + \frac{(-13)^2}{29.0} + \frac{(-3)^2}{29.0}$$
$$+ \frac{(+7)^2}{29.0} + \frac{(+8)^2}{29.0} + \frac{(+12)^2}{29.0} + \frac{(-3)^2}{29.0} + \frac{(+5)^2}{29.0} = 22.07.$$

There are $k = 12$ categories, and hence $k - 1 = 11$ degrees of freedom. In the chi-square distribution with 11 degrees of freedom, we see that 22.07 falls between the tabled values of 21.920 and 24.724. These are

TABLE 14.1
Number of Deaths Observed Before, During, and After the Birth Month

	Number of months before the birthmonth						Birth month	Number of months after the birth month				
	6	5	4	3	2	1		1	2	3	4	5
Observed number of deaths	24	31	20	23	34	16	26	36	37	41	26	34
Expected number of deaths	29	29	29	29	29	29	29	29	29	29	29	29
"Excess" number of deaths	−5	+2	−9	−6	+5	−13	−3	+7	+8	+12	−3	+5
	$n = 348$ people							$n/12 = 29.0$				

source: Phillips (1972), p. 58.

$\chi_{11}^2(.025)$ and $\chi_{11}^2(.01)$, respectively, so that the P value interval for the test statistic is $.01 < P < .025$. We would reject the hypothesis that the hypothetical distribution (1/12 for each month) is correct at the 5% level. These data provide evidence that at least for famous people the death rate has some relation to birth date; a possible explanation is that such a person looks forward to a birthday enough to affect health.

Combining Categories

Lucky Louis has brought the dice for your weekly game. You wish to examine each of the dice, suspecting that either the dice have been loaded in favor of 6, or else that they are fair. You select one die and roll it 60 times, obtaining the results in Table 14.2. In formal terms the null hypothesis is $p_1 = p_2 = \cdots = p_6 = 1/6$. The value of the chi-square statistic is

$$\frac{(5-10)^2}{10} + \frac{(10-10)^2}{10} + \frac{(10-10)^2}{10} + \frac{(9-10)^2}{10}$$

$$+ \frac{(9-10)^2}{10} + \frac{(17-10)^2}{10}$$

$$= \frac{25 + 0 + 0 + 1 + 1 + 49}{10}$$

$$= 7.6.$$

There are $6 - 1 = 5$ degrees of freedom. From the table of the chi-square distribution with 5 degrees of freedom (Appendix IV) we see that the test statistic is lower than the 10% significance point $\chi_5^2(.10) = 9.236$. That is, the P value interval for the test statistic is $P > .10$ and the result is not particularly significant. However, this χ^2 test does not take into account the fact that you have one particular alternative in mind, namely, that the die may be loaded in favor of 6.

TABLE 14.2

Results of Rolling a Die

Face	1	2	3	4	5	6	Total
Frequency	5	10	10	9	9	17	60

Hypothetical data.

Since a die is constructed so that the sum of the numbers of spots on opposite faces is seven, if 6 is favored, it must be the case that 1 will turn up less frequently. Accordingly, we categorize the outcome of the die into the three events: [1], [2, 3, 4, or 5], and [6]. The frequencies are given in Table 14.3. The value of the chi-square statistic for Table 14.3 is

$$\frac{(5-10)^2}{10} + \frac{(38-40)^2}{40} + \frac{(17-10)^2}{10} = \frac{25}{10} + \frac{4}{40} + \frac{49}{10}$$
$$= 2.5 + 0.1 + 4.9 = 7.5.$$

With this new grouping of values there are $3 - 1 = 2$ degrees of freedom. The test statistic falls between the 1% significance point $[\chi_2^2(.01) = 9.210]$ and the 2.5% point $[\chi_2^2(.025) = 7.378]$. Thus $.01 < P < .025$ and the result is significant. The hypothesis that the die is fair is rejected at the 5% level.

TABLE 14.3

Results of Rolling a Die

Face	1	2,3,4,5	6	Total
Frequency	5	38	17	60

source: Table 14.2.

These are various ways of categorizing the outcomes. For each there is a power function for the corresponding test, that is, the probability of rejecting the null hypothesis when it is not true; the power depends on the true probabilities of the outcomes. In the above example the test with 2 degrees of freedom is more powerful than that with 5 degrees of freedom when the discrepancy from the null hypothesis only involves the 1 and 6 faces.

14.2   Chi-Square Tests of Independence

In Chapter 6 we studied frequency tables in the case in which individuals were classified simultaneously on two categorical variables. Independence in 2×2 tables was defined in Section 6.1 and in larger two-way tables in Section 6.2. In this section we consider testing the null hypothesis that in a *population* two variables are independent on the basis of a *sample* drawn from that population. Even though the variables are independent in the population, they may not be (and in fact probably will not be) independent in a sample.

Two-by-Two Tables

In Section 12.4 data are presented from one of Asch's experiments (discussed in Section 1.1). The numbers of subjects making one or more errors in the set of trials with group pressure and in the control set and the numbers making no errors in each set constituted the entries. (See Table 12.5.) These data were used to test the null hypothesis that the probability of an individual making one or more errors is the same for individuals under group pressure as for individuals not under group pressure. Stated another way, the null hypothesis is that the probability of error does not depend on whether group pressure is applied; this is a hypothesis of independence.

As another example, consider some data in Exercise 10.27. At each of thee different dates, a sample of 1000 registered voters was drawn; at each time each respondent was asked which of two potential candidates he or she would favor. The results for October 1991 and January 1992 are summarized in Table 14.4. In the underlying population, let the proportion favoring Bush be p_1 at the time of the first poll and p_2 at the time of the second poll. The null hypothesis that the proportion favoring one candidate does not depend on the date of polling is $H_0: p_1 = p_2$. In this example the estimates of p_1 and p_2 are $\hat{p}_1 = 0.523$ and $\hat{p}_2 = 0.502$, respectively, based on sample sizes $n_1 = n_2 = 1000$.

TABLE 14.4

Favored Candidates in Two Polls

		Date		
		October 1991	*January 1992*	*Total*
Candidate	*Bush*	523	502	1025
	Clinton	477	498	975
	Total	1000	1000	2000

Hypothetical data.

The estimate of the standard deviation of the difference between two sample proportions when the null hypothesis is true is

$$\sqrt{\hat{p}\hat{q}\left(\frac{1}{n_1} + \frac{1}{n_2}\right)} = \sqrt{0.5125 \times 0.4875 \left(\frac{1}{1000} + \frac{1}{1000}\right)}$$
$$= 0.02235.$$

The test statistic (Section 12.4) is

$$z = \frac{\hat{p}_1 - \hat{p}_2}{\sqrt{\hat{p}\hat{q}(1/n_1 + 1/n_2)}} = \frac{0.021}{0.02235} = 0.940, \quad (14.6)$$

which is to be referred to the standard normal table. In this example a two-tailed test is called for, because a change could go either way; at the .05 level, the hypothesis would be rejected if (14.6) were greater in numerical value than 1.96.

An equivalent test procedure is to compare the square of (14.6) with the corresponding significance point of the chi-square distribution with 1 degree of freedom. Thus, $0.940^2 = 0.884$ is to be compared with $1.96^2 = 3.84$.

The computation of the criterion can be put into a form similar to the χ^2-statistic in Section 14.1 [that is, (14.5)]. The "expected" numbers are calculated on the basis of the hypothesis of independence being true. If voter preference and date of polling are independent, then $p_1 = p_2$ and one estimate of the common value of p_1 and p_2 describes preferences at both points in time. This estimate is obtained from the *total* number of individuals who favored Bush, that is, $\hat{p} = 1025/2000 = 0.5125$.

On the basis of this estimate of the proportion favoring Bush, the expected numbers favoring Bush on the two dates are $n_1\hat{p} = 1000 \times 0.5125 = 512.5$ and $n_2\hat{p} = 512.5$. Similarly, the expected numbers favoring Clinton on the two dates are $n_1\hat{q} = 1000 \times 0.4875 = 487.5$ and $n_2\hat{q} = 487.5$. These results may be tabulated in a table of expected values (Table 14.5). The marginal totals in Table 14.5 are, by construction, the same as the marginals in Table 14.4.

TABLE 14.5
Expected Numbers for Table 14.4

		Date		Total
		October 1991	*January 1992*	
Candidate	*Bush*	$e_{11} = 512.5$	$e_{12} = 512.5$	1025
	Clinton	$e_{21} = 487.5$	$e_{22} = 487.5$	975
	Total	$n_1 = 1000$	$n_2 = 1000$	2000

The expected values represent independence; they are the numbers of individuals who would have responded if the proportion favoring Bush were the same (0.5125) at the first and second time points. In contrast, the observed values (Table 14.4) are the actual responses of the 2000 individuals. The test statistic summarizes the differences between these two sets of values into a single number. The bigger the differences (and bigger the test statistic) the more the hypothesis of independence is contradicted by the data. If the differences are smaller, then the test statistic will be smaller and the hypothesis of independence is tenable.

The differences between observed numbers and expected numbers are given in Table 14.6. Note that the marginals are 0 and entries are ± 10.5. The X^2 statistic is

$$X^2 = \frac{10.5^2}{512.5} + \frac{(-10.5)^2}{512.5} + \frac{(-10.5)^2}{487.5} + \frac{10.5^2}{487.5}$$
$$= 0.883.$$

It will be seen that this is the same value as that obtained before (except for rounding error). This form for X^2 can be generalized to larger two-way tables.

General Notation and Procedure for 2 × 2 Tables

In general, we may represent the observed counts (or observed frequencies) as n_{ij}, where the first subscript denotes the row in the 2 × 2

TABLE 14.6
Differences Between Observed and Expected Values for Table 14.4

		Date		Total
		First	*Second*	
Candidate	*Bush*	$523 - 512.5 = 10.5$	$502 - 512.5 = -10.5$	0
	Clinton	$477 - 487.5 = -10.5$	$498 - 487.5 = 10.5$	0
	Total	0	0	0

table, and the second subscript denotes the column. Thus the four frequencies in Table 14.4 are represented as n_{11}, n_{12}, n_{21}, and n_{22}. (See Table 14.7.)[2] The total numbers of observations in the rows are represented as $n_1.$ and $n_2.$, respectively, and the column totals are $n._1$ and $n._2$, respectively. The total sample size is n.

TABLE 14.7

Table of Observed Frequencies

		B		
		B_1	B_2	*Total*
A	A_1	n_{11}	n_{12}	$n_1.$
	A_2	n_{21}	n_{22}	$n_2.$
	Total	$n._1$	$n._2$	n

The expected values for the four cells, e_{11}, e_{12}, e_{21}, and e_{22}, may be arranged in a similar table. (See Table 14.8.) From the preceding section, it can be seen that the expected value in the upper left-hand corner (e_{11}) is the overall proportion of individuals in the top row of the table ($n_1./n = 0.5125$) multiplied by the number of observations in the first column ($n._1$), that is, $e_{11} = n_1. \times n._1/n = 1025 \times 1000/2000 = 512.5$.

TABLE 14.8

Table of Expected Frequencies

		B		
		B_1	B_2	*Total*
A	A_1	e_{11}	e_{12}	$n_1.$
	A_2	e_{21}	e_{22}	$n_2.$
	Total	$n._1$	$n._2$	n

The same pattern gives expected values for all four cells. That is,

$$e_{ij} = \frac{n_i. n._j}{n}. \tag{14.7}$$

The expected value in each cell of Table 14.8 is obtained by multiplying the *row total* for that row by the *column total* for that column and dividing by the *total sample size*. As another example, note that $e_{21} = n_2. \times n._1/n = 975 \times 1000/2000 = 487.5$. The process is repeated for

---

[2]These values are represented by *a*, *b*, *c*, and *d* in Table 6.16 for simplicity. Different letters of the alphabet do not expand as easily to larger tables, however.

each cell in the table. The row and column totals of the expected values are the same as the row and column totals of the observed values, that is, $n_{i.}$ and $n_{.j}$.

The cell-by-cell differences, $n_{ij} - e_{ij}$, are summarized into the test statistic

$$X^2 = \frac{(n_{11} - e_{11})^2}{e_{11}} + \frac{(n_{12} - e_{12})^2}{e_{12}} + \frac{(n_{21} - e_{21})^2}{e_{21}} + \frac{(n_{22} - e_{22})^2}{e_{22}}$$

$$= \sum\sum \frac{(n_{ij} - e_{ij})^2}{e_{ij}}. \qquad (14.8)$$

The double summation indicates that the terms are summed across both the rows and columns of Tables 14.7 and 14.8, that is, terms are added for all 4 cells. For random samples when the null hypothesis is true, X^2 has approximately a chi-square distribution with 1 degree of freedom. The 1 degree of freedom is consistent with the fact that any one difference $n_{ij} - e_{ij}$ determines the other three entries since the differences sum to 0 in each row and column. (See Table 14.6.)

Another Application. In the political poll described in the preceding section one sample of individuals was asked their opinions at one point in time and a *separate sample* was surveyed at another time point. In contrast, Section 6.1 describes data in which a single sample of individuals has been classified on the basis of two characteristics. For example, Table 6.7 summarizes data for 25 college students who have been classified by gender and by college status (graduate or undergraduate). Suppose that these frequencies are obtained from a random sample of students from a large university. We may ask whether gender and college status are independent or, on the other hand, whether "maleness" (or "femaleness") is significantly associated with graduate status and the opposite gender is more prevalent among undergraduates.

This question can be answered using the same chi-square test for independence in a 2×2 table. To illustrate, Table 14.9 presents data for a sample of white youngsters in elementary schools in a southern town, collected by Patterson, Kupersmidt, and Vaden (1990).[3] The children were classified according to whether they lived in a one-parent or two-parent home. Also, the youngsters were classified by whether or not they came from a "low-income" household, that is, whether the family received some form of public assistance. If these data are a random sample of white youngsters from the community, they can be analyzed to see if number of parents and income level are independent or, on the other hand, if coming from a one-parent (or two-parent) home is significantly associated with having a low-income family.

---

[3]Further information from this study of household composition is given in Exercises 14.25–14.27.

TABLE 14.9

Family Structure and Income for a Sample of White Elementary Pupils

		Number of parents		Total
		Two	*One*	*Total*
Income	*Low*	42	72	$n_{1.} = 114$
	Not low	339	75	$n_{2.} = 414$
	Total	$n_{.1} = 381$	$n_{.2} = 147$	$n = 528$

source: Patterson, Kupersmidt, and Vaden (1990).

The expected values for these data—the cell counts that would have been obtained if income and number of parents were independent—are given in Table 14.10. These are obtained from (14.7). For example, $e_{11} = n_{1.} \times n_{.1}/n = 114 \times 381/528 = 82.26$ and $e_{12} = n_{1.} \times n_{.2}/n = 114 \times 147/528 = 31.74$. The test statistic (14.8) is the weighted sum of cell-by-cell differences squared. For these data

$$X^2 = \frac{(42 - 82.26)^2}{82.26} + \frac{(72 - 31.74)^2}{31.74} + \frac{(339 - 298.74)^2}{298.74} + \frac{(75 - 115.26)^2}{115.26}$$

$$= 90.26.$$

TABLE 14.10

Expected Values for Data of Table 14.9

		Number of parents		Total
		Two	*One*	*Total*
Income	*Low*	82.26	31.74	114
	Not low	298.74	115.26	414
	Total	381	147	528

This figure exceeds the .01 significance point of the χ^2-distribution with 1 degree of freedom, $\chi_1^2(.01) = 6.637$, and we conclude that income and number of parents are not independent in the population from which these youngsters were drawn.

To see the relationship more clearly, we may estimate the proportion of families in each category in the population, that is, the probability of each combination of income and number of parents. The population proportions may be represented as p_{ij} with row proportions $p_{i.}$ and column proportions $p_{.j}$ as in Table 14.11. The estimates (Table 14.12) are obtained by dividing each entry in Table 14.9 by the total sample

TABLE 14.11
Proportions in a Population

		B		
		B_1	B_2	*Total*
A	A_1	p_{11}	p_{12}	$p_{1\cdot}$
	A_2	p_{21}	p_{22}	$p_{2\cdot}$
	Total	$p_{\cdot 1}$	$p_{\cdot 2}$	1

TABLE 14.12
Sample Proportions for Data of Table 14.9

		Number of parents		
		Two	*One*	*Total*
Income	*Low*	0.08	0.14	0.22
	Not low	0.64	0.14	0.78
	Total	0.72	0.28	1

size *n*. (For example, $\hat{p}_{11} = 42/528 = 0.08$.) These show that there are 8 times as many "not low income" homes as low income homes when there are two parents present, while there is about the same proportion of "low" and "not low" homes when there is only a single parent present. In other words, low-income households are significantly more prevalent among single-parent families.

Theoretical Basis for Test.   The theoretical concept on which this test is based is the *independence of events*, discussed in Section 7.7. If an individual is drawn at random from a population with proportions given in Table 14.10, the probability of drawing an individual from the pair of categories A_i and B_j is $\Pr(A_i \text{ and } B_j) = p_{ij}$. The dichotomies low income or not low and two-parents or one-parent are examples of categorizations A_1 and A_2 and B_1 and B_2. The probability of drawing an individual with characteristic A_i is $p_{i\cdot}$ and the probability of an individual with B_j is $p_{\cdot j}$. These are called the *marginal probabilities* for A_i and B_j, respectively.

By (7.4) *A* and *B* are independent in the population if $\Pr(A_i \text{ and } B_j) = \Pr(A_i) \times \Pr(B_j)$, that is, if

$$p_{ij} = p_{i\cdot}p_{\cdot j}. \tag{14.9}$$

The null hypothesis of independence is that (14.9) is true for all pairs *i* and *j*. In words, *A* and *B* are independent if the probability in *each cell* cell of the table is equal to the product of the marginal probability for its row and the marginal probability for its column. [In a 2×2 table, when (14.9) is true for one pair *i* and *j* it is true for every pair.]

The hypothesis is tested on the basis of a random sample of n observations with frequencies n_{ij} and row and column proportions $\hat{p}_{i\cdot} = n_{i\cdot}/n$ and $\hat{p}_{\cdot j} = n_{\cdot j}/n$ (illustrated in Table 14.12). If the null hypothesis of independence is true, then p_{ij} is estimated by $\hat{p}_{i\cdot} \times \hat{p}_{\cdot j}$ and the estimate of the *number* of observations in the cell if the null hypothesis is true is $n\hat{p}_{i\cdot}\hat{p}_{\cdot j}$. This is exactly the expected value e_{ij} given by (14.7).[4] The test statistic (X^2) summarizes the difference between the actual frequencies and the frequencies that would be expected if the two dichotomies were independent, and is compared with values of the chi-square distribution with 1 degree of freedom.

Two-Way Tables in General

Table 7.1 is a 12×3 table relating to the 1969 draft lottery. The 12 rows correspond to the months of the year; the 3 columns, to high-, medium-, or low-priority numbers. Under an assumption of adequate mixing of the 366 date-bearing capsules, about one-third of the days in each month should fall into each of the high, medium, and low categories.

Table 14.13 shows the corresponding expected values. These were obtained from Table 7.1 in the same way as for a 2×2 table, that is,

TABLE 14.13
Expected Values for Draft Lottery

| | Priority numbers | | | |
Month	1 to 122	123 to 244	245 to 366	Total
January	10.33	10.33	10.33	31
February	9.67	9.67	9.67	29
March	10.33	10.33	10.33	31
April	10.00	10.00	10.00	30
May	10.33	10.33	10.33	31
June	10.00	10.00	10.00	30
July	10.33	10.33	10.33	31
August	10.33	10.33	10.33	31
September	10.00	10.00	10.00	30
October	10.33	10.33	10.33	31
November	10.00	10.00	10.00	30
December	10.33	10.33	10.33	31
Total	122	122	122	366

source: Table 7.1.

---

[4]To prove this recall that $n\hat{p}_{i\cdot} = n_{i\cdot}$ and $\hat{p}_{\cdot j} = n_{\cdot j}/n$.

using (14.7) for each of the 36 cells. Table 14.14 gives the discrepancies $n_{ij} - e_{ij}$. The value of the X^2-statistic for testing independence of month and priority is the sum of the values of $(n_{ij} - e_{ij})^2/e_{ij}$, that is,

$$X^2 = \frac{(-1.33)^2}{10.33} + \frac{(-2.67)^2}{9.67} + \cdots + \frac{(-6.33)^2}{10.33} = 37.2.$$

Since the discrepancies of observed and expected values sum to 0 in each row *and* in each column of Table 14.14, the number of degrees of freedom is $(12 - 1)(3 - 1) = 11 \times 2 = 22$. From Appendix IV, the .05 significance point from the chi-square distribution with 22 degrees of freedom is $\chi^2_{22}(.05) = 33.924$. Since $37.2 > 33.924$, the results are statistically significant at the 5% level $(.01 < P < .025)$. The hypothesis of independence of month and priority is rejected; there is evidence that the three classes of priority numbers were not spread fairly among the different months.

TABLE 14.14
Values of $n_{ij} - e_{ij}$ for Draft Lottery

| | Priority numbers | | | |
Month	1 to 122	123 to 244	245 to 366	Total
January	−1.33	1.67	−0.33	0
February	−2.67	2.33	0.33	0
March	−5.33	−0.33	5.67	0
April	−2.00	−2.00	4.00	0
May	−1.33	−3.33	4.67	0
June	1.00	−3.00	2.00	0
July	1.67	−3.33	1.67	0
August	2.67	−3.33	0.67	0
September	0.00	5.00	−5.00	0
October	−1.33	4.67	−3.33	0
November	2.00	2.00	−4.00	0
December	6.67	−0.33	−6.33	0
Total	0	0	0	0

source: Tables 7.1 and 14.13.

In this example the 366 dates of the year were classified simultaneously according to two classifications, month and priority number. The determination of the priority numbers was analogous to shuffling a deck of 366 cards, each card containing a date of the year; the first 122 cards were dealt to the priority category 1 to 122, the next 122 cards to category 123 to 244, and the last 122 cards to the third category. The mechanism is similar to dealing bridge hands; all marginal totals are specified in advance.

A situation that is more common is drawing one sample from one population. Each individual may be cross-classified simultaneously into one of the categories $A_1, A_2, \ldots, A_r$ and into one of the categories $B_1, B_2, \ldots, B_c$. On the basis of a sample of n, one may test the hypothesis that these classifications are independent in the population, that is, that the corresponding variables A and B are independent. For example, we may classify an individual according to educational level (non-high school graduate, high school graduate, college graduate, post-college study) and political party affiliation (Democrat, Republican, Other). The hypothesis to be tested is that educational level and political party affiliation are independent in the population.

A two-way table that summarizes a sample of observations that has been classified on two dimensions (e.g., Table 7.1) is referred to as a *contingency table*.

General Procedure

The same chi-square procedure may be used to test for independence when several samples have been classified into categories and when a single sample has been cross-classified on two dimensions. In general we represent the number of rows in a two-way table by r and the number of columns by c. An $r \times c$ table of observed frequencies has the form of Table 14.15 where n_{ij} represents the number of observations is row i, column j of the table, $n_{i\cdot}$ represents the row total in row i, and $n_{\cdot j}$ represents the column total in column j. Table 14.16 is an analogous table of population probabilities $\Pr(A_i \text{ and } B_j) = p_{ij}$.

TABLE 14.15

Cross-Classification of Sample

		B				
		B_1	B_2	$\cdots$	B_c	*Total*
	A_1	n_{11}	n_{12}	$\cdots$	n_{1c}	$n_{1\cdot}$
	A_2	n_{21}	n_{22}	$\cdots$	n_{2c}	$n_{2\cdot}$
A	$\vdots$	$\vdots$	$\vdots$		$\vdots$	$\vdots$
	A_r	n_{r1}	n_{r2}	$\cdots$	n_{rc}	$n_{r\cdot}$
	Total	$n_{\cdot 1}$	$n_{\cdot 2}$	$\cdots$	$n_{\cdot c}$	n

The classifications (or variables) A and B are said to be independent if $\Pr(A_i \text{ and } B_j) = \Pr(A_i) \times \Pr(B_j) = p_{i\cdot} p_{\cdot j}$ for all cells in the table, that is, for all values of i and j. In a sample of n observations, the row and column probabilities are estimated by $\hat{p}_{i\cdot} = n_{i\cdot}/n$ and $\hat{p}_{\cdot j} = n_{\cdot j}/n$, respectively. If the hypothesis of independence is true, then the number

TABLE 14.16
Table of Probabilities

		\multicolumn{5}{c}{B}			
		B_1	B_2	$\cdots$	B_c	*Total*
	A_1	p_{11}	p_{12}	$\cdots$	p_{1c}	$p_1.$
	A_2	p_{21}	p_{22}	$\cdots$	p_{2c}	$p_2.$
A	$\vdots$	$\vdots$	$\vdots$		$\vdots$	$\vdots$
	A_r	p_{r1}	p_{r2}	$\cdots$	p_{rc}	$p_r.$
	Total	$p._1$	$p._2$	$\cdots$	$p._c$	1

of observations that would be *expected* in row i, column j is $n\hat{p}_{i\cdot} \times \hat{p}_{\cdot j}$, which is

$$e_{ij} = \frac{n_{i\cdot}n_{\cdot j}}{n}.$$

The χ^2 test is based on comparing each observed value n_{ij} with its corresponding e_{ij}. The test statistic is written

$$X^2 = \sum\sum \frac{(n_{ij} - e_{ij})^2}{e_{ij}}$$

where the double summation is used to indicate that i is summed from 1 to r and j is summed from 1 to c. In total the X^2-statistic is a sum of $r \times c$ terms.

The sampling distribution of this statistic when the hypothesis of independence is true is approximated by the chi-square distribution χ^2_f with $f = (r-1)(c-1)$ degrees of freedom. The approximation is good if each e_{ij} is at least 5. (The error of approximation is not large if only a few, say 20% or fewer, of the e_{ij}'s are less than 5.) The hypothesis of independence is rejected at level α if the value of the test statistic exceeds $\chi^2_f(\alpha)$, where $f = (r-1)(c-1)$.

The number of degrees of freedom f is $(r-1)(c-1)$ because that is the number of *independent* quantities among the rc quantities $n_{ij} - e_{ij}$. In a table of quantities $n_{ij} - e_{ij}$, all the marginals are zero. When $(r-1)(c-1)$ entries in the table, say all but those in the last row or column, are filled in, the omitted entries are determined by the fact that all the marginals are zero.

Combining Categories

Just as in goodness-of-fit tests (Section 14.1), the investigator may exercise some judgment in assigning categories. From a table in which the categories are quite finely divided, the investigator may combine some categories or may delete some categories (which reduces the sample size). The powers of tests of independence will depend on the definitions of categories.

For example, Table 14.17 gives the voting records of professors from various colleges. Are the probabilities of voting Democratic the same in colleges of different kinds? The overall chi-square test statistic, with 2 degrees of freedom, is 12.98 ($P < .005$). Combining the first and second columns for comparison with the third produces Table 14.18; the value of the chi-square test statistic for this table, based on 1 degree of freedom, is 12.85 ($P < .005$).

It should be noted that the value of X^2 for the 2×2 table is almost as great as for the 2×3 table; the lack of independence (that is, difference in Democrat probabilities) is due mainly to the difference between "Other" (that is, mixed sponsorship) and the combined category of purely private and purely public schools. This result is verified by observing that the percentages in the first two columns of Table 14.17 are about the same.

Further examples of the effects of combining categories are given in Exercises 14.20 and 14.31.

TABLE 14.17

2×3 Table: 1948 Voting Record of Professors from Different Types of Colleges

		Type of school				
		Public	*Private*	*Other*		
Vote	*Democrat*	402 (71%)	493 (72%)	331 (63%)	1226 (69%)
	Not democrat	164 (29%)	192 (28%)	195 (37%)	551 (31%)
		566 (100%)	685 (100%)	526 (100%)	1777 (100%)

source: Lazarsfeld and Thielens (1958), p. 28.

TABLE 14.18

2×2 Table: Public and Private Schools versus Others

	Public or private	*Other*	
Democrat	895 (72%)	331 (63%)	1226 (69%)
Not democrat	356 (28%)	195 (37%)	551 (31%)
	1251 (100%)	526 (100%)	1777 (100%)

source: Table 14.17.

14.3   Measures of Association

When two categorical variables are not independent, they are dependent or associated. How do we measure the degree of association? As we shall see, the appropriate definition of a measure of association depends to some extent on the purpose for which it is to be used. The measures to be described apply to both samples and populations. The value of a measure of association in a sample usually varies statistically from the value of the measure in the population sampled. Even when a sample measure leads to rejection of the hypothesis of independence in the population (that is, is "statistically significant"), it may not indicate a *practically* significant degree of association. For instance, to predict one variable from another for a useful purpose may require a high degree of dependence.

The Phi Coefficient

A measure of association useful for 2×2 tables is introduced in Section 6.1: the *phi coefficient*. In Chapter 6 the observed frequencies are represented by a, b, c, and d for simplicity while in this chapter we use n_{11}, n_{12}, n_{21}, and n_{22}, respectively, in order to generalize to larger tables. Row totals are represented as $n_{i.}$ and column totals as $n_{.j}$ (Table 14.7). Using this notation the phi coefficient is

$$\phi = \frac{n_{11}n_{22} - n_{12}n_{21}}{\sqrt{(n_{11} + n_{12})(n_{21} + n_{22})(n_{11} + n_{21})(n_{12} + n_{22})}}$$

$$= \frac{n_{11}n_{22} - n_{12}n_{21}}{\sqrt{n_{1.}n_{2.}n_{.1}n_{.2}}}.$$

The value of ϕ is in the range from -1 to 1, with values closer to unity (in either direction) indicating stronger degrees of association between characteristics A and B. If variables A and B are *ordered* (for example, high-low dichotomies) then the magnitude of ϕ indicates the strength of association and the sign indicates the direction of association as well.

For the data of Table 14.9, the value of ϕ is

$$\frac{42 \times 75 - 339 \times 72}{\sqrt{114 \times 414 \times 382 \times 147}} = -0.413.$$

Note that $n\phi^2$ is identical to the test statistic X^2 that provides a simple yes/no decision about whether the two variables are independent in the population. The ϕ coefficient indicates the *extent* or *degree* to which there is association (non-independence) between the variables.[5]

---

[5]A continuity adjustment for X^2 for small values of n_{ij} is given in Appendix 14A.

Given a measure of association and a corresponding test of the hypothesis of nullity of that association, one can make a simultaneous assessment of both "statistical significance" and "practical significance." The situation is similar to that discussed in Section 11.1, where we considered the relation of hypothesis tests to confidence intervals for a mean. The four cases that may arise are classified in Table 14.19.

TABLE 14.19
Strength of Association and Acceptance or Rejection of Hypothesis

| | | Hypothesis of independence | |
		Accepted	*Rejected*
Measure of	*Small*	(i)	(ii)
association	*Large*	(iii)	(iv)

In case (i), the sample measure of association is small, that is, near zero. The hypothesis of independence is accepted with some conclusiveness, since the investigator has confidence that the value of the measure of association in the population is near zero. A confidence interval for it would include only values that are so close to zero as to be equivalent to zero in practical terms.

In case (ii), the sample measure of association is small, yet the hypothesis of independence is rejected. The investigator is confident that the population value is different from zero, yet this value may be so small that the relationship between variables cannot be exploited. If the sample size is large, there is a good chance that the value of the chi-square test statistic will lead to rejection of the hypothesis of independence, even though the association between variables is not large; this situation leads to case (ii).

In case (iii), the hypothesis of independence is accepted in spite of the fact that the sample measure of association is large. This large result may be due to a misleading sample and thus would not be reproducible. It may occur when the sample size is small, and hence the variability inherent in the sample measure of association is great.

In case (iv), the sample measure of association is large and the hypothesis of independence is rejected. The result is both statistically significant and practically significant. The result is reproducible and useful.

A Coefficient Based on Prediction

Another measure of association is based on the idea of using one variable to predict the other. Consider the cross-classification of exercise

TABLE 14.20

Exercise and Health Status

| | | Health status | | |
		Good	*Poor*	*Total*
Exercise	*Exerciser*	92	14	106
category	*Non-exerciser*	25	71	96
	Total	117	85	202

Hypothetical data.

and health in Table 14.20. How well does exercise group predict health status? If one of the 202 persons represented in this table is selected at random, our best guess of the health status—if we don't know anything about the person—is to say that the person is in the good-health group because more of the people are in that group (117, compared with 85 for the poor-health group); the good-health category is the *mode*. If we make this prediction for each of the 202 persons, we shall be right in 117 cases and wrong in 85 cases. However, if we take the person's exercise level into account, we can improve our prediction. If we know the person exercises regularly, our best guess is still that the person is in the good-health group, for 92 of the regular exercisers are in the good-health group, compared with only 14 in the nonexercise group. On the other hand, if we know the person is not an exerciser, we should guess that the person is in the poor-health group; in this case we would be correct 71 times and incorrect 25 times. Our total number of errors in predicting all 202 health conditions for both exercisers and nonexercisers is $14 + 25 = 39$, compared with 85 errors if we do not use the exercise category in making the prediction.

A coefficient of association that measures the improvement in prediction of the column category due to using the row classification is

$$\lambda_{c \cdot r} = \frac{\left(\begin{array}{c}\text{number of errors} \\ \text{not using the rows}\end{array}\right) - \left(\begin{array}{c}\text{number of errors} \\ \text{using the rows}\end{array}\right)}{\left(\begin{array}{c}\text{number of errors} \\ \text{not using the rows}\end{array}\right)} \quad (14.10)$$

$$= \frac{\begin{array}{c}\text{reduction in errors when using} \\ \text{the rows to predict columns}\end{array}}{\text{number of errors not using the rows}},$$

the decrease in errors divided by the number of errors made when not using the rows. The letter λ is lowercase Greek *lambda*. The symbol $\lambda_{c \cdot r}$ may be read "lambda c dot r." The subscript $c \cdot r$ refers to predicting the column category using the row category. For the example, we have

$$\lambda_{c \cdot r} = \frac{85 - (14 + 25)}{85} = \frac{85 - 39}{85} = \frac{46}{85} = 0.54.$$

Another equivalent expression for this coefficient

$$\lambda_{c \cdot r} = \frac{\begin{pmatrix} \text{number of} \\ \text{correct predictions} \\ \text{using rows} \end{pmatrix} - \begin{pmatrix} \text{number of correct} \\ \text{predictions not} \\ \text{using rows} \end{pmatrix}}{(\text{number}) - \begin{pmatrix} \text{number of correct predictions} \\ \text{not using rows} \end{pmatrix}}. \qquad (14.11)$$

This shows that $\lambda_{c \cdot r}$ can also be interpreted in terms of the increase in the number of correct predictions. For the example, this is

$$\lambda_{c \cdot r} = \frac{(92 + 71) - 117}{202 - 117} = \frac{163 - 117}{85} = \frac{46}{85} = 0.54.$$

Formulas for the coefficient λ based on prediction of the row category from the column classification are obtained from (14.10) and (14.11) by interchanging "columns" and "rows".

The values of $\lambda_{r \cdot c}$ and $\lambda_{c \cdot r}$ range between 0 and 1. The value 1 is attained when there are no errors in using the cross-classification for predictions; for $\lambda_{c \cdot r}$ this happens when in each row all the frequencies but one are zero; similarly $\lambda_{r \cdot c}$ equals 1 when each column contains only one frequency that is not zero. When two classifications are independent, the coefficient is zero, for in the case of independence the modal category is the same for each row, and the column classification gives no help in predicting. However, the coefficient can be zero even if the classifications are not independent. In Table 14.21,

$$\lambda_{c \cdot r} = \frac{(63 + 54) - 117}{85} = \frac{117 - 117}{85} = 0.$$

The modal category is good health for both exercisers and nonexercisers, so that $\lambda_{c \cdot r} = 0$; yet the proportion of exercisers in the good-health

TABLE 14.21
Exercise and Health Status

| | | Health status | | |
		Good	Poor	Total
Exercise	*Exerciser*	63	43	106
category	*Non-exerciser*	54	42	96
	Total	117	85	202

Hypothetical data.

group is $63/117 = 0.54$, which is not equal to $43/58 = 0.51$, the proportion of exercisers in the poor-health group.

A disadvantage of the coefficients of association based on predicting one variable from the other is that they do not treat the variables symmetrically. An alternative is to take a weighted average of the two coefficients:

$$\lambda = \dfrac{\begin{pmatrix}\text{reduction in errors}\\ \text{when using rows}\\ \text{to predict columns}\end{pmatrix} + \begin{pmatrix}\text{reduction in errors}\\ \text{when using columns}\\ \text{to predict rows}\end{pmatrix}}{\begin{pmatrix}\text{number of errors}\\ \text{in predicting columns}\\ \text{without using rows}\end{pmatrix} + \begin{pmatrix}\text{number of errors}\\ \text{in predicting rows}\\ \text{without using columns}\end{pmatrix}}.$$

This is called the *coefficient of mutual association*. Let p be the reduction in errors when using rows to predict columns, q the number of errors in predicting columns without using rows, r the reduction in errors when using columns to predict rows, and s the number of errors in predicting rows without using columns. Then

$$\lambda_{r \cdot c} = \frac{p}{q}, \quad \lambda_{c \cdot r} = \frac{r}{s}, \quad \lambda = \frac{p + r}{q + s}.$$

An interpretation for λ can be made by supposing that on half the occasions we shall predict columns and on half the occasions rows. Then the average number of errors is $(q + s)/2$, and the average reduction in errors is $(p + r)/2$. Then we take as the coefficient of mutual association the ratio

$$\frac{\text{average reduction in errors}}{\text{average number of errors}}.$$

Algebraically this is

$$\frac{(p + r)/2}{(q + s)/2} = \frac{p + r}{q + s} = \lambda.$$

The coefficient λ will be zero when and only when both p and r are zero, that is, when and only when both $\lambda_{r \cdot c}$ and $\lambda_{c \cdot r}$ are zero. In particular, $\lambda = 0$ in the case of independence. As with $\lambda_{r \cdot c}$ and $\lambda_{c \cdot r}$, the coefficient λ can be zero even in cases in which the classifications are not independent. For Table 14.21, $\lambda_{r \cdot c} = 0$ and $\lambda_{c \cdot r} = 0$, and so $\lambda = 0$, but the proportions based on the column totals are $63/117 = 0.54$ and $43/85 = 0.51$, which are not equal.

The following example illustrates computation of the coefficient for a larger two-way table. Consider the data of Table 14.22 in which 6800 newly-built homes have been classified according to type of dwelling and number of bedrooms. How well can one predict number of bed-

TABLE 14.22

Type of Newly-Built Dwelling and Numbers of Bedrooms

		Number of bedrooms				
		1	*2*	*3*	*4 or more*	*Total*
Type of dwelling	*Apartment unit*	1768	807	189	47	2811
	Attached house	946	1387	746	53	3132
	Single house	115	438	288	16	857
	Total	2829	2632	1223	116	6800

Hypothetical data.

rooms from the type of dwelling? The coefficient based on predicting columns from rows is

$$\lambda_{c \cdot r} = \frac{(6800 - 2829) - [(2811 - 1768) + (3132 - 1387) + (857 - 438)]}{6800 - 2829}$$

$$= \frac{3971 - (1043 + 1745 + 419)}{3971}$$

$$= \frac{3971 - 3207}{3971} = \frac{764}{3971} = 0.192.$$

The coefficient based on predicting rows from columns is

$$\lambda_{r \cdot c} = \frac{(6800 - 3132) - [(2829 - 1768) + (2632 - 1387) + (1223 - 746) + (116 - 53)]}{6800 - 3132}$$

$$= \frac{3668 - (1061 + 1245 + 477 + 66)}{3688}$$

$$= \frac{3668 - 2849}{3668} = \frac{819}{3668} = 0.223.$$

The coefficient of mutual association is

$$\lambda = \frac{819 + 764}{3668 + 3971}$$

$$= \frac{1583}{7639} = 0.207.$$

The value of this index indicates there is a weak association between type of dwelling and number of bedrooms in this set of 6800 new housing units.

A Coefficient Based on Ordering*

If both categorical variables are based on ordinal scales a measure of association may take this fact into account. The measure is appropriate, for example, if variables are ranked as high or low (or high, medium, or low) on some characteristics or above or below a certain criterion level, for example, pass or fail scores on a particular test.

Two-by-Two Tables   In Table 6.44, 75 communities were cross-classified on rate of juvenile delinquency and population density. A measure of association for such data can be constructed as follows: Consider two of the 75 communities, A and B. A's population density may be classified as either high or low; B's population density may be either high or low. Thus A and B can be ranked according to population density. The population density of A is either higher than that of B (A High, B Low), equal to that of B (both High or both Low), or lower than that of B (A Low, B High). Similarly, A and B can be ranked according to rate of juvenile delinquency. There are nine possibilities if both orderings are considered (Table 14.23).

TABLE 14.23

Possible Orderings of Communities A and B on Two Variables

	Rate of juvenile delinquency	Population density	Agreement
1	A's < B's	A's < B's	Yes
2	A's < B's	A's = B's	· · ·
3	A's < B's	A's > B's	No
4	A's = B's	A's < B's	· · ·
5	A's = B's	A's = B's	· · ·
6	A's = B's	A's > B's	· · ·
7	A's > B's	A's < B's	No
8	A's > B's	A's = B's	· · ·
9	A's > B's	A's > B's	Yes

In the actual data (Table 6.44) the number of communities classified as Low for both variables is 30, and the number of communities classified as High for both variables is 36. This give rise to 30 × 36 pairs of communities in which the orderings on the two variables are the same. That is, each of the 30 low-low communities may be compared with each of the 36 high-high communities and the comparisons will result in the same conclusion: this is a low-high pair.

From Table 6.44 it can be seen that the orderings disagree for 5 × 4 pairs. The value of the coefficient based on ordering, γ (lower-case

Greek *gamma*) is

$$\gamma = \frac{30 \times 36 - 5 \times 4}{30 \times 36 + 5 \times 4} = \frac{1080 - 20}{1080 + 20} = \frac{1060}{1100} = 0.96.$$

In general, the formula for γ in the case of a 2×2 table with ordered categories and with frequencies a, b, c, and d is

$$\gamma = \frac{ad + bc}{ad + bc}.$$

In the more general notation using n_{11}, n_{12}, n_{21}, and n_{22} for a, b, c, and d, respectively, the coefficient is

$$\gamma = \frac{n_{11} n_{22} - n_{12} n_{21}}{n_{11} n_{22} + n_{12} n_{21}}.$$

Coefficient γ for 2×2 tables is equivalent to Yule's Q-statistic described in some textbooks.

Notice that the quantity $n_{11} n_{22} - n_{12} n_{21}$ is the same as the numerator of ϕ in Section 14.3. The two coefficients will be 0 under the same conditions. However, ϕ can only reach a maximum of 1 if the row totals $n_{1\cdot}$ and $n_{2\cdot}$ are the same two values as the column totals $n_{\cdot 1}$ and $n_{\cdot 2}$. (See Exercises 14.42–14.44.) Since this restriction does not hold for γ, it is the preferred index of association between ordinal variables if the marginal totals are not equal.

The theory underlying γ is as follows. Suppose that a pair of individuals is selected at random from among the $n_{11} n_{22} + n_{12} n_{21}$ pairs of individuals who are either the same on both variables or differ on both variables. Then $n_{11} n_{22}/(n_{11} n_{22} + n_{12} n_{21})$ is the probability that the chosen pair will be put in the same order by both variables, while $n_{12} n_{21}/(n_{11} n_{12} + n_{12} n_{21})$ is the probability that the chosen pair will be put in opposite orders by the two variables. The coefficient γ is the difference between these two probabilities.

Larger Two-Way Tables A drawback of the chi-square test statistic and of the measure of association based on prediction is that they fail to take into account any ordering of the categories. Table 14.24 gives frequencies for a hypothetical sample of 100 students enrolled in four-year colleges for which both variables are ordinal. The location of the college may be viewed as "distance from home town," either 0 or greater than 0; the numbers of years completed in that college are ordered from less than 1 year, to 1 to 3 years, to completing all 4 years. Table 14.25 has the same six frequencies but the second and third columns of frequencies have been interchanged. The X^2-statistic and the measure based on prediction have the same values in the two tables, yet if A_1 and A_2 and B_1, B_2,

TABLE 14.24

Location of College Attended and Years Completed

		Years completed			
		Less than 1	*1–3*	*Graduated*	*Total*
Location of college (distance)	*Home town*	30 (50%)	25 (42%)	5 (8%)	60 (100%)
	Away from home	3 (8%)	17 (42%)	20 (50%)	40 (100%)
	Total	33 (33%)	42 (42%)	25 (25%)	100 (100%)

Hypothetical data.

TABLE 14.25

A 2 × 3 Table

		B			
		B_1	B_2	B_3	*Total*
A	A_1	30	5	25	60
	A_2	3	20	17	40
	Total	33	25	42	100

and B_3 are ordered, the positive association seems higher in Table 14.24. One method that distinguishes between the two tables is the coefficient based on ordering. The definitional formula for γ is, in words,

$$\gamma = \frac{\left(\begin{array}{c}\text{number of pairs} \\ \text{of individuals for} \\ \text{which both orderings} \\ \text{are the same}\end{array}\right) - \left(\begin{array}{c}\text{number of pairs} \\ \text{of individuals for} \\ \text{which the orderings} \\ \text{are different}\end{array}\right)}{\left(\begin{array}{c}\text{total number of pairs of individuals} \\ \text{for which neither ordering is a tie}\end{array}\right)}.$$

To illustrate the idea, we consider Table 14.26, with only six individuals. We list the 15 possible pairs of individuals in Table 14.27. The number of pairs with the same ordering is 5, and the number with different orderings is 1; hence the value of γ is

$$\gamma = \frac{5 - 1}{5 + 1} = \frac{4}{6} = 0.67.$$

TABLE 14.26
Data for Six Students on Two Ordinal Variables

Individual	Location	Years
A	Away	Less than 1
B	Home	Less than 1
C	Away	Graduated
D	Away	1–3
E	Home	1–3
F	Home	Less than 1

Hypothetical data.

TABLE 14.27
Ranking of Persons on Two Variables

Pairs of persons	Relationship of distance of first person to distance of second person	Relationship of years completed by first person to years completed by second person	Same order?	Opposite order?
AB	>	=		
AC	=	<		
AD	=	<		
AE	>	<		×
AF	>	=		
BC	<	<	×	
BD	<	<	×	
BE	=	<		
BF	=	=		
CD	=	>		
CE	>	>	×	
CF	>	>	×	
DE	>	=		
DF	>	>	×	
EF	=	>		
			5	1

source: Table 14.26.

However, the coefficient can be computed directly from cross-classification of the six students in Table 14.28. The same reasoning as that given for the special case of a 2 × 2 table applies. Relative to the $n_{11} = 2$ students in the (Home, Less-than-1) category, there are $n_{22} + n_{23} = 1 + 1 = 2$ students in categories that are higher on both

TABLE 14.28

Cross-Classification of Six Students

		Years completed			
		Less than 1	*1–3*	*Graduated*	*Total*
Location	*Home town*	$n_{11} = 2$	$n_{12} = 1$	$n_{13} = 0$	3
	Away from home	$n_{21} = 1$	$n_{22} = 1$	$n_{23} = 1$	3
	Total	3	2	1	6

source: Table 14.26.

variables; this fact yields $2 \times 2 = 4$ pairs of students that are ordered the same way on both variables. The $n_{12} = 1$ student in the (Home, 1–3) category can be paired with the $n_{23} = 1$ student in the (Away, Graduated) category to yield $1 \times 1 = 1$ pair of students ordered the same way on both variables. No other pairs of students are ordered similarly. The $n_{12} = 1$ student can be paired with the $n_{21} = 1$ student in the (Away, Less-than-1) category to form $1 \times 1 = 1$ pair of students ordered differently on the two variables. This gives $4 + 1 = 5$ pairs of students ordered the same way and 1 pair ordered differently on the two variables. The value of γ is $(5 - 1)/(5 + 1) = 4/6 = 0.67$. Table 14.29 is a worksheet for this calculation. At the bottom of column (4) is the sum of the numbers in that column; this sum is the number of pairs of students having the orderings the same. Similarly, the sum at the bottom of the column (5) is the number of pairs having different orderings. (The last three rows are unnecessary.)

TABLE 14.29

Worksheet for Calculations of γ for Table 14.28

(1)	(2)	(3)	(4)	(5)
	Sum of frequencies below and to the right of this frequency	*Sum of frequencies below and to the left of this frequency*	*Pairs with same orders*	*Pairs with different orders*
Frequency	*frequency*	*frequency*	$(1) \times (2)$	$(1) \times (3)$
$n_{11} = 2$	$1 + 1 = 2$	0	4	0
$n_{12} = 1$	1	1	1	1
$n_{13} = 0$	0	0	0	0
$n_{21} = 1$	0	0	0	0
$n_{22} = 1$	0	0	0	0
$n_{23} = 1$	0	0	0	0
			5	1

Returning to Table 14.24, we see that for each of the 30 students in the category (Home, Less-than-1) there are $17 + 20 = 37$ students in a category that is higher on both variables. This yields $30 \times 37 = 1110$ pairs of students for which the variables give the same ordering (entered on the first line of Table 14.30). For each of the 25 students in the category (Home, 1–3), there are 20 students who are ranked higher on both variables. This yields $25 \times 20 = 500$ pairs of students for which the ordering is the same for the variables. For each of these 25 students in the category (Home, 1–3), there are also 3 students who are ranked higher in distance but lower in years completed. This yields $25 \times 3 = 75$ pairs for which the one variable gives a different ordering of the students than does the other variable. For each of the 5 students in the (Home, Graduated) category, there are $3 + 17 = 20$ persons who are ranked higher in distance from home but lower on years completed. This yields $5 \times 20 = 100$ pairs with different orderings. The number of pairs with the same ordering is $1110 + 500 = 1610$. The number of pairs with different orderings is $75 + 100 = 175$. The difference is $1610 - 175 = 1435$. The total number of relevant pairs is $1610 + 175 = 1785$. The value of γ is $1435/1785 = 0.80$.

The value of γ for Table 14.25 is

$$
\gamma = \frac{[30 \times (20 + 17) + 5 \times 17] - [5 \times 3 + 25 \times (3 + 20)]}{[30 \times (20 + 17) + 5 \times 17] + [5 \times 3 + 25 \times (3 + 20)]}
$$

$$
= \frac{(1110 + 85) - (15 + 575)}{(1110 + 85) + (15 + 575)}
$$

$$
= \frac{605}{1785} = 0.34,
$$

indicating a weaker but still positive association.

TABLE 14.30

Worksheet for Calculation of γ for Table 14.24

(1)	(2)	(3)	(4) = (1) × (2)	(5) = (1) × (3)
$n_{11} = 30$	$17 + 20 = 37$	0	1110	0
$n_{12} = 25$	20	3	500	75
$n_{13} = 5$	0	$3 + 17 = 20$	0	100
			1610	175
		(4) − (5):	1435	
		(4) + (5):	1785	
		γ:	0.80	

Summary

A test of goodness of fit is a test of whether a specified distribution fits the distribution of a sample.

A test of independence is a test of whether the joint distribution of two variables is the product of their marginal distributions.

Tests of goodness of fit and independence are made by computing the frequencies expected under the respective hypotheses and comparing these with observed frequencies. The statistics used have distributions that are approximated by chi-square distributions.

Measures of association indicate the *extent* of departure from independence in a two-way table. The phi (ϕ) coefficient is a correlation-like statistic generally applicable to 2 × 2 tables. The lambda ($\lambda_{c \cdot r}$) coefficient may be used for larger tables. It is based on the extent to which an individual's classification on one variable can be known by knowledge of his or her classification on the other variable. The gamma (γ) statistic indicates the extent of association between two ordinal variables, that is, for which the categories reflect greater or lesser amounts of a characteristic.

Appendix 14A   *The Continuity Adjustment*

The continuity adjustment for the chi-square test statistic for independence in a 2 × 2 table consists of replacing $(n_{11}n_{22} - n_{12}n_{21})^2$ by

$$\left(|n_{11}n_{22} - n_{12}n_{21}| - \frac{n}{2} \right)^2 .$$

Note that this "makes it harder to reject." This continuity correction agrees with replacing $|\hat{p}_1 - \hat{p}_2|$ by

$$|\hat{p}_1 - \hat{p}_2| - \frac{1}{2} \left(\frac{1}{n_1} + \frac{1}{n_2} \right)$$

in the test statistic for comparing two proportions (Appendix 12A).

Exercises

14.1   In each of the following situations test the null hypothesis that the coin is fair at the 5% significance level. For each answer, conclude that the coin is "fair" or "not fair" and give a *P* value associated with the test.

(a) The coin is tossed 10 times; heads turn up 6 times;

(b) The coin is tossed 10 times; heads turn up 8 times;

(c) The coin is tossed 25 times; heads turn up 15 times;

(d) The coin is tossed 25 times; heads turn up 18 times;

(e) The coin is tossed 25 times; heads turn up 20 times;

14.2   Do Exercise 11.19 by means of a chi-square test.

14.3   Using the data in Exercise 10.27, test the null hypothesis at the 5% level that Bush and Clinton were equally favored in October 1991. Take $n = 1000$.

14.4   Using the data of Exercise 10.27, test the null hypothesis at the 5% level that Bush and Clinton were equally favored in January 1992. Use $n = 1000$.

14.5   The study by Ware and Stuck (1985) of the portrayal of males and females in computer magazines (Exercise 12.37) also counted the appearances of boys and girls. The frequencies are given in Table 14.31 together with the proportion of each age-gender group in the population as reported by the U.S. Census Bureau.

TABLE 14.31

Number of Men, Women, Boys, and Girls Portrayed in Computer Magazines

	Men	Women	Boys	Girls
Observed number	$n_1 = 465$	$n_2 = 196$	$n_3 = 38$	$n_4 = 28$
Proportion from U.S. census	$p_{10} = 0.34$	$p_{20} = 0.37$	$p_{30} = 0.14$	$p_{40} = 0.14$

source: Ware & Stuck (1985), p. 208.

Use a chi-square test to decide whether individuals chosen for portrayal in computer magazines represent a random sample of American men, women, boys, and girls. Give a P value interval for your result.

14.6   A certain community has a population that is 72% non-Hispanic white, 22% African American, and 6% of Hispanic origin. The 40-member police force includes 34 white officers and 4 African American officers. Use a chi-square test at the 5% level to see whether the composition of the force is consistent with a random recruitment model. Give a P value interval for your result.

14.7   Table 14.32 gives the number of wins for each of the first six post positions at Waterford Park, Chester, West Virginia, for the 118 races from the beginning of the season to March 11, 1969. Test at the 5%

TABLE 14.32
Number of Wins, by Post Position

Post position	1	2	3	4	5	6
Number of wins	30	17	21	27	12	11

source: Pittsburgh Press, March 11, 1969.

level the hypothesis that the probability of winning is the same for all post positions. Give a *P* value interval for your result.

14.8 Test at the 10% level the hypothesis that the faces of the die have the same probability of turning up, based on the data in Table 14.33 for 100 rolls of the die.

TABLE 14.33
Results of 100 Rolls of a Die

Face	1	2	3	4	5	6	Total
Frequency	21	19	20	15	14	11	100

Hypothetical data.

14.9 A die was tossed 300 times, giving the results in Table 14.34. Give a *P* value interval to express the degree of evidence that the die is not fair.

TABLE 14.34
Results of Tossing a Die

Face	1	2	3	4	5	6	Total
Frequency	46	51	44	57	38	64	300

Hypothetical data.

14.10 A random sample of 200 digits is tabulated in Table 14.35. Test the hypothesis that this random sample is from a uniform distribution. Express your result in the form of a *P* value interval.

TABLE 14.35
Distribution of a Random Sample of Digits

Digit	0	1	2	3	4	5	6	7	8	9	Total
Frequency	10	20	19	21	21	15	21	22	25	26	200

Hypothetical data.

14.11   (continuation) For the data in Table 14.34, do the following:

(a) Partition the table into Odd Faces vs. Even Faces.

(b) Compute the corresponding chi-square statistic.

(c) Test the hypothesis that the probability of an Odd Face is 1/2 at the 5% level.

14.12   (continuation) For the data in Table 14.35, do the following:

(a) Partition the table into 0 versus others.

(b) Compute the corresponding chi-square statistic.

(c) Test the hypothesis that the probability of 0 is 0.1.

14.13   W. E. McGrath (1986) analyzed data on the size and expenditures of a sample of 49 academic libraries. One variable was the number of volumes added to the library's collection during the most recent fiscal year. Another was the type of college or university, that is, whether the school is a publicly or privately supported institution. The following cross-tabulation resulted:

		Type of library		
		Public	*Private*	*All*
Volumes added	*Fewer than 100,000*	20	11	31
	More than 100,000	10	8	18
	Total	30	19	49

source: McGrath (1986), p. 34.

Use these data as a random sample of the nation's academic libraries and test at the 5% level the hypothesis of independence between type of institution and volumes added to the library. In your answer, give the test statistic, the significance point from the appropriate probability distribution, and your conclusion about whether the type of institution and volumes added are or are not independent.

14.14   The study by McGrath (1986) also examined the relationship of the number of library staff members with volumes added to the library's collection. The following cross-tabulation resulted:

		Number of staff		
		Under 200	*Over 200*	*All*
Volumes added	*Fewer than 100,000*	19	12	31
	More than 100,000	2	16	18
	Total	21	28	49

source: McGrath (1986), p. 33.

Use these data as a random sample of the nation's academic libraries and test at the 1% level the hypothesis of independence between number of staff members and volumes added to the library. In your answer, give the test statistic, the significance point from the appropriate probability distribution, and your conclusion about whether the number of staff members and volumes added are or are not independent.

14.15 Do Exercise 12.30 using a chi-square test statistic.

14.16 Do Exercise 12.31 using a chi-square test statistic.

14.17 Compute the phi (ϕ) index of association for the data in Exercise 12.30 and in Exercise 12.31.

(a) Is the association strong, moderate, or weak in each case?

(b) How do the two measures compare in magnitude?

(c) How would you explain the similarity or difference between them?

14.18 Table 14.36 gives the distribution of a sample of primary householders by income and age in Suburbs A and B of a metropolitan area.

(a) Compute the age distribution in each suburb. Which suburb is "younger"?

(b) Compute the income distribution in each suburb. Which suburb is "richer"?

(c) Consider Table 14.36 as three separate 3 × 2 tables, one for each age group. Compute the chi-square statistic for each of these three tables. Are there any significant differences in income between the two suburbs, once you have "controlled" for age?

TABLE 14.36
Income and Age in Suburbs A and B

| | | Below 25 | | 25–45 | | Above 45 | |
		A	B	A	B	A	B
Income level	Low	52	12	13	6	6	8
	Medium	29	7	25	13	6	5
	High	33	7	12	6	25	26

Hypothetical data.

14.19 As shown in Table 14.37, African Americans were 2% of all officers in the Armed Forces and 3% of those in Vietnam as of June 30, 1967. About 1000 in a total of 8000 black officers were serving in Vietnam. African Americans were 10 percent of all enlisted men and 12 percent of those

TABLE 14.37
African-American Officers and Enlisted Men in the Armed Forces, June 20, 1967

| | | | Percent |
	Total	Black	Black
		(Numbers in thousands)	
Total	3365	305	9
Officers	384	8	2
Outside Vietnam	342	7	2
Inside Vietnam	43	1	3
Enlisted men	2981	297	10
Outside Vietnam	2536	246	10
Inside Vietnam	444	51	12

source: U.S. Department of Defense (*Social and Economic Conditions of Negroes in the United States*, October 1967, BLS Report No. 332, U.S. Department of Labor, Bureau of Labor Statistics, p. 84; Current Population Reports, Series P-23, No. 24, U.S. Bureau of the Census).

in Vietnam—51,000 in Vietnam in a total of 297,000 black enlisted men in the Armed Forces.

The percentages of black soldiers in Vietnam were higher than would be expected on the basis of the overall percentages of blacks among both officers and enlisted men. Could this be explained on the basis of chance alone? Study this question, as follows.

(a) For Officers, make a 2 × 2 table, Outside Vietnam/Inside Vietnam by black/Other. Compute the value of the chi-square statistic. Give the achieved level of significance (*P* value).

(b) Do the same for Enlisted Men.

14.20 Consider the data on the portrayal of men and women in popular computer magazines in Table 12.17 (Exercise 12.37).

(a) Is the hypothesis that the same distribution obtains in the portrayals of men and women tenable? (Do the chi-square test of independence at the 1% level of significance.)

(b) Decide whether the distributions are different if only the two most popular roles are distinguished in the cross-tabulation and all others are combined. Do this by computing the chi-square test statistic for Table 14.38; give a *P* value interval for your result.

14.21 Consider the data of Exercise 6.7 on damage to property caused by high school students.

(a) Test the overall hypothesis that the number of times individuals have damaged other people's property is independent of gender. Use $\alpha = .05$.

(continued . . .)

TABLE 14.38

Illustrations of Males and Females in Three Most Popular Roles

Role	Men	Women	Both
Seller	124	81	205
Manager	101	13	114
All others	240	102	342
Total	465	196	661

source: Table 12.17.

(b) Compare those who "never" damaged other people's property with those who damaged property one or more times. Do this by conducting a chi-square test on the data table of part (b) of Exercise 6.7.

14.22 An experiment was performed to test whether male moths' ability to steer toward a source of odorous sex pheromone depends on air movement. A male was considered to have steered successfully if, having started at one end of a flight tunnel, he flew through a hoop near the source at the other end. The results are summarized in Table 14.39. Test the hypothesis that flying through the hoop is independent of the pheromone plume.

TABLE 14.39

Number of Male Moths Flying Through Hoop and Outside of Hoop for Three Test Conditions

	Number flying through hoop	Number flying outside hoop
1. Pheromone plume in moving air	17	3
2. Pheromone plume in still air	16	4
3. No pheromone plume in still air (control)	3	17

source: Reprinted with permission from Farkas, S. R., & Shorey, H. H. (1972). Chemical trail-following by flying insects: A mechanism for orientation to a distant odor source. *Science, 178*, 67–68. Copyright © 1972 by the American Association for the Advancement of Science.

14.23 Beginning in 1992, a sample of 16- to 19-year old males was interviewed as part of a long-term study of the relationship between alcohol consumption and delinquency. In the interview subjects reported whether they belonged to a gang and the frequency of committing each of six selected violent acts, for example, attacking someone with the intent of

TABLE 14.40

Violent Acts Committed by Sample of 16- to 19-year-old Males

| | | \multicolumn{3}{c}{Number of commissions} | | |
		None	*1–24*	*25+*	*Total*
Gang afiliation	*Never*	194	217	34	445
	Previously	19	89	30	138
	Currently	3	22	15	40

source: Unpublished data provided by the Research Institute on Addictions, Buffalo, New York, printed with permission of John W. Welte, Senior Scientist.

hurting or killing them. A total was computed that indicated the number of times the violent acts had been committed during the previous 12 months; results are given in Table 14.40.

(a) Compute the percentage of each gang-affiliation group who reported no violent acts, 1–24 violent acts, or more than 24 violent acts committed during the previous 12 months. Compute the percentages for all 623 respondents.

(b) Compute the chi-square test of independence of gang affiliation and violent acts. In your answer give the test statistic, the number of degrees of freedom on which the statistic is based, and a P value interval for the test.

(c) Summarize your conclusion verbally. Include a statement about whether gang affiliation and violent acts are or are not independent and further information to explain your conclusion. For example, if affiliation and violence are independent, then what is the "common" distribution of violent acts for all three groups? If the gang-affiliation groups have different patterns (affiliation and violence not independent) then what distinguishes one group from another (e.g., is one group more inclined to nonviolence or to violence than another?)

14.24   Consider the results of one of the early experiments in the Asch study on conformity under pressure in reporting comparative length of line segments, as summarized in Table 12.5.

(a) Compute the value of the chi-square test statistic for these data.

(b) Compute ϕ.

(c) Compute $n\phi^2$ in order to verify that it equals the chi-square test statistic. (It can be shown algebraically that this is true in general. Here you are asked only to verify it numerically.)

14.25   Table 14.9 gives data on family structure and income for homes of a sample of white elementary-school pupils. Table 14.41 gives the same tabulation for families of African American students obtained in the same study.

(a) Compute the value of the chi-square statistic for testing the independence of family structure and income for the data of Table 14.41.

(b) Compute the value of ϕ for these data.

(c) Compute $n\phi^2$ in order to verify that it equals the chi-square test (within rounding error).

TABLE 14.41

Family Structure and Income for Sample of African-American Elementary Pupils

		Number of parents		
		Two	*One*	*Total*
Income	*Low*	39	172	211
	Not low	67	62	129
	Total	106	234	340

source: Patterson, Kupersmidt, and Vaden (1990).

14.26 (continuation) Compare the values of chi-square and ϕ for the data of Table 14.9 with those obtained from Table 14.41. What differences, if any, do you see between the families of white and African-American elementary students in this community in

(a) the proportion of one-parent and two-parent families,

(b) the proportion of low-income and not-low-income homes,

(c) the degree of association between family structure and income. How would you explain the similarities or differences?

14.27 (continuation) The classification of African-American and white pupils as low income or not-low income in the Patterson, Kupersmidt, and Vaden (1990) study was made by asking the child's teacher whether or not the family received "public assistance," that is, whether the child had free or reduced price lunches at school and whether the child's family lived in subsidized housing. What improvements can you suggest in how families could be classified according to income? How would you implement these improvements in an actual study of elementary school children and their families?

14.28 The data of Table 14.42 are part of a study of the association between complications in pregnancy of mothers and behavior problems in children. The comparison is between mothers of children who had been referred by their teachers as behavior problems and mothers of children not so referred. For each mother it was recorded whether she had lost any children prior to the birth of the child. The birth order of the child was recorded and used as a control.

TABLE 14.42

Cross-Classification of Mothers by Referral of Child, Prior Infant Loss, and Birth Order of Child

Birth order	Behavioral referral?	Prior infants Loss	No loss
2	Yes	20	82
	No	10	54
3–4	Yes	26	41
	No	16	30
≥ 5	Yes	27	22
	No	14	23

source: Reproduced from W. G. Cochran, "Some Methods for Strengthening the Common Chi-Square Tests," *Biometrics* 10: 417–451, 1954, with the permission of the Biometric Society.

(a) Compute the phi coefficient for each of the three 2 × 2 tables.

(b) Compute the chi-square statistic for each of the three tables, and give a *P* value interval for each.

14.29 From Table 14.43, study the association of estrogen treatment with the condition of tumor, as follows. Study separately each of the four 2 × 2 tables for Estrogen Treatment and Condition of Tumor; this controls for Age and Parity. (A nulliparous woman has never given birth; a parous woman has given birth at least once.)

(a) Compute the phi coefficient for each of the four tables.

(b) Compute the chi-square statistic for each of the four tables.

TABLE 14.43

Condition of Tumor, by Age, Parity, and Estrogen Treatment

		Age < 50 Benign	Malignant	Age ≥ 50 Benign	Malignant
Nulliparous	Estrogen	9	14	9	14
	No estrogen	64	27	24	38
Parous	Estrogen	30	9	14	14
	No estrogen	189	71	39	93

source: Black and Leis (1972), Table I, p. 1602.

14.30 P. G. Norton and E. V. Dunn (1985) investigated whether snoring was related to various diseases. The results for heart disease for a sample of 2484 respondents is given in Table 14.44.

TABLE 14.44

Heart Disease and Snoring

		Frequency of snoring				
		Non-snorers	*Occasional snorers*	*Snore nearly every night*	*Snore every night*	*All*
Heart disease	*Absent*	1355	603	192	224	2374
	Present	24	35	21	30	110
	All	1379	638	213	254	2484

source: Norton & Dunn (1985). Reprinted in Hand et al. (1994), p. 19.

Use these data to examine the relationship between snoring and heart disease as follows:

(a) Compute the column distributions of heart disease for each snoring classification.

(b) Compute the chi-square test of independence of snoring and heart disease; use the 5% level of significance. In your answer include

- the table of expected frequencies,
- the value of the test statistic,
- the significance point from the appropriate chi-square distribution,
- your conclusion about the independence or non-independence of snoring and heart disease.

(c) Compute $\lambda_{r \cdot c}$, the coefficient of association based on predicting rows (heart disease) from columns (amount of snoring).

14.31 (continuation) Construct a new table from Table 14.44 by combining all snorers into a single classification so that the resulting table has just two columns (non-snorers and snorers). Repeat the steps in the preceding exercise for this 2 × 2 table. How do these results compare with those obtained in Exercise 14.30?

14.32 (continuation) Both variables in Table 14.44 have ordinal scales; heart disease "present" indicates more than "absent" and the snoring classifications are ordered from "none" to "occasional" to "almost every night" to "every night."

(a) Compute the γ coefficient based on ordering for this table. How strong is the tendency for greater amounts of snoring to be associated with the presence of heart disease (weak, moderate, or strong)?

(b) Compare the γ coefficient with $\lambda_{r \cdot c}$ obtained in Exercise 14.30. Are they similar in magnitude or quite different?

(c) Compute the γ coefficient for the 2 × 2 table obtained in Exercise 14.31 by combining all snoring classifications.

14.33   Consider the data of Table 6.49, "Cross-Classification of 40 Years, by Yield of Hay, Rainfall, and Temperature."

(a) Calculate $\lambda_{\text{yield·temperature}}$ for the years of light rainfall and for the years of heavy rainfall.

(b) Test the independence of temperature and yield for the years of light rainfall and for the years of heavy rainfall.

14.34   For the data in Table 6.49, calculate the coefficient based on ordering: (a) for the years of light rainfall, (b) for the years of heavy rainfall.

14.35   Analyze the data in Table 6.25 as follows.

(a) Compute the coefficient of association based on predicting the "leader's rating" from the "degree of change in understanding."

(b) Compute the coefficient of association based on predicting the "degree of change in understanding" from the "leader's rating."

(c) Compute the coefficient of mutual association, based on both types of prediction.

(d) Compute the coefficient of association based on ordering.

(e) Interpret the association in these data as measured by the coefficients in parts (a), (b), (c), and (d).

14.36   Analyze the data in Table 6.24, as follows.

(a) Compute the coefficient of association based on predicting the number of volumes from the length of programs.

(b) Compute the coefficient of association based on predicting the length of programs from the number of volumes.

(c) Compute the coefficient of mutual association, based on both types of prediction.

(d) Compute the coefficient of association based on ordering.

(e) Interpret the association in these data as measured by the coefficients in parts (a), (b), (c), and (d).

14.37   Calculate the phi coefficient for each of the three tables of Table 6.32.

14.38   Calculate the phi coefficient for each of three tables of Table 6.50.

14.39   Test the hypothesis that the heights in Table 2.12 are a sample from a normal distribution with a mean of 52.5 inches and standard deviation of 2 inches, as follows.

(a) Standardize the endpoints of the class intervals by subtracting the mean of 52.5 and dividing the results by the standard deviation of 2. Make the first interval $-\infty$ to -3.0 (in standardized units) and the last 3.0 to $+\infty$.

(continued . . .)

(b) Use Appendix I to find the proportion of the standard normal distribution that lies between the standardized endpoints.

(c) Multiply each result of (b) by the sample size to obtain expected frequencies.

(d) Compare the expected frequencies with the observed frequencies by means of the chi-square statistic, and give a corresponding P value interval.

14.40   Test the hypothesis that the heights in Table 8.7 are a sample from a normal distribution with a mean of 54.5 inches and a standard deviation of 2 inches. Use steps corresponding to (a) to (d) of Exercise 14.39.

14.41   Table 14.45 resulted from a study of auto weight and safety. Test the hypothesis of independence between auto weight and accident frequency.

TABLE 14.45
Accident Frequency, by Weight Class

Auto weight class	Observed accident frequency	Registration distribution
Under 3000 pounds	162	21.04%
3000–4000 pounds	318	46.13%
4000–5000 pounds	689	31.13%
Over 5000 pounds	35	1.70%
	1204	100.00%

source: Yu, Wrather, and Kozmetsky (1975), p. 8.

14.42   Consider the hypothetical cross-classification of individuals on two characteristics A and B given in Table 14.46.

TABLE 14.46
Table of Observed Frequencies

		B		
		B_1	B_2	Total
A	A_1	6	3	9
	A_2	0	6	6
	Total	6	9	15

Hypothetical data.

(a) Calculate the ϕ index of association for the table.

(b) Calculate the γ index of association for the table.

(c) Which of the coefficients is larger? How would individuals have to be distributed in the table in order for the ϕ index to be 1?

14.43 Consider the hypothetical cross-classification of individuals on two characteristics A and B given in Table 14.47.

(a) Calculate the ϕ index of association for the table.

(b) Calculate the γ index of association for the table.

(c) Construct several other tables from these data by adding the 5 individuals in the upper right and lower left cells (n_{12} and n_{21}) into the upper left and lower right cells instead. What distributions, if any, will give a ϕ value of 1?

TABLE 14.47
Table of Observed Frequencies

		B		
		B_1	B_2	*Total*
A	A_1	4	2	6
	A_2	3	6	9
	Total	7	8	15

Hypothetical data.

14.44 The preceding two exercises illustrate the fact that the γ index of association is often larger than ϕ. Algebraically it can be shown that γ is always greater than or equal to ϕ. Construct two 2×2 tables with the same total n that have the same value for γ (not 0 or 1) but different values of ϕ. [Hint: try $n = 17$.]

15

Simple Regression Analysis

Introduction

In this chapter we return to the statistical relationship between two quantitative variables. In Chapter 5 the correlation coefficient is described as a symmetric index of strength of association. In this chapter we examine the *directional* relationship of two variables. In many instances one variable may have a direct effect on the other or may be used to predict the other. For example, sodium intake may affect blood pressure; rainfall influences crop yield; SAT scores may predict college grade averages; and parents' heights may predict offsprings' heights.

In a statistical relationship, when one variable is used to predict or "explain" values of a second variable, there is some discrepancy between the actual values of the second variable and the predictions, that is, some imperfection in the prediction. This chapter develops ways of expressing statistical relationships, and then introduces methods for estimating those relationships from a sample of data. Finally the chapter discusses ways to measure and interpret variability around the predicted or explained values. The correlation coefficient is one useful measure for this purpose.

15.1 Functional Relationship

A variable y is said to be a *function* of a variable x if to any value of x there corresponds one and only one value of y, that is, if we know that $x = x_0$, a specified numerical value of x, then we know that $y = y_0$, a specific value of y. We symbolize a functional relationship by writing $y = f(x)$, where f represents the function. For example, if f stands for the function "squaring," then $y = f(x)$ is $y = x^2$. Table 15.1 gives five values of x and y when y is x^2. The variable x is called the *independent variable*; the variable y is called the *dependent variable* because it is considered to depend on x. If x is the height from which a ball is dropped and y is the time the ball takes to fall to the ground, then y is functionally related to x because the law of gravity determines y in terms of x.

TABLE 15.1
Some Values of the Function $y = x^2$

x	-1	0	$\frac{1}{2}$	1	2
y	1	0	$\frac{1}{4}$	1	4

Figure 15.1 is a graph of a functional relationship. To the value x_0 there corresponds a value of y, which is labeled y_0.

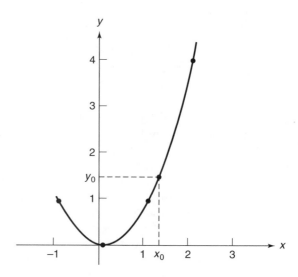

Figure 15.1 *Graph of the functional relationship $y = x^2$.*

A *linear relationship* is a functional relationship of the form $y = a + bx$, where a and b are numbers. For example, $y = 3 + 2x$ and $y = 23.7 + 37.4x$ are linear relationships. Figure 15.2 is the graph of a linear functional relationship; it is a straight line. The number a is called the *intercept* because it is the height at which the line intercepts the y axis, that is, the value of y when $x = 0$. Table 15.2 is a table of four values of x and y when y is linearly related to x. The variable y increases by b as x increases by 1. An example of a linear relationship is that between Centigrade and Fahrenheit temperatures, $F = 32 + 1.8\,C$, where F and 32 denote degrees Fahrenheit and C denotes degrees Centigrade.

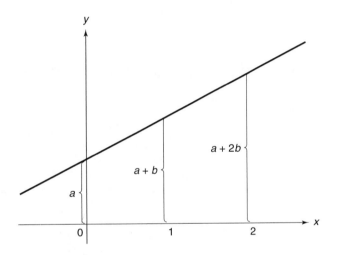

Figure 15.2 *Graph of the linear relationship $y = a + bx$.*

A relationship that is not linear, but is nevertheless smooth, can be *approximated* as a straight line, as in Figure 15.3. Although the real relationship is not linear, the linear description may be adequate in a particular interval of interest.

TABLE 15.2
Some Values of a Linear Function

x	y
0	a
0.5	$a + 0.5b$
1	$a + b$
2	$a + 2b$

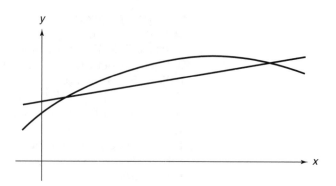

Figure 15.3   *A curve approximated by a straight line.*

15.2   Statistical Relationship

Often the relationship between x and y is not an exact, mathematical relationship, but rather several y values corresponding to a given x value scatter about a value that depends on the x value. For example, although not all persons of the same height have exactly the same weight, their weights bear some relation to that height. On the average, people who are 6 feet tall are heavier than those who are 5 feet tall; the mean weight in the population of 6-footers exceeds the mean weight in the population of 5-footers.

This relationship is modeled statistically as follows: For every value of x there is a corresponding population of y values. The population mean of y for a particular value of x is denoted by $\mu(x)$. As a function of x it is called the *regression function.* For many variables encountered in statistical research, the regression function is a linear function of x, and thus may be written as $\mu(x) = \alpha + \beta x$. The quantities α and β are parameters that define the relationship between x and $\mu(x)$. In conducting a regression analysis, we use a sample of data to estimate the values of these parameters so that we can understand this relationship. The population of y values at a particular x value also has a variance, denoted σ^2; the usual assumption is that the variance is the same for all values of x.

The data required for regression analysis are observations on the pair of variables (x, y). Variable x may be uncontrolled or "naturally occurring" as in the case of observing a sample of n individuals with their heights x and their weights y, or it may be controlled, as in an experiment in which persons are trained as data processors for different lengths of time x, and one measures the accuracy of their work y.

The focus of regression analysis is on making inferences about α, β, and σ^2. Samples of data are used to estimate the magnitude of these parameters and test hypotheses about them. Of particular interest is the hypothesis H_0: $\beta = 0$. If this null hypothesis is true then $\mu(x) = \alpha + 0 \times x = \alpha$, the same number for all values of x. This means that the values of y do not depend on x, that is, there is no statistical relationship between x and y. If H_0 is rejected, then the existence of a statistical relationship between x and y is confirmed.

15.3 Least-Squares Estimates

The data consist of n pairs of numbers $(x_1, y_1), (x_2, y_2), \ldots, (x_n, y_n)$; they can be plotted as a *scatter plot*, such as those described in Section 5.1. Figure 15.4 is a scatter plot of hypothetical data for two variables x and y with a "best fit" line drawn through the points (as in Figure 5.2 also).

   Y values directly on the line (in contrast to observed values y_i) will be represented as $\hat{y}_i$. In regression analysis we seek to determine the equation of that line that gives y_i values as close as possible to the data values y_i. The equation will be of the type $\hat{y}_i = A + Bx_i$, where A is the "y-intercept" of the best fit line and B is the "slope" of the line. The values of A and B will be our estimates of α and β, respectively, permitting us to obtain confidence intervals and to test hypotheses about the parameters of interest.

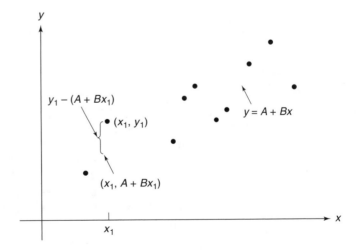

Figure 15.4 *Scatter plot. Deviation from the line $y = A + Bx$ is shown for* (x_1, y_1).

How can we choose a line of best fit from all the lines that can be drawn through the points in a scatter plot? Of course, this question is equivalent to asking how we can choose an appropriate y intercept and slope, A and B. One approach is to assume for the moment that trial values A and B have been chosen and the corresponding estimates $\hat{y}_1, \hat{y}_2, \ldots, \hat{y}_n$ of the values $y_1, y_2, \ldots, y_n$ have been computed. To decide whether the values A and B are good choices, we determine the distance between the $\hat{y}$'s and the y's. The differences $y_i - \hat{y}_i$ are squared and summed, giving the criterion

$$\sum(y_i - \hat{y}_i)^2.$$

Since $\hat{y}_i = A + Bx_i$, this is

$$\sum[y_i - (A + Bx_i)]^2.$$

The *principle of least squares* says to take as the estimates of α and β the values of A and B that minimize this criterion. The quantity $y_i - (A + Bx_i)$ is the vertical distance from the line $y = A + Bx$ to the point (x_i, y_i). (See Figure 15.4.) The criterion is the sum of the squares of these vertical distances. The least-squares estimates, which will be denoted by a and b, are the values of A and B that minimize the criterion. These values are

$$b = \frac{s_{xy}}{s_x^2} \tag{15.1}$$

and

$$a = \bar{y} - b\bar{x}, \tag{15.2}$$

where

$$s_{xy} = \frac{\sum(x_i - \bar{x})(y_i - \bar{y})}{n - 1}$$

is the covariance between x and y (see Section 5.1) and s_x^2 is the variance of x. The statistics a and b are the *least-squares estimates*. The slope b is called the sample *regression coefficient* or *regression weight*.

The *estimated regression line*, the estimate of the function $\mu(x)$, is $\hat{y} = a + bx$. It is interesting to write this using $\bar{y}$ instead of a; that is, $\hat{y} = a + bx = (\bar{y} - b\bar{x}) + bx = \bar{y} + b(x - \bar{x})$. The predicted value of y_i is thus

$$\hat{y}_i = \bar{y} + b(x_i - \bar{x}).$$

The vertical distance from the data point (x_i, y_i) to the line $\hat{y} = \bar{y} + b(x - \bar{x})$ is $y_i - [\bar{y} + b(x_i - \bar{x})]$. (See Figure 15.4.) The estimate of the variance σ^2 depends on these distances and is[1]

$$s_{y \cdot x}^2 = \frac{\sum (y_i - \hat{y}_i)^2}{n - 2}.$$

The numerator of $s_{y \cdot x}^2$ is the sum of squared distances of the observations from the regression line. Substituting for $\hat{y}_i$ gives us a relatively simple way to obtain $s_{y \cdot x}^2$ from s_x^2, s_y^2, and s_{xy}. That is,

$$
\begin{aligned}
\sum (y_i - \hat{y}_i)^2 &= \sum [y_i - \bar{y} - b(x_i - \bar{x})]^2 \\
&= \sum [(y_i - \bar{y})^2 - 2b(y_i - \bar{y})(x_i - \bar{x}) + b^2(x_i - \bar{x})^2] \\
&= (n - 1)(s_y^2 - 2bs_{xy} + b^2 s_x^2) \\
&= (n - 1)\left[s_y^2 - 2\frac{s_{xy}^2}{s_x^2} + \left(\frac{s_{xy}}{s_x^2}\right)^2 s_x^2 \right] \\
&= (n - 1)\left(s_y^2 - \frac{s_{xy}^2}{s_x^2} \right),
\end{aligned}
$$

where $(n - 1)s_y^2 = \sum (y_i - \bar{y})^2$; we have used the definition of b in (15.1). The estimate of σ^2 can be written

$$s_{y \cdot x}^2 = \left(s_y^2 - \frac{s_{xy}^2}{s_x^2} \right) \frac{n - 1}{n - 2}. \tag{15.3}$$

The divisor $n - 2$ is the number of degrees of freedom, obtained as the number of observations minus the number of coefficients estimated. The standard deviation $s_{y \cdot x}$ is an estimate of σ; it is called the *standard error of estimate*. The computations are illustrated in the following example.

EXAMPLE   One of the questions about peaceful uses of atomic energy is the possibility that radioactive contamination poses health hazards. Since World War II, plutonium has been produced at the Hanford, Washington, facility of the Atomic Energy Commission. Over the years, appreciable quantities of radioactive wastes have leaked from their open-pit storage areas into the nearby Columbia River, which flows through parts of Oregon to the Pacific. As part of the assessment of the consequences of this

---

[1]The conventional symbol $s_{y \cdot x}^2$, which is the estimate of variance, should not be confused with s_{xy}, which is the covariance between x and y.

contamination on human health, investigators calculated, for each of the nine Oregon counties having frontage on either the Columbia River or the Pacific Ocean, an "index of exposure." This index of exposure was based on several factors, including distance from Hanford and average distance of the population from water frontage. The cancer mortality rate, cancer mortality per 100,000 person-years (1959–1964), was also determined for each of these nine counties. Table 15.3 shows the index of exposure and the cancer mortality rate for the nine counties, given as mortality per 100,000 person years.

TABLE 15.3

Radioactive Contamination and Cancer Mortality

County	Index of exposure	Cancer mortality per 100,000 person-years
Clatsop	8.34	210.3
Columbia	6.41	177.9
Gilliam	3.41	129.9
Hood River	3.83	162.3
Morrow	2.57	130.1
Portland	11.64	207.5
Sherman	1.25	113.5
Umatilla	2.49	147.1
Wasco	1.62	137.5

source: Fadeley (1965).

Figure 15.5 is a scatter plot of the data in Table 15.3. Each point corresponds to one county. The straight line in the figure is the estimated regression line of y (cancer mortality) on x (index of exposure). Summary statistics needed for the computation are given in Table 15.4.

TABLE 15.4

Computations for Regression Example

$$n = 9$$

$\sum x_i = 41.56$	$\sum y_i = 1,416.1$
$\bar{x} = 4.618$	$\bar{y} = 157.34$
$\sum x_i^2 = 289.4222$	$\sum y_i^2 = 232,498.97$

$$\sum x_i y_i = 7439.37$$

$$\sum(x_i - \bar{x})^2 = 97.5074 \quad \sum(x_i - \bar{x})(y_i - \bar{y}) = 900.13 \quad \sum(y_i - \bar{y})^2 = 9683.50$$

source: Table 15.3.

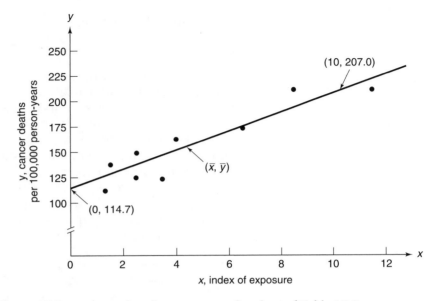

Figure 15.5   *Scatter plot of cancer mortality data of Table 15.3.*

The numerator of the sample covariance is

$$(n - 1)s_{xy} = \sum(x_i - \bar{x})(y_i - \bar{y})$$

$$= \sum x_i y_i - \frac{\left(\sum x_i\right)\left(\sum y_i\right)}{n}$$

$$= 7439.37 - \frac{41.56 \times 1416.1}{9}$$

$$= 900.13.$$

The numerator of the variance of x is

$$(n - 1)s_x^2 = \sum(x_i - \bar{x})^2$$

$$= \sum x_i^2 - \frac{\left(\sum x_i\right)^2}{n}$$

$$= 289.4222 - \frac{(41.56)^2}{9} = 97.5074.$$

The sample regression coefficient is

$$b = \frac{s_{xy}}{s_x^2} = \frac{(n - 1)s_{xy}}{(n - 1)s_x^2} = \frac{900.13}{97.507} = 9.23.$$

The estimate of the intercept is

$$a = \bar{y} - b\bar{x} = 157.34 - 9.23 \times 4.62 = 114.7.$$

The estimated regression line is

$$\hat{y} = 114.7 + 9.24x.$$

One concludes that a zero exposure rate ($x = 0$) would suggest a death rate of about 115 and that, as the exposure index increases by one unit, the death rate increases by about 9.23 per 100,000.

Two points are required to draw the line in Figure 15.5. One of these can be taken to be $(0, a)$, which is $(0, 114.7)$. Another point may be obtained by substituting into the equation of the estimated regression line a large value of x and computing the corresponding value of $\hat{y}$. For example, $x = 10$ gives $\hat{y} = 114.7 + 9.23 \times 10 = 207.0$. The corresponding point, $(10, 207.0)$, is shown in Figure 15.5.

The residual sum of squares, which is a sum of squares of deviations, is computed from

$$\sum (y_i - \hat{y}_i)^2 = (n - 1)s_y^2 - \frac{[(n - 1)s_{xy}]^2}{(n - 1)s_x^2}$$

$$= 9683.50 - \frac{(900.13)^2}{97.507} = 1374.06.$$

The estimate of the variance is

$$s_{y \cdot x}^2 = \frac{1374.06}{9 - 2} = \frac{1374.06}{7} = 196.294;$$

the corresponding standard deviation is $s_{y \cdot x} = \sqrt{196.294} = 14.01$. This standard deviation is used in statistical inference.

The example above shows that the elements of simple regression analysis—the least-squares estimates of α and β, the sum of squared deviations from the regression line, and the variance of the observations about the regression line ($s_{y \cdot x}^2$)—can be obtained from a few summary statistics. The basic ingredients are the sample means, $\bar{x}$ and $\bar{y}$, the sample variances s_x^2 and s_y^2 as defined in Section 4.3, and the sample covariance s_{xy}, defined originally in Section 5.1. These are also the ingredients for inferential procedures in regression analysis, discussed in the section that follows.

15.4   Statistical Inference for β

Suppose we drew repeated samples with the same set of x's; these samples would be of the form

$$(x_1, y_1), \ldots, (x_n, y_n),$$

$$(x_1, y_1'), \ldots, (x_n, y_n'),$$

$$(x_1, y_1''), \ldots, (x_n, y_n''),$$

and so forth. For example, we could think of $x_1, \ldots, x_n$ as being fixed numbers of hours of training in typing; each sample arises from n different people who give rise to n different words-per-minute scores.

For each sample we could compute the value of b, the estimate of β. This procedure would generate the sampling distribution of b. The mean of this sampling distribution is β; that is, b is an unbiased estimate of β. The standard deviation of this distribution, the *standard error* of b, denoted by σ_b, is

$$\sigma_b = \frac{\sigma}{\sqrt{\sum(x_i - \bar{x})^2}}.$$

The more spread out the x's are, the smaller is σ_b. Since $s_{y \cdot x}^2$ estimates σ^2, the estimate of the standard error of b is

$$s_b = \frac{s_{y \cdot x}}{\sqrt{\sum(x_i - \bar{x})^2}}. \tag{15.4}$$

If the y's are normally distributed, the sampling distribution of b is also normally distributed. Since the mean of the sampling distribution is β and the standard deviation is σ_b, the statistic $(b - \beta)/\sigma_b$ is normally distributed with mean 0 and variance 1.

These results provide a basis for testing hypotheses and constructing confidence intervals for β. The general form of the hypothesis is H_0: $\beta = \beta_0$, where β_0 is a specified constant. Most research using regression analysis is asking whether there is *any* relationship of y with x in the population. Thus the most common application is when $\beta_0 = 0$ and the null hypothesis is H_0: $\beta = 0$ and the alternative is H_1: $\beta \neq 0$. If H_0 is rejected, then variables x and y are statistically related. If H_0 is accepted, we conclude that there is no statistical relationship of x with y. One-sided alternatives may also be tested. For example, if it is hypothesized that there is a positive association of x with y, then the hypotheses are H_0: $\beta \leq 0$ and H_1: $\beta > 0$.

Like other tests of significance presented in this book, the test statistic depends on whether σ is known or σ is unknown and must be estimated from the data. If σ is known, the test statistic is $z = (b - \beta_0)/\sigma_b$. Significance points are taken from the table of the standard normal distribution. For example, H_0 would be rejected making a two-tailed test at the 5% level if the absolute value of z is greater than 1.96.[2]

---

[2] The procedures here are valid if n is large, even if the original y variable is not normally distributed.

It is much more common that σ is not known and must be estimated. If $s_{y \cdot x}$ is used as the estimate of σ, then the estimate of σ_b is s_b given by (15.4). With this estimate of σ_b the test statistic is

$$t = \frac{b - \beta_0}{s_b}. \tag{15.5}$$

Significance points are obtained from the t-distribution with $n - 2$ degrees of freedom (the degrees of freedom for $s_{y \cdot x}$). H_0 is rejected against the two-sided alternative at significance level[3] α if $|t| \geq t_{n-2}(\alpha/2)$.

In the example $s_b = 14.01/\sqrt{97.5074} = 1.419$. To test whether there is any association of x with y, that is $H_0: \beta = 0$, the test statistic is $t = b/s_b = 9.23/1.419 = 6.50$. If the hypothesis is tested at the 1% level, then the significance points for the two-tailed test are $\pm t_7(.005) = \pm 3.499$. The value of the test statistic exceeds this and we conclude that the two variables are related (positively) in the population of counties represented by this sample.

When σ^2 is unknown, a confidence region for β with confidence coefficient $1 - \alpha$ is

$$b - s_b t_{n-2}(\alpha/2) < \beta < b + s_b t_{n-2}(\alpha/2). \tag{15.6}$$

In the example, the .99 interval has endpoints $9.23 \pm 1.419 \times 3.499 = 9.23 \pm 4.97 = (4.26, 14.20)$. Thus, one has confidence that β is at least 4.26 units but no more than 14.20 units, where a unit is the increase in mortality associated with a one-degree increase in the exposure index.

EXAMPLE

Figure 5.3 consists of data and a scatter plot of two measures for a sample 10 wooden beams, $x =$ moisture content, and $y =$ beam strength. The scatter plot indicates a negative association of the two variables. Either a computer program or hand computations will produce the following basic summary statistics:

Means	$\bar{x} = 9.880$	$\bar{y} = 11.88$
Sums of Deviations	$\sum(x_i - \bar{x})^2 = 7.856$	$\sum(y_i - \bar{y})^2 = 5.310$
Variances	$s_x^2 = 0.8729$	$s_y^2 = 0.5900$
Standard Deviations	$s_x = 0.9343$	$s_y = 0.7681$
Sum of Cross-Products	$\sum(x_i - \bar{x})(y_i - \bar{y}) = -4.904$	
Covariance	$s_{xy} = -0.5449$	

Using (15.1) the least-squares estimate of the regression coefficient is $b = -0.5449/0.8729 = -0.6242$; the negative relationship is confirmed for the sample.

---

[3] The level of significance should not be confused with the intercept α in $\mu(x) = \alpha + \beta x$.

The test of significance and confidence interval both require the estimated standard error given by (15.4); the standard error in turn requires the variance of y about the regression line, given by (15.3). The variance is

$$s_{y \cdot x}^2 = \left(0.5900 - \frac{0.5449^2}{0.8729} \right) \left(\frac{10 - 1}{10 - 2} \right) = 0.2811,$$

and the standard deviation of y about the regression line is $s_{y \cdot x} = 0.5302$. The standard error is

$$s_b = 0.5302 / \sqrt{7.856} = 0.1892.$$

We shall test the null hypothesis that there is no regression of beam strength on moisture content. The null and alternative hypotheses are H_0: $\beta = 0$ and H_1: $\beta \neq 0$, respectively. The test statistic (15.5) is $t = -0.6242/0.1892 = -3.299$. This value is compared to percentage points of the t-distribution with $n - 2 = 8$ degrees of freedom. The P interval for the two-tailed test is $.01 < P < .02$. (The test statistic is between 2.896 and 3.355, the values that separate the most extreme 1% and 0.5% of the distribution *in each tail* from the central portion.) Thus there is solid evidence that greater moisture content is associated with reduced beam strength.

A 90% confidence interval for β may be constructed using (15.6). The value required from the t-distribution is $t_8(.05) = 1.860$. The interval is from $-0.6242 - (1.860 \times 0.1892)$ to $-0.6242 + (1.860 \times 0.1892)$ or $(-0.98, -0.27)$. This is the range of plausible values for β; we can state with 90% confidence that every one-unit increment in moisture content is associated with at least a 0.27-unit decrement in beam strength and perhaps as much as a 0.98-unit decrement.

15.5   The Correlation Coefficient: A Measure of Linear Relationship

In Section 5.1 the *correlation coefficient*

$$r = \frac{s_{xy}}{s_x s_y}$$

is discussed as a measure of association between two quantitative variables. It is a statistic describing one aspect of n pairs of measurements on two variables x and y. Although the measure is symmetric in the two variables, it can be interpreted in terms of how well one variable y can be predicted from the other variable x.

In Chapter 14 we define a measure of association for categorical variables that has the form

$$\frac{\left(\begin{array}{c}\text{Number of errors}\\ \text{in prediction of } y\\ \text{without using } x\end{array}\right) - \left(\begin{array}{c}\text{Number of errors}\\ \text{in prediction of } y\\ \text{using } x\end{array}\right)}{\left(\begin{array}{c}\text{Number of errors in prediction}\\ \text{of } y \text{ without using } x\end{array}\right)}.$$

With quantitative variables, the notion that replaces "number of errors" is "sum of squared deviations" of the predicted values of y from the actual values. These deviations are "errors" in the sense that they tell us how much error there would be if we tried to predict the y value from the x's. Thus a measure of association for quantitative variables is defined as

$$\frac{\left(\begin{array}{c}\text{Sum of squares of}\\ \text{deviations in predicting}\\ y \text{ without using } x\end{array}\right) - \left(\begin{array}{c}\text{Sum of squares of}\\ \text{deviations in predicting}\\ y \text{ using } x\end{array}\right)}{\left(\begin{array}{c}\text{Sum of squares of deviations in}\\ \text{predicting } y \text{ without using } x\end{array}\right)}.$$

Without the use of x_i the best prediction for an observed value y_i is the mean $\bar{y}$ in the sense that the sum of squared deviations from $\bar{y}$ is minimal, that is, there is no other value v for which $\sum (y_i - v)^2$ is smaller. The sum of squared deviations without using information from the x variable is $\sum (y_i - \bar{y})^2$.

With the use of x_i the predicted value of y_i is $\hat{y}_i = a + bx_i = \bar{y} + b(x_i - \bar{x})$, and the sum of squared deviations is $\sum (y_i - \hat{y}_i)^2$. The measure of association is

$$\frac{\left[\sum (y_i - \bar{y})^2\right] - \left[\sum (y_i - \hat{y}_i)^2\right]}{\sum (y_i - \hat{y})^2}. \tag{15.7}$$

In terms of the sample variances and covariance, (15.7) is

$$\frac{\sum (y_i - \bar{y})^2 - \sum (y_i - \hat{y}_i)^2}{\sum (y_i - \bar{y})^2} = \frac{(n-1)s_y^2 - (n-1)(s_y^2 - s_{xy}^2/s_x^2)}{(n-1)s_y^2}$$

$$= 1 - 1 + \frac{s_{xy}^2}{s_y^2 s_x^2} = \frac{s_{xy}^2}{s_x^2 s_y^2}.$$

In Section 5.1 the sample correlation is defined as $r = s_{xy}/s_x s_y$. Thus the measure of association is r^2, the square of the correlation between x and y.

The correlation coefficient is itself a useful measure of association. In addition we see that the *square* of correlation can be interpreted as the proportional reduction in the sum of squared deviations achieved by using the best linear function of x to predict y. It is commonly referred to as the *proportion of variation in y explained by (or attributable to) x*.

Since the absolute value of r cannot exceed 1, the squared correlation has possible values from 0 to 1. If $r = 0$, then x is of no help in predicting y. At the other extreme, if $r = 1$ (and $r^2 = 1$), then x predicts y exactly. This can only occur if each point (x_i, y_i) falls exactly on the regression line; all of the variation in y can be explained by variability in x. In between these extremes, a weak correlation (e.g., in the range 0 to 0.3) is accompanied by a small proportion of explained variation (0 to 0.09), while a strong correlation (e.g., between 0.6 and 1.0) is accompanied by a substantially larger proportion of explained variation (0.36 to 1.0).

The correlation coefficient is positive or negative depending on whether the regression line has a positive or negative slope, that is, whether it runs from lower left to upper right or from upper left to lower right in the scatter plot, respectively. Thus both r and r^2 are useful measures of association for regression analysis. The correlation tells the *direction* of association and its square tells the *extent* to which y is predictable from x. For the cancer mortality data, the correlation coefficient is $r = 900.13/\sqrt{97.507 \times 9683.52} = 0.926$. This value is very high, and the proportion of variation in mortality attributable to radioactive exposure is also high, $0.926^2 = 0.857$ (that is, 85.7%).

The Bivariate Normal Distribution and Test of Significance for a Correlation

An example of a two-way frequency table of pairs of quantitative variables is Table 15.7 (page 588), which gives the frequencies of average heights of pairs of parents and heights of grown children. For example, there were 48 children who had a height of 69.2 inches (within 0.5 inch) and whose parents had an average height of 68.5 inches. A three-dimensional histogram can be constructed by raising a column over each square with height equal to the frequency. For instance, over the square from 68 to 69 in one direction and 68.7 to 69.7 in the other, erect a height of 48. The frequency table can be described by such a collection of columns. A particular smooth surface fitted over this is a bivariate normal distribution. This is a theoretical relative frequency surface. The normal distribution of one variable x depends on its mean μ_x and its standard deviation σ_x, and the normal distribution of the other variable y depends on its mean μ_y and its standard deviation σ_y. The *bivariate* normal distribution depends on these four parameters and the population correlation coefficient ρ, which is the population analog of r.

The mode of the normal distribution (that is, the maximum value of the relative frequency function) is the point (μ_x, μ_y). The distribution of y values for a fixed value of x is a normal distribution with a mean that bears a linear relation to x and a variance not depending on x. Similarly, the distribution of x values for a fixed value of y is normal with a mean linearly related to y and a variance independent of y. Another remarkable property of the bivariate normal distribution is that x and y are independent if and only if $\rho = 0$.

These ideas lead to a test of significance for the correlation coefficient. Specifically, we may test the hypothesis that the population correlation is null, that is, H_0: $\rho = 0$ against the alternative hypothesis H_1: $\rho \neq 0$. If the null hypothesis is rejected, the conclusion is that there is a nonzero correlation between x and y in the population. If H_0 is accepted, the conclusion is that there is no correlation in the population. Further, if x and y have a bivariate normal distribution, then a population correlation of zero means that the two variables are statistically independent. Of course, one-sided alternatives may be tested as well if, for example, it is hypothesized that two variables have a positive correlation in the population (H_0: $\rho \leq 0$; H_1: $\rho > 0$) or that two variables have a negative correlation in the population (H_0: $\rho \geq 0$; H_1: $\rho < 0$).

The test statistic for the correlation is

$$t = r\sqrt{\frac{n-2}{1-r^2}}. \tag{15.8}$$

It can be seen that if the sample correlation is positive, t will be positive, and if r is negative then t will be negative. Similarly a large sample correlation (with either sign) will produce a large test statistic. The t-statistic is compared to percentage points from the t-distribution with $n - 2$ degrees of freedom. If r is large enough, given the sample size, then t will exceed its respective significance point and H_0 is rejected.

The correlation coefficient for the data of Table 15.3 is $r = 0.926$ based on 9 observations. The research was conducted initially because scientists hypothesized that exposure to radioactive material is *positively* related to the incidence of cancer. Test of a one-sided alternative is consistent with this reasoning; the hypotheses are H_0: $\rho \leq 0$ and H_1: $\rho > 0$. The test statistic is

$$t = 0.926\sqrt{\frac{9-2}{1-0.926^2}} = 6.49.$$

The .01 upper-tail significance point from the t-distribution with 7 degrees of freedom is $t_7(.01) = 2.998$. Thus H_0 is rejected and we conclude that there is a positive correlation between exposure to radioactive material and cancer mortality in the population of counties represented by this sample.

It may be noted that, except for rounding error, the value of the test statistic is identical to that for testing that the regression coefficient is zero, discussed in Section 15.4. The statistic for testing that the correlation is zero [(15.8)] is algebraically equivalent to the statistic for testing that the regression of y on x is null [(15.5)] and for testing that the regression of x and y is null as well.

Summary

Regression analysis is a set of statistical techniques for analyzing the relationship between two numerical variables. One variable is viewed as the outcome or *dependent variable* (usually represented by y) and the other is the antecedent or *independent variable* (usually represented by x). The purpose of regression analysis is to understand the direction and extent to which y values can be predicted by corresponding values of x.

The *regression line* is a line of best fit through the swarm of points (x_i, y_i) in the scatter plot. The slope of the regression line indicates the direction of association of x and y (positive or negative). The degree of association is indicated by $r^2 = s_{xy}^2/(s_x^2 s_y^2)$. The coefficients of the line, including the slope, are usually estimated by the *method of least-squares* based on minimizing the sum of squared differences between the observed y values and those predicted from x.

The correlation coefficient is a measure of linear relationship between x and y. The square of the correlation is the proportion of variability in variable y that is explained by a linear function of variable x. The most common test of significance in regression analysis is a test of the null hypothesis that the population correlation is 0, or equivalently, that the slope of the regression line is 0.

Exercises

15.1 Make a scatter plot of the data in Table 15.5. Use the data to find the regression line of variable y on variable x as follows:

(a) Compute the estimated slope and y-intercept for the regression line;

(b) Use any two real or hypothetical x values to obtain predicted y values, and place these two points on the scatter plot;

(c) Connect the two points with a straight edge.

TABLE 15.5
Data Set

x	1	2	5	8	9
y	5	2	3	4	1

15.2   Make a scatter plot of the data in Table 15.6. Use the data to find the regression line of variable y on variable x as follows:

(a) Compute the estimated slope and y-intercept for the regression line;

(b) Use the two real or hypothetical x values to obtain predicted y values, and place these two points on the scatter plot;

(c) Connect the two points with a straight edge.

TABLE 15.6
Data Set

x	1	2	5	8	9
y	1	4	3	2	5

15.3   Figure 15.6 gives two population regression lines showing the relationship between hours of study for the final examination and average grade on the examination. Line *CD* is for an art course; line *EF* is for a psychology course. What do these lines imply about the relationships between the variables in the two cases?

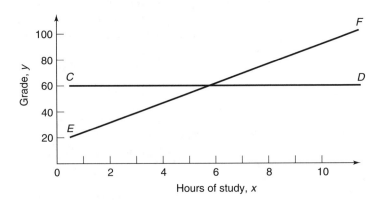

Figure 15.6   *Grades and hours of study in two courses.*

15.4   In an investigation of hours of study and grade in two large lecture courses in history, the data and sample regression lines shown in Figure 15.7 were obtained. The observed sample regression lines are very nearly the same in the two courses. Relatively how reliable are conclusions about the population regression line in the two cases? Compare the usefulness of x in predicting y in the two cases.

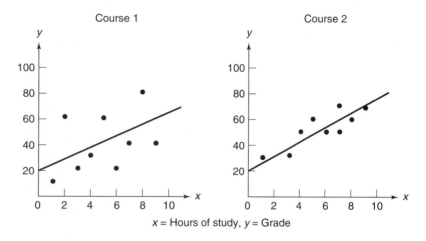

Figure 15.7   *Grades and hours of study in two courses.*

15.5   In the study from which "regression" takes its name, Galton obtained the heights of 205 couples together with the heights of their grown children. The frequencies of heights of children and their parents are given in Table 15.7. (These data are for both males and females. It was desirable to adjust for the sex difference in height. Since the mean height for males was 8% greater than that for females, all female heights were multiplied by 1.08 to make them equivalent, on the average, to male heights.)

(a) Give the relative frequencies of average heights of the 205 couples (that is, the "marginal distribution").

(b) Give the relative frequencies of heights of the 928 children.

(c) Give the distribution of the heights of the children, given that the average height of the two parents was 68.5 inches (that is, the "conditional distribution").

(d) Plot the medians of the average height of the two parents against the height of the children.

(e) Plot the medians of the height of the children against the average height of the two parents.

(continued ...)

TABLE 15.7
Number of Grown Children of Various Heights Born to 205 Couples of Various Heights (Units: Inches)

Heights of grown children	Below 64.5	64.5	65.5	66.5	67.5	68.5	69.5	70.5	71.5	72.5	Above 72.5	Totals	Medians of rows
Below 62.2	1	1	1			1		1				5	
62.2		1		3	3							7	
63.2	2	4	9	3	5	7	1	1				32	66.3
64.2	4	4	5	5	14	11	16					59	67.8
65.2	1	1	7	2	15	16	4	1	1			48	67.9
66.2	2	5	11	17	36	25	17	1	3			117	67.7
67.2	2	5	11	17	38	31	27	3	4			138	67.9
68.2	1		7	14	28	34	20	12	3	1		120	68.3
69.2	1	2	7	13	38	48	33	18	5	2		167	68.5
70.2			5	4	19	21	25	14	10	1		99	69.0
71.2			2		11	18	20	7	4	2		64	69.0
72.2			1		4	4	11	4	9	7	1	41	70.0
73.2						3	4	3	2	2	3	17	
Above 73.2							5	3	2	4		14	
Total number of:													
children	14	23	66	78	211	219	183	68	43	19	4	928	
couples	1	5	12	20	33	49	41	22	11	6	5	205	
Medians of columns		65.8	66.7	67.2	67.6	68.2	68.9	69.5	69.9	72.2			

Heading "Average height of the two parents" spans the columns Below 64.5 through Above 72.5.

source: Galton (1885).

In (d) and (e), draw on the graphs the 45° line $y = x$, the line representing equality of height of children and the average height of their two parents.

15.6 If you have not already done so, complete Exercise 5.8 in Chapter 5. Once you have obtained the results of that Exercise, compute the y-intercept and slope of the regression line of noise level on average speed. What is the predicted value of noise level on a section of road whose average speed is 35 mph? 40 mph? 45 mph?

15.7 Complete Exercise 5.9 in Chapter 5. Once you have obtained the results of that Exercise, compute the y-intercept and slope of the regression line of noise level on the square of average speed. From these results, what is the predicted value of noise level on a section of road whose average speed is 35 mph? 40 mph? 45 mph?

15.8 Exercise 5.10 in Chapter 5 gives data on competitiveness and productivity among interviewers in two sections of a state employment agency. Compute the slope of the regression line of productivity on competitiveness in each section. How do these two slopes compare to one another? How would you explain any differences between them in terms of the behavior of the employees in the two sections?

15.9   Table 15.8 gives the collection size and number of full-time staff in a sample of 22 large college and university libraries. Use these data to examine the number of staff members who are employed to manage collections of different sizes, as follows:

(a) Present the data in the form of a scatter plot in which the independent variable (x) is size of the collection and the dependent variable (y) is the number of staff members.

(b) Use the summary statistics in Table 15.8 to obtain the least-squares estimates of the intercept (α) and slope (β) of the regression equation. Choose two x values as reference points, obtain the "predicted" values of y for those x's, and draw the regression line connecting those points.

15.10   (continuation) Test the null hypothesis that the slope of the regression line in the preceding exercise is *no greater than zero* in a hypothetical population of large university libraries. In your answer

(a) State the null and alternative hypotheses in statistical form;

(b) Give the formula and show your computations for the standard error of estimate ($s_{y \cdot x}$), the estimated standard error of the regression coefficient (s_b), and the test statistic (t);

(c) Give a P value interval for the test statistic. Would the null hypothesis be accepted or rejected at the 1% level of significance?

15.11   (continuation) Administrators considering a major upgrade in a library's holdings must consider how many additional staff members are required to manage the larger collection. To answer this question, report (once more) the estimate of β, that is, the number of additional staff members associated with a 100,000-volume increment in holdings. Also, compute a 95% confidence interval for β. Write a statement giving the most likely number of additional staff persons required to manage the larger collection and the minimum and maximum as indicated by the confidence interval.

15.12   (continuation) Compute the sum of squares of the observations about the regression line, $\sum(y_i - \hat{y}_i)^2$. From this value and $\sum(y_i - \bar{y})^2$, compute the percentage of variation in y (number of staff members) attributable to x (size of the collection) using (15.7). Use these results plus any results from Exercise 15.10 or 15.11 to explain whether the size of the library's collection is a *strong* predictor of size of staff.

15.13   (continuation) Compute the correlation coefficient between size of the library collection and number of staff members using the definition of r given in Section 5.1. Verify that the percentage of explained variation in the preceding exercise is equal to the square of r.

TABLE 15.8
Collection Size and Number of Full-Time Staff in College and University Libraries (December 1988)

Institution	Number of volumes (100,000's)	Total staff
University of Illinois	73.8	536
University of Michigan	61.3	677
UCLA	58.1	711
University of Chicago	48.6	349
University of Minnesota	46.5	484
Ohio State University	42.5	454
Princeton University	40.7	375
University of Washington	37.2	447
University of Arizona	36.2	328
University of Pennsylvania	35.0	368
Stanford University	31.9	407
University of Virginia	30.0	343
University of Florida	27.0	405
University of Rochester	26.0	231
Northwestern University	25.7	288
University of Missouri	24.1	235
University of Hawaii	23.6	247
Johns Hopkins University	23.3	308
University of Delaware	23.0	220
Wayne State University	22.3	266
Syracuse University	22.2	271
MIT	21.8	265

source: Snyder & Hoffman (1993), p. 431.

$$n = 22$$

$$\sum x_i = 780.80 \qquad \sum y_i = 8215.0$$
$$\sum x_i^2 = 32{,}144.10 \qquad \sum y_i^2 = 3{,}454{,}309.0$$
$$\sum (x_i - \bar{x})^2 = 4432.80 \qquad \sum (y_i - \bar{y})^2 = 386{,}753.3$$
$$\sum x_i y_i = 326{,}842.90$$
$$\sum (x_i - \bar{x})(y_i - \bar{y}) = 35{,}285.08$$

15.14   Table 11.13 (Exercise 11.9) lists the best times of 10 British sprinters running 100-meter and 200-meter distances.

(a) Make a scatter plot of these values with 100-meter times as the x variable and 200-meter times as the y variable.

(b) Compute the correlation coefficient between the two sets of running times.

(continued ...)

(c) Assuming that this sample is representative of a large population of sprinters, test the hypothesis at the 5% level that the population correlation is greater than zero. In your answer give

- the null and alternative hypotheses in statistical form,
- the computations and final value of the test statistic,
- the significance point from the appropriate t distribution,
- the conclusion that H_0 is accepted or rejected,
- a short verbal summary stating whether these two measures are correlated in the population or not; if so, was the degree of association in the sample weak, moderate, or strong?

15.15 Figure 5.3 shows a scatter plot and measures taken on a sample of 10 wooden beams; variable y is an index of the strength of the beam. The published report [Draper & Stoneman (1966), reprinted in Hand *et al.* (1994), p. 200] also provided a measure of the specific gravity of each beam. For the 10 beams in the order listed in Figure 5.3, the values are 0.499, 0.558, 0.604, 0.441, 0.550, 0.528, 0.418, 0.480, 0.406, and 0.467. Represent specific gravity as variable x and beam strength as variable y; summary statistics are given in Table 15.9.

TABLE 15.9
Summary Statistics for Sample of Wooden Beams

$n = 10$	
$\sum x_i = 4.951$	$\sum y_i = 118.76$
$\sum (x_i - \bar{x})^2 = 0.0378$	$\sum (y_i - \bar{y})^2 = 5.310$
$\sum (x_i - \bar{x})(y_i - \bar{y}) = 0.4089$	

Use these data to

(a) draw a scatter plot showing the relationship of beam strength and specific gravity,

(b) compute the basic summary statistics, $\bar{x}$, $\bar{y}$, s_x^2, s_y^2, and s_{xy},

(c) compute the least squares estimate of the intercept and slope of the regression line of strength on specific gravity,

(d) compute the sample correlation between beam strength and specific gravity.

15.16 (continuation) Use the data of the previous exercise; test at the 5% level the null hypothesis that the slope of the regression line in the population of beams from which the sample was drawn is equal to zero. In your answer give

(a) the null and alternative hypotheses in statistical form,

(b) the computations and final result for the standard error of the slope and the test statistic,

(c) the 5% significance points from the appropriate t-distribution,

(d) the conclusion to accept or reject H_0, and an accompanying sentence stating whether you found no association, a positive association, or a negative association of specific gravity with beam strength.

15.17   (continuation) Use the data of the two preceding exercises to construct a 95% confidence interval for the slope of the regression line.

15.18   Table 15.10 gives the IQs of monozygotic (identical) twins raised in separate homes.

(a) Compute the correlation r between $x =$ IQ of first born and $y =$ IQ of second born.

(b) Compute r^2. The percentage $100r^2$ is called the *heritability coefficient*. Its complement $100(1 - r^2)$ is taken as a measure of the contribution of environmental effects on the development of intelligence.

15.19   Table 2.22 gives agility scores and written test scores of 28 firefighter applicants. If you completed Exercise 5.13, you have already drawn scatter plots that show the associations of "Scaled Agility Score" with "Written Test Score" for the 14 male and 14 female applicants. Use the data to

(a) compute the coefficient of correlation between these two variables for the 14 male applicants,

(b) compute the percentage of variation shared by these two variables for male applicants, that is, $100r^2$,

(c) test the hypothesis that the correlation between the two variables is zero in the (hypothetical) population of possible male applicants. Express your results as a P value interval for the test statistic.

15.20   (continuation) Repeat the preceding exercise for the 14 female applicants. Compare the results obtained for male and female applicants in terms of

(a) the direction of association,

(b) the magnitude of association, that is, the percentage of shared variation,

(c) the P value for the test of significance. Explain any noteworthy discrepancies that you find.

15.21   Table 5.14 gives data on available seat miles and the load factor for a sample of 10 airlines during one quarter of a year of operation. Use these data to compute the regression equation for predicting load factor from available seat miles. In addition to the estimated slope and y intercept,

TABLE 15.10

IQs of Monozygotic Twins Raised Apart (n = 34 pairs of twins)

Case	First born	Second born
Sm 1	22	12
Sm 2	36	34
Sm 4	13	10
Sm 5	30	25
Sm 6	32	28
Sm 7	26	17
Sm 8	20	24
Sm 10	30	26
Sm 11	29	35
Sm 12	26	20
Sm 13	28	22
Sm 14	21	27
Sm 15	13	4
Sf 5	32	33
Sf 7	30	34
Sf 9	27	24
Sf 10	32	18
Sf 11	27	28
Sf 12	22	23
Sf 13	15	9
Sf 14	24	33
Sf 15	6	10
Sf 16	23	21
Sf 17	38	27
Sf 18	33	26
Sf 19	16	28
Sf 20	27	25
Sf 22	4	2
Sf 23	19	9
Sf 24	41	40
Sf 25	40	38
Sf 26	12	9
Sf 27	13	22
Sf 29	29	30

source: Reprinted by permission of Oxford University Press from Shields, J. (1962). *Monozygotic twins*. London: Oxford.

compute the standard error of the slope and a 90% confidence interval for the slope of the regression line.

15.22 Table 5.18 (Exercise 5.25) gives correlations among several physiological measures for a sample of mothers, their older daughters ("adolescent"), and their younger daughters ("child"). Test the hypothesis that the following correlations are greater than zero in the population; the hypotheses are H_0: $\rho \leq 0$ and H_1: $\rho > 0$. In your answer give the correlation coefficient, the value of the test statistic, and a P value interval from the appropriate t-distribution. (If the exact distribution you need is not in the table, use the tabled distribution with the next lower number of degrees of freedom.)

(a) The correlation between the systolic blood pressure (SBP) of mothers and the systolic blood pressure of their older daughters;

(b) The correlation between the systolic blood pressure of mothers and the systolic blood pressure of their younger daughters.

15.23   (continuation) Repeat the preceding exercise for the correlation between

(a) the diastolic blood pressure (DBP) of mothers and the diastolic blood pressure of their older daughters,

(b) the diastolic blood pressure of mothers and the diastolic blood pressure of their younger daughters,

(c) the heart rate (HR) of mothers and the heart rate of their older daughters,

(d) the heart rate (HR) of mothers and the heart rate of their younger daughters,

(e) the body mass index (BMI) of mothers and the body mass index of their older daughters,

(f) the body mass index of mothers and the body mass index of their younger daughters.

15.24   (continuation) Construct a table that summarizes the results of the preceding two exercises in an easy-to-read, well-labeled format. For each result include the correlation coefficient and the P value interval. *Circle* those correlation coefficients for which the null hypothesis would be rejected at the 1% level of significance. Write a brief report that summarizes the general findings and any patterns revealed by this analysis of inter-generational correlations.

15.25   Suppose that a manufacturing process were performed at each of five different temperatures and the strength of the resulting product measured. The data might have appeared as in Table 15.11.

(a) Compute the residual sum of squares $\sum(y_i - \hat{y}_i)^2$ that results when $\hat{y}_i = 0.3x_i - 87$.

(b) Compute the residual sum of squares that results when $\hat{y}_i = 0.3x_i - 86$.

TABLE 15.11

Temperature and Strength of Product

x, Temperature	y Strength
480	57
500	63
520	71
540	75
560	82

Hypothetical data.

TABLE 15.12

Summary Statistics for Data of Table 15.11

$n = 5$

$\sum x_i = \quad 2{,}600$	$\sum y_i = \quad 348$
$\sum x_i^2 = 1{,}356{,}000$	$\sum y_i^2 = 24{,}608$

$\sum x_i y_i = 182{,}200$

$\bar{x} = \quad 520$	$\bar{y} = \quad 69.6$
$s_x^2 = \quad 1{,}000$	$s_y^2 = \quad 96.8$

$s_{xy} = \quad 310$

(c) When $\hat{y}_i = a + bx_i$, where a and b are the least-squares estimates, then $\sum(y_i - \hat{y}_i)^2 = \sum(y_i - \bar{y})^2 - [\sum(x_i - \bar{x})(y_i - \bar{y})]^2 / \sum(x_i - \bar{x})^2$. Use the results in Table 15.12 to compute this value. Note that it is considerably less than the results of (a) and (b).

16

Comparison of Several Populations

Introduction

Throughout this book we have stressed the basic statistical concept of *variability*. When some measurement, such as height or aptitude for a particular job, is made on several individuals, the values vary from person to person. The variability of a quantitative scale is measured by its variance. If the set of individuals is stratified into more homogeneous groups, the variance of the measurements within the more homogeneous groups will be less than that of the measurements in the entire group; that is what "more homogeneous" means. For example, the variance of the heights of pupils in an elementary school is usually greater than the variance of heights of pupils in just the first grade, the variance in the second grade, and the variance in each of the other grades. At the same time, the *average* height of pupils also varies from grade to grade.

The facts that the within-grade variances are less than the overall variance, and that the averages vary between grades correspond to two ways of looking at the same phenomenon. The total variability (of heights) is made up of two components: the variability of individuals within groups (grades) and the variability of means between groups (grades). At the extreme, all of the variability of a measured variable may be within the groups and none of it between groups, that is, the means of the subgroups are equal.

The *analysis of variance* is a set of statistical techniques for studying variability from different sources and comparing them to understand the relative importance of each of the sources. It is also used to make inferences about the population through tests of significance, including the very important comparison of the means of two or more separate populations. Thus, the analysis of variance is the most straightforward way to examine the association between a categorical variable ("groups") and a numerical variable (the measure on which the means are based).

Suppose that we wish to compare the means of three populations on some measured dependent variable. The population means are represented as μ_1, μ_2, and μ_3. A hypothesis that may be tested is

$$H_0: \mu_1 = \mu_2 = \mu_3,$$

that is, that the means of the three populations are equal. The hypothesis is rejected if the population means are different in any way, for example, if one of the means differs from the other two or if all three means are different.

A procedure for comparing the means of two populations is discussed in Chapter 12. The techniques described in this chapter will lead to exactly the same conclusion when only two populations are compared, but also generalize to three or more groups.

The procedure for testing H_0 through the analysis of variance parallels that for other tests of significance. Samples of data are obtained and a test statistic is computed. In this application the test statistic reflects the extent to which variation among the sample means is greater than variation among observations within the groups. If the test statistic is large enough (that is, if it exceeds the corresponding significance point) then H_0 is rejected. If the sample means are close together and hence the test statistic is small, then H_0 is accepted.

When only two groups are being compared, it is obvious from knowledge of the sample means which group has a significantly higher mean than the other. When there are three or more groups, however, rejection of H_0 only means that *at least* one population mean is different from the others. It is necessary to follow the test of significance by additional analysis-of-variance procedures to determine *which* of the population means are different from which other means. Some of these follow-up procedures are described in Section 16.2.

The "groups" compared through analysis of variance may be created in a number of different ways. In many studies the data are already classified into groups, such as states, countries, gender groups, medical diagnostic categories, or religious affiliations.

In other instances, the statistician may define the groups based on one or more measured characteristics. For example, socioeconomic classifications may be created by combining individuals' educational levels, incomes, and employment statuses according to some set of rules. The result might be, say, three or four or five separate socioeconomic categories. Or a medical researcher might classify patients as "high risk" or "low risk" for some disease based on a number of indicators, for example, age, family history, and health behaviors. In conducting surveys, groups are frequently formed from the respondents' answers, for example, all people who support a particular political issue, those who are opposed, and those who have no opinion.

In an experiment, the groups are defined by the various experimental conditions or "treatments," such as instructional methods, psychological or social interventions, or dosages or different forms of a drug. In these situations, individuals are assigned to the conditions by the researcher. When possible, the assignment should be done randomly so that the only difference between the groups is the experimental intervention. Thus, any differences in outcomes can be explained by differences in the treatments received and not by other factors that are irrelevant to the study.

EXAMPLE 1   A psychologist is interested in comparing the effects of three different informational "sets" on children's ability to memorize words. Eighteen 7-year-old children were randomly assigned to one of three groups. In condition 1, the children were shown a list of 12 words and were asked to study them in preparation for a recall test. In condition 2, the children were told that the words comprised three global categories, flowers, animals, and foods, and were asked to study the list. In condition 3, the children were told that the words comprised 6 more detailed categories. After studying the word list for 10 minutes, each youngster was asked to list all of the words that he/she could remember. The number of correctly recalled words was the measured outcome variable of the study.

EXAMPLE 2   A study was performed to see if school grades are related to television viewing habits among high-school juniors. Respondents were monitored for 15 weekdays during the year and classified into 4 groups according to the average amount of television they viewed on those days (0–0.5 hour; 0.5–1.5 hours; 1.5–3.0 hours; more than 3 hours). Each student's grade-average (GPA) was recorded for all courses taken during the year. Mean GPAs were compared among the four television viewing groups.

EXAMPLE 3   In the study *The Academic Mind* by Lazarsfeld and Thielens, a total of 2451 social science faculty members from 165 of the larger American colleges and universities were interviewed in order to assess the impact of the McCarthy era on social science faculties. At each college, the number of "academic freedom incidents" was counted. These were incidents mentioned by more than one respondent as an attack on the academic freedom of the faculty. They ranged from small-scale matters, such as a verbal denunciation of a professor by a student group, to large-scale matters, including a Congressional investigation. It was of interest to examine whether and how the institutional basis of a school's support and control affected the number of "incidents" occurring there. Hence, each college was classified as publicly controlled, privately controlled, or controlled by some other institution. (Teachers' colleges and schools controlled by a religious institution were included in the "other" category.) The distributions of numbers of "incidents" in the different types of institutions were studied.

EXAMPLE 4   A large manufacturing firm employs high-school dropouts, high-school graduates, and individuals who attended college as production-line workers. The company management speculated that job proficiency was related to educational attainment. If so, only high-school graduates, or perhaps only individuals who had attended college, would be hired in the future. To test their idea, the job performance of a sample of employees was rated by their supervisors on an extensive rating scale that yielded possible proficiency scores ranging from 0 to 200. The mean ratings of the three education groups (high-school dropouts, high-school graduates, college attendees) were compared by the analysis of variance.

In each example the major research question concerns differences among the means of two or more separate groups. The analysis of variance answers this question by comparing the variability of scores *within* a particular population to *between-group* variability, that is, the difference between the mean of one population and the mean of another.[1]

The comparison of within- and between-group variability is illustrated in Figure 16.1 with hypothetical distributions for Example 4. Suppose that the population of all high-school dropouts has an average job proficiency of 97, all high-school graduates an average proficiency of 100, and all college attendees an average of 103. Let us also assume that the standard deviation of proficiency ratings is about 15 points in each of the three populations. The distributions of proficiencies are shown in Figure 16.1a.

In this hypothetical situation, the variability within each group is so large that the slight differences in average group scores are of no con-

---

[1]Although rules of grammar would dictate that the word "among" be used for three or more groups, standard statistical terminology is "between-group variability" regardless of the number of groups.

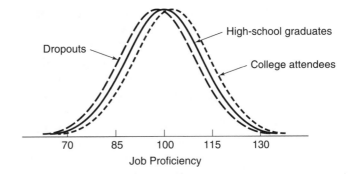

Figure 16.1a *Hypothetical job ratings of three education groups, most variation within groups.*

sequence. An employer would not be justified in imposing a minimum educational prerequisite on the basis of such small differences in averages. If just those individuals who had graduated from high school or more were hired, the variability in job proficiency would still be almost as large as it is with high-school dropouts included.

On the other hand, if between-group differences are an important part of the variability in scores, then it is useful to distinguish among the populations. This situation is illustrated in Figure 16.1b. Proficiency ratings range from below 40 points to above 160 points and differences among the means of the three groups are pronounced. Thus if one group were to be eliminated (e.g., high-school dropouts) variability in job performance would be reduced considerably, and the average job proficiency of all remaining employees would be raised. In this example, if high-school graduates were also eliminated, variability in the remaining scores would be reduced further. Thus the classification of

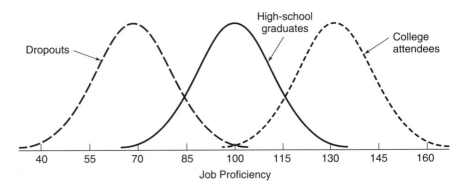

Figure 16.1b *Hypothetical job ratings of three education groups, substantial between-group variation.*

employees by educational attainment is of consequence to the prospective employer.

In practice a *sample* of observations is drawn from each of the populations and the analysis of variance is used to examine the variability of scores in the samples. Every observation is a member of one of several nonoverlapping groups, representing one distinct population. Each observation also has a score on the measured outcome or dependent variable from which sample means ($\bar{x}_1, \bar{x}_2, \bar{x}_3, \ldots$) and sample variability are computed.

In general, if variability among the sample means is small relative to the variability among individuals within the groups, then the test of significance will tell us that the hypothesis of equal population means is tenable. This is likely to occur if the population distributions are like those pictured in Figure 16.1a. On the other hand, if variability among the group means is large relative to variability among individuals within the groups, the test of significance should tell us that the hypothesis of equal means is rejected. This is likely to occur if the population distributions are like those pictured in Figure 16.1b.

The full discussion of the analysis of variance begins in Section 16.1. Like the *t*-test for comparing two means, certain conditions must be met in order for the analysis of variance to provide an appropriate way to test for mean differences. As with the *t*-test, the mean should be a suitable measure of location. This requirement implies the measured outcome variable should have nearly equal intervals and that its distribution should be unimodal and (roughly) symmetric. Likewise, the variance (or standard deviation) should be a suitable measure of variability. The formal statistical assumptions for analysis of variance are given last in Section 16.1, together with recommended approaches for comparing groups if those assumptions are not met.

16.1 One-Way Analysis of Variance

A Complete Example

The analysis of variance indicates whether population means are different by comparing the variability among sample means with variability among individual observations within groups. The data for Example 1 illustrate this principle. Table 16.1 gives the numbers of words memorized by 18 children given three different information sets. The set of scores for six children given a particular information set is considered as

TABLE 16.1

Words Memorized by 18 Children With Three Different Information Sets

	No information	*3 categories*	*6 categories*
	2	9	4
	4	10	5
	3	10	6
	4	7	3
	5	8	7
	6	10	5
Sums	24	54	30
Sample means	4	9	5
Total of 3 samples: 108; mean of 3 samples: 6			

Hypothetical data.

a sample from a hypothetical population of scores of all seven-year-old children who might have that information set. The scores are graphed in Figure 16.2.

The means of the three samples are 4, 9, and 5, respectively. Figure 16.2 shows these as the centers of three samples; there is clear variability from group to group. The variability in the entire pooled sample of 18 is shown on the last line.

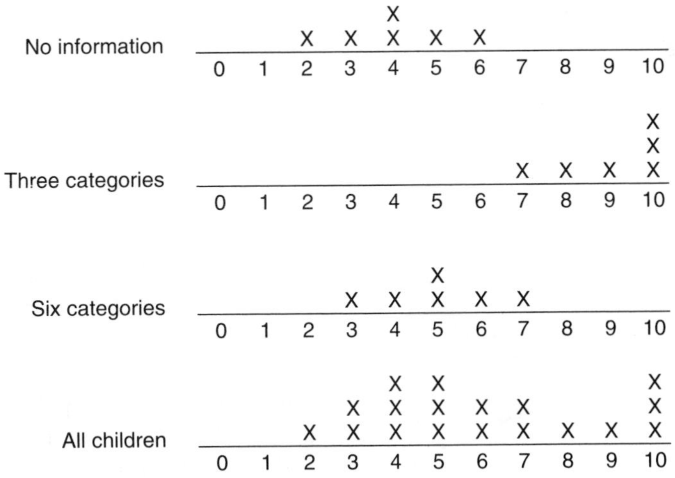

Figure 16.2   *Words memorized by 18 children with three different information sets.*

In contrast to this rather typical pattern, Tables 16.2 and 16.3 are illustrations of extreme cases. In Table 16.2 every observation in Group A is 3, every observation in Group B is 5, and every observation in Group C is 8. There is no variation within groups, but there is variation between groups. In Table 16.3 the mean of each group is 3. There is no variation among the group means, although there is variability within each group. Neither extreme can be expected to occur in an actual data set. In actual data, one needs to make an assessment of the relative sizes of the between-group and within-group variability.

TABLE 16.2

No Variation Within Groups

| | Group | | |
	A	B	C
	3	5	8
	3	5	8
	3	5	8
	3	5	8
Means:	3	5	8

Hypothetical data.

TABLE 16.3

No Variation Between Groups (Group Means Equal)

| | Group | | |
	A	B	C
	3	3	1
	5	6	4
	1	2	3
	3	1	4
Means:	3	3	3

Hypothetical data.

To test the hypothesis of equal means, a measure of variability among the groups means is computed and compared to a measure of variability within the groups. Each of these is a variance-like measure, and the test statistic (F) is the ratio of the two measures of variability. The ratio is compared to percentage points of the F-distribution to see if H_0 is rejected.

Variability Within Groups

The measure of variability among observations within the groups is called the *Mean Square Within Groups* and is denoted MS_W. The Mean Square Within Groups is the variance of all the scores, but computed separately for each of the three samples and then combined. Like the sample variance discussed in Chapter 4, the Mean Square Within Groups has a sum of squared deviations in the numerator and the corresponding number of degrees of freedom in the denominator. The numerator is called the *Sum of Squares Within Groups* and the denominator is called the *degrees of freedom within groups.*

The Sum of Squares Within Groups (SS_W) is the sum of squared deviations of individual scores from their subgroup means. The degrees of freedom within groups (df_W) is the total number of degrees of freedom of the deviations within the groups.

Table 16.4 shows the computation of the sum of squared deviations of each of the 18 scores from their respective group means. For each individual in each group, the table gives the deviation of that individual's score from the respective group mean and the square of the deviation. At the foot of the table is the sum of squared deviations for each group. The sum of these is $SS_W = 10 + 8 + 10 = 28$.

TABLE 16.4

Computation of Within-Groups Sum of Squares by Use of Deviations from Group Mean

	Group 1	Group 2	Group 3
	$(2-4)^2 = (-2)^2 = 4$	$(9-9)^2 = 0^2 = 0$	$(4-5)^2 = (-1)^2 = 1$
	$(4-4)^2 = 0^2 = 0$	$(10-9)^2 = 1^2 = 1$	$(5-5)^2 = 0^2 = 0$
	$(3-4)^2 = (-1)^2 = 1$	$(10-9)^2 = 1^2 = 1$	$(6-5)^2 = 1^2 = 1$
	$(4-4)^2 = 0^2 = 0$	$(7-9)^2 = (-2)^2 = 4$	$(3-5)^2 = (-2)^2 = 4$
	$(5-4)^2 = 1^2 = 1$	$(8-9)^2 = (-1)^2 = 1$	$(7-5)^2 = 2^2 = 4$
	$(6-4)^2 = 2^2 = 4$	$(10-9)^2 = 1^2 = 1$	$(5-5)^2 = 0^2 = 0$
Sum	10	8	10

source: Table 16.1.

The Sum of Squares Within Groups is made up of three sample sums of squares. Each involves 6 squared deviations, and hence, each has $6 - 1 = 5$ degrees of freedom. In total there are $df_W = 5 + 5 + 5 = 15$ degrees of freedom within groups. Thus the Mean Square Within Groups is $MS_W = 28/15 = 1.867$.

Variability Between Groups

The measure of variability among the group means is called the *Mean Square Between Groups* and is denoted MS_B. The Mean Square Between Groups is similar to a variance but is computed from the three subgroup means, 4, 9, and 5. The Mean Square Between Groups also has a sum of squared deviations in the numerator and a corresponding number of degrees of freedom in the denominator. The numerator is called the *Sum of Squares Between Groups* and the denominator is called the *degrees of freedom between groups.*

The Sum of Squares Between Groups (SS_B) is the sum of squared deviations of the subgroup means from the overall or "pooled" mean. The degrees of freedom between groups (df_B) is the number of independent deviations summarized in SS_B.

The pooled mean of all 18 observations is 6 (from Table 16.1), which is also the average of the values 4, 9, and 5 (see footnote 2). The squared deviations of the group means from the pooled mean are

$$(4 - 6)^2 = 4, \qquad (9 - 6)^2 = 9, \qquad \text{and} \qquad (5 - 6)^2 = 1.$$

Their sum is $4 + 9 + 1 = 14$.

This is not quite the final Sum of Squares Between Groups, however. Before dividing by the degrees of freedom between groups, one additional adjustment is necessary. The explanation is somewhat complex, but the calculation is simple. To test the null hypothesis of equal means, SS_B is going to be compared to SS_W by converting each to a mean square and then dividing one mean square by the other. However, SS_B cannot be meaningfully compared to SS_W because SS_B reflects variability among *means* while SS_W reflects variability among individual measurements. As shown in Chapter 9, the sampling variability of the mean of N observations is only $1/N$ as large as the variability of individual observations. Thus to put SS_B on a comparable scale to SS_W, it is multiplied by the number of observations in each subgroup, that is, 6. The Sum of Squares Between Groups in the 3-group example is $SS_B = 6 \times 14 = 84$.

The Sum of Squares Between Groups is comprised of three deviations from the pooled mean. Thus the number of degrees of freedom is $df_B = 3 - 1 = 2$, and the Mean Square Between Groups is $MS_B = 84/2 = 42$.

The Test Statistic and Conclusion

The test statistic to test H_0: $\mu_1 = \mu_2 = \mu_3$ is the ratio of MS_B/MS_W, or $42/1.867 = 22.50$. This ratio seems to indicate that the variability among groups is much greater than that within groups. However, we know that different samples would give us different values of the ratio even if the three population means are equal. Thus, the calculated ratio of 22.50 is compared to values from the F-distribution to see if the test statistic is large enough for 3 samples and 18 observations to reject H_0. The number of degrees of freedom for the numerator mean square is

[2]The pooled mean is the simple average of the subgroup means here because the three groups have the same number of observations (6). If the sample sizes are different, the pooled mean is a weighted average of the subgroup means.

$df_B = 2$. The number of degrees of freedom for the denominator mean square is $df_W = 15$.

If we are using a 1% significance level, the significance point from the table of the F-distribution with 2 and 15 degrees of freedom is 6.36. Thus the calculated ratio of 22.50 is very significant and we conclude that there are real differences in the average number of words memorized depending on the amount of organizing information that is provided.

The follow-up question remains, specifically: Which of the three means are significantly different from which others? Procedures for addressing this question are described in Section 16.2.

Summary Table.   The results of the computations are usually summarized in an Analysis of Variance Table like Table 16.5. In addition to the between and within sums of squares and degrees of freedom, an additional row is usually added that includes the total of these two sources of variability. It is an algebraic fact that the Sum of Squares Between Groups plus the Sum of Squares Within Groups is equal to the sum of squares of deviations of all observations from the overall mean. From Table 16.1 the *Total Sum of Squares* is $(2 - 6)^2 + (4 - 6)^2 + \cdots + (7 - 6)^2 + (5 - 6)^2 = 112$, which is exactly the sum of SS_B and SS_W, namely, $28 + 84 = 112$; the Total Sum of Squares is denoted by SS_T. There is only a single value in the column labeled "F" because there is only one test statistic, even though the table has several rows.

TABLE 16.5
Analysis of Variance Table

Source of variation	Sums of squares	Degrees of freedom	Mean squares	F
Between groups	84	2	42	22.50
Within groups	28	15	1.867	
Total	112	17		

source: Table 16.1.

The Algebra of ANOVA

The statistical notation needed to explain the analysis of variance is summarized in Table 16.6. The number of groups being compared is represented by k. The null hypothesis is

$$H_0: \mu_1 = \mu_2 = \cdots = \mu_k,$$

that is, the k population means are equal.

TABLE 16.6

Notation for Analysis of Variance

		Group			
		1	*2*	$\cdots$	*k*
Population	Mean	μ_1	μ_2	$\cdots$	μ_k
	Variance	σ_1^2	σ_2^2	$\cdots$	σ_k^2
Sample	Observations	x_{11}	x_{21}		x_{k1}
		x_{12}	x_{22}		x_{k2}
		$\vdots$	$\vdots$		$\vdots$
	Sample size	n_1	n_2	$\cdots$	n_k
	Mean	$\bar{x}_1$	$\bar{x}_2$	$\cdots$	$\bar{x}_k$
	Variance	s_1^2	s_2^2	$\cdots$	s_k^2

The number of sampled observations in a "typical" group, group g, is denoted by n_g and the total number of observations is $n = n_1 + n_2 + \cdots + n_k$. Each observation has two subscripts, the first indicating the group to which the observation belongs and the second indicating the observation number within that group. Thus, x_{gi} represents the ith observation in group g. For instance, in Table 16.1, $x_{11} = 2, x_{12} = 4$, $\ldots, x_{21} = 9, \ldots$ and so on.

The sample means of the subgroups are represented by $\bar{x}_1, \bar{x}_2, \ldots, \bar{x}_k$. The mean in the "typical" group, group g, is the sum of observations in group g divided by the number of observations in group g, that is,

$$\bar{x}_g = \frac{1}{n_g} \sum_i x_{gi}.$$

The notation $\sum_i x_{gi}$ indicates that all of the data in a particular group are summed, that is, the x's are summed over values of the i subscript. Thus for group 1 in Table 16.1,

$$\sum_i x_{1i} = x_{11} + x_{12} + \cdots + x_{16}$$

$$= 2 + 4 + \cdots + 6 = 24.$$

The overall mean $\bar{x}$ is simply the total of all observations divided by n,

$$\bar{x} = \frac{\sum_g \sum_i x_{gi}}{n}.$$

The notation $\sum_g \sum_i x_{gi}$ indicates that all the observed values in all groups are summed, that is, terms are summed over values of both subscripts.

In Table 16.1 the number of groups is $k = 3$, and n_1, n_2, and n_3 are all 6. The means are $\bar{x}_1 = 4$, $\bar{x}_2 = 9$, and $\bar{x}_3 = 5$, and the overall mean is $\bar{x} = 6$.

The ANOVA Decomposition

To test H_0, variability among group means is compared with variability among observations within groups. That is, we are "explaining" why observed values on a measured variable differ from one another or, equivalently, why scores differ from the overall mean $\bar{x}$. The "explanation" has two parts: first, scores differ from the mean of their subgroup, and second, the mean of a particular subgroup differs from the mean of other groups and thus from the overall mean as well.

These two explanations can be assembled into an algebraic expression showing the "decomposition" of a single typical score x_{gi} as

$$(x_{gi} - \bar{x}) = (\bar{x}_g - \bar{x}) + (x_{gi} - \bar{x}_g). \tag{16.1}$$

Expression (16.1) shows that the deviation of a score from the overall mean $\bar{x}$ can be decomposed as the sum of two parts, the deviation of the group mean from the overall mean and the deviation of an observation from its group mean.

The terms on the right-hand side of (16.1) are the fundamental elements of "between-group" and "within-group" variability, respectively. By squaring and summing the terms in (16.1) we obtain exactly the sum-of-squares partition essential for the analysis of variance. That is, the sum of squares of $x_{gi} - \bar{x}$ is equal to the sum of squares of the first term on the right-hand side, namely, $\bar{x}_g - \bar{x}$, plus the sum of squares of the second term $x_{gi} - \bar{x}_g$. (See Exercise 16.22 for proof.)

Table 16.7 illustrates these sums of squares for each group (1, 2, and 3) in the three-group example. The left-hand side of each expression is the sum of squared deviations of scores in the group from the overall mean $\bar{x}$. The first term on the right-hand side of each expression contains the squared deviation of the group mean $\bar{x}_g$ from the overall mean. The second term on the right-hand side is the sum of squared deviations of scores x_{gi} from the mean in their particular group.

The "Total" line of Table 16.7 shows the sum of these sums of squares for all 3 groups. The summation notation $\sum_g$ indicates that the terms are summed across groups, that is, summed over values of g, the group subscript.

The three sums of squares in the "Total" line are exactly SS_T, SS_B, and SS_W as computed for the three-group example. That is, the partition of sums of squares in the "Total" line of Table 16.7 is identical to

$$SS_T = SS_B + SS_W.$$

TABLE 16.7
Sums of Squares

Group 1:	$\sum_i (x_{1i} - \bar{x})^2 = n_1(\bar{x}_1 - \bar{x})^2 + \sum_i (x_{1i} - \bar{x}_1)^2$
Group 2:	$\sum_i (x_{2i} - \bar{x})^2 = n_2(\bar{x}_2 - \bar{x})^2 + \sum_i (x_{2i} - \bar{x}_2)^2$
Group 3:	$\sum_i (x_{3i} - \bar{x})^2 = n_3(\bar{x}_3 - \bar{x})^2 + \sum_i (x_{3i} - \bar{x}_3)^2$
Total:	$\sum_g \sum_i (x_{gi} - \bar{x})^2 = \sum_g n_g(\bar{x}_g - \bar{x})^2 + \sum_g \sum_i (x_{gi} - \bar{x}_g)^2$

These three terms are exactly the sums of squares of the three terms in the decomposition of a single score represented by (16.1). Numerical results are given for the three-group example in Table 16.5. The Sum of Squares Between Groups (SS_B) and Sum of Squares Within Groups (SS_W) are the numerators of the Mean Square Between Groups and the Mean Square Within Groups, respectively, that form the *F*-statistic.

This partition shows us why the technique is called the *analysis of variance.* Total variation in the data, represented by SS_T, is decomposed or "analyzed" into two parts, one a measure of differences among the group means and the other a measure of variation within groups. The components are compared to determine if differences among the means are substantially greater than variability among individuals within the groups.

The components of the partition plus the *F*-statistic are usually summarized in an Analysis of Variance table like Table 16.5. The general form is shown in Table 16.8. The "Source of Variation" column lists the components in the ANOVA decomposition. The Between-groups and Within-groups rows give the sum of squares for each component, the number of degrees of freedom associated with each component, and the mean squares. There is only a single *F*-statistic and it is listed in the Between-group row. The "Total" row of the table gives the total sum of squares and total number of degrees of freedom.

TABLE 16.8
Analysis of Variance Table

Source of variation	*Sum of squares*	*Degrees of freedom*	*Mean squares*	*F*
Between groups	SS_B	df_B	$MS_B = SS_B/df_B$	$F = MS_B/MS_W$
Within groups	SS_W	df_W	$MS_W = SS_W/df_W$	
Total	SS_T	df_T		

Computing Formulas

The expressions in Table 16.7 and used in the three-group example are not the most efficient for actually computing the three basic sums of squares. The usual computational procedures are given below.

Total Sum of Squares.   The quantity

$$SS_T = \sum_g \sum_i (x_{gi} - \bar{x})^2,$$

is more efficiently calculated as

$$SS_T = \sum_g \sum_i x_{gi}^2 - \frac{1}{n} \left(\sum_g \sum_i x_{gi} \right)^2. \qquad (16.2)$$

The first term on the right-hand side of (16.2) involves squaring every original data value and summing the squares. The second term involves summing all the original values, squaring the sum and dividing by the total sample size.

For the data of Table 16.1, the left-hand term is

$$2^2 + 4^2 + 3^2 + \cdots + 3^2 + 7^2 + 5^2 = 760.$$

The right-hand term is

$$(2 + 4 + 3 + \cdots + 3 + 7 + 5)^2/18 = 108^2/18 = 648,$$

and $SS_T = 760 - 648 = 112$.

The number of degrees of freedom is the number of independent deviations of the original observations from the overall mean. If the total sample size is represented by n, the total degrees of freedom is $df_T = n - 1$. For the data of Table 16.1, $n = 18$ and $df_T = 17$.

Between-group Sum of Squares and Mean Square.   The quantity

$$SS_B = \sum_g n_g (\bar{x}_g - \bar{x})^2,$$

is usually calculated as

$$SS_B = \sum_g \frac{\left(\sum_i x_{gi} \right)^2}{n_g} - \frac{1}{n} \left(\sum_g \sum_i x_{gi} \right)^2. \qquad (16.3)$$

The first term on the right-hand side of (16.3) involves summing the values in each group, squaring the sum, dividing by the number of observations in the group, and summing these terms across all groups. The second term is the same as the second term on the right-hand side of (16.2).

For the data of Table 16.1, the left-hand term is

$$\frac{24^2}{6} + \frac{54^2}{6} + \frac{30^2}{6} = 96 + 486 + 150 = 732.$$

The second term is 648, and $SS_B = 732 - 648 - 84$.

The number of degrees of freedom between groups is the number of independent deviations of subgroup means from the overall mean. If the number of groups is represented by k, then $df_B = k - 1$. For the data of Table 16.1, $k = 3$ and $df_B = 2$. The Mean Square Between Groups is $MS_B = SS_B/df_B$. For the data of Table 16.1, $MS_B = 42.0$.

Within-group Sum of Squares and Mean Square.   Since the Total Sum of Squares is simply the sum of the Sum of Squares Between Groups and the Sum of Squares Within Groups, one of the three sums of squares can be obtained by simple addition or subtraction. The most efficient procedure is to subtract SS_B from SS_T:

$$SS_W = SS_T - SS_B. \tag{16.4}$$

For the data of Table 16.1, this $SS_W = 760 - 732 = 28$.

The number of degrees of freedom within groups is the number of independent deviations of observed values from their subgroup means. If there are n_g observations in subgroup g, then the number of independent deviations in that group is $n_g - 1$. Summing these across the k groups,

$$df_W = \sum_g (n_g - 1) = n - k,$$

and the Mean Square Within Groups is $MS_W = SS_W/df_W$. For the data of Table 16.1 with $n = 18$ and $k = 3$, $df_W = 15$ and $MS_W = 1.867$. This is the final term needed to calculate the F-ratio, $F = 42/1.867 = 22.50$.

The F-statistic is compared to significance points of the F-distribution with df_B and df_W degrees of freedom. For the example, the 5% significance point is $F_{2,15}(.05) = 3.68$, and the 1% significance point is $F_{2,15}(.01) = 6.36$. In either case, H_0 is rejected. If we were using a P value approach, we would report $P < .005$ for the test because 22.50 is greater than the largest value tabled for the F-distribution with 2 and 15 degrees of freedom (7.7008).

An Example with Unequal Sample Sizes

In Example 2 (Table 16.9) high-school juniors have been classified according to four levels of weekday television viewing, and each student's grade-average (GPA) was recorded as well. The purpose of the study is to determine if mean GPAs are related to the amount of time spent watching television. In statistical terms the null hypothesis is

$$H_0: \mu_1 = \mu_2 = \mu_3 = \mu_4.$$

The number of groups in this example is $k = 4$. Unlike the previous example, the numbers of observations differ from group to group with $n_1 = 3$, $n_2 = 4$, $n_3 = 3$, and $n_4 = 2$. The total sample size is $n = 3 + 4 + 3 + 2 = 12$. Table 16.9 gives the GPAs for all 12 students plus the sum of the observed values and the mean for each television viewing group.

TABLE 16.9

Grade Averages of Four Groups of High-School Juniors

| | Daily amount of television viewing | | | |
	$g = 1$ *0–0.5 hours*	$g = 2$ *0.5–1.5 hours*	$g = 3$ *1.5–3.0 hours*	$g = 4$ *More than 3 hours*
	$x_{11} = 3.5$	$x_{21} = 3.5$	$x_{31} = 2.5$	$x_{41} = 2.0$
	$x_{12} = 4.0$	$x_{22} = 3.0$	$x_{32} = 3.0$	$x_{42} = 1.0$
	$x_{13} = 3.0$	$x_{23} = 2.5$	$x_{33} = 2.0$	
		$x_{24} = 3.0$		
$\sum_i x_{gi}:$	$\sum_i x_{1i} = 10.5$	$\sum_i x_{2i} = 12.0$	$\sum_i x_{3i} = 7.5$	$\sum_i x_{4i} = 3.0$
$n_g:$	3	4	3	2
$\bar{x}_g:$	3.5	3.0	2.5	1.5

Hypothetical data.

The overall mean is the total of all observations divided by n, that is,

$$\bar{x} = \frac{\sum_g \sum_i x_{gi}}{n}$$

$$= \frac{10.5 + 12.0 + 7.5 + 3.0}{12} = 2.75.$$

The sample means, 3.5, 3.0, 2.5, and 1.5 are different from one another and decrease systematically as the amount of television viewing increases.[3] The test of H_0, however, is to determine if the sample means are sufficiently different that we conclude there is also a difference among the respective *population* means.

The basic computational formulas are given by (16.2), (16.3), and (16.4). The Total Sum of Squares (16.2) has two terms of which the first

---

[3]The ANOVA F ratio does not consider the order of the groups, however. That is, the F-statistic would be the same for the same data even if the groups were permuted in order so that the means were, say, 3.0, 1.5, 3.5, and 2.5.

is the sum of squares of all the data, that is,

$$\sum_g \sum_i x_{gi}^2 = 3.5^2 + 4.0^2 + 3.0^2 + \cdots + 2.0^2 + 2.0^2 + 1.0^2$$

$$= 98.00.$$

The second term in SS_T is

$$\frac{1}{n} \sum_g \sum_i x_{gi} = \frac{1}{12} \times (10.5 + 12.0 + 7.5 + 3.0)^2$$

$$= 90.75,$$

and $SS_T = 98.00 - 90.75 = 7.25$. The total number of degrees of freedom in $n - 1 = 11$.

The Sum of Squares Between Groups (16.3) also has two terms. The first term is the sum of k components, one for each group, that is,

$$\sum_g \frac{\left(\sum_i x_{gi}\right)^2}{n_g}.$$

For group 1, the sum of values in the group is squared and divided by n_1, that is,

$$\frac{\left(\sum_i x_{1i}\right)^2}{n_1} = \frac{10.5^2}{3} = 36.75.$$

For groups 2, 3, and 4,

$$\frac{\left(\sum_i x_{2i}\right)^2}{n_2} = 36.00; \qquad \frac{\left(\sum_i x_{3i}\right)^2}{n_3} = 18.75; \qquad \frac{\left(\sum_i x_{4i}\right)^2}{n_4} = 4.50.$$

Thus the first term in SS_B is $36.75 + 36.00 + 18.75 + 4.50 = 96.00$. The second term in SS_B is the same as the second term in SS_T (90.75) so that $SS_B = 96.00 - 90.75 = 5.25$.

The number of degrees of freedom between groups is $df_B = k - 1 = 4 - 1 = 3$. The Mean Squares Between Groups is $MS_B = 5.25/3 = 1.75$.

The Sum of Squares Within Groups (16.4) can be obtained by subtraction. For the example $SS_W = 7.25 - 5.25 = 2.00$. The number of degrees of freedom within groups is the number of independent deviations of the data from their subgroup means, that is, $df_W = 2 + 3 + 2 + 1 = 8$; this is equivalent to the total number of observations minus the total number of groups, $n - k$. The Mean Square Within Groups is $MS_W = 2.00/8 = 0.25$.

These results can be assembled into an analysis of variance table like Table 16.10. In addition to the sums of squares, degrees of freedom, and mean squares, the F-ratio has been entered in the table. This is $F = MS_B/MS_W = 1.75/.25 = 7.00$.

TABLE 16.10

Analysis of Variance Table for Television Viewing and GPAs

Source of variation	Sum of squares	Degrees of freedom	Mean square	F
Between viewing groups	5.25	3	1.75	7.00
Within viewing groups	2.00	8	0.25	
Total	7.25	11		

source: Table 16.9.

The Mean Square Between Groups is seven times as large as the Mean Square Within Groups, suggesting that there is substantial between-group variation. To decide formally if H_0 is rejected, the F-ratio is compared to values of the F-distribution with 3 and 8 degrees of freedom. At the 5% significance level, the significance point is $F_{3,8}(.05) = 4.07$. The test statistic exceeds this value and H_0 is rejected. We conclude that in the population, average GPAs differ according to the amount of television individuals watch. It is clear from the sample means that GPAs decreased with an increase in television viewing. However, to estimate the *magnitude* of the impact of watching more television on grades, or to state definitively that any particular *pair* of sample means is significantly different, further analysis is needed. This is discussed in Section 16.2, "Which Groups Differ from Which, and by How Much?"

Had we chosen to work at the 1% significance level, the significance point would have been $F_{3,8}(.01) = 7.59$. In this case the test statistic $F = 7.00$ is not large enough to cause us to reject H_0 and we would conclude that amount of television viewing is not related to average GPAs. In other words, the differences that were observed in the sample were not sufficient to cause us to conclude that there are differences among population means. A P value interval for this test of significance is $.01 < P < .025$.

Obviously the choice of α—the probability of a Type I error—can have an important effect on the results of a statistical analysis. It should be given careful consideration, taking into account the potential costs of a statistical error, *before* the analysis is conducted. If a 5% chance of making a Type I error is unacceptably high, then a 1% level of significance (or lower still) should be adopted.

More About the Analysis of Variance

Validity Conditions for the F-test   Like the tests of significance and confidence intervals for a single mean and for the difference of two means, the analysis of variance rests on a set of assumptions that should be considered before the procedure is applied. The most inflexible assumption is that the n observations must be independent. Sampling observations at random is an important step in assuring that this condition is met. In addition, steps should be taken to assure that the response on the measured outcome variable of no respondent is in any way affected by other respondents. If observations are humans, they should not have the opportunity to hear, see, or otherwise be influenced by other subjects' answers or behavior. If the assumption of independence is not met, then the analysis of variance tests of significance are not generally valid.

Second, the sampling distribution of the subgroup means should be nearly normal. This condition is usually met in analysis of variance, especially if the subgroup n's are moderate to large, since the Central Limit Theorem assures us that means based on large sample sizes are nearly normally distributed. The normality condition will also be met if the distribution of the underlying measured variable is normal, whether or not the sample sizes are large. It is always advisable to make histograms of the data to see whether there are any gross irregularities in the distributions, however. If the sample sizes are small and the distribution of the measured variable is highly non-normal, then the analysis of variance of ranks (Section 16.3) should be used in place of the ANOVA methods presented in Section 16.1.

Third, the F-test is based on an assumption that the population variances are all equal, that is, $\sigma_1^2 = \sigma_2^2 = \cdots = \sigma_k^2$. (If the values are equal, they may be represented without different subscripts, that is, just σ^2). This condition is especially important if the sample sizes $(n_1, n_2, \ldots, n_k)$ are not equal. Sample variances should be computed prior to conducting the analysis of variance to see if they are in the same general range as one another. If they appear to be very different, a formal test of equality of the σ^2's may be conducted (see Chapter 13).

If the test indicates that the variances are not homogeneous, several options may be available. For one, the data may be *transformed* to a scale on which the variances are more equal. For example, this might involve analyzing the logarithms of the original observed values, the square roots of the observed values, or some other function of the data. A number of more advanced statistics texts discuss this approach in depth, for example, Ott (1993). As an alternative, the analysis of variance of ranks presented in Section 16.3 of this book may be used instead.

Comparing Two Groups   In Chapter 12 a t-test is presented for comparing the means of two separate populations, μ_1 and μ_2. Obviously the analysis of variance can also be used to compare means when the number of groups is $k = 2$.

The reader need not be concerned that these ways of comparing μ_1 and μ_2 will yield two different results.

When there are two groups, the F-ratio produced by ANOVA is algebraically identical to the *square* of the t-statistic when the variances are considered equal, that is, $F = t^2$. Further, the significance point from the F table is exactly the square of the significance point from the corresponding t-distribution[4] at the same value of α. Thus the decision about H_0 will be identical regardless of whether a t-test is conducted or the analysis of variance is used to compare two means.

When there are more than two groups, only the analysis of variance can be used to test equality of the set of k means. In this case there is no simple relationship of the F-test and its significance points to any t-statistic or t-distribution.

How Does Analyzing Variability Tell Us About Differences Among Means?

A closer examination of the Mean Square Within Groups and the Mean Square Between Groups gives us further understanding of why the F-ratio works as it does. Table 16.7 shows the components of the Sums of Squares for three groups that make up these Mean Squares.

It can be seen that the Sum of Squares Within Groups is the sum of three quantities of the form $\sum_i (x_{gi} - \bar{x}_g)^2$, one for each group. Each is the sum of squared deviations of the observed values from the subgroup mean. If the sum of squared deviations in group g is divided by $n_g - 1$, we obtain the sample variance in group g, that is,

$$s_g^2 = \frac{\sum_i (x_{gi} - \bar{x}_g)^2}{n_g - 1}.$$

This sample variance is an estimate of the common population variance σ^2. (See Chapter 13.)

The Mean Square Within Groups is a weighted average of the three sample variances, that is,

$$MS_W = \frac{\sum_g \sum_i (x_{gi} - \bar{x}_g)^2}{\sum (n_g - 1)}$$

$$= \frac{(n_1 - 1) \times s_1^2 + (n_2 - 1) \times s_2^2 + \cdots + (n_k - 1) \times s_k^2}{(n_1 - 1) + (n_2 - 1) + \cdots + (n_k - 1)}.$$

In fact, if the sample sizes are equal, then MS_W is simply the average of the subgroup variances. For example, if there are three groups and $n_1 = n_2 = n_3$, then

$$MS_W = \frac{s_1^2 + s_2^2 + s_3^2}{3}.$$

---

[4]Assuming that a two-tailed t-test is being conducted. The reader may wish to verify from Appendices III and V that for degrees of freedom f, $t_f^2(\alpha/2) = F_{1,f}(\alpha)$.

Since each variance estimates σ^2 individually, their average also estimates σ^2 based on the observations in all three groups. That is, the Mean Square Within Groups is an estimate of the population variance of the measured outcome variable.

Now let us examine the Mean Square Between Groups. *First* assume that the null hypothesis is true and the k population means are equal, that is, $\mu_1 = \mu_2 = \cdots = \mu_k = \mu$. Each group mean $\bar{x}_g$ is the mean of a sample of n_g observations from a population with population mean μ and population variance σ^2. We know from Section 10.1 that means of random samples of n_g observations have a sampling distribution with variance equal to the variance of the measured variable divided by n_g. That is, the variance of $\bar{x}_g$ is σ^2/n_g.

To understand the implications of this for MS_B, let us assume for the moment that the sample sizes are equal ($n_1 = n_2 = \cdots = n_k = n$). The preceding discussion implies that if we were to draw several random samples from a population, calculate the sample means, and compute the sample variance of these means, this final result would estimate σ^2/n. That is, if we compute

$$\frac{\sum_g (\bar{x}_g - \bar{x})^2}{k - 1},$$

this value is an estimate of σ^2/n. If we multiply this value by n, the result is an estimate of σ^2, that is,

$$\frac{n \sum_g (\bar{x}_g - \bar{x})^2}{k - 1} = \frac{\sum_g n(\bar{x}_g - \bar{x})^2}{k - 1}.$$

This expression is exactly the Mean Square Between Groups (MS_B). Nothing of consequence is any different if the sample sizes are different, that is, if n_g replaces n in the right-hand expression above.

What does this mean in terms of the analysis of variance? This result demonstrates that *if H_0 is true and the population means are equal*, then MS_B is an estimate of the population variance σ^2. Since the Mean Square Within Groups is *always* an estimate of σ^2, F is the ratio of two estimates of the same value. That is, if H_0 is true, the F-ratio MS_B/MS_W should be close to 1.0.

Second, consider how this discussion would differ if H_0 is false. If the k population means are different, then we can expect the three sample means to be different (no matter how large the sample sizes are). This tends to inflate MS_B. Sample means vary, not just because they are repeated samples from a single population, *but also* because they reflect differences among population means. The sample variance of the k means estimates a value greater than σ^2/n and the Mean Square Between Groups estimates a value greater than σ^2. In this situation the F-ratio MS_B/MS_W should be large, that is, greater than 1.0.

In sum, the Mean Square Within Groups is an estimate of σ^2. The Mean Square Between Groups is an estimate of σ^2 if H_0 is true and an estimate of a value larger than σ^2 if H_0 is false. Thus the F-ratio will tend to be close to 1.0 if H_0 is true and large if H_0 is false.

Even if the population means are equal, however, the F-ratio computed for a particular sample may be greater than 1.0. Thus to test the hypothesis of equal population means, the F-statistic is compared with a significance point from the table of the F-distribution with $k - 1$ and $n - k$ degrees of freedom. If the test statistic exceeds the tabled significance point, we conclude that it arose from populations with different means and H_0 is rejected. If the test statistic does not exceed the significance point, then H_0 is accepted and any differences among the population means are considered to be negligible.

16.2   Which Groups Differ from Which, and by How Much?

The F-test gives information about all the means $\mu_1, \mu_2, \ldots, \mu_k$ simultaneously. If the hypothesis of equal means is accepted, the conclusion is that the data do not indicate differences among the population means. On the other hand, if the null hypothesis is rejected, the conclusion is that there are some differences; then the researcher may want to know which specific means are significantly different from which others and the direction and the magnitudes of the differences. This section begins by reviewing the methods for comparing two means described in Chapter 12, and then expands those ideas to the comparison of three or more groups.

Comparing Two Means

If $k = 2$ and only two means are being compared, the question of which is greater and by how much is a simple matter. Either the analysis-of-variance F-ratio or the t-test described in Chapter 12 indicates whether the null hypothesis is to be rejected.[5] The F-ratio is MS_B / MS_W; the t-statistic is

$$t = \frac{\bar{x}_1 - \bar{x}_2}{s_p \sqrt{\dfrac{1}{n_1} + \dfrac{1}{n_2}}},$$

---

[5] The reader is reminded that the F-ratio for comparing two groups is the square of the t-statistic and that both tests will lead to the same conclusion about H_0.

where $\bar{x}_1$ and $\bar{x}_2$ are the sample means based on n_1 and n_2 observations, respectively, and s_p^2 is the pooled within-group variance.[6] The t-statistic is compared to significance points from the t-distribution with $n_1 + n_2 - 2 = n - 2$ degrees of freedom. If the hypothesis of equal means is rejected, however, the *direction* and *magnitude* of the difference still remain to be described.

The most basic statistic for this purpose is the mean difference $\bar{x}_1 - \bar{x}_2$. The difference is positive if $\bar{x}_1$ is greater and negative if $\bar{x}_2$ is greater. If consumers of the statistical analysis are familiar with the scale of the measured variable, then this difference will be enough to convey the effect of interest. It leads to such conclusions as "group 1 watched television for 40 minutes more than group two, on the average," or "these packages of cookies weigh an average of 2 ounces more than those packages of cookies."

Often, however, the scale of the measured variable is not as familiar to consumers of a statistical analysis as hours of TV viewing or the size of a package of cookies. The scale might be scores on a personality test or classroom test, physiological measures, or observations of animal or human behavior where significant mean differences are harder to interpret. In this situation, the difference may be expressed as an *effect size*, that is, the number of standard deviations that separate the group means.

The effect size is

$$\frac{\bar{x}_1 - \bar{x}_2}{s},$$

where s^2 is an estimate of the population variance on the measured variable, and s is its square root, the standard deviation.[7] In a t-test of the difference of two means (Section 12.2) the estimate of σ^2 is the pooled variance s_p^2, so that the standard deviation is its square root, s_p. In conducting an analysis of variance, the estimate of σ^2 is the Mean Square Within Groups. Thus if two means are compared in an analysis of variance, the standard deviation for the effect size is $s = \sqrt{MS_W}$.

As an example, suppose that two means have been found to be significantly different using and F-test. Suppose that group 1 has $n_1 = 12$ and $\bar{x}_1 = 19.6$ while group 2 has $n_2 = 10$ and $\bar{x}_2 = 14.1$; also Mean Square Within Groups is $MS_W = 25.0$. The effect size is

$$\frac{19.6 - 14.1}{\sqrt{25}} = \frac{5.5}{5} = 1.1;$$

---

[6]A comparison of the definitions of s_p^2 and MS_W will reveal that these two statistics are identical; they simply have different names in different contexts.

[7]Note that the denominator of the effect size (s) is an estimate of the population standard deviation (σ); it is *not* an estimate of the standard error of $\bar{x}_1 - \bar{x}_2$.

the mean for group 1 is 1.1 standard deviations greater than the mean for group 2.

Even if we are not familiar with the original measurement scale, a mean difference of more than a full standard deviation is substantial in most contexts.[8] On the other hand a mean difference of, say, 1/10 of a standard deviation or even 2/10 of a standard deviation is small. Effect sizes are a relatively common way of expressing mean differences in many fields and can also be used to make some rough comparisons among the results of different studies. It may be, for example, that our study of a particular mouthrinse produced a 1.1 standard deviation decrease in plaque formation while Proctor and Lever's study, performed under similar conditions, produced only 1/2 of a standard deviation decrease. Even if the standard deviations of our study and Proctor and Lever's study were different, the effectiveness of our mouthrinse is roughly twice that of P&L's.

The sample difference $\bar{x}_1 - \bar{x}_2$ and the effect size $(\bar{x}_1 - \bar{x}_2)/s$ are both estimates of the population mean difference, expressed on two different scales. They are useful following a significant F-test to understand the direction and magnitude of differences that were obtained. The effect size is similar to the t-statistic for comparing two means, except that the t-statistic incorporates the sample n's for determining statistical significance. Thus, an effect size is a simple expression of the size of a mean difference, while the t-statistic reveals whether a difference is large *given* the size of the sample.

Confidence Intervals

A confidence interval for the difference of the two means may be computed as described in Section 12.2. The endpoints of the interval are obtained by subtracting an amount from the sample mean difference, and adding the same amount, so that the resulting interval contains the population mean difference with a prespecified confidence level. The $1 - \alpha$ confidence interval for $\mu_1 - \mu_2$ is

$$\left[\bar{x}_1 - \bar{x}_2 - t_{n-2}(\alpha/2)s\sqrt{\frac{1}{n_1} + \frac{1}{n_2}}, \quad \bar{x}_1 - \bar{x}_2 + t_{n-2}(\alpha/2)s\sqrt{\frac{1}{n_1} + \frac{1}{n_2}} \right].$$

As with the effect size, s^2 is the estimate of σ^2, that is, s_p^2 from the t-test or MS_W from the analysis of variance; s is its square root.

A 95% confidence interval for the hypothetical data above requires the tabled value $t_{20}(.025) = 2.086$. The amount to be subtracted from

---

[8]One way to form such a judgment is by comparison to the normal distribution. A difference of one standard deviation in the center of the normal curve, for example from $z = -0.5$ to $z = 0.5$, encompasses 38% of the entire population.

and added to $\bar{x}_1 - \bar{x}_2$ is

$$2.086 \times 5.0\sqrt{\frac{1}{10} + \frac{1}{12}} = 4.47,$$

and the resulting interval has lower and upper limits

$$5.50 - 4.47 = 1.03 \quad \text{and} \quad 5.50 + 4.47 = 9.97,$$

respectively.

Comparing More Than Two Means

When three or more groups are compared by the analysis of variance, several specific comparisons may be made after the overall hypothesis of equality is rejected. For example, if there are three groups and H_0 is rejected, μ_1 may be compared with μ_2, μ_1 may be compared with μ_3, and μ_2 may be compared with μ_3.

It is up to the researcher to decide which comparisons to make. The decision rests partially on the design of the research. For example, if group 1 is a control group and groups 2 and 3 are two different experimental conditions, then it would be sensible to compare μ_2 with μ_1 and μ_3 with μ_1, that is, both experimental conditions with the control. If students from four different universities are being compared on mean scores on the Law School Admissions Test, every school's mean might be routinely compared to every other school's mean. In the latter case, the number of pairwise comparisons among k means is $k \times (k - 1)/2$.

The procedure for *any one* of these comparisons is very much like comparing two groups as described in Section 16.2, but one additional factor needs to be considered: the probability of making a Type I error when performing several tests of significance from the same data set. If, indeed, all the μ's were equal, so that there were no real differences, the probability that any particular *one* of the pairwise differences would exceed the corresponding t-value is α. However, the probability of making at least one Type I error out of *two or more* pairwise comparisons is greater than this. That is, when many differences are tested, the probability that some will appear to be "significant" when the corresponding population means are equal is greater than the nominal significance level α. The more comparisons that are made, the greater the probability of making at least one Type I error.

How can a researcher protect against too high a Type I error rate? One widely used approach is based on the *Bonferroni inequality*. The Bonferroni inequality states that the probability of making *at least one Type I error out of a given set of comparisons* is less than or equal to the sum of the α's used for the separate comparisons. Thus, for example, if we make 2 comparisons among 3 group means and use an α level of

.05 for each comparison, the probability of making at least one Type I error is no greater than .10. This overall α level for the pair or "family" of tests is called the *familywise (or experimentwise) Type I error rate*.

The Bonferroni inequality can be put to use to keep the familywise error rate acceptably small. Suppose that we wish to make m specific pairwise comparisons. We can then decide on a reasonable familywise error rate (α) and divide this value by m to obtain a significance level to be used for each comparison separately; call this result α^*. If α^* is used for each of m comparisons, the probability of making at least one Type I error out of the set is not greater than $m \times \alpha^* = \alpha$.

We illustrate this procedure with the data of Example 1 (see Tables 16.1 and 16.5). Suppose that we wish to ask specifically whether each mean is significantly different from each other mean. Since the number of groups is 3, this will involve making $m = 3 \times (3 - 1)/2 = 3$ comparisons: μ_1 with μ_2, μ_1 with μ_3, and μ_2 with μ_3. If we want the overall Type I error probability to be at most .03, say, then the significance level we shall use for testing each comparison is $\alpha^* = .03/3 = .01$.

For Example 1 (Table 16.5) the estimate of σ^2 is $MS_W = 1.867$ based on 15 degrees of freedom. The significance points for any one comparison are taken from the table of the t-distribution with 15 degrees of freedom.[9] With $\alpha^* = .01$, the significance points for a two-tailed test are $\pm t_{15}(.005)$, that is, ± 2.947.

The sample means from Example 1 are $\bar{x}_1 = 4$, $\bar{x}_2 = 9$, and $\bar{x}_3 = 5$, based on $n_1 = n_2 = n_3 = 6$ observations. The test statistic for comparing μ_1 with μ_2 is

$$t = \frac{\bar{x}_1 - \bar{x}_2}{\sqrt{MS_W} \times \sqrt{\frac{1}{n_1} + \frac{1}{n_2}}}$$

$$= \frac{4 - 9}{1.366 \times \sqrt{\frac{1}{6} + \frac{1}{6}}} = -6.34.$$

This value is smaller than -2.947 and we conclude that $\bar{x}_1$ and $\bar{x}_2$ are significantly different. Providing a 3-category information set to children results in greater word memorization than providing no organizing information.

Applying the same procedure to compare μ_1 and μ_3 gives a nonsignificant t-value of -1.27 and we conclude that these two means are not different. It may be that a very detailed information set is too difficult for children to process and so does not help with word

---

[9]Note that since the analysis of variance requires an assumption that the three population variances are equal, MS_W and its associated degrees of freedom from *all 3* groups are used here even though the means of just two groups are being compared.

memorization. Finally, the comparison of μ_2 with μ_3 gives $t = 5.07$, which exceeds $+2.947$, and we conclude that a 3-category information set is superior to the 6-category set.

In sum, the follow-up tests revealed that μ_2 is greater than both μ_1 and μ_3, while μ_1 and μ_3 are indistinguishable. The study reveals that 3 categories of organizing information facilitates word memorization more than providing no organizing information and more than providing 6 categories of information; the latter two conditions produce results that are equally poor.

This procedure of comparing the 3 pairs of means could also have been done without a preliminary test of the "overall" hypothesis H_0: $\mu_1 = \mu_2 = \mu_3$. In fact, the set of 3 comparisons constitutes a test of the same hypothesis in the sense that if all 3 comparisons are not significant, the conclusion is that the data do not contradict the overall null hypothesis; H_0 is accepted. This fact suggests that the familywise Type I error probability should be similar to the significance level of the overall F-test.

One-tailed tests may be conducted if the study originated with *several specific hypotheses* about which groups would have higher or lower means than other groups (and not if the sample means were used to suggest these "hypotheses" after the data were collected). Two-tailed tests are necessary if the researcher is simply examining the means after the F-test to see which are significantly different. This would certainly be the case if all possible pairs $[k \times (k - 1)/2$ comparisons] are being tested.

General Procedure.   In general the test statistic for comparing any two means μ_g and μ_b is computed from $\bar{x}_g$ and $\bar{x}_b$ based on n_g and n_b observations, respectively, it is

$$t = \frac{\bar{x}_g - \bar{x}_b}{\sqrt{MS_W} \times \sqrt{\dfrac{1}{n_g} + \dfrac{1}{n_b}}}.$$

This general form allows for the possibility that the two groups being compared have different numbers of observations. The test statistic is compared with percentage points of the t-distribution with $n - k$ degrees of freedom.

The magnitudes of significant differences are estimated by subtracting the sample means or by computing effect sizes. For Example 1, the condition-2 mean exceeded the condition-1 mean by $9 - 4 = 5$ words, on average, and condition 2 exceeded condition 3 by an average of $9 - 5 = 4$ words. To help understand whether these differences are important, each of these may be divided by the standard deviation $\sqrt{MS_W} = 1.366$. The two mean differences are equivalent to 3.66

standard deviations and 2.93 standard deviations, respectively, which indicate large differences in word memorization. In general, an effect size for comparing $\bar{x}_g$ with $\bar{x}_h$ is

$$\frac{(\bar{x}_g - \bar{x}_h)}{\sqrt{MS_W}}.$$

Confidence Intervals

The Bonferroni inequality can also be used to obtain simultaneous confidence intervals for two or more comparisons. The confidence will be *at least* $1 - \alpha$ that *all* of the intervals contain their respective population mean differences. The confidence interval for the difference between μ_g and μ_h, when this is one of a set of m confidence intervals, is

$$\left[\bar{x}_g - \bar{x}_h - t_{n-k}(\alpha^*/2)\sqrt{MS_W\left(\frac{1}{n_g} + \frac{1}{n_h}\right)}, \right.$$

$$\left. \bar{x}_g - \bar{x}_h + t_{n-k}(\alpha^*/2)\sqrt{MS_W\left(\frac{1}{n_g} + \frac{1}{n_h}\right)} \right],$$

where $\alpha^* = \alpha/m$.

We illustrate with the television viewing data of Table 16.9. To assess the impact of television viewing on school grades, we shall compare the mean GPA for youngsters who watch little or no television (group 1) with the means for groups 2, 3, and 4. Thus the number of contrasts is $m = 3$. If we require .85 confidence for the set of 3 intervals, then $\alpha = .15$, $\alpha^* = .15/3 = .05$, and each interval is obtained with confidence level $1 - .05 = .95$.

The Mean Square Within Groups for the study (Table 16.10) is $MS_W = 0.25$. Altogether the study involved $n = 12$ subjects in $k = 4$ groups, so that the degrees of freedom within groups is $12 - 4 = 8$. The confidence intervals require the value from the t-distribution with 8 degrees of freedom that has .025 probability of being exceeded. From Appendix III this is $t_8(.025) = 2.306$.

To compare μ_1 with μ_2, the sample means are $\bar{x}_1 = 3.5$ based on $n_1 = 3$ observations, and $\bar{x}_2 = 3.0$ based on $n_2 = 4$ observations. The mean difference is $\bar{x}_1 - \bar{x}_2 = 0.5$, and the confidence interval is from

$$0.5 - 2.306 \times \sqrt{0.25\left(\frac{1}{3} + \frac{1}{4}\right)} \quad \text{to} \quad 0.5 + 2.306 \times \sqrt{0.25\left(\frac{1}{3} + \frac{1}{4}\right)},$$

or from -0.38 to 1.38. Because the interval contains zero, there is no evidence that watching small amounts of television (less than $1\frac{1}{2}$ hours per day) is associated with reduced school grades.

To compare μ_1 with μ_3 we require $\bar{x}_3 = 2.5$ based on $n_3 = 3$ observations. The mean difference is $\bar{x}_1 - \bar{x}_3 = 1.0$, and the confidence interval is from

$$1.0 - 2.306 \times \sqrt{0.25 \left(\frac{1}{3} + \frac{1}{3} \right)} \quad \text{to} \quad 1.0 + 2.306 \times \sqrt{0.25 \left(\frac{1}{3} + \frac{1}{3} \right)},$$

or from 0.06 to 1.94. The interval does not contain 0 so we are confident that watching between $1\frac{1}{2}$ and 3 hours of television per day is associated with reduced school grades. The reduction is *at least* 6/100 of a letter grade but may be substantially larger.

The comparison with μ_4 requires $\bar{x}_4 = 1.5$ based on $n_4 = 2$ observations. The mean difference is $\bar{x}_1 - \bar{x}_4 = 2.0$, and the resulting confidence interval is from 0.95 to 3.05. Watching more than 3 hours of television per day is associated with a reduction in school grades of almost an entire letter grade *at the minimum*, and perhaps as much as three whole letter grades.[10] Of course, a nonexperimental study of this sort does not tell us whether grades are reduced because of watching too much television or whether students watch more television because their grades are low and they are frustrated with school work.

The Bonferroni inequality is used widely among researchers to control Type I error rates when several comparisons are made. There are numerous other procedures for controlling the familywise α in particular situations, for example, when the means of several experimental groups are compared to a control group. These methods are generally called *multiple comparisons* and are described in more advanced statistics books, for example, Ott (1988).

16.3 Analysis of Variance of Ranks*

In the previous sections we compared the distributions of a variable in several populations by focusing on the means of the samples. The *F*-test relies on the assumptions that the variable has a normal distribution in the population and that the population variances are the same. The locations of distributions can also be compared through an analysis of *ranks*. This approach can be applied even when the data are ordinal, and does not require the assumption of normality or equal variances. The null hypothesis is that the locations of the populations are the same, and the alternative hypothesis is that they are not.

---

[10]We will not bother with effect sizes in this example because most of us are familiar with the GPA scale. We know, for example, that a one-unit difference is the difference between an "A average" and a "B average" or between a "C average" and a "D average" and can be very meaningful.

The test procedure was developed by Kruskal and Wallis (1952). Table 16.11 gives some data from their original article: the daily outputs of three bottle-cap machines. We shall use these 12 values to test whether, in general, the three machines produce equal numbers of bottle caps.

TABLE 16.11

Daily Outputs of Three Bottle-Cap Machines

Machine A:	340, 345, 330, 342, 338
Machine B:	339, 333, 344
Machine C:	347, 343, 349, 355

source: Kruskal and Wallis (1952).

Sometimes the data we collect are already in the form of ranks. For example, ratings of an artistic or athletic performance may be in terms of best (rank 1), second best (rank 2), and so on. Or we may ask whether students' grades on an exam are related to the order in which they finish the test, with the first "completer" ranked 1, the second completer 2, and so on. If this is the case, then the Kruskal-Wallis analysis can be computed from the raw data.

More often, raw data are not in the form of ranks and the first step is to rank-order the data values for the entire sample. Thus, the lowest value in Table 16.11 (330) receives rank 1, the next lowest (333) receives rank 2, and so on up to the highest value (355) that receives rank 12. The ranks are given in Table 16.12.

TABLE 16.12

Ranks in the Combined Sample

	Ranks	*Sum of ranks*	*Mean rank*	*Sample size*
Machine A	5, 9, 1, 6, 3	24	4.80	5
Machine B	4, 2, 8	14	4.67	3
Machine C	10, 7, 11, 12	40	10.00	4
Sum		78		12

source: Table 16.11.

The test applies analysis of variance formulas to these ranks, replacing scores x_{gi} by r_{gi}, the rank in the whole sample of the ith observation in the gth group. The formulas simplify somewhat, however, because ranks are just consecutive integers from 1 to n. The Sum of Squares

Between Groups is the same as (16.3) but computed from ranks,[11] that is,

$$SS_B = \sum_g \frac{\left(\sum_i r_{gi}\right)^2}{n_g} - n\frac{(n+1)^2}{4}.$$

The Total Sum of Squares is the same as (16.2) but computed from ranks, that is,

$$SS_T = \frac{n(n+1)(2n+1)}{6} - n\left(\frac{n+1}{2}\right)^2 = \frac{n(n+1)(n-1)}{12}.$$

In the analysis of ranks, SS_T is divided by the total degrees of freedom, $n-1$, to obtain a "Total Mean Square." This is

$$MS_T = \frac{SS_T}{n-1} = \frac{n(n+1)}{12}.$$

Unlike the analysis of variance, the test statistic for the analysis of ranks is the ratio of the Sum of Squares Between Groups to the Total Mean Square, that is,

$$H = \frac{\sum_g \left(\sum_i r_{gi}\right)^2 / n_g - n(n+1)^2/4}{n(n+1)/12}$$

$$= \frac{12}{n(n+1)} \sum_g \frac{\left(\sum_i r_{gi}\right)^2}{n_g} - 3(n+1).$$

The null hypothesis of no difference between the locations of the three populations is equivalent to random sampling, that is, that the ranks have been allocated at random to the k groups. When this hypothesis is true, the sampling distribution of H is approximately χ^2 with $k-1$ degrees of freedom if the sample sizes are large.[12]

In the example,

$$H = \frac{12}{12 \times 13}\left(\frac{24^2}{5} + \frac{14^2}{3} + \frac{40^2}{4}\right) - 3 \times (12+1) = 5.66.$$

The test statistic is compared to significance points from the χ^2-distribution with $k - 1 = 2$ degrees of freedom. If we choose $\alpha = .10$, the significance point is $\chi_2^2(.10) = 4.605$ and H_0 is re-

---

[11]These results use the fact that $1 + 2 + 2 + \cdots + n = n(n+1)/2$ and thus the mean is $(n+1)/2$. Also $1^2 + 2^2 + 3^2 + \cdots + n^2 = n(n+1)(2n+1)/6$.

[12]Although the sample sizes in the example are rather small, we will continue for purposes of illustration.

jected ($.05 < P < .10$). We conclude that the three machines do not have equal productivity, with Machine B producing the fewest bottle caps and Machine C the most. If the test statistic did not exceed the significance point, we would conclude that the productivity was the same for all three machines.

Summary

The analysis of variance comprises a set of procedures for comparing two or more population means. The total variability of the data is allocated to two sources: variability among the group means and variability of the observations within the groups. Two measures of variability are computed, Mean Square Between Groups and Mean Square Within Groups, respectively. The hypothesis that the k population means are equal is tested by comparing the magnitude of these two mean squares. The F-ratio is the ratio of Mean Square Between Groups to Mean Square Within Groups. The F-distribution provides significance points for deciding if the ratio is large enough to reject the hypothesis of equal means.

A finding that the means are not equal is followed by further statistical analysis to discover which of the k group means differ from which others, and the direction and magnitude of the difference. Since this may involve conducting several follow-up tests of significance, or computing several confidence intervals, it is important to control for the overall "familywise" probability of a Type I error. This may be done using the "Bonferroni inequality" that assures that the overall Type I error rate is never greater than the sum of the Type I error rates for all of the tests conducted.

The analysis of variance is based on the conditions that the sample means have a normal distribution and that the populations being compared have the same variance σ^2. When these conditions are not met, or when the raw data are ordinal, the Kruskal-Wallis analysis of ranks may be used instead of the more common analysis of variance of means.

Exercises

16.1 In the March 1993 issue of *The Journal of Psychology*, Persinger, Ballance, and Moland reported data on the relationship of snow fall and admissions to three Canadian hospitals for heart attacks during the months of November through March, 1983 through 1986. Table 16.13

is adapted from their report. It lists the mean and standard deviation of the daily number of admissions for cardiac emergencies under 4 conditions of snowfall: days on which there was no snowfall, days on which the snowfall was more than 1 but less than 2 standard deviations above average, days on which the snowfall was more than 2 but less than 3 standard deviations above average, and days on which the amount of snowfall was more than 3 standard deviations above average.

Since the effects of poor weather may not be felt until several days afterward, the mean number of admissions was computed on the day of the snowfall (lag 0), 1 day later (lag 1), 2 days later (lag 2), and 3 days later (lag 3). For each lag, the researchers used the analysis of variance to see if the mean number of cardiac admissions is related to the severity of the snowfall. All together, 64 snowfalls were studied so that the sample size for each comparison is $n = 64$.

TABLE 16.13

Means, Standard Deviations, and F-statistics for Comparing Average Number of Hospital Admissions Across Four Snow Conditions

	Amount of Snowfall								
	0		1 SD		2 SD		3 SD		
Lag	*Mean*	*SD*	*Mean*	*SD*	*Mean*	*SD*	*Mean*	*SD*	*F*
0	2.4	1.5	2.1	1.2	2.6	1.7	2.6	1.3	0.82
1	2.4	1.5	2.5	1.4	2.8	1.7	1.9	1.0	1.14
2	2.4	1.5	2.2	1.5	2.8	1.2	2.5	1.6	0.81
3	2.4	1.5	2.0	1.4	2.1	1.7	2.6	1.5	1.88

source: Persinger, M. A., Ballance, S. E., & Moland, M. (1993). Snow fall and heart attacks. *The Journal of Psychology, 127*, 243–252. Reprinted with permission of the Helen Dwight Reid Educational Foundation. Published by Heldref Publications, 1319 Eighteenth St., N.W., Washington, D.C. 20036. Copyright © 1993.

(a) What are the numbers of degrees of freedom for each comparison of the four amounts of snowfall?

(b) What is the significance point for the comparison of mean numbers of admissions for the four amounts of snowfall, if $\alpha = .05$?.

(c) Is the hypothesis of equal mean numbers of admissions for the 4 amounts of snowfall accepted or rejected for the days on which the snowfall occurred (lag-0)?

16.2   (continuation) Determine if the hypothesis of equal mean numbers of admissions for the 4 amounts of snowfall is accepted or rejected for 1, 2, and 3 days after the snowfall (also at the 5% level).

(a) Find a *P* value interval for each of the four test statistics obtained for comparing the 4 amounts of snowfall.
(continued . . .)

(b) Summarize the results of the entire study in a short succinct paragraph. Include a statement about which "lag" shows the strongest association of snowfall with admissions (from the P values).

16.3 A retail business has two stores and is often transferring merchandise between the two. There are four reasonable routes between the stores. Table 16.14 shows the numbers of minutes needed to make the trip on six different occasions for each route. Is there reason to think that in the long run it will matter which route is taken? Test the hypothesis of no difference in the true mean time it takes to drive from store to store along the four different routes, at the 5% significance level.

TABLE 16.14
Times (Minutes) by Four Different Routes

		Sum	*Mean*
Route 1:	39, 36, 43, 37, 33, 40	228	38
Route 2:	34, 41, 38, 45, 35, 29	222	37
Route 3:	36, 46, 47, 42, 48, 51	270	45
Route 4:	53, 57, 54, 52, 58, 50	324	54

Hypothetical data.

16.4 (continuation) Compare each of the routes with each of the other routes using a t-test for equality of means. Use a familywise α level of 0.12. Summarize your findings in one or two succinct sentences. For each pairwise comparison that is significant, compute the difference of the two respective sample means and include the "advantage" (in minutes) of taking the shorter of the two routes.

16.5 (See Example 3 of the Introduction.) Figure 16.3 gives the distribution of the number of academic freedom incidents by type of control of college. Only 46 "moderately large" schools (student bodies of 2500 to 9000) are included to improve comparability. Table 16.15 gives some statistics derived from the distributions. These are to be used in making a test of the hypothesis that the mean number of incidents is the same for the three populations from which the schools were drawn.

(a) Compute MS_B and MS_W.

(b) Test at the 1% level the hypothesis that the population means are identical.

(c) What conclusion do you draw?

16.6 (See Example 4 of the Introduction.) The job performance of a sample of 22 employees in a large manufacturing firm was rated by their supervisors. The rating form was extensive and resulted in a total proficiency score for each employee ranging from 0 to 200. Employees were

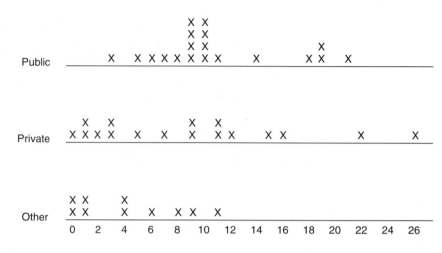

Figure 16.3 *Distributions of number of academic freedom incidents in three types of colleges (Source: Lazarsfeld and Thielens (1958)).*

TABLE 16.15

Summary Statistics for Distributions of Number of Academic Freedom Incidents in Three Types of Colleges

Type of control	Sum	Sum of squares	Number of colleges
Public	207	2711	19
Private	153	2287	17
Other	44	336	10

source: Figure 16.3.

classified into one of three educational attainment groups: high-school noncompleters, high-school graduates, and college attendees. The ratings are given in Table 16.16. Use these data to test the hypothesis of equal mean proficiencies for the three levels of education.

(a) State H_0 using statistical notation.

(b) Use the analysis-of-variance F-test to compare the mean proficiencies of the three education groups, with $\alpha = .01$.

(c) Make an analysis-of-variance source table summarizing your findings, and put the significance point from the F-distribution below the table.

(d) Summarize the result of this test in words. In your summary state whether H_0 is accepted or rejected and what this means in terms of average job proficiencies. What recommendation would you make to the employer about hiring new employees *based on these results*?

TABLE 16.16
Job Proficiency Ratings of Employees Classified by Educational Attainment

Noncompleters	High-school	College
174	146	182
161	182	159
186	149	170
122	138	180
102	180	192
178	176	104
	155	152
	114	
	190	

Hypothetical data.

16.7 (continuation) Explain why the sample means are or are not significantly different in the preceding exercise by making a line graph of the data for each group, as in Figure 16.3. Examine the magnitude of variability within the groups compared to variability among the group means. What does this comparison show?

16.8 (continuation) Using the data in Table 16.16, compare the mean job proficiency of high-school noncompleters with high-school graduates, and high-school graduates with college attendees using a familywise α level of 0.02. Summarize your findings in words. Are the results of these pairwise comparisons consistent with the conclusion you reached about mean differences in Exercise 16.6?

16.9 Table 2.21 gives final grades for 24 students in a graduate statistics course, and also the number of previous semesters of statistics that each student had taken. Classify those students according to the number of prior statistics courses (0, 1, 2, or 3) and compare the mean final grades for the 4 groups using an analysis-of-variance F-test. Summarize your results in a written report as follows:

(a) State the null hypothesis using statistical notation.

(b) Make a summary table giving the n's and means for all 4 groups.

(c) Make an analysis-of-variance source table that includes the value of the test statistic.

(d) Give a P value interval corresponding to your F-statistic as a footnote to the table.

(e) Using the 5% level of significance, state in words whether H_0 is accepted or rejected and what this means in terms of average course grades and prior statistics courses.

16.10 (continuation) From your now-extensive knowledge of statistics courses, select a set of pairwise comparisons that you think are "meaningful." For example, is the most important experience simply that of having one or more prior courses? If so, you might want to compare the mean final grades of students who had 1 course, 2 courses, and 3 courses, respectively, to the grades of those who had none. If you think that each additional prior course is helpful, you might want to compare the mean grades of those who had 1 course to those who had none, those who had 2 courses to those who had 1, and those who had 3 courses to those with 2.

Choose your own set of comparisons and justify your choice in words. Choose an appropriate familywise α level. Conduct the pairwise tests and report your statistical results including the α^* value and significance points you used for each comparison, the value of the test statistic for each comparison, and your conclusions for each comparison. In one or two brief sentences, summarize your findings about the relationship of prior statistics courses and final grade in this particular graduate course.

16.11 In 1995, Frisancho et al. reported a study of the aerobic capacity of five groups of adults living in La Paz, Bolivia: High Altitude Rural Natives (HARN), High Altitude Urban Natives (HAUN), Bolivians of foreign ancestry who were acclimatized to high altitude since birth (AHAB), Bolivians of foreign ancestry who were acclimatized to high altitude during growth (AHAG), and non-Bolivians acclimatized to high altitude during adulthood (AHAA). Each participant performed an intensive exercise routine following which a number of energy-related measures were taken. The results for heart rate (HR) and for "maximal aerobic capacity" (VO_2STPD) for the male participants are summarized in Table 16.17. Aerobic capacity is the amount of oxygen intake per unit of body weight; following strenuous exercise, lower heart rates and higher aerobic capacity are preferable.

Compare the mean heart rate of the five groups through the analysis of variance. The entries in the ANOVA source table can be obtained as follows:

(a) Compute the Sum of Squares Between Groups from the means in Table 16.17 using the definition of SS_B in the top line of (16.3). Compute the degrees of freedom and the Mean Square Between Groups in the usual manner.

(b) The Mean Square Within Groups is given with Table 16.17. Compute the number of degrees of freedom within groups. For completeness, compute the Sum of Squares Within Groups by multiplication and enter its value in the ANOVA table.

(c) Compute the F-statistic. Obtain a P interval by comparing the test statistic to significance points of the F-distribution with the closest

TABLE 16.17
Results of Aerobic Testing for Bolivian Males

Group	n_g	Heart Rate Mean	SD^a	Maximal Aerobic Capacity Mean	SD^a
HARN	39	167.28	13.49	48.24	5.93
HAUN	32	183.59	10.35	40.00	4.24
AHAB	33	179.76	11.26	40.43	4.77
AHAG	25	178.84	14.80	42.95	6.05
AHAA	25	171.75	13.05	36.69	7.70
All participants[b]	154	175.95		42.12	

source: Adapted from Frisancho et al., Developmental, genetic, and environmental components of aerobic capacity at high altitude. *American Journal of Physical Anthropology*, p. 435, ©1995. Reprinted by permission of John Wiley & Sons, Inc.
[a]The published report gives a standard error for each mean from which standard deviations were computed for this table.
[b]The Mean Square Within Groups was computed from the standard deviations given above. For Heart Rate, $MS_W = 158.64$; for Maximal Aerobic Capacity, $MS_W = 33.05$.

(lower) denominator degrees of freedom. State whether the hypothesis of equal means is or is not rejected at the 5% level of significance.

16.12 (continuation) The cardiovascular functioning of high-altitude natives has often been of interest to scientists. Use the heart rate data of Table 16.17 to examine the contrast of each group except HARN with the mean of the HARN group. For each contrast, compute the difference between the sample means, a test of the hypothesis of no difference between population means (use a common variance estimate), and a P interval from the table of the t-distribution with the closest (lower) degrees of freedom. Assuming a familywise error rate of .10, it is reasonable to test each contrast at the 2.5% level; mark those contrasts that are statistically significant at this level.

16.13 (continuation) Repeat Exercises 16.11 and 16.12 for maximal aerobic capacity. Construct a table that gives the mean difference and P interval for all 8 contrasts, that is, for both heart rate and maximal aerobic capacity. Write a brief summary of your analysis that includes answers to the following questions:

- Are some of the groups significantly different from high-altitude rural natives in terms of heart rate or maximal aerobic capacity, while other groups are not?
- For groups that are significantly different, how large is the difference and which group is in the preferred direction?
  (continued ...)

- Does there appear to be an advantage associated with living in a high-altitude *rural* area? With living in a high-altitude since birth, whether or not you are a native? To having Bolivian ancestry?

16.14 Managers of a large retail chain are concerned because sales are dropping. They wonder if it is because the sales staff is dressing too informally, thus detracting from the image that the company would like to project. They choose a random sample of salespersons and classify their appearance as "formal and neat," "casual but neat," or "unkempt." The average daily number of sales is recorded for each salesperson over a one-month period, yielding the data in Table 16.18.

TABLE 16.18
Average Daily Numbers of Sales for Salespersons Classified by Dress Style

Formal	Casual	Unkempt
23	28	17
18	33	25
16	37	24
20	34	19
21	29	22

Hypothetical data.

Perform an analysis-of-variance *F*-test of these data to test the hypothesis of equal mean numbers of sales for the three dress styles, using $\alpha = .01$. Is the hypothesis accepted or rejected?

16.15 (continuation) For the data of Table 16.18

(a) Compute the effect size for the the comparison of each dress style with each other dress style;

(b) Construct a confidence interval for the difference in mean numbers of sales between each dress style and each other dress style, using a familywise confidence level of .94.

(c) Summarize the results of this study by stating which dress style is associated with the greatest average number of sales, which other dress styles it is superior to, and by how much, that is, the mean differences and the effect sizes.

16.16 (continuation) If the managers of the retail chain were considering imposing a dress code to implement their findings, it might be preferable if the study were conducted as an *experiment*. How would you redesign the same study so that it is an experiment? What would be the primary advantages and disadvantages of doing so?

16.17 Refer to Exercise 16.3 (Table 16.14). Construct a confidence interval for the mean difference of each of the routes compared to each of the other routes using a familywise confidence level of 0.88.

16.18 In January and February of 1991, during the Persian Gulf War, a comparison was made of white and minority students' attitudes toward American participation in the war (Lee, 1993). A questionnaire was administered to 47 minority students and 104 white students representing four different racial/ethnic groups who were enrolled in a psychology course at State University of New York at Stony Brook. The author concluded that "The minority students were less in favor and more opposed to American involvement in the Gulf War than the nonminority students were" (p. 710).

This conclusion was based on two attitude measures, one of which is "patriotic attitudes related to the war." For this scale, the mean for minority students was $\bar{x}_1 = 16.13$ and the mean for white students was $\bar{x}_2 = 23.90$; the common within-group variance was $s_p^2 = MS_W = 104.719$.

Use these results to confirm the author's conclusion, as follows:

(a) Compute the t-statistic for comparing the population means of white and minority students. Find the significance point for a two-tailed test at the 1% level and decide whether the hypothesis of equal means is accepted or rejected.

(b) Compute the Mean Square Between Groups from the sample n's and sample means, and compute the analysis-of-variance F-statistic. Show your work.

(c) Confirm that the F-statistic and the significance point from the F-distribution are the squares of the respective t-values from part (a), and that the same conclusion is reached.

16.19 (continuation) Describe the magnitude of the difference between white and minority students found by Lee as follows:

(a) Express the mean difference in patriotic attitudes as an effect size. Would you say that this difference is small, moderate, or large? Why?

(b) Construct a 90% confidence interval for the mean difference. Once you have obtained lower and upper limits for the interval, re-express each of these differences as an effect size.

16.20 Art galleries from five large cities have entered pieces from their collections in a nationwide contest. The judges ranked all 20 pieces from 1 (best) to 20 (worst); the ranks are given in Table 16.19. Use the analysis of variance of ranks to test the null hypothesis that all five galleries have art works of equal average quality. Use the 5% level of significance.

TABLE 16.19

Ranks of 20 Pieces of Art from Galleries in Five Cities

City	Ranks
YN	1, 3, 6, 9
FS	2, 11, 14
C	4, 7, 8, 13
B	10, 15, 18
AL	5, 12, 16, 17, 19, 20

Hypothetical data from hypothetical cities.

16.21   Use the analysis of variance of ranks to make a 5%-level test of equality of locations for the data of Exercise 16.14. Do you reach the same overall conclusion as you did using an analysis-of-variance F-test?

16.22   Expression (16.1) shows that the deviation of a single score from the overall mean can be expressed as the sum of two components, the deviation of the group mean from the overall mean and the deviation of the observation from its group mean, respectively, that is, $(x_{gi} - \bar{x}) = (\bar{x}_g - \bar{x}) + (x_{gi} - \bar{x}_g)$. It is stated without proof that the sum of squares of the $(x_{gi} - \bar{x})$ is equal to the sum of squares of the $(\bar{x}_g - \bar{x})$ plus the sum of squares of the $(x_{gi} - \bar{x}_g)$, that is,

$$\sum_g \sum_i (x_{gi} - \bar{x})^2 = \sum_g n_g(\bar{x}_g - \bar{x})^2 + \sum_g \sum_i (x_{gi} - \bar{x}_g)^2.$$

Demonstrate this fact algebraically. [Hint: Expression (16.1) is of the form $a = b + c$. Squaring both sides gives $a^2 = b^2 + 2bc + c^2$. The sum-of-squares partition is of the form $\sum a^2 = \sum b^2 + \sum c^2$. Thus the problem requires demonstrating that $\sum 2bc = 0$.]

17

Sampling from Populations: Sample Surveys

Introduction

Much empirical data arises from experiments, in which the investigator interacts in some way with the units of observation and actually influences the conditions of the units leading to the measurements. Many other sets of data result from simply *observing*, that is, making a *survey*. It is to such investigations that we now turn our attention. Usually one cannot observe every individual in the population, and often this would not even be desirable, for many individuals are similar. One does not need to eat the whole bowl to learn how the soup tastes; a spoonful will suffice, provided that the soup has been adequately stirred. The "spoonful" is a *sample* from the bowl (*population*), and "stirring" corresponds to drawing a random sample.

Sample surveys are especially useful in business and economics, market research, sociology, and industrial quality control. A survey often involves an interview, a questionnaire, or some sort of inspection. Care must be taken in working out the details of administration of the interview, questionnaire, or inspection. The planning of a survey should involve two kinds of professionals working together—a subject-matter specialist (economist, sociologist, production engineer) and a statistician.

Suppose that one wants to draw a random sample from among the households in a county to find out how many hours a day children watch television. Individual children are the "units of observation." Households are the "sampling units." The list of households on the county tax roll is a "frame." The "population sampled" consists of the children in the county who live in households on the tax roll. The "target population" is the set of all children who live in the county.

In general, the *target population* is the population about which it is desired to make inferences. Any difference between the *population sampled* and the target population is a potential source of bias. The *frame* is a physical list of the sampled population. Each *sampling unit* contains one, more than one, or no units of observation.

Many problems would arise in carrying out the survey of TV-viewing habits. How should the survey be designed? For how many days should each child be observed? If it is decided to monitor the TV viewing of the children for two weeks, which two weeks should be chosen? Should different two-week periods be chosen for different subsamples of children? How can one ensure that the parents keep an accurate record of how long their children watch TV? If there is more than one child in a household, should all of them be included in the sample? Should some inducement to participate be offered?

When a survey involves an expenditure of time or effort on the part of those selected to be in the sample, there are almost always some individuals who refuse to respond. When the sampling unit is a household, an interviewer may find no one home when he or she calls a household designated to be in a sample. Such failures to obtain information from each unit meant to be sampled can introduce some bias into the results when those not responding differ systematically from those responding. Bias resulting from this source is termed *nonresponse bias*. It means, in effect, that the population sampled differs from the target population, inasmuch as the population sampled consists only of those individuals in the target population who are willing to respond or can be induced to do so. If, with respect to the characteristic of interest in the study, the population of persons who do not respond differs from those who do, there is a nonresponse bias. A statistical method for adjusting for nonresponse bias will be discussed briefly. In order for any such adjustment to be possible, however, there must be a second try, in which at least some of those who did not respond the first time do in fact respond.

Another type of bias is *interviewer bias*, which results when different persons in the sample are questioned by different interviewers. This may occur because interviewers interpret the instructions given by the director of the survey in slightly different ways, or simply because people react in different ways to different interviewers. Care should be taken to minimize these effects. Sometimes statistical methods, such as analysis of variance, can be employed after the fact to assess the extent of interviewer bias and adjust for it.

Our emphasis in this chapter will be on explanation of several sampling methods, and we shall focus attention on the problem of assessing the accuracy and precision of estimators resulting from the various methods. For such assessment to be possible, the sample drawn must be a *probability sample*, a sample drawn in such a way that the investigator knows the probability that any individual unit of observation will be included in the sample. We shall discuss several sampling methods: simple random sampling, stratified random sampling, cluster sampling, systematic sampling with a random start, and systematic subsampling with random starts. In each case some random device plays a role in determining which members of the population shall be included in the sample.

When sampling involves interviewing people selected in a probability sample, the amount of work involved can be tremendous. W. E. Deming (1950, page 10), a leading sampling expert, puts it this way: "A probability-sample will send the interviewer through mud and cold, over long distances, up decrepit stairs, to people who do not welcome an interviewer; but such cases occur only in their correct proportions. Substitutions are not permitted: the rules are ruthless."

The reader interested in learning more about designing and conducting sample surveys, as well as analyzing the data that result, is referred to books by Cochran (1977), Deming (1950), Hanson, Hurwitz, and Madow (1953), and Levy and Lemeshow (1991).

17.1 Simple Random Sampling

A *simple random sample* of size n is a sample of n drawn in such a way that all sets of n units in the population have the same chance of being in the sample. (There are C_n^N such sets, where N is the population size and C_n^N was defined in Section 9.3. The fact that all these sets are equally likely to be chosen implies in particular that all units have the same chance of being included.) A *frame*, a list of all units with one of the numbers from 1 to N assigned to each, is constructed. A random sample can then be drawn.

The basic ideas of drawing a random sample were presented in Section 7.8. If the sample size is not very large, random numbers can be taken from a table of random numbers such as Appendix VI. Since the frame of N units is finite, the sampling is done without replacement. A computer can be used to obtain a stream of numbers that behave like random numbers (often called pseudorandom numbers). The set of n random numbers determines the n sampling units to be included in the sample. Characteristics of the sampled units are then obtained by interview, written questionnaire, or direct measurement.

In the case of a numerical variable, let the numbers $y_1, y_2, \ldots, y_N$ represent the values of the variable for the N individuals in the population. The population mean μ is $\sum y_k / N$ (Section 8.3), and the population variance is $\sigma^2 = \sum (y_k - \mu)^2 / N$. In a finite population the total, such as total energy used, may be of primary interest rather than the mean, such as average amount of energy used. The total $\sum y_k$ is, of course, N times the mean, that is, $N\mu$.

Let the values of the variable for the n individuals in the sample be $x_1, x_2, \ldots, x_n$, where x_i is the value of the variable for the ith individual *in the sample*, $i = 1, 2, \ldots, n$. (The symbol x_1 bears no special relation to y_1; if $n = 3$ and the sample of three individuals consists of the individuals numbered 13, 17, and 8 in the frame, then $x_1 = y_{13}$, $x_2 = y_{17}$, and $x_3 = y_8$.) The mean of the population μ is estimated by the mean of the sample $\bar{x} = \sum x_i / n$. Then the total $N\mu$ is estimated by $N\bar{x}$. Note that

$$N\bar{x} = \frac{N}{n} \sum x_i;$$

this is the sample total $\sum x_i$ scaled up to population size by multiplying by N/n.

The sample mean $\bar{x}$ is an unbiased estimate of the population mean μ (Section 10.1), that is, $E(\bar{x}) = \mu$ (whether sampling is with or without replacement). The variance of the sample mean for sampling without replacement (Section 10.7) is

$$\sigma_{\bar{x}}^2 = \frac{N-1}{N-1} \frac{\sigma^2}{n}.$$

The finite population correction factor for the variance $(N - n)/(N - 1)$ is approximately $(N - n)/N$, or $1 - n/N$. When the fraction sampled n/N is small, this correction is negligible.

The statistic $\bar{x}$ has a sampling distribution that is approximated by a normal distribution when n is large. Tests and confidence intervals are based on this fact.

EXAMPLE

A survey of a school of 800 students is conducted to ascertain smoking behavior. In a random sample of 40 students (sampling without replacement), the mean number of cigarettes smoked per day is $\bar{x} = 12$ cigarettes and the sample standard deviation is $s = 6$ cigarettes. The estimated standard deviation of the sample mean is

$$\sqrt{\frac{N-n}{N-1} \cdot \frac{s^2}{n}} = \sqrt{\frac{800-40}{800-1} \cdot \frac{6}{\sqrt{40}}} = \sqrt{0.951} \times 0.917$$

$$= 0.975 \times 0.917 = 0.947.$$

A 95% confidence interval for the mean number of cigarettes smoked per student is

$$(12 - 1.96 \times 0.947, \quad 12 + 1.96 \times 0.947) = (10.14, \ 13.86).$$

Adjustment for Nonresponse Bias

Often a nonresponse can be converted into a response. If no one in a household is at home at the time the interviewer calls, the interviewer can return (a "call-back"). If a questionnaire mailed to a respondent is not returned, the respondent can be tried by telephone. Typically, the second attempt is more expensive than the first (for example, the households may be more scattered), and the response on the second round may not be complete.

Suppose that of a sample of 100 taken in March to estimate the unemployment rate in the county there were 10 nonrespondents. Of the 90 respondents, 5 were unemployed, and the unemployment rate was estimated as $\hat{p}_1 = 5/90 = 0.056$. This is an estimate of the unemployment rate *in the population of respondents*. The 10 nonrespondents were revisited in April and asked if they were employed in March; 8 of them responded, 3 saying that they were unemployed in March. The estimate of the unemployment in the population of nonrespondents is $\hat{p}_2 = 3/8 = 0.375$. The overall unemployment rate p is $0.9p_1 + 0.1p_2$, in terms of the rates in the population of respondents and the population of nonrespondents. The corresponding estimate is $0.9\hat{p}_1 + 0.1\hat{p}_2 = 0.9 \times 0.056 + 0.1 \times 0.375 = 0.088$. This is considerably higher than the initial estimate of 5.6%, due to the high rate of unemployment among the nonrespondents.

17.2   Stratified Random Sampling

Randomness in drawing a sample, which is essential to obtaining unbiased estimates, results in sampling variability. In some situations the variability can be reduced without introducing bias by using other information about the population. Suppose that a sample of engineers employed in a large corporation is to be drawn to estimate the mean salary of all engineers. An individual's salary depends heavily on his or her corporate function—whether the position is supervisory or nonsupervisory. If the listing of engineers (a frame) is such that the function of each is identified, it is possible to draw a random sample of each type of engineer and estimate the average salary of each type of engineer in the entire corporation. A weighted average of these two estimates yields an estimate of the average salary of all engineers. The sampling variability of this procedure is less than that of simple random sampling because the variability within the set of supervisory engineers and within the set of nonsupervisory engineers is less than the variability among all the engineers.

A *stratum* is a subpopulation. A set of *strata* is a collection of subsets of individuals in the population such that each individual belongs to one and only one such subset. To use "stratified sampling," it is essential that a frame be available for each stratum; this implies that the sizes of the strata are known.

Two Strata

To begin the development of the theory of stratified random sampling, let us assume two strata: supervisory and nonsupervisory. Let N denote the size of the total population of engineers and μ the mean salary in this population. Suppose that some N_1 of these N have supervisory positions; let μ_1 denote the mean income in this stratum. The other N_2 $(= N - N_1)$ engineers have nonsupervisory positions, and their mean income is μ_2. The parameter μ is equal to $(N_1\mu_1 + N_2\mu_2)/N$. The parameters μ, μ_1, and μ_2 are, of course, unknown to the investigator, but N, N_1, and N_2 are known from information on the sampling frame (the list of engineers).

For the purpose of explaining the principles, suppose that the engineers are numbered in such a way that the numbers $1, 2, \ldots, N_1$ denote the supervisory engineers and the numbers $N_1 + 1, N_1 + 2, \ldots, N_1 + N_2$ denote the nonsupervisory engineers. Here y_i is the outcome of the ith engineer. The mean for supervisory engineers is

$$\mu_1 = \frac{y_1 + y_2 + \cdots + y_{N_1}}{N_1};$$

the mean for nonsupervisory engineers is

$$\mu_2 = \frac{y_{N_1+1} + y_{N_1+2} + \cdots + y_{N_1+N_2}}{N_2}.$$

The overall mean is

$$\mu = \frac{y_1 + y_2 + \cdots + y_{N_1} + y_{N_1+1} + y_{N_1+2} + \cdots + y_{N_1+N_2}}{N_1 + N_2}$$

$$= \frac{N_1\mu_1 + N_2\mu_2}{N}.$$

That is, the mean of all engineers is the weighted mean of the two types of engineers,

$$\mu = \frac{N_1}{N}\mu_1 + \frac{N_2}{N}\mu_2. \tag{17.1}$$

The population has been *stratified* by occupational position (supervisory versus nonsupervisory). To estimate μ by stratified random sampling, one takes a random sample of specified size n_1 from the N_1 supervisory engineers and a random sample of size n_2 from the N_2 nonsupervisory engineers. The estimates of the strata means μ_1 and μ_2 are the sample means $\hat{\mu}_1 = \bar{x}_1$ and $\hat{\mu}_2 = \bar{x}_2$ of the samples from the two strata. As an estimate of μ, we take the weighted average corresponding to (17.1), namely

$$\hat{\mu} = \frac{N_1}{N}\hat{\mu}_1 + \frac{N_2}{N}\hat{\mu}_2 = \frac{N_1}{N}\bar{x}_1 + \frac{N_2}{N}\bar{x}_2. \tag{17.2}$$

The mean of the sampling distribution of $\hat{\mu}$ is

$$E[\hat{\mu}] = E\left[\frac{N_1}{N}\bar{x}_1 + \frac{N_2}{N}\bar{x}_2\right]$$

$$= E\left[\frac{N_1}{N}\bar{x}_1\right] + E\left[\frac{N_2}{N}\bar{x}_2\right]$$

$$= \frac{N_1}{N}\mu_1 + \frac{N_2}{N}\mu_2 = \mu,$$

by (17.1); thus $\hat{\mu}$ is an unbiased estimate of μ. Note that, in order to form the estimate, one has to know the weights N_1/N and N_2/N.
From Section 10.7 we know that

$$\mathrm{Var}(\bar{x}_g) = \frac{N_g - n_g}{N_g - 1}\frac{\sigma_g^2}{n_g}$$

for $g = 1, 2$, where $\text{Var}(\bar{x}_g)$ means $\sigma^2_{\bar{x}_g}$ and σ^2_g is the variance within the gth stratum. That is, $\sigma^2_1 = \sum(y_k - \mu_1)^2/N_1$ for members of population 1 and $\sigma^2_2 = \sum(y_k - \mu_2)^2/N_2$ for members of population 2. The variance of the estimate $\hat{\mu}$ is

$$\text{Var}(\hat{\mu}) = \text{Var}\left(\frac{N_1}{N}\bar{x}_1 + \frac{N_2}{N}\bar{x}_2\right) \tag{17.3}$$

$$= \text{Var}\left(\frac{N_1}{N}\bar{x}_1\right) + \text{Var}\left(\frac{N_2}{N}\bar{x}_2\right)$$

$$= \left(\frac{N_1}{N}\right)^2 \text{Var}(\bar{x}_1) + \left(\frac{N_2}{N}\right)^2 \text{Var}(\bar{x}_2)$$

$$= \left(\frac{N_1}{N}\right)^2 \frac{N_1 - n_1}{N_1 - 1}\frac{\sigma^2_1}{n_1} + \left(\frac{N_2}{N}\right)^2 \frac{N_2 - n_2}{N_2 - 1}\frac{\sigma^2_2}{n_2}.$$

If the fractions sampled, n_1/N_1 and n_2/N_2, are small, we can ignore the finite population correction and write

$$\text{Var}(\hat{\mu}) \doteq \left(\frac{N_1}{N}\right)^2 \frac{\sigma^2_1}{n_1} + \left(\frac{N_2}{N}\right)^2 \frac{\sigma^2_2}{n_2}. \tag{17.4}$$

EXAMPLE   Suppose an automobile manufacturer employs $N_1 = 100$ supervisory engineers and $N_2 = 1000$ nonsupervisory engineers. In a random sample (without replacement) of $n_1 = 10$ supervisory engineers, the mean annual salary is \$135,000 and the standard deviation is $s_1 = \$20,000$, whereas in a sample of $n_2 = 50$ nonsupervisory engineers the mean is $\bar{x}_2 = \$80,000$ and the standard deviation is $s_2 = \$10,000$. Then the unbiased estimate of the average salary of the 1100 engineers is

$$\frac{N_1}{N}\bar{x}_1 + \frac{N_2}{N}\bar{x}_2 = \frac{100}{1100}\$135,000 + \frac{1000}{1100}\$80,000$$

$$= \$85,000.$$

The estimated variance of the estimated mean is

$$\frac{N_1 - n_1}{N_1 - 1}\frac{s^2_1}{n_1} + \frac{N_2 - n_2}{N_2 - 1}\frac{s^2_2}{n_2}$$

$$= \frac{100 - 10}{100 - 1}\cdot\frac{(\$20,000)^2}{10} + \frac{1000 - 50}{1000 - 1}\cdot\frac{(\$10,000)^2}{50}$$

$$= 38,266,000$$

dollars squared or a standard deviation of \$6,186.

The estimate $\hat{\mu}$ is unbiased no matter what n_1 and n_2 are (as long as each is positive). Since the variance, however, depends on the sample sizes as well as population sizes, n_1 and n_2 will be selected to make (17.3) or (17.4) small. When the sample sizes are taken so that $n_1/n_2 = N_1/N_2$, the procedure is called *proportional stratified sampling*. If the total sample size is n, the stratum sample sizes are $n_1 = (N_1/N)n$ and $n_2 = (N_2/N)n$, and the approximate variance (17.4) is

$$\text{Var}(\hat{\mu}_{\text{prop}}) = \frac{1}{n}\left(\frac{N_1}{N}\sigma_1^2 + \frac{N_2}{N}\sigma_2^2\right),$$

which is $1/n$ times the same weighted average of the stratum variances that μ is of the stratum means. An analysis of variance, which is detailed below, shows that $\text{Var}(\hat{\mu}_{\text{prop}})$ is never greater than σ^2/n, where σ^2 is the variance of the whole population, and is less than σ^2/n when $\mu_1 \neq \mu_2$. Thus proportional stratified random sampling is preferred over simple random sampling.

Figure 17.1 suggests the distributions of salaries in the two strata and the entire population. The variability in each of the strata is less than the variability in the whole population, and the weighted average of the strata variabilities is less than the variability of the whole population. It is shown in Section 12.2 how the weighted average of the variances of sets of measurements is smaller than the variance of all the measurements.

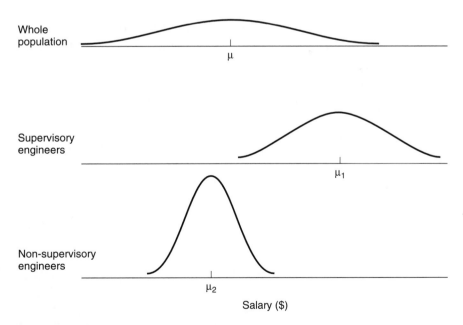

Figure 17.1 *Distributions in strata and in whole population.*

If the variances of the strata are known (while the means of the strata are unknown), the allocation of the sample can be improved. The basic idea is that, if one stratum has a very small variance, even if the sample from that stratum is quite small, the sample mean from that stratum will have a small variance. It turns out that the (approximate) variance (17.4) is minimized subject to $n_1 + n_2 = n$, fixed, by taking

$$n_1 = \frac{\sigma_1 N_1}{\sigma_1 N_1 + \sigma_2 N_2} n, \qquad n_2 = \frac{\sigma_2 N_2}{\sigma_1 N_1 + \sigma_2 N_2} n.$$

That is, n_i is proportional to $\sigma_i N_i$. For instance, if $n = 70$, $N_1 = 2N_2$, and $\sigma_1 = 3\sigma_2$, then $\sigma_1 N_1 = 6\sigma_2 N_2$, and $n_1 = 60$ and $n_2 = 10$. Stratified random sampling with this allocation of total sample size is called *Neyman sampling* because it was developed by the famous Polish-American statistician Jerzy Neyman. Neyman sampling is an improvement over proportional sampling if the variances in the strata are different (and known).

More than Two Strata

The efficiency of stratification may be increased by increasing the number of strata. The ideas presented for the case of two strata hold for more strata.

In the case of k strata, let y_{gi} be the value of the variable for the ith individual in the gth stratum, $i = 1, 2, \ldots, N_g$, $g = 1, 2, \ldots, k$. The population mean is written in this notation as

$$\mu = \frac{1}{N} \sum_g \sum_i y_{gi};$$

the population variance is

$$\sigma^2 = \frac{1}{N} \sum_g \sum_i (y_{gi} - \mu)^2.$$

The mean in the gth stratum is

$$\mu_g = \frac{1}{N_g} \sum_i y_{gi};$$

the variance in the gth stratum is

$$\sigma_g^2 = \frac{1}{N_g} \sum_i (y_{gi} - \mu_g)^2.$$

Making a decomposition like that used in analysis of variance (Section 16.1), we can write the population variance in terms of the stratum means and variances as

$$N\sigma^2 = \sum_g \sum_i (y_{gi} - \mu)^2$$

$$= \sum_g \sum_i (y_{gi} - \mu_g)^2 + \sum_g N_g(\mu_g - \mu)^2$$

$$= \sum_g N_g\sigma_g^2 + \sum_g N_g(\mu_g - \mu)^2.$$

The indicated decomposition is shown in an analysis of variance table for the population in Table 17.1.

TABLE 17.1

Decomposition of Population Sum of Squares in Terms of Stratum Means and Variances

Source of variation	Sum of squares
Between strata	$\sum_g N_g(\mu_g - \mu)^2$
Within strata	$\sum_g \sum_i (y_{gi} - \mu_g)^2 = \sum_g N_g\sigma_g^2$
Population	$\sum_g \sum_i (y_{gi} - \mu)^2 = N\sigma^2$

A random sample of size n_g is taken from the gth stratum, yielding a sample mean of $\bar{x}_g$, $g = 1, \ldots, k$. The estimate of the population mean based on sampling from k strata is the weighted average of the estimates of the strata means,

$$\hat{\mu} = \sum_g \frac{N_g}{N} \bar{x}_g.$$

Note that the weights are the population proportions. The variance of $\hat{\mu}$ is

$$\text{Var}(\hat{\mu}) = \sum_g \left(\frac{N_g}{N}\right)^2 \frac{N_g - n_g}{N_g - 1} \frac{\sigma_g^2}{n_g},$$

which, ignoring the finite population correction factors, is approximately

$$\text{Var}(\hat{\mu}) \doteq \sum_g \left(\frac{N_g}{N}\right)^2 \frac{\sigma_g^2}{n_g}. \tag{17.5}$$

Proportional sampling occurs when an overall sample of size n is allocated to the strata in proportion to the strata sizes, that is, $n_g =$

$(N_g/N)n$. Then the approximate variance (17.5) is

$$\text{Var}(\hat{\mu}_{\text{prop}}) = \frac{1}{n} \sum_g \frac{N_g}{N} \sigma_g^2.$$

This is the Within Strata sum of squares in Table 17.1 divided by $n \times N$. Since the variance of the estimate based on simple random sampling is the population sum of squares divided by $n \times N$, the reduction in variance due to proportional sampling is the Between Strata sum of squares divided by $n \times N$, that is,

$$\frac{1}{n} \sum_g \frac{N_g}{N}(\mu_g - \mu)^2.$$

This decrease in the variability of the estimate is made large by choosing homogeneous strata, in which case the strata means will be very different. (We note that for proportional sampling, $\hat{\mu} = \sum_g n_g \bar{x}_g/n$, which is the mean of all the observations.)

If the strata variances are known and some are different, a further increase in efficiency can be made. The optimal allocation of a total sample number n (Neyman sampling) is proportional to $N_g \sigma_g$:

$$n_g = \frac{N_g \sigma_g}{N \bar{\sigma}} n, \qquad g = 1, \ldots, k,$$

where $\bar{\sigma} = \sum_g (N_g/N)\sigma_g$, the weighted average of strata standard deviations. The variance of the estimate, say $\hat{\mu}_{\text{Neyman}}$, is

$$\text{Var}(\hat{\mu}_{\text{Neyman}}) = \sum_g \left(\frac{N_g}{N}\right)^2 \frac{\sigma_g^2}{n\sigma_g N_g/(N\bar{\sigma})}$$

$$= \frac{\bar{\sigma}}{n} \sum_g \frac{N_g}{N} \sigma_g = \frac{\bar{\sigma}^2}{n}.$$

The saving over proportional sampling is

$$\frac{1}{n}\left(\sum_g \frac{N_g}{N}\sigma_g^2 - \bar{\sigma}^2\right) = \frac{1}{n} \sum_g \frac{N_g}{N}(\sigma_g - \bar{\sigma})^2.$$

This shows that such a saving is due to the variation among standard deviations of the strata.

Usually the strata variances are not known exactly; they may be guessed on the basis of studies made earlier in time or based on related characteristics. For any method of stratified random sampling, the variance σ_g^2 can be estimated by the sample variance s_g^2 of the n_g observations from the stratum (if $n_g \geq 2$). Substituting s_g^2 for σ_g^2 into (17.5) or the preceding exact expression yields an estimate of the variance of the estimate of μ.

17.3   Cluster Sampling

Cluster sampling refers to sampling "clusters" of potential respondents and then sampling respondents in the clusters in the sample. To determine the total number of unemployed in a city, for example, one might consider city blocks as the clusters and households as the "respondents." A sample of city blocks is taken, using a map of the city to number the blocks. In each block the households are enumerated, and a random sample of households is taken in each block in the sample. The *total* number of unemployed in a sampled block is estimated from the sample of households in that block. In turn, the total number of unemployed in the city is estimated from these estimated block totals. An unbiased estimate of the population total results from this procedure. An unbiased estimate of the population mean is obtained by dividing the estimated total by the number of units in the population.

Sampling variability arises from two sources: the sampling of clusters and the sampling of units within clusters. The formula for the variance of an estimate depends on the rule for sample size within the sampled cluster. For example, the sample sizes may be the same in all clusters or they may be a fixed proportion of the cluster sizes. It is beyond the scope of this book to develop these formulas. See Cochran (1977), for example.

What are the advantages of cluster sampling? When the clusters represent geographically compact sets of units, as in the above illustration, with cluster sampling the interviewers may spend more time in interviewing than traveling. Also, a frame for clusters (for instance, blocks) may be available, making enumeration of clusters feasible, while preparation of a frame for the entire population of units is not practical.

In a sense the clusters are strata, since each individual or unit belongs to one and only one cluster. The greater flexibility here results because clusters (or strata) are themselves sampled.

17.4   Systematic Sampling with a Random Start

The idea of systematic sampling is to take every tenth name on a list, or check every fifth car passing a toll booth, or review every twentieth file folder in a drawer. The method is appealing because it is easy to carry out and it spreads the "sample" out through the population. However, it is clearly not random. To add an element of randomness—which is necessary to obtain unbiased estimates—one may select the starting point at random.

The systematic sampling procedure begins with the construction of a frame, as before. Suppose we want a sample of size n out of a population of N, and suppose that $N/n = l$, approximately an integer. The number l is called the *sampling interval*. If we want a sample of 300 families out of 3000, the sampling interval is 10. We choose an integer at random from $1, 2, \ldots, l$. Suppose we get 7. Then we choose the individuals numbered $7, 7 + l, 7 + 2l, \ldots, 7 + (n-1)l$ as our sample; our observations will be

$$x_1 = y_7, \quad x_2 = y_{7+l}, \quad \ldots, x_n = y_{7+(n-1)l}.$$

The estimate of the population mean is the sample mean $\bar{x} = \sum x_i/n$.

The sample mean is unbiased (because the start is random), but the precision of $\bar{x}$ depends on how the characteristic under observation varies as we go through the frame. If the population is the 365 days of the year, the frame is the calendar. When the sampling interval is 7 and $n = 52$, we get a systematic sample that is based on the same day of the week over the entire year. The method is good for estimating the average hours of daylight per day over the year, but poor for estimating the average hours of work per day over the year.

Suppose again that we are sampling households and that the frame lists houses in the following order: Avenue A from west to east; Avenue B from west to east on the north side of the street, Avenue B from west to east on the south side of the street, etc. The blocks are oblong and narrow between avenues, so that all houses face on avenues. Suppose that the ordering in the frame corresponds to the cost of the houses; house 1 is least expensive and house 36 most expensive (Figure 17.2). Then systematic sampling gives us a sample of households that is varied and representative as far as cost of house (and consequently family income) is concerned. In this case, systematic sampling performs roughly like stratified sampling, where the strata are defined in terms of income.

On the other hand, if the variation in the population is related to the sampling interval, then systematic sampling can be much less precise than simple random sampling; at worst it can be equivalent to having only a sample of size 1. If the corner houses are most expensive, and our sample interval is 3, then we get a very nonrepresentative sample: either all corner houses or all middle-of-the-block houses.

Let us denote by m the starting number, the integer chosen at random from $1, 2, \ldots, l$. The quantity m is a random variable that takes on the values $1, 2, \ldots, l$, each with probability $1/l$. Only l different samples are possible. These are

$$x_1 = y_m, \quad x_2 = y_{m+l}, \quad \ldots, x_n = y_{m+(n-1)l},$$

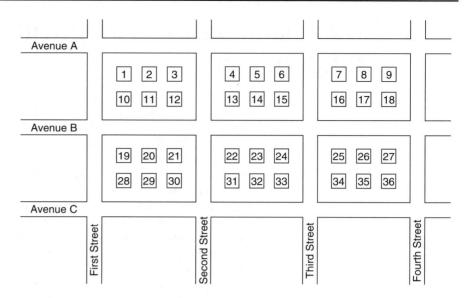

Figure 17.2 *Frame for systematic sampling.*

where m ranges over $1, 2, \ldots, l$. Although all individuals have the same chance of being included in the sample, not all possible *sets* of n have the same chance: l sets have probability $1/l$ each. All the other sets have zero probability. The possible values of $\bar{x}$ are

$$
\begin{aligned}
\bar{y}_1 &= \frac{y_1 + y_{1+l} + \cdots + y_{1+(n-1)l}}{n}, \\
\bar{y}_2 &= \frac{y_2 + y_{2+l} + \cdots + y_{2+(n-1)l}}{n}, \\
&\vdots \\
\bar{y}_l &= \frac{y_l + y_{2l} + \cdots + y_{nl}}{n};
\end{aligned}
\tag{17.6}
$$

each of these values has probability $1/l$.

In effect, the observed $\bar{x}$ is a random sample of 1 from the population of l values, $\bar{y}_1, \ldots, \bar{y}_l$. The mean of the sampling distribution of $\bar{x}$ is

$$
E(\bar{x}) = \frac{1}{l} \sum_b \bar{y}_b = \frac{1}{nl} \sum_k y_k,
$$

which is μ if $nl = N$ (and is approximately μ in any case). The variance is $\sum_b (\bar{y}_b - \mu)^2 / l$, which is unknown. It cannot be estimated from the sample because the sample is effectively a sample of 1 from the population (17.6). The investigator cannot assess variability, carry out tests of significance, or construct confidence intervals.

17.5   Systematic Subsampling with Random Starts

A simple modification of systematic sampling makes it possible for one to estimate an appropriate variance. The point is to carry out the procedure with several random starts.

Suppose that the population is of size $N = 3000$ and a sample of $n = 300$ is wanted. With systematic sampling we would compute the sampling interval as $3000/300 = 10$ and observe every 10th individual, randomly selecting a starting number from $1, 2, \ldots, 10$. Suppose that, instead of taking *one* sample of size $n = 300$, we take, for example, 5 *subsamples* of size $n/5 = 300/5 = 60$. For samples of size 60, the sampling interval is $N/60 = 3000/60 = 50$. We randomly obtain five different starting numbers from $1, 2, \ldots, 50$. Suppose that the first starting number is 17. Then the first subsample of 60 consists of the individuals numbered $17, 67, 117, \ldots, 2967$, and we observe the values $y_{17}, y_{67}, y_{117}, \ldots, y_{2967}$. Suppose that the second starting number is 23. The second subsample gives the values $y_{23}, y_{73}, y_{123}, \ldots, y_{2973}$. Continuing in this way, we get five systematic subsamples. Let the means of the subsamples be denoted $\bar{x}_1, \bar{x}_2, \bar{x}_3, \bar{x}_4, \bar{x}_5$. Then the mean of all $5 \times 60 = 300$ observations can be written as the mean of the subsample means,

$$\bar{x} = \frac{\bar{x}_1 + \bar{x}_2 + \cdots + \bar{x}_5}{5}.$$

This statistic is an unbiased estimate of the population mean. The sample variance of subsample means,

$$s_{\bar{x}_i}^2 = \frac{\sum(\bar{x}_i - \bar{x})^2}{4},$$

is an unbiased estimate of the variance $\sigma_{\bar{x}_i}^2$ of subsample means. The subsample means constitute a sample of size 5 from a population of subsample means. This population is of size 50, for the possible values of the subsample means are

$$\frac{y_1 + y_{51} + \cdots + y_{2951}}{60},$$

$$\frac{y_2 + y_{52} + \cdots + y_{2952}}{60},$$

$$\vdots$$

$$\frac{y_{50} + y_{100} + \cdots + y_{3000}}{60}.$$

Hence the variance of $\bar{x}$ is

$$\sigma_{\bar{x}}^2 = \frac{\sigma_{\bar{x}_i}^2}{5} \frac{50 - 5}{50 - 1},$$

which is estimated by

$$s_{\bar{x}}^2 = \frac{s_{\bar{x}_i}^2}{5} \frac{50 - 5}{50 - 1}.$$

Here the "population" is a population of subsample means; this population is usually approximately normal, and inferences can be based on the normal distribution.

Summary

Usually, for budgetary or other reasons, an investigator knows that only a sample of relatively limited size is affordable. Then the design of the survey becomes especially important.

Often an investigator partitions the population into subpopulations, or *strata*. Making the allocation of observations proportional to the strata sizes is called *proportional sampling*.

Proportional sampling is always better than simple random sampling from the population as a whole. If the strata variances are known, the best method of stratified sampling is *Neyman sampling*, in which the sample sizes are proportional to the products of the strata sizes and the strata standard deviations.

Systematic sampling has also been discussed.

Exercises

17.1   Define the events $A = (1, 2, 3)$, $B = (4, 5, 6)$, and $C = (7, 8, 9)$. If the procedure used in constructing the table in Appendix VI is random, the (conditional) probability of each event is $1/3$, given that the digit is not 0. Use as your "sample" the first 15 nonzero digits in the line of the table, numbered according to the day of the month of your birthday. For example, if your birthday is on the 13th of some month, use the 13th line of the table.

(a) Calculate the chi-square statistic for goodness of fit.

(b) Test the null hypothesis that the (conditional) probabilities are $1/3$ at the 5% level.

17.2   Show that for simple random sampling of a sample of size $n = 2$ without replacement from a population of size $N = 3$, the variance of the sample mean is

$$\frac{\sigma^2}{4} = \frac{3-2}{3-1}\frac{\sigma^2}{2} = \frac{N-n}{N-1}\frac{\sigma^2}{n}.$$

17.3   In the TV-viewing example, if more than one child in a household is included in the sample, do you think it would be valid to consider the corresponding observations as statistically independent? Why or why not?

17.4   From (17.4), deduce the conditions under which n_1 and n_2 would be chosen to be equal.

17.5   [*Note:* The assumption is made throughout this exercise that sampling (from a finite population) is done with replacement, so that there need be no correction for finite population. It is assumed that the stratum sizes are known, so that proportional random sampling can be used.]

Suppose an investigator wishes to estimate the income of a population of engineers by drawing a sample. The investigator believes that a stratified sample would be better than a simple random sample, but is not sure whether it would be better to stratify by occupational position (supervisory versus nonsupervisory) or by department (production versus development). Table 17.2 shows a hypothetical *population* of 10 incomes (to the nearest thousand dollars) in two departments for supervisory and nonsupervisory engineers. Your computations in answer to the following questions illustrate the answer to this investigator's question.

TABLE 17.2
Incomes of Engineers ($1000's)

	Production	*Development*
Supervisory	83, 80, 69	96
Nonsupervisory	62, 59, 51	33, 30, 29

Hypothetical data.

(a) Compute the standard deviation of the sampling distribution of the mean income (to the nearest dollar) for samples of size 5 in proportional stratified sampling from this population: (i) stratified by occupational position, and (ii) stratified by department.

(b) Show (algebraically) what you would use as an estimate of mean income in the population on the basis of the above samples.

(c) Compute the standard deviation of the sampling distribution of the mean income for simple random sampling.

(d) Describe the information about stratified sampling that is illustrated by your computations.

(e) Graph on one line the incomes of supervisory engineers and on a parallel line the incomes of nonsupervisory engineers. Graph on a pair of parallel lines the incomes of production and development engineers, respectively.

17.6   An investigator is interested in the average amount a large garage charges for automobile repairs. The repair bills from the garage are categorized according to the size of car on which the repair was made, giving the breakdown in Table 17.3.

TABLE 17.3
Repair Bills for Autos

Category	Number of receipts	Average charge	Standard deviation within category
Large cars	100	$110.00	$42.00
Small cars	150	$75.00	$34.00

Hypothetical data.

(a) What is the average repair bill from the garage?

(b) Suppose that the investigator does not know the average charge and variance within each category, but only the number of receipts in each category. The investigator randomly selects 10 receipts from "large cars" and finds that their total is $1480, then randomly selects 10 receipts from "small cars" and finds that their total is $711. What is the best estimate of the overall average repair bill?

(c) What is the variance of this estimate? (The investigator cannot compute this, but since we have more information, we can.)

17.7   An investigator proposed a systematic sample of residential apartment buildings for a study of neighborhood incomes. Since all buildings had one family per floor and the same number of floors, the proposal was to choose a floor randomly, then to sample the *same* floor in each building. (Note that we can think of the families as listed in order with a serial number, so that our sample includes every third family, which makes the sampling interval equal to 3.)

(a) Assuming a random starting point for this sampling, list all the different possible samples of 3 that could be drawn by systematic sampling in the two situations shown in Table 17.4.

(b) Compute the standard error of the mean (to the nearest dollar) for samples of 3 in the above two situations.

(c) Define the sample mean whose standard error is computed in (b).

(d) In Situations A and B of Table 17.4, compute the standard error of the mean, assuming simple random sampling from the population, for samples of size 3.

(e) Discuss your conclusions in regard to the efficiency of systematic sampling.

TABLE 17.4

Incomes ($1000s) of Residents on Each Floor for Three Apartment Buildings

Situation A

	Building		
	1	*2*	*3*
First floor	16	32	40
Second floor	16	32	40
Third floor	16	32	40

Situation B

	Building		
	1	*2*	*3*
First floor	40	40	40
Second floor	32	32	32
Third floor	16	16	16

Hypothetical data.

17.8   Table 17.5 is a table of hours of study time per day for a Stanford student on 28 successive days.

TABLE 17.5

Hours of Study Time per Day of 28 Successive Days

4 6 5 4 3 1 2 6 6 4 3 5 2 2 4 6 5 5 3 2 3 6 6 6 4 5 3 1
$\bar{x} = 4,\ \sum x_i^2 = 520,\ \sum(x_i - \bar{x})^2 = 72$

(a) What is the variance of the estimate of the mean number of hours studied per day, using systematic sampling with a random start and a sampling interval of 7 days (sample size of 4)?

(b) What is the variance of the estimate of the mean hours studied per day, using random sampling with sample size 4?

(c) Compare the results in parts (a) and (b), and explain.

Answers to Selected Exercises

Chapter 1

1.9 (a) Examples: Better students read more pages; Students who read more pages of the book also read other books that were more helpful; Students who exaggerate the number of pages read find alternate ways to get better scores on the test (e.g., help from friends).

1.10 Both the experiment and nonexperimental evaluation should involve a random sample of individuals from the population of tooth-brushers. Experiment: Randomly assign "C" and "P" to members of the sample, using a single generic label on all tubes so that subjects do not know which product they are using. Nonexperimental study: Survey a sample of individuals and ask them which toothpaste they use. (Disregard subjects who use toothpastes other than "C" and "P".)

Chapter 2

2.1 (a) ordinal, (b) ratio, (c) nominal, (d) ordinal, (e) interval, (f) ratio.

2.2 discrete: (a), (d), (e); continuous: (b), (c), (f), (g).

2.7   (a)

Class interval	Frequency	Relative frequency
0– 4	1	0.033
5– 9	6	0.200
10–14	9	0.300
15–19	6	0.200
20–24	2	0.067
25–29	1	0.033
30–34	1	0.033
35–39	1	0.033
40–44	2	0.067
45–49	0	0.0
50–54	0	0.0
55–59	0	0.0
60–64	1	0.033

2.9   (c) 18¢ (in the interval $18.00 - 18.99$).

2.12   (a) 5,886,481, (b) 13.04% (divide by total n), (e) 14,803,995, (f) 32.80%.

2.19   Two- to five-year olds.

1	334444444
1	5555555555555666666666666666666666667777777777788888888889999
2	001112
2	
3	
3	
4	
4	
5	

Chapter 3

3.1   (a) 25, (b) 4, (c) 6.25.

3.3   (b) Trial 1: $\bar{x} = 21.0$; Trial 5: $\bar{x} = 8.25$.

3.5   (a) Mean = 8.96 inches; Median = 9.3 inches. Distribution is skewed (slightly) to the left.

3.7   From Table 2.10, median = 150 pounds. Median estimated by interpolation from Table 2.11 is 150.4.

3.9   For males: $\bar{x} = -0.746$; median = -0.71.

3.10 (a) Mode (= 2) is most descriptive; 9 of the 12 families have 2 children.

3.13 (a) 3, (b) 3, (c) 2.84.

3.14 (a) Distribution is skewed to the right.

3.16 (b) 23 hours per week.

3.19 (a) 20.0, (b) 0.0, (c) 34.8, (d) -5.22.

3.21 $67.80.

3.22 (a) 83.8%, (b) Median for no-VCR students on weekdays is 3 hours per day.

3.25 (a) $\sum x_i = 55$; $\sum z_i = 0$,

(c) $\frac{1}{2}(\sum x_i)^2 = 1512.5$,

(e) $\sum(x_i + 1) = 65$; $\sum(1 - z_i) = 10$,

(g) $\sum(z_i + 2)^2 = 50$; $\sum(z_i^2 + 2) = 30$,

(i) $\sum \frac{1}{n}x_i = 5.5$.

3.26 (b) $\sum(x_i - y_i)^2 = 19{,}668$; $\left[\sum(x_i - y_i)\right]^2 = 108{,}900$,

(d) $\sum x_i \sum y_i = 21{,}175$.

Chapter 4

4.2 $+4$.

4.3 Doctor A: $s = 11.27$.

4.4 Doctor A: first quartile = 11.5; third quartile = 25; interquartile range = 13.5.

4.7 $s = 3.67$.

4.8 Figure 4.1: $s^2 = 7.70$.

4.10 (b) $\bar{x} = 2.8$; $s = 0.84$, (c) Mean deviation = 0.64.

4.13 (a) $\bar{x} = 204.89$; $s = 45.85$.

(b) The two-standard deviation rule of thumb estimates that most of the observations will be between 113.19 and 296.59. All of the data points in the exercise are between these limits.

4.16 SU: range is 11 (computed by $18 - 7$); $s = 2.04$.

4.20 $\bar{x} = 73.77$; $s = 2.22$.

4.25   (a) Mean = 18.00; Standard deviation = 3.30.

(c) Mean = 162.0; Standard deviation = 33.0.

(e) Mean = 4.3; Standard deviation = 0.825.

4.26   (a) −1.0, (b) 2.0.

4.28   (b) Mean = 0; Standard deviation = 1.0.

Chapter 5

5.1   (a) +, (b) −, (c) 0, (d) +, (e) +.

5.3   (a) 2.5, (b) 1.581, (c) 2.5, (d) 1.581, (e) −2.25, (f) −0.9.

5.5   (b) 0.849.

5.7   Correlation for Figure 5.3: −0.759.

5.8   (b) −0.818.

5.11   (b) The pair (1.4, 1.02) appears to be an outlier. (c) There is a curvilinear relationship of x with y. (d) The correlation is 0.090 with the outlier removed. It is weak because there is little *linear* association of x with y.

5.15   (b) $r_s = 0.636$.

5.16 $r_s = 0.788$ [using (5.2)]; It seems that students who leave the examination earlier tend to get higher grades than those who leave later.

5.18   The results are the same as those for Exercise 5.3.

5.20

	$x_i - 6$		$3y_i + 1.5$
Mean:	1.42		
Variance:			
Standard deviation:	1.30		6.30
Covariance:		−3.450	
Correlation:			

5.22   (b) 0.477, (c) Similar.

5.24   (a) 0.50, (c) Mathematics performance: IQ.

Chapter 6

6.1 (a)

2	0	2
1	2	3
3	2	5

1	1	2
2	1	3
3	2	5

0	2	2
3	0	3
3	2	5

6.2

0	4	4
4	0	4
4	4	8

1	3	4
3	1	4
4	4	8

2	2	4
2	2	4
4	4	8

3	1	4
1	3	4
4	4	8

4	0	4
0	4	4
4	4	8

(a) $\phi = -1.00,$ $\phi = -.05,$ $\phi = 0.0,$ $\phi = 0.5,$ $\phi = 1.0.$

6.5 (a) 0.84, (b) 0.62.

6.6 (a)

	No	Yes	Total
Low	32 (58%)	23 (42%)	55 (100%)
High	64 (33%)	128 (67%)	192 (100%)
Total	96	151	247

$\phi = 0.21$

(b)

	No	Yes	Total
Low	83 (49%)	87 (51%)	170 (100%)
High	13 (17%)	64 (83%)	77 (100%)
Total	96	151	247

$\phi = 0.30$

6.7 (a)

	Never	1–2 Times	3–11 Times	12 or more	Total
Male	37.3%	46.8%	11.8%	4.0%	100.0%
Female	71.5%	23.2%	5.0%	0.0%	100.0%
Total	54.8%	34.7%	8.3%	2.1%	100.0%

(b)

	Never	1 or more	Total
Male	37.3%	62.7%	100.0%
Female	71.5%	28.5%	100.0%
Total	54.8%	45.2%	100.0%

6.9

	T	C	OS	DK	Total
Squirrels	6.0%	12.0%	18.1%	63.9%	100.0%
Bears	9.2%	11.1%	23.5%	56.2%	100.0%
Cats	11.6%	14.5%	21.7%	52.2%	100.0%

6.13   (a)

	Not convicted	Convicted: probation	Convicted: prison	Total
Robbery	8.1%	2.0%	89.9%	100.0%
Larceny	23.6%	47.7%	28.8%	100.0%
Embezzlement	11.8%	44.0%	44.2%	100.0%
Forgery	12.5%	41.8%	45.7%	100.0%
Drug	15.1%	9.8%	75.1%	100.0%
Total	14.7%	23.9%	61.5%	100%

(i) 14.7%; 23.9%; 61.5%.

(ii) Robbery; Larceny; Larceny.

6.14   (a) (i) 65.5%, (ii) 21.9%, (iii) 41.7%, (iv) 32.6%, (v) 58.3%, (vi) 16.5%.

(b) Percent Voting Democratic, by Age and Productivity.

		Productivity		
		Low	Medium	High
Age	Less than 41	69%	79%	79%
	41–50	50%	63%	72%
	More than 50	42%	50%	54%

6.17   (a) 30%; 48%, (b) $\phi = 0.18$.

6.19

		Number of magazines read					
		0	1	2	3	4	Total
Number of children	0	29%	33%	30%	6%	2%	100%
	1	27%	33%	30%	8%	2%	100%
	2	25%	35%	30%	7%	3%	100%
	3 or more	25%	31%	31%	8%	5%	100%
All nos. of children		27%	33%	30%	7%	3%	100%

6.21   0.91; 1.49; 1.62.

Chapter 7

7.2	(a) a person is either wealthy or a college graduate, (b) a person is a wealthy college graduate, (c) a person is either wealthy or a college graduate, but not both, (d) a person is not a college graduate, (e) a person is not wealthy, (f) a person is neither wealthy nor a college graduate.
7.4	(a) $\Pr(C)$, (b) $\Pr(D)$.
7.6	(a) 0.6, (b) 0.1, (c) 0.4.
7.8	(a) independent, (b) dependent, (c) dependent.
7.10	(a) 0.0121, (b) 0.0118.
7.12	0.0965.
7.15	1/3.
7.17	1/6.
7.19	0.06.
7.21	(a) $\Pr(D \mid S) = 0.8$, (b) $\Pr(D \mid \bar{S}) = 0.05$.
7.23	(a) 0.064, (b) 0.124.
7.25	brown: 3/4; blue 1/4.
7.26	1/3.
7.28	$1/2, 1/4, 1/8, \ldots$
7.31	0.84.
7.32	0.81.
7.35	0.5625.

Chapter 8

8.1	(a) 0.42, (b) 0.58, (d) 0.95.
8.2	(a) 0.50, (b) 0.50, (c) 0.25, (d) 0.50.
8.4	(a) $E(y) = 2.50$; $\sigma_y = 0.50$.
8.5	$E(x) = 1.844$; $\sigma_x = 0.953$
8.6	Expected value: $1844.

8.7   (a) 0.05, (c) 0.20.

8.8   (a) .4505, (b) .0256, (c) .4032, (d) .8413.

8.10   (a) .0968, (b) .9032, (c) .2266, (d) .9429.

8.11   (a) .1178.

8.12   (a) $z = 0.25$, (d) $z = -0.39$.

8.13   (a) $z(.40)$, (d) $z(.65)$.

8.14   (a) $z_0 = 2.33$, (b) $z_0 = 1.23$, (e) $z_0 = 0.80$.

8.16   (a) .9332, (c) .5328, (e) .0495.

8.18   (a) .9332, (c) .0062.

8.19   (b) 21.4 years to 22.8 years (to the nearest 1/10 year).

8.23   F: $-\infty$ to -1.04,
D: -1.04 to -0.39,
C: -0.39 to $+0.39$,
B: $+0.39$ to $+1.04$,
A: 1.04 to ∞.

8.27   (a) Assuming that the only whole-number scores are obtained, with 0 as the lowest possible score

F: 0 to 64,
D: 65 to 70,
C: 71 to 77,
B: 78 to 83,
A: 84 and above.

Chapter 9

9.3   (2)   (3)
(a) .9192, .9452,
(b) .1587, .2119,
(c) .2743, .3446,
(d) .1935, .2898.

9.5   0.3125.

9.7   0.2009.

9.8   0.9377.

9.11   0.3214.

9.13

x :	0	1	2	3
prob.:	0.0156	0.1406	0.4219	0.4219

9.15   No. The cell proportions are not equal to the products of the corresponding marginal entries.

9.18   (a) 0.9844, (b) 0.8438, (c) 0.9844, (d) 1.0.

9.20   Mean: 2.25; Standard deviation: 0.75.

9.22   Mean: 0.75; Standard deviation: 0.0137.

9.25   (a) 19.4, (b) 2.1, (c) .5, (d) .1587, (e) .9987.

9.27   12.22.

9.28   .0222.

Chapter 10

10.1   Point estimate: $10,300; Interval: ($10,104, $10,496).

10.3   Point estimate: $10,300; Interval: ($10,136, $10,464).

10.5   Point estimate: 161 lb.; Interval: (158.4 lb, 163.6 lb).

10.7   Point estimate: 90 mm; Interval: (86.1 mm, 93.9 mm).

10.9   (a) 2.764, (d) 1.708.

10.10   (a) 0.95, (c) 0.99, (e) 0.98.

10.12   (a) (208.4, 219.6).

10.13   (10.6, ?) using percentage points from t_{30}.

10.14   Interval: (85.9, ?).

10.18   (a) (29.3, 38.7), (c) 1/3 yes; 40% no.

10.20   Expected attendance is 42.

10.21   (0.075, ?).

10.23   Interval: (0.322, ?).

10.25   (a) (0.593, 0.747), (d) Less than 1/2 - no.

10.27   Bush, October 1991: 99% interval is (0.482, 0.564).

10.29 $(353, 1945)$ $[r = 2.5$, obtained from $z(.05) = 1.64$, is rounded down to 2].

10.32   Mean weight loss: 6.6 lbs; 80% interval: $(5.2, ?)$.

10.34   (a) $(-16.2, 0.2)$, (b) yes.

10.36   (48 hours, 52 hours).

Chapter 11

11.1   (a) $H_0: \mu \leq 500$, $H_1: \mu > 500$, (b) Standard normal, (c) 1.645, (d) $z = 1.80$, (e) Reject H_0.

11.2 P is approximately 0.0359, (a) Yes, (b) No.

11.3   (a) $H_0: \mu \geq 500$, $H_1: \mu < 500$, (c) -2.33.

11.5   (a) $H_0: \mu = 43$, $H_1: \mu \neq 43$, (b) t_{24}, (c) -2.797 and 2.797, (d) $t = 2.08$, (e) Accept H_0.

11.6 $.02 < P < .05$; H_0 would be rejected for any α value of .05 or greater; this range includes .10 and .05 but not .01.

11.7   (a) $H_0: \mu \geq 98.2°$, $H_1: \mu < 98.2°$, (c) -1.699.

11.9   (a) $H_0: \mu \leq 10$, $H_1: \mu > 10$, (b) t_9; significance point is 1.383, (c) $t = 6.3$, (d) Reject H_0.

11.10   (a) $H_0: \mu \leq 1000$, $H_1: \mu > 1000$; Use t_{16}; significance point is 1.746. The test statistic is $t = 1.98$; Reject H_0.

11.11   (a) $.025 < P < .05$.

11.14   (a) $H_0: \mu_d \geq 0$ $H_1: \mu_d < 0$ where $d_i = y_i - x_i$, (b) The 5% significance point from t_{60} is -1.671 and from t_{120} it is -1.658, (c) $t = -4.77$, (d) Reject H_0, (e) The program is effective.

11.17 $H_0: \mu_d \leq 0$ $H_1: \mu_d > 0$ where $d_i = y_i - x_i$ (after $-$ before); The 1% significance point from t_8 is 2.896; The test statistic is $t = 2.78$; Accept H_0, the study does not provide evidence of an average increase in anxiety.

11.19 $z = 2.5$; Reject H_0.

11.22   (a) $H_0: p \leq .5$ $H_1: p > .5$, (b) $N(0, 1)$; significance point $= 1.645$, (c) $z = 3.4$, (d) Reject H_0; more than $1/2$ of the population supports the proposition.

11.24 (a) $H_0: p \geq 0.46$, $H_1: p < 0.46$; The 5% significance point from the standard normal distribution is -1.645; The test statistic is $z = -1.5$; Accept H_0.

11.27 Test statistic: $z = 0.83$; Significance points: ± 1.96; Accept H_0.

11.29 (a) P approximately .0681, (c) P approximately .0340, (d) $.025 < P < .05$, (f) From t_{120}, $P < .01$.

11.31 $P < .025$; Reject H_0 for any α value of .025 or greater.

11.33 (a) $H_0: p = 0.12$, $H_1: p \neq 0.12$; z-statistic for test of a proportion. (c) $H_0: \mu \leq 21$, $H_1: \mu > 21$; t-test for a population mean.

11.34 (a) Yes, if the test statistic falls between the .05 and .01 significance points, (c) Yes. The 1% significance level requires a larger test statistic than the 5% level.

11.35 (a) Yes, (b) No.

11.38 (a) Reject H_0, (b) Accept H_0, (f) Insufficient information.

11.39 (a) 0.25, (c) 0.82, (d) 0.99.

11.41 (a) 0.14.

Chapter 12

12.1 (a) $H_0: \mu_1 \leq \mu_2$, $H_1: \mu_1 > \mu_2$ (one-tailed), (b) 4.6; 3.059, (c) $z = 1.50$; significance point $= 1.645$, (d) Accept H_0.

12.3 P approximately .0668, (a) Yes, (b) No.

12.4 $(-1.40, 10.60)$; the mean for those who take the course may be the same as or lower than the mean for noncourse takers; the maximum plausible benefit is 10.6 points, on average, or 0.88 standard deviations.

12.5 (a) $H_0: \mu_1 = \mu_2$, $H_1: \mu_1 \neq \mu_2$, (b) 5; 24.881, (c) $z = 0.20$, (d) P approximately .8414, (e) Accept H_0 with $\alpha = .05$ and with $\alpha = .01$.

12.7 (a) $H_0: \mu_1 = \mu_2$, $H_1: \mu_1 \neq \mu_2$; Significance points are ± 1.645; WEIGHT: $\sigma_{\bar{x}_1 - \bar{x}_2} = \sqrt{12.1^2/9984 + 13.3^2/240} = 0.867$; $z = (81.9 - 83.2)/0.867 = -1.50$; Accept H_0. SYSTOLIC BLOOD PRESSURE: $z = 5.13$; Reject H_0. (b) SYSTOLIC BLOOD PRESSURE: Effect size $= 0.44$ standard deviations.

12.10 $H_0: \mu_1 \geq \mu_2$, $H_1: \mu_1 < \mu_2$, where population 1 had no prior coursework and population 2 had some previous coursework; $n_1 = 10$, $n_2 = 14$, significance point $-t_{22}(.05) = -1.717$; $s_p^2 = 1,426.6$; $t = (244.7 - 251.9)/\sqrt{1,426.6 \times (1/10 + 1/14)} = -0.46$; Accept H_0. Prior course work does not generally increase mean performance.

12.11 (a) 21.97, (b) 1.555, (c) $t = 2.06$, (d) $t_{35}(.10) = 1.306$ (by interpolation), (e) Reject H_0.

12.13 (a) $t = 2.40$, (b) $\pm t_{40}(.025) = \pm 2.021$, (c) Reject H_0.

12.15 $.02 < P < .05$; H_0 would be rejected for any α value of $.05$ or greater.

12.16 (a) $H_0: \mu_1 = \mu_2$, $H_1: \mu_1 \neq \mu_2$, (b) $t = 2.72$, (c) t_{156}; significance points from t_{120} are ± 1.980.

12.19 (a) $H_0: \mu_1 \leq \mu_2$, $H_1: \mu_1 > \mu_2$ where μ_1 is the population mean for women with no children, and μ_2 for women with children, (b) $t = 1.34$, (d) $.05 < P < .10$ using t_{120}.

12.20 $(-1.23, 11.39)$ [using percentage points from t_{120}].

12.21 (a) $n_1 = n_2 = 100$, $\bar{x}_1 = 11.09$ (B.U.), $\bar{x}_2 = 11.83$ (S.U.); $s_1^2 = 5.01$, $s_2^2 = 4.14$, $s_p^2 = 4.58$; (b) $H_0: \mu_1 = \mu_2$, $H_1: \mu_1 \neq \mu_2$; (c) $t = -2.45$; (d) $.01 < P < .02$ ($2.326 < t < 2.5746$; two-tailed test).

12.24 (a) Reject H_0, (b) Accept H_0, (c) Reject H_0.

12.26 (b) Let μ_1 = mean for non-English speaking group, and μ_2 for English-speaking. Interval for $\mu_2 - \mu_1$: $(4.45, 28.09)$.

12.27 (b) 26,581, (c) 64, (d) $s_p^2 = 415$.

12.29 (b) $\hat{p}_1 = 0.023$, $\hat{p}_2 = 0.014$, $z = 3.1$.

12.30 $\hat{p}_A = 0.88$, $\hat{p}_B = 0.40$, $\hat{p} = 0.64$, $\sigma_{\hat{p}_1 - \hat{p}_2} = 0.096$, $z = (0.88 - 0.40)/0.096 = 5.00$; Significance points are $\pm z(.005) = \pm 2.58$; Reject H_0.

12.32 (b) $\hat{p}_{HDA} = 0.429$, $\hat{p}_{LDA} = 0.619$, $z = -1.23$; (d) Accept H_0; (e) No greater proportion of LDA patients complete a treatment plan.

12.35 MALES: $(-0.244, -0.136)$.

12.36 $\hat{p}_1 = 0.000164$, $\hat{p}_2 = 0.000547$, $\hat{p} = 0.000356$, $\sigma_{\hat{p}_1 - \hat{p}_2} = 0.595 \times 10^{-4}$; $z = (0.000164 - 0.000547)/(0.595 \times 10^{-4}) = -6.4$; Significance point (one-tailed test) is $-z(.01) = -2.33$; Reject H_0.

12.37 MANAGERS: (a) $H_0: p_1 = p_2$, $H_1: p_1 \neq p_2$, (b) $N(0, 1)$, (c) $\hat{p}_1 = 0.320$, $\hat{p}_2 = 0.091$, $z = 1.8$, Accept H_0.

12.38 MANAGERS: $P = .0718$.

12.41 Common median = 96; Above common median: $\hat{p}_1 = 12/21 = 0.57$, $\hat{p}_2 = 2/6 = 0.33$; $z = 2 \times \sqrt{21(6)/(21 + 6)} \times (0.57 - 0.33) = 1.03$; Significance points are $\pm z(.025) = \pm 1.96$; Accept H_0.

Chapter 13

13.1   (a) 5.992, (b) 0.103, (c) 9.488, (d) 0.711.

13.3   (a) (4.97, 16.01), (b) Accept.

13.5   (4.88, 17.2).

13.7   (1.76, 2.68).

13.9   (a) $H_0: \sigma \geq 100$, $H_1: \sigma < 100$; Test statistic $\chi^2 = 23.91$; significance point from χ^2_{40} is 26.509; Reject H_0.

13.10   (a) 2.3479, (b) 1.9367, (c) 7.5594, (d) 5.3903.

13.13 $H_0: \sigma^2_1 = \sigma^2_2$, $H_1: \sigma^2_1 \neq \sigma^2_2$; Test statistic $F = 0.50$; Significance points from $F_{10,20}$ are 0.293 and 2.7737; Accept H_0.

13.15   Use $F_{30,25}$; Significance points are 0.532 and 1.9192; the value 1.17 is between these, so H_0 is accepted.

13.17   (a) $P < .005$, (c) $P > .10$, (e) $.01 < P < .02$.

13.19   (b) $F = 0.335$, (c) $F_{14,14}$; Significance points from $F_{12,12}$ are 0.204 and 4.9062; from $F_{15,15}$ (liberal) they are 0.246 and 4.0698; Accept H_0 in either case.

13.21 $s/\bar{x}$; Doctor A: 0.58.

Chapter 14

14.1   (a) $\chi^2 = 0.4$, Significance point from χ^2_1 is 3.843, coin is fair, $P > 10$; (e) $\chi^2 = 9.0$; coin is not fair; $P < .005$.

14.3 $\chi^2 = 2.12$; Significance point from χ^2_1 is 3.843, Accept H_0.

14.5 $\chi^2 = 305.2$, $P < .005$ using χ^2_3; Reject $H_0: p_1 = p_2 = p_3 = p_4$.

14.7 $\chi^2 = 15.4$, significance point from χ^2_5 is 11.07, probability not the same at all post positions.

14.9 $\chi^2 = 8.84$; $P > .10$ using χ^2_5; die is fair.

14.11   (a) 128 odd; 172 even, (b) $\chi^2 = 6.45$, (c) Reject H_0.

14.13 $\chi^2 = 0.38$, significance point from χ^2_1 is 3.843; Accept H_0, type of library and volumes added are independent.

14.15 $\chi^2 = 25.0$, $P < .005$ using χ^2_1; type of syringe is related to reaction.

14.18   (a) Suburb A is younger (see table).

		Suburb	
		A	B
	Below 25	114 (56.7%)	26 (28.9%)
Age	*25–45*	50 (24.9%)	25 (27.8%)
	Above 45	37 (18.4%)	39 (43.3%)

(c) 0.04, 0.03, 0.31; no significant differences.

14.20   (b) $\chi^2 = 27.7$; $P < .005$ using χ^2_2.

14.23   (b) $\chi^2 = 78.4$; 4 degrees of freedom; $P < .005$.

14.24   (a) $\chi^2 = 40.45$, (b) $\phi = -0.6819$.

14.25   (b) $\phi = -0.350$.

14.28

Birth order		*2*	*3–4*	*≥ 5*
(a)	ϕ	0.05038	0.04090	0.17114
(b)	χ^2	0.421	0.189	2.519
	P	$> .10$	$> .10$	$> .10$

14.30   (a) Proportions with heart disease: 0.02, 0.05, 0.10, 0.12,

(b) Expected frequencies:

		Non-snorers	*Occasional snorers*	*Nearly every night*	*Every night*
Heart	*Present*	61.07	28.25	9.43	11.25
disease	*Absent*	1317.93	609.75	203.57	242.75

(c) $\lambda_{r \cdot c} = 0$.

14.32   (a) $\gamma = -0.56$; moderate association of snoring and heart disease.

14.35   (a) $\lambda_{c \cdot r} = 0$, (b) $\lambda_{r \cdot c} = 0.025$, (c) $\lambda = 0.0139$, (d) $\gamma = 0.358$, (e) slight association is detected by the coefficient based on ordering.

14.37 $\phi_{RB} = 0.50$, $\phi_{RC} = 0.42$, $\phi_{CB} = 0.26$.

14.39   (d) $\chi^2 = 390.17$, $P < .005$ using χ^2_{13}.

14.41 $\chi^2 = 408.10$.

14.42   (a) $\phi = 0.67$, (b) $\gamma = 1$, (c) $n_{12} = n_{21} = 0$.

Chapter 15

15.1   (a) $b = -0.2$, $a = 4.0$.

15.3   *CD:* grade independent of hours of study. *EF:* positive correlation between grade and hours of study; slope is about 5 grade points per hour of study.

15.6 $b = -0.0306$, $a = 1.48$, $\hat{y}$ for 35 miles per hour is 0.41.

15.8   SECTION A: $b = 0.175$.

15.9   (b) $b = 7.96$.

15.10   (a) $H_0: \beta \leq 0$, $H_1: \beta > 0$, (b) $s_{y \cdot x} = 72.76$, $s_b = 1.093$, $t = 7.28$, (c) using t_{20}, $P < .005$.

15.11   Interval: $(5.68, 10.24)$.

15.13 $r = 0.852$.

15.14   (b) $r = 0.837$, (c) $H_0: \rho \leq 0$, $H_1: \rho > 0$, $t = 0.837\sqrt{8}/\sqrt{1 - 0.837^2} = 4.33$, significance point from t_8 is 1.86, Reject H_0.

15.15   (c) $b = 10.81$.

15.16   (b) $s_b = 1.72$, $t = 6.29$.

15.17 $(6.85, \ ?)$.

15.18   (a) $r = 0.793$.

15.21 $b = 0.266$, $s_b = 0.109$.

15.22   (a) $r = 0.10$; $t = 0.80$; $P > .10$ using t_{60}.

15.23   (b) $r = 0.46$; $t = 4.14$; $P < .005$.

15.25   (a) 5, (b) 4, (c) 2.8.

Chapter 16

16.1   (a) 3, 60, (b) 2.7581, (c) No.

16.3 $F = 18.32$; Significance point from $F_{3,20}$ is 3.0984, Reject H_0.

16.4   Significance points for 6 two-tailed tests with $\alpha^* = .02$, from t_{20}, are ± 2.528. For the comparison of μ_1 with μ_2, $t = 0.39$; difference not significant.

16.5   (a) $MS_B = 138.82$; $MS_W = 35.08$.

16.6   (a) $H_0: \mu_1 = \mu_2 = \mu_3$, (b) $F = 127.606/837.113 = 0.15$; Significance point from $F_{2,19}$ is 5.9259; Accept H_0.

16.8   NONCOMPLETERS–HIGH-SCHOOL GRADUATES: $t = -5.06/15.25 = -0.33$; Significance points for 2 two-tailed contrasts with $\alpha^* = .01$, from t_{19}, are ± 2.861; difference not significant.

16.11   (a) $SS_B = 5928.248$; $df_B = 4$; $MS_B = 1482.06$. (b) $df_W = 149$, $SS_W = 23{,}637.36$, (c) $F = 9.34$, $P < .005$ using $F_{4,100}$; Reject H_0.

16.12   HAUN–HARN: $\bar{x}_2 - \bar{x}_1 = 16.31$; $t = 16.31/3.00 = 5.43$, $P < .01$ using t_{120} for a two-tailed test.

16.15   FORMAL–CASUAL: (a) $-12.6/3.28 = -3.84$, (b) For 3 comparisons, confidence level for each is .98; interval is from $-12.6 - (2.681 \times 2.075)$ to $-12.6 + (2.681 \times 2.075)$, or $(-18.2, -7.0)$.

16.18   (a) $t = -4.32$, significance points from t_{120} are ± 2.617,

         (b) $\bar{x} = 21.48$, $MS_B = 1954.32$.

16.19   (a) Effect size: -0.76, moderate-to-large.

16.20 $H = 9.164$, significance point from χ_4^2 is 9.488; null hypothesis accepted.

Chapter 17

17.4 $\sigma_1 N_1 = \sigma_2 N_2$.

17.5   (a) (i) 6.10, (ii) 10.21, (c) 9.48.

17.7   (a) Situation A: $\{16, 32, 40\}$ is the only possible sample. Situation B: $\{40, 40, 40\}$, $\{32, 32, 32\}$, and $\{16, 16, 16\}$ are the only possible samples.

         (b) Situation A: standard error is 0. Situation B: standard error is 9.98.

References

American Airlines (1988). [Call workload data for a reservations office]. Unpublished raw data.

Anderson, T. W., & Sclove, S. L. (1986). *The statistical analysis of data* (2nd ed.). Palo Alto, CA: The Scientific Press.

Asch, S. (1951). Effects of group pressure upon the modification and distortion of judgement. In H. Guetzkow (Ed.), *Groups, leadership, and men.* Pittsburgh, PA: Carnegie Press. Also in E. E. Maccoby, T. M. Newcomb, & E. L. Hartley (Eds.) (1958), *Readings in social psychology.* New York: Holt, Rinehart, and Winston.

Bayes, T. (1958). An essay towards solving a problem in the doctrine of chances. *Biometrika,* **45**, 293–315. (Reprinted from *Philosophical Transactions of the Royal Society,* **53** (1763), 370–418).

Black, M. M., & Leis, H. P. (1972). Mammary carcinogenesis: Influence of parity and estrogens. *New York State Journal of Medicine,* **72**, 1601–1605.

Blair, S. N., Kohl, H.W., III, Paffenberger, R. S., Jr., Clark, D. G., Cooper, K. H., and Gibbons, L. W. (1989). Physical fitness and all-cause mortality. *Journal of the American Medical Association,* **262**, 2395–2401.

Blau, P. M. (1955). *The dynamics of bureaucracy.* Chicago: University of Chicago Press.

Bowditch, H. P. (1877). *The growth of children.* Report of the Board of Health of Massachusetts, VIII.

British Amateur Athletic Board (1989). *British athletics 1989.* United Kingdom: The Amateur Athletic Association and the National Union of Track Statisticians (ISBN 0-85134-092-X).

Carlin, J. E. (1962). *Lawyers on their own: A study of individual practitioners in Chicago.* New Brunswick, NJ: Rutgers University Press.

Ciancio, S. G., & Mather, M. L. (1990). A clinical comparison of two electric toothbrushes with different mechanical actions. *Clinical Preventive Dentistry,* **12**.

675

Clark, K. B., & Clark, M. P. (1958). Racial identification and preference in Negro children. In E. E. Maccoby, T. M. Newcomb, & E. L. Hartley (Eds.), *Readings in social psychology* (3rd ed., pp. 602–611). New York: Holt, Rinehart, and Winston.

Cochran, W. G. (1954). Some methods for strengthening the common χ^2 tests. *Biometrics*, **10**, 417–451.

Cochran, W. G. (1977). *Sampling techniques* (3rd ed.). New York: Wiley.

Deming, W. E. (1950). *Some theory of sampling.* New York: Wiley. (Reprinted in 1966, New York: Dover).

Draper, N. R., & Stoneman, D. M. (1966). Testing for the inclusion of variables in linear regression by a randomisation technique. *Technometrics*, **8**, 695–699.

Drew, D. R., & Dudek, C. L. (1965). *Investigation of an internal energy model for evaluating freeway level of service.* College Station: Texas A&M University, Texas Transportation Institute.

Dunn, J. E., Jr., & Greenhouse, S. W. (1950). *Cancer diagnostic tests: Principles and criteria for development and educaiton* (PHS Publication No. 9). Washington, D. C.: U.S. Government Printing Office.

Fadeley, R. C. (1965). Oregon malignancy pattern physiographically related to Hanford, Washington, radioisotope storage. *Journal of Environmental Health*, **27**, 883–897.

Farkas, S. R., & Shorey, H. H. (1972). Chemical trail-following by flying insects: A mechanism for orientation to a distant odor source. *Science*, **178**, 67–68.

Finn J. D. (1969). Multivariate analysis of repeated measures data. *Multivariate Behavioral Research*, **4**, 391–413.

Finn, J. D. (1979). *Quality of schooling: A process approach.* Final report submitted to the Spencer Foundation.

Fisher, R. A. (1960). *The design of experiments* (8th ed.). Edinburgh, Scotland: Oliver and Boyd.

Fisher, R. A. (1970). *Statistical methods for research workers* (14th ed.). Edinburgh, Scotland: Oliver and Boyd.

Francis, T., Jr., *et al.* (1957). *Evaluation of 1954 field trial of Poliomyelitis vaccine: Final report.* Ann Arbor: University of Michigan, Poliomyelitis Vaccine Evaluation Center.

Frets, G. P. (1921). Heredity of head form in man. *Genetica*, **3**, 193–384.

Frisancho, A. R., Frisancho, H. G., Milotich, M., Brutsaert, T., Albalak, R., Spielvogel, H., Villena, M., Vargas, E., & Soria, R. (1995). Developmental, genetic, and environmental components of aerobic capacity at high altitude. *American Journal of Physical Anthropology*, **96**, 431–442.

Galle, O. R., Gove, W. R., & McPherson, J. M. (1972). Population density and pathology: What are the relations for man? *Science*, **176**, 23–30.

Gallup, G. (1972, January 20). Nixon's rating at [State of the Union] message time. *San Francisco Chronicle.* Syndicated Column, Field Enterprises, Inc. Courtesy of Field Newspaper Syndicate.

Galton, F. (1885). Regression towards mediocrity in hereditary stature. *Journal of the Anthropological Institute*, **15**, 246–263.

Gebski, V., Leung, O., McNeil, D. R., & Lunn, A. D. (1992). *The SPIDA user's manual.* Sydney, Australia: Statistical Computing Laboratory.

Gentile, J. R., Voelkl, K. E., Monaco, N., & Mt. Pleasant, J. (1995). Recall after relearning by fast and slow learners. *Journal of Experimental Education*, **63**, 185–197.

Hammond, E. C., & Horn, D. (1958). Smoking and death rates: Report on forty-four months of follow-up of 187,783 men: II. Death rates by cause. *Journal of the American Medical Association*, **166**, 1294–1308.

Hand, D. J., Daly, F., Lunn, A. D., McConway, K. J., & Ostrowski, E. (Eds.). (1994). *A handbook of small data sets*. London: Chapman & Hall.

Hanson, M. H., Hurwitz, W. H., & Madow, W. G. (1953). *Sample survey methods and theory: Vol. 1. Methods and applications*. New York: John Wiley & Sons.

Hoel, P. G., Port, S. C., & Stone, C. J. (1972). *Introduction to probability theory*. Boston: Houghton Mifflin.

Hooker, R. H. (1907). The correlation of the weather and crops. *Journal of the Royal Statistical Society*, **70**, 1–42.

Hyman, H. H., Wright, C. R., & Hopkins, T. K. (1962). *Applications of methods of evaluations: Four studies of the encampment for citizenship*. Berkeley, CA: University of California Press.

Iannotti, R. J., & Bush, P. J. (1992). Perceived vs. actual friends' use of alcohol, cigarettes, marijuana, and cocaine: Which has the most influence? *Journal of Youth and Adolescence*, **21**, 375–389.

Iannotti, R. J., Zuckerman, A. E., Blyer, E. M., O'Brien, R. W., Finn, J., & Spillman, D. (1994). Comparison of dietary intake methods with young children. *Psychological Reports*, **74**, 883–889.

Kottmeyer, W. (1959). *Teacher's guide for remedial reading*. St. Louis, MO: Webster Publishing.

Kruskal, W. H., & Wallis, W. A. (1952). Use of ranks in one-criterion analysis of variance. *Journal of the American Statistical Association*, **47**, 583–621.

Lazarsfeld, P. F. (1940). *Radio and the printed page*. New York: Duell, Sloan, and Pierce. (Reprinted in 1971, New York: Arno Press).

Lazarsfeld, P. F., Berelson, B., & Gaudet, H. (1968). *The people's choice* (3rd ed). New York: Columbia University Press.

Lazarsfeld, P. F., & Thielens, W. (1958). *The academic mind*. Glencoe, IL: Free Press.

Lee, Y. T. (1993). Reactions of American minority and nonminority students to the Persian Gulf War. *The Journal of Social Psychology*, **133**, 707–713.

Levy, P. S., & Lemeshow, S. (1991). *Sampling of populations: Methods and Applications*. New York: John Wiley & Sons.

The Literary Digest (1963, October 31). **122**, 5–6.

The Literary Digest (1963, November 14). **122**, 7–8.

The Market Research Society. (1972). *Public opinion polling in the 1970 election*. London: Author.

McCarthy, P. J. (1949). Election predictions. In F. Mosteller, H. Hyman, P. J. McCarthy, E. S. Marks, & D. B. Turman (Eds.), *The pre-election polls of 1948*. New York: Social Science Research Council.

McGrath, W. E. (1986). Levels of data in the study of library practice: Definition, analysis, inference and explanation. In G. G. Allen & F. C. A. Exon (Eds.), *Research and the practice of librarianship: An international symposium* (pp. 29–40). Perth, Australia: Western Australian Institute of Technology.

McNemar, Q. (1947). Note on the sampling error of the difference between correlated proportions or percentages. *Psychometrika*, **12**, 153–157.

Mosteller, F., & Wallace, D. L. (1964). *Inference and disputed authorship: The federalist*. Reading, MA: Addison–Wesley.

Nanji, A. A., & French, S. W. (1985). Relationship between pork consumption and cirrhosis. *The Lancet*, **i**, 681–683.

Norton, P. G., & Dunn, E. V. (1985). Snoring as a risk factor for disease: An epidemiological survey. *British Medical Journal*, **291**, 630–632.

O'Brien, R. W., & Iannotti, R. J. (1994). How maternal characteristics influence differences between mothers' and teachers' ratings of Type A behavior in Black preschool children. *Behavioral Medicine*, **19**, 162–168.

Ott, R. L. (1993). *An introduction to statistical methods and data analysis* (4th ed.). Belmont, CA: Duxbury Press.

Pallas, A. M., Natriello, G., & McDill, E. L. (1989). The changing nature of the disadvantaged population: Current dimensions and future trends. *Educational Researcher*, **18**(5), 16–22.

Patterson, C. J., Kupersmidt, J. B., & Vaden, N. A. (1990). Income level, gender, ethnicity, and household composition as predictors of children's school based competence. *Child Development*, **61**, 485–494.

Pearson, E. S., & Hartley, H. O. (Eds.). (1954). *Biometrika tables for statisticians* (Vol. 1). Cambridge, England: Cambridge University Press.

Persinger, M. A., Ballance, S. E., & Moland, M. (1993). Snow fall and heart attacks. *The Journal of Psychology*, **127**, 243–252.

Phillips, D. (1972). Deathday and birthday: An unexpected connection. In J. M. Tanur, *et al.* (Eds.), *Statistics: A guide to the unknown*. San Francisco: Holden-Day.

The RAND Corporation (1955). *A million random digits with 100,000 normal deviates*. New York: The Free Press.

Rosenblatt, J. R., & Filliben, J. J. (1971). Randomization and the draft lottery. *Science*, **171**, 306–308.

San José Mercury News (1989, July 26). 1A.

Schneer, J. A., & Retiman, F. (1993). Effects of alternate family structures on managerial career paths. *Academy of Management Journal*, **36**, 830–843.

Shields, J. (1962). *Monozygotic twins*. London: Oxford University Press.

Shulte, A. P. (1970). The effects of a unit in probability and statistics on students and teachers of ninth grade General Mathematics. *The Mathematics Teacher*, **63**, 56–64.

Singer, S. I., Levine, M., Rowley, J., & Bazargan, M. (1987). *Amherst Youth Board needs assessment survey*. Unpublished report, State University of New York at Buffalo, Research Center for Children and Youth.

Snyder, T. D., & Hoffman, C. M. (1993). *Digest of education statistics* (NCES Publication No. 93–292). Washington, DC: U. S. Department of Education, Office of Educational Research and Improvement, National Center for Education Statistics.

Somers, R. H. (1959). *Young Americans abroad: A study of the selection and evaluation procedure of the experiment in international living*. Unpublished report, Columbia University, Bureau of Applied Social Research, New York.

Stevens, S. S. (1951). Mathematics, measurement, and psychophysics. In S. S. Stevens (Ed.), *Handbook of experimental psychology*. New York: Wiley.

Stewart, J. E., Marcus, M., Christenson, P. D., & Lin, W. L. (1994). Comprehensive treatment among dental school patients with high and low dental anxiety. *Journal of Dental Education*, **58**, 697–700.

Student (Gosset, W. S.) (1908). The probable error of a mean. *Biometrika*, **6**, 1–25.

Tennessee State Education Department (1988). [Years experience for teachers participating in the Student-Teacher Achievement Ratio (STAR) project]. Unpublished raw data.

Thucydides (1954). *History of the Peloponnesian War* (R. Warner, Trans.). Baltimore, MD: Penguin Books.

Time (1971, December 27). 42.

Tukey, J. W. (1977). *Exploratory data analysis.* Reading, MA: Addison-Wesley.

U. S. Bureau of the Census (1961). *Historical statistics of the United States: Colonial times to 1957.* Washington, DC: U. S. Department of Commerce.

U. S. Bureau of the Census (1964). *U. S. census of population: 1960: Vol 1. Characteristics of the population.* Washington, DC: U. S. Department of Commerce.

U. S. Bureau of the Census (1993). *Statistical abstracts of the United States.* Washington, DC: U. S. Department of Commerce.

U. S. Department of Education (1989). *National education longitudinal study of 1988* (NCES Publication No. 90–464). Washington, DC: Office of Educational Research and Improvement, National Center for Education Statistics.

U. S. Office for Civil Rights (1978). [Biennial survey of special education placements in American public schools]. Unpublished raw data.

U. S. Public Health Service (1964). *Smoking and health: Report of the advisory committee to the Surgeon General of the Public Health Service* (PHS Publication No. 1103). Washington, DC: U. S. Department of Health, Education, and Welfare.

Ware, M. C., & Stuck, M. F. (1985). Sex-role messages vis-à-vis microcomputer use: A look at the pictures. *Sex Roles,* **13**, 205–214.

Weiss, D., Whitten, B., & Leddy, D. (1972). Lead content of human hair (1871–1971). *Science,* **178**, 69–70.

Westinghouse Learning Corporation/Ohio Univ. (1969). *The impact of Head Start: An evaluation of the effects of Head Start on children's cognitive and affective development.* Washington, DC: Office of Economic Opportunity.

Whelpton, P. K., & Kiser, C. V. (1950). *Social and psychological factors affecting fertility: Vol. 1. The household survey in Indianapolis.* New York: Milbank Memorial Fund.

Williams, J. (1992). *Academic Libraries: 1990* (NCES Publication No. 93-004). Washington, DC: U. S. Department of Education, Office of Educational Research and Improvement, National Center for Education Statistics. (ERIC Document Reproduction Service No. ED 355 943).

Wiltsey, R. G. (1972). *Doctoral use of foreign languages: A survey: Part II. Supplementary tables.* Princeton, NJ: Educational Testing Service, Graduate Record Examinations Board.

Wise, P. S., & Cramer, S. H. (1988). Correlates of empathy and cognitive style in early adolescence. *Psychological Reports,* **63**, 179–192.

Witkin, H. A. (1971). *The embedded figures test.* Palo Alto, CA: Consulting Psychologists Press.

Yu, P. L., Wrather, C., & Kozmetsky, G. (1975). *Auto weight and public safety: A statistical study of transportation hazards* (Research Rep. No. 233). Austin: University of Texas, Center for Cybernetic Studies.

Zehna, P. W., & Barr, D. R. (1970). *Tables of common probability distributions.* Monterey CA: United States Naval Postgraduate School.

Appendices

Appendix I   Table of the Standard Normal Distribution

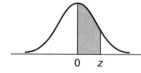

The entries in this table are the probabilities that a standard normal variate will have a value between 0 and z.

z	.00	.01	.02	.03	.04	.05	.06	.07	.08	.09
0.0	.0000	.0040	.0080	.0120	.0160	.0199	.0239	.0279	.0319	.0359
0.1	.0398	.0438	.0478	.0517	.0557	.0596	.0636	.0675	.0714	.0753
0.2	.0793	.0832	.0871	.0910	.0948	.0987	.1026	.1064	.1103	.1141
0.3	.1179	.1217	.1255	.1293	.1331	.1368	.1406	.1443	.1480	.1517
0.4	.1554	.1591	.1628	.1664	.1700	.1736	.1772	.1808	.1844	.1879
0.5	.1915	.1950	.1985	.2019	.2054	.2088	.2123	.2157	.2190	.2224
0.6	.2257	.2291	.2324	.2357	.2389	.2422	.2454	.2486	.2517	.2549
0.7	.2580	.2611	.2642	.2673	.2704	.2734	.2764	.2794	.2823	.2852
0.8	.2881	.2910	.2939	.2967	.2995	.3023	.3051	.3078	.3106	.3133
0.9	.3159	.3186	.3212	.3238	.3264	.3289	.3315	.3340	.3365	.3389
1.0	.3413	.3438	.3461	.3485	.3508	.3531	.3554	.3577	.3599	.3621
1.1	.3643	.3665	.3686	.3708	.3729	.3749	.3770	.3790	.3810	.3830
1.2	.3849	.3869	.3888	.3907	.3925	.3944	.3962	.3980	.3997	.4015
1.3	.4032	.4049	.4066	.4082	.4099	.4115	.4131	.4147	.4162	.4177
1.4	.4192	.4207	.4222	.4236	.4251	.4265	.4279	.4292	.4306	.4319
1.5	.4332	.4345	.4357	.4370	.4382	.4394	.4406	.4418	.4429	.4441
1.6	.4452	.4463	.4474	.4484	.4495	.4505	.4515	.4525	.4535	.4545
1.7	.4554	.4564	.4573	.4582	.4591	.4599	.4608	.4616	.4625	.4633
1.8	.4641	.4649	.4656	.4664	.4671	.4678	.4686	.4693	.4699	.4706
1.9	.4713	.4719	.4726	.4732	.4738	.4744	.4750	.4756	.4761	.4767
2.0	.4772	.4778	.4783	.4788	.4793	.4798	.4803	.4808	.4812	.4817
2.1	.4821	.4826	.4830	.4834	.4838	.4842	.4846	.4850	.4854	.4857
2.2	.4861	.4864	.4868	.4871	.4875	.4878	.4881	.4884	.4887	.4890
2.3	.4893	.4896	.4898	.4901	.4904	.4906	.4909	.4911	.4913	.4916
2.4	.4918	.4920	.4922	.4925	.4927	.4929	.4931	.4932	.4934	.4936
2.5	.4938	.4940	.4941	.4943	.4945	.4946	.4948	.4949	.4951	.4952
2.6	.4953	.4955	.4956	.4957	.4959	.4960	.4961	.4962	.4963	.4964
2.7	.4965	.4966	.4967	.4968	.4969	.4970	.4971	.4972	.4973	.4974
2.8	.4974	.4975	.4976	.4977	.4977	.4978	.4979	.4979	.4980	.4981
2.9	.4981	.4982	.4982	.4983	.4984	.4984	.4985	.4985	.4986	.4986
3.0	.4987	.4987	.4987	.4988	.4988	.4989	.4989	.4989	.4990	.4990
3.1	.4990	.4991	.4991	.4991	.4992	.4992	.4992	.4992	.4993	.4993
3.2	.4993	.4993	.4994	.4994	.4994	.4994	.4994	.4995	.4995	.4995
3.3	.4995	.4995	.4995	.4996	.4996	.4996	.4996	.4996	.4996	.4997
3.4	.4997	.4997	.4997	.4997	.4997	.4997	.4997	.4997	.4997	.4998
3.5	.4998	.4998	.4998	.4998	.4998	.4998	.4998	.4998	.4998	.4998

For $z = 4.0$, 5.0, and 6.0, the areas are 0.49997, 0.4999997, and 0.499999999, respectively.

Appendix II Table of Binomial Probabilities

					PROBABILITY OF HEAD p				
n	r	.01	.05	.10	1/6	.25	1/3	.49	.50
2	0	.9801	.9025	.8100	.6944	.5625	.4444	.2601	.2500
	1	.0198	.0950	.1800	.2778	.3750	.4444	.4998	.5000
	2	.0001	.0025	.0100	.0278	.0625	.1111	.2401	.2500
3	0	.9703	.8574	.7290	.5787	.4219	.2963	.1327	.1250
	1	.0294	.1354	.2430	.3472	.4219	.4444	.3823	.3750
	2	.0003	.0071	.0270	.0694	.1406	.2222	.3674	.3750
	3	.0000	.0001	.0010	.0046	.0156	.0370	.1176	.1250
4	0	.9606	.8145	.6561	.4823	.3164	.1975	.0677	.0625
	1	.0388	.1715	.2916	.3858	.4219	.3951	.2600	.2500
	2	.0006	.0135	.0486	.1157	.2109	.2963	.3747	.3750
	3	.0000	.0005	.0036	.0154	.0469	.0988	.2400	.2500
	4	.0000	.0000	.0001	.0008	.0039	.0123	.0576	.0625
5	0	.9510	.7738	.5905	.4019	.2373	.1317	.0345	.0312
	1	.0480	.2036	.3280	.4019	.3955	.3292	.1657	.1562
	2	.0010	.0214	.0729	.1608	.2637	.3292	.3185	.3125
	3	.0000	.0011	.0081	.0322	.0879	.1646	.3060	.3125
	4	.0000	.0000	.0004	.0032	.0146	.0412	.1470	.1562
	5	.0000	.0000	.0000	.0001	.0010	.0041	.0282	.0312
6	0	.9415	.7351	.5314	.3349	.1780	.0878	.0176	.0156
	1	.0571	.2321	.3543	.4019	.3560	.2634	.1014	.0938
	2	.0014	.0305	.0984	.2009	.2966	.3292	.2436	.2344
	3	.0000	.0021	.0146	.0536	.1318	.2195	.3121	.3125
	4	.0000	.0001	.0012	.0080	.0330	.0823	.2249	.2344
	5	.0000	.0000	.0001	.0006	.0044	.0165	.0864	.0938
	6	.0000	.0000	.0000	.0000	.0002	.0014	.0138	.0156
7	0	.9321	.6983	.4783	.2791	.1335	.0585	.0090	.0078
	1	.0659	.2573	.3720	.3907	.3115	.2048	.0604	.0547
	2	.0020	.0406	.1240	.2344	.3115	.3073	.1740	.1641
	3	.0000	.0036	.0230	.0781	.1730	.2561	.2786	.2734
	4	.0000	.0002	.0026	.0156	.0577	.1280	.2676	.2734
	5	.0000	.0000	.0002	.0019	.0115	.0384	.1543	.1641
	6	.0000	.0000	.0000	.0001	.0013	.0064	.0494	.0547
	7	.0000	.0000	.0000	.0000	.0001	.0005	.0068	.0078

Appendix II Table of Binomial Probabilities (Continued)

		PROBABILITY OF HEAD p							
n	r	.01	.05	.10	1/6	.25	1/3	.49	.50
8	0	.9227	.6634	.4305	.2326	.1001	.0390	.0046	.0039
	1	.0746	.2793	.3826	.3721	.2670	.1561	.0352	.0312
	2	.0026	.0515	.1488	.2605	.3115	.2731	.1183	.1094
	3	.0001	.0054	.0331	.1042	.2076	.2731	.2273	.2188
	4	.0000	.0004	.0046	.0260	.0865	.1707	.2730	.2734
	5	.0000	.0000	.0004	.0042	.0231	.0683	.2098	.2188
	6	.0000	.0000	.0000	.0004	.0038	.0171	.1008	.1094
	7	.0000	.0000	.0000	.0000	.0004	.0024	.0277	.0312
	8	.0000	.0000	.0000	.0000	.0000	.0002	.0033	.0039
9	0	.9135	.6302	.3874	.1938	.0751	.0260	.0023	.0020
	1	.0830	.2985	.3874	.3489	.2253	.1171	.0202	.0176
	2	.0034	.0629	.1722	.2791	.3003	.2341	.0776	.0703
	3	.0001	.0077	.0446	.1302	.2336	.2731	.1739	.1641
	4	.0000	.0006	.0074	.0391	.1168	.2048	.2506	.2461
	5	.0000	.0000	.0008	.0078	.0389	.1024	.2408	.2461
	6	.0000	.0000	.0001	.0010	.0087	.0341	.1542	.1641
	7	.0000	.0000	.0000	.0001	.0012	.0073	.0635	.0703
	8	.0000	.0000	.0000	.0000	.0001	.0009	.0153	.0176
	9	.0000	.0000	.0000	.0000	.0000	.0001	.0016	.0020
10	0	.9044	.5987	.3487	.1615	.0563	.0173	.0012	.0010
	1	.0914	.3151	.3874	.3230	.1877	.0867	.0114	.0098
	2	.0042	.0746	.1937	.2907	.2816	.1951	.0494	.0439
	3	.0001	.0105	.0574	.1550	.2503	.2601	.1267	.1172
	4	.0000	.0010	.0112	.0543	.1460	.2276	.2130	.2051
	5	.0000	.0001	.0015	.0130	.0584	.1366	.2456	.2461
	6	.0000	.0000	.0001	.0022	.0162	.0569	.1966	.2051
	7	.0000	.0000	.0000	.0002	.0031	.0163	.1080	.1172
	8	.0000	.0000	.0000	.0000	.0004	.0030	.0389	.0439
	9	.0000	.0000	.0000	.0000	.0000	.0003	.0083	.0098
	10	.0000	.0000	.0000	.0000	.0000	.0000	.0000	.0010

$\Pr(\text{exactly } r \text{ heads in } n \text{ tosses}) = C_r^n p^r q^{n-r},$

$$C_r^n = \frac{n!}{r!(n-r)!}.$$

For values of p greater than .50, use the rule $\Pr(\text{exactly } r \text{ heads in } n \text{ tosses if heads probability is } p) = \Pr(\text{exactly } n - r \text{ heads in } n \text{ tosses if heads probability is } 1 - p).$

Appendix III   Percentage Points of Student's t-Distributions (f = number of degrees of freedom)

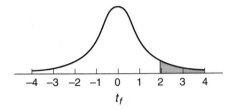

f	PROPORTION OF AREA IN ONE TAIL				
	.10	.05	.025	.01	.005
1	3.078	6.314	12.706	31.821	63.657
2	1.886	2.920	4.303	6.964	9.925
3	1.638	2.353	3.182	4.541	5.841
4	1.533	2.132	2.776	3.747	4.604
5	1.476	2.015	2.571	3.365	4.032
6	1.440	1.943	2.447	3.143	3.707
7	1.415	1.895	2.365	2.998	3.499
8	1.397	1.860	2.306	2.896	3.355
9	1.383	1.833	2.262	2.821	3.250
10	1.372	1.812	2.228	2.764	3.169
11	1.363	1.796	2.201	2.718	3.106
12	1.356	1.782	2.179	2.681	3.054
13	1.350	1.771	2.160	2.650	3.012
14	1.345	1.761	2.145	2.624	2.977
15	1.341	1.753	2.132	2.602	2.947
16	1.337	1.746	2.120	2.584	2.921
17	1.333	1.740	2.110	2.567	2.898
18	1.330	1.734	2.101	2.552	2.878
19	1.328	1.729	2.093	2.540	2.861
20	1.325	1.725	2.086	2.528	2.845
21	1.323	1.721	2.080	2.518	2.831
22	1.321	1.717	2.074	2.508	2.819
23	1.320	1.714	2.069	2.500	2.807
24	1.318	1.711	2.064	2.492	2.797
25	1.316	1.708	2.060	2.485	2.788
26	1.315	1.706	2.056	2.479	2.779
27	1.314	1.703	2.052	2.473	2.771
28	1.312	1.701	2.048	2.467	2.763
29	1.311	1.699	2.045	2.462	2.756
30	1.310	1.697	2.042	2.457	2.750
40	1.303	1.684	2.021	2.423	2.704
60	1.296	1.671	2.000	2.390	2.660
120	1.289	1.658	1.980	2.358	2.617
∞	1.282	1.645	1.960	2.326	2.576

For a one-tailed test with significance level α refer to the column headed by α in the table. For a two-tailed test refer to the column headed by $\alpha/2$; for example, if the significance level is 5% use the column headed .025.

Appendix IV Percentage Points of Chi-Square Distributions (f = number of degrees of freedom)

					AREA IN RIGHTHAND TAIL					
f	.995	.99	.975	.95	.90	.10	.05	.025	.01	.005
1	0.000	0.000	0.001	0.004	0.016	2.706	3.843	5.025	6.637	7.882
2	0.010	0.020	0.051	0.103	0.211	4.605	5.992	7.378	9.210	10.597
3	0.072	0.115	0.216	0.352	0.584	6.251	7.815	9.348	11.344	12.837
4	0.207	0.297	0.484	0.711	1.064	7.779	9.488	11.143	13.277	14.860
5	0.412	0.554	0.831	1.145	1.610	9.236	11.070	12.832	15.085	16.748
6	0.676	0.872	1.237	1.635	2.204	10.645	12.592	14.440	16.812	18.548
7	0.989	1.239	1.690	2.167	2.833	12.017	14.067	16.012	18.474	20.276
8	1.344	1.646	2.180	2.733	3.490	13.362	15.507	17.534	20.090	21.954
9	1.735	2.088	2.700	3.325	4.168	14.684	16.919	19.022	21.665	23.587
10	2.156	2.558	3.247	3.940	4.865	15.987	18.307	20.483	23.209	25.188
11	2.603	3.053	3.816	4.575	5.578	17.275	19.675	21.920	24.724	26.755
12	3.074	3.571	4.404	5.226	6.304	18.549	21.026	23.337	26.217	28.300
13	3.565	4.107	5.009	5.892	7.041	19.812	22.362	24.735	27.687	29.817
14	4.075	4.660	5.629	6.571	7.790	21.064	23.685	26.119	29.141	31.319
15	4.600	5.229	6.262	7.261	8.547	22.307	24.996	27.488	30.577	32.799
16	5.142	5.812	6.908	7.962	9.312	23.542	26.296	28.845	32.000	34.267
17	5.697	6.407	7.564	8.682	10.085	24.769	27.587	30.190	33.408	35.716
18	6.265	7.015	8.231	9.390	10.865	25.989	28.869	31.526	34.805	37.156
19	6.843	7.632	8.906	10.117	11.651	27.203	30.143	32.852	36.190	38.580
20	7.434	8.260	9.591	10.851	12.443	28.412	31.410	34.170	37.566	39.997
21	8.033	8.897	10.283	11.591	13.240	29.615	32.670	35.478	38.930	41.399
22	8.643	9.542	10.982	12.338	14.042	30.813	33.924	36.781	40.289	42.796
23	9.260	10.195	11.688	13.090	14.848	32.007	35.172	38.075	41.637	44.179
24	9.886	10.856	12.401	13.848	15.659	33.196	36.415	39.364	42.980	45.558
25	10.519	11.523	13.120	14.611	16.473	34.381	37.652	40.646	44.313	46.925
26	11.160	12.198	13.844	15.379	17.292	35.563	38.885	41.923	45.642	48.290
27	11.807	12.878	14.573	16.151	18.114	36.741	40.113	43.194	46.962	49.642
28	12.461	13.565	15.308	16.928	18.939	37.916	41.337	44.461	48.278	50.993
29	13.120	14.256	16.147	17.708	19.768	39.087	42.557	45.722	49.586	52.333
30	13.787	14.954	16.791	18.493	20.599	40.256	43.773	46.979	50.892	53.672
31	14.457	15.655	17.538	19.280	21.433	41.422	44.985	48.231	52.190	55.000
32	15.134	16.362	18.291	20.072	22.271	42.585	46.194	49.480	53.486	56.328
33	15.814	17.073	19.046	20.866	23.110	43.745	47.400	50.724	54.774	57.646
34	16.501	17.789	19.806	21.664	23.952	44.903	48.602	51.966	56.061	58.964
35	17.191	18.508	20.569	22.465	24.796	46.059	49.802	53.203	57.340	60.272
40	20.706	22.164	24.433	26.509	29.050	51.805	55.758	59.342	63.691	66.766
50	27.991	29.707	32.357	34.764	37.689	63.167	67.505	71.420	76.154	79.490
60	35.535	37.485	40.482	43.188	46.459	74.397	79.082	83.298	88.379	91.952
70	43.275	45.442	48.758	51.739	55.329	85.527	90.531	95.023	100.425	104.215
80	51.172	53.540	57.153	60.392	64.278	96.578	101.879	106.629	112.329	116.321
90	59.196	61.754	65.647	69.126	73.291	107.565	113.145	118.136	124.116	128.299
100	67.328	70.065	74.222	77.930	82.358	118.498	124.342	129.561	135.807	140.169

source: Zehna and Barr (1970), Table 2, pp. 11–14.

Appendix V   Upper Percentage Points of F-Distributions (m = numerator degrees of freedom, n = denominator degrees of freedom)

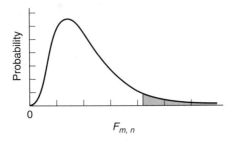

		AREA IN RIGHT-HAND TAIL				
m	n	.10	.05	.025	.01	.005
1	1	39.8635	161.4476	647.7890	4052.1806	16210.7221
	2	8.5263	18.5128	38.5063	98.5025	198.5013
	3	5.5383	10.1280	17.4434	34.1162	55.5520
	4	4.5448	7.7086	12.2179	21.1977	31.3328
	5	4.0604	6.6079	10.0070	16.2582	22.7848
	6	3.7759	5.9874	8.8131	13.7450	18.6350
	7	3.5894	5.5914	8.0727	12.2464	16.2356
	8	3.4579	5.3177	7.5709	11.2586	14.6882
	9	3.3603	5.1174	7.2093	10.5614	13.6136
	10	3.2850	4.9646	6.9367	10.0443	12.8265
	11	3.2252	4.8443	6.7241	9.6460	12.2263
	12	3.1765	4.7472	6.5538	9.3302	11.7542
	13	3.1362	4.6672	6.4143	9.0738	11.3735
	14	3.1022	4.6001	6.2979	8.8616	11.0603
	15	3.0732	4.5431	6.1995	8.6831	10.7980
	16	3.0481	4.4940	6.1151	8.5310	10.5755
	17	3.0262	4.4513	6.0420	8.3997	10.3842
	18	3.0070	4.4139	5.9781	8.2854	10.2181
	19	2.9899	4.3807	5.9216	8.1849	10.0725
	20	2.9747	4.3512	5.8715	8.0960	9.9439
	25	2.9177	4.2417	5.6864	7.7698	9.4753
	30	2.8807	4.1709	5.5675	7.5625	9.1797
	40	2.8354	4.0847	5.4239	7.3141	8.8279
	60	2.7911	4.0012	5.2856	7.0771	8.4946
	100	2.7564	3.9361	5.1786	6.8953	8.2406
	1000	2.7106	3.8508	5.0391	6.6603	7.9145

Appendix V
F-Distributions
(Continued)

m = numerator degrees of freedom
n = denominator degrees of freedom

		AREA IN RIGHT-HAND TAIL				
m	n	.10	.05	.025	.01	.005
2	1	49.5000	199.5000	799.5000	4999.4999	19999.4992
	2	9.0000	19.0000	39.0000	99.0000	199.0000
	3	5.4624	9.5521	16.0441	30.8165	49.7993
	4	4.3246	6.9443	10.6491	18.0000	26.2843
	5	3.7797	5.7861	8.4336	13.2739	18.3138
	6	3.4633	5.1433	7.2599	10.9248	14.5441
	7	3.2574	4.7374	6.5415	9.5466	12.4040
	8	3.1131	4.4590	6.0595	8.6491	11.0424
	9	3.0065	4.2565	5.7147	8.0215	10.1067
	10	2.9245	4.1028	5.4564	7.5594	9.4270
	11	2.8595	3.9823	5.2559	7.2057	8.9122
	12	2.8068	3.8853	5.0959	6.9266	8.5096
	13	2.7632	3.8056	4.9653	6.7010	8.1865
	14	2.7265	3.7389	4.8567	6.5149	7.9216
	15	2.6952	3.6823	4.7650	6.3589	7.7008
	16	2.6682	3.6337	4.6867	6.2262	7.5138
	17	2.6446	3.5915	4.6189	6.1121	7.3536
	18	2.6239	3.5546	4.5597	6.0129	7.2148
	19	2.6056	3.5219	4.5075	5.9259	7.0935
	20	2.5893	3.4928	4.4613	5.8489	6.9865
	25	2.5283	3.3852	4.2909	5.5680	6.5982
	30	2.4887	3.3158	4.1821	5.3903	6.3547
	40	2.4404	3.2317	4.0510	5.1785	6.0664
	60	2.3933	3.1504	3.9253	4.9774	5.7950
	100	2.3564	3.0873	3.8284	4.8239	5.5892
	1000	2.3079	3.0047	3.7025	4.6264	5.3265
m	n	.10	.05	.025	.01	.005
3	1	53.5932	215.7073	864.1630	5403.3519	21614.7406
	2	9.1618	19.1643	39.1655	99.1662	199.1664
	3	5.3908	9.2766	15.4392	29.4567	47.4672
	4	4.1909	6.5914	9.9792	16.6944	24.2591
	5	3.6195	5.4095	7.7636	12.0600	16.5298
	6	3.2888	4.7571	6.5988	9.7795	12.9166
	7	3.0741	4.3468	5.8898	8.4513	10.8824
	8	2.9238	4.0662	5.4160	7.5910	9.5965
	9	2.8129	3.8625	5.0781	6.9919	8.7171
	10	2.7277	3.7083	4.8256	6.5523	8.0807
	11	2.6602	3.5874	4.6300	6.2167	7.6004
	12	2.6055	3.4903	4.4742	5.9525	7.2258
	13	2.5603	3.4105	4.3472	5.7394	6.9258

Appendix V
F-Distributions
(Continued)

m = numerator degrees of freedom
n = denominator degrees of freedom

m	n	.10	.05	.025	.01	.005
				AREA IN RIGHT-HAND TAIL		
3	14	2.5222	3.3439	4.2417	5.5639	6.6804
	15	2.4898	3.2874	4.1528	5.4170	6.4760
	16	2.4618	3.2389	4.0768	5.2922	6.3034
	17	2.4374	3.1968	4.0112	5.1850	6.1556
	18	2.4160	3.1599	3.9539	5.0919	6.0278
	19	2.3970	3.1274	3.9034	5.0103	5.9161
	20	2.3801	3.0984	3.8587	4.9382	5.8177
	25	2.3170	2.9912	3.6943	4.6755	5.4615
	30	2.2761	2.9223	3.5894	4.5097	5.2388
	40	2.2261	2.8387	3.4633	4.3126	4.9758
	60	2.1774	2.7581	3.3425	4.1259	4.7290
	100	2.1394	2.6955	3.2496	3.9837	4.5424
	1000	2.0893	2.6138	3.1292	3.8012	4.3048

m	n	.10	.05	.025	.01	.005
4	1	55.8330	224.5832	899.5833	5624.5832	22499.5824
	2	9.2434	19.2468	39.2484	99.2494	199.2497
	3	5.3426	9.1172	15.1010	28.7099	46.1946
	4	4.1072	6.3882	9.6045	15.9770	23.1545
	5	3.5202	5.1922	7.3879	11.3919	15.5561
	6	3.1808	4.5337	6.2272	9.1483	12.0275
	7	2.9605	4.1203	5.5226	7.8466	10.0505
	8	2.8064	3.8379	5.0526	7.0061	8.8051
	9	2.6927	3.6331	4.7181	6.4221	7.9559
	10	2.6053	3.4780	4.4683	5.9943	7.3428
	11	2.5362	3.3567	4.2751	5.6683	6.8809
	12	2.4801	3.2592	4.1212	5.4120	6.5211
	13	2.4337	3.1791	3.9959	5.2053	6.2335
	14	2.3947	3.1122	3.8919	5.0354	5.9984
	15	2.3614	3.0056	3.8043	4.8932	5.8029
	16	2.3327	3.0069	3.7294	4.7726	5.6378
	17	2.3077	2.9647	3.6648	4.6690	5.4967
	18	2.2858	2.9277	3.6083	4.5790	5.3746
	19	2.2663	2.8951	3.5587	4.5003	5.2681
	20	2.2489	2.8661	3.5147	4.4307	5.1743
	25	2.1842	2.7587	3.3530	4.1774	4.8351
	30	2.1422	2.6896	3.2499	4.0179	4.6234
	40	2.0909	2.6060	3.1261	3.8283	4.3738
	60	2.0410	2.5252	3.0077	3.6490	4.1399
	100	2.0019	2.4626	2.9166	3.5127	3.9634
	1000	1.9505	2.3808	2.7986	3.3380	3.7390

Appendix V
F-Distributions
(Continued)

m = numerator degrees of freedom
n = denominator degrees of freedom

		AREA IN RIGHT-HAND TAIL				
m	n	.10	.05	.025	.01	.005
5	1	57.2401	230.1619	921.8479	5763.6494	23055.7973
	2	9.2926	19.2964	39.2982	99.2993	199.2996
	3	5.3092	9.0135	14.8848	28.2371	45.3916
	4	4.0506	6.2561	9.3645	15.5219	22.4564
	5	3.4530	5.0503	7.1464	10.9670	14.9396
	6	3.1075	4.3874	5.9876	8.7459	11.4637
	7	2.8833	3.9715	5.2852	7.4604	9.5221
	8	2.7264	3.6875	4.8173	6.6318	8.3018
	9	2.6106	3.4817	4.4844	6.0569	7.4712
	10	2.5216	3.3258	4.2361	5.6363	6.8724
	11	2.4512	3.2039	4.0440	5.3160	6.4217
	12	2.3940	3.1059	3.8911	5.0643	6.0711
	13	2.3467	3.0254	3.7667	4.8616	5.7910
	14	2.3069	2.9582	3.6634	4.6950	5.5623
	15	2.2730	2.9013	3.5764	4.5556	5.3721
	16	2.2438	2.8524	3.5021	4.4374	5.2117
	17	2.2183	2.8100	3.4379	4.3359	5.0746
	18	2.1958	2.7729	3.3820	4.2479	4.9560
	19	2.1760	2.7401	3.3327	4.1708	4.8526
	20	2.1582	2.7109	3.2891	4.1027	4.7616
	25	2.0922	2.6030	3.1287	3.8550	4.4327
	30	2.0492	2.5336	3.0265	3.6990	4.2276
	40	1.9968	2.4495	2.9037	3.5138	3.9860
	60	1.9457	2.3683	2.7863	3.3389	3.7599
	100	1.9057	2.3053	2.6961	3.2059	3.5895
	1000	1.8530	2.2231	2.5792	3.0355	3.3730
m	n	.10	.05	.025	.01	.005
6	1	58.2044	233.9860	937.1111	5858.9860	23437.1102
	2	9.3255	19.3295	39.3315	99.3326	199.3330
	3	5.2847	8.9406	14.7347	27.9107	44.8385
	4	4.0097	6.1631	9.1973	15.2069	21.9746
	5	3.4045	4.9503	6.9777	10.6723	14.5133
	6	3.0546	4.2839	5.8198	8.4661	11.0730
	7	2.8274	3.8660	5.1186	7.1914	9.1553
	8	2.6683	3.5806	4.6517	6.3707	7.9520
	9	2.5509	3.3738	4.3197	5.8018	7.1339
	10	2.4606	3.2172	4.0721	5.3858	6.5446
	11	2.3891	3.0946	3.8807	5.0692	6.1016
	12	2.3310	2.9961	3.7283	4.8206	5.7570
	13	2.2830	2.9153	3.6043	4.6204	5.4819

Appendix V
F-Distributions
(Continued)

m = numerator degrees of freedom
n = denominator degrees of freedom

		AREA IN RIGHT-HAND TAIL				
m	n	.10	.05	.025	.01	.005
6	14	2.2426	2.8477	3.5014	4.4558	5.2574
	15	2.2081	2.7905	3.4147	4.3183	5.0708
	16	2.1783	2.7413	3.3406	4.2016	4.9134
	17	2.1524	2.6987	3.2767	4.1015	4.7789
	18	2.1296	2.6613	3.2209	4.0146	4.6627
	19	2.1094	2.6283	3.1718	3.9386	4.5614
	20	2.0913	2.5990	3.1283	3.8714	4.4721
	25	2.0241	2.4904	2.9685	3.6272	4.1500
	30	1.9803	2.4205	2.8667	3.4735	3.9492
	40	1.9269	2.3359	2.7444	3.2910	3.7129
	60	1.8747	2.2541	2.6274	3.1187	3.4918
	100	1.8339	2.1906	2.5374	2.9877	3.3252
	1000	1.7800	2.1076	2.4208	2.8200	3.1138
m	n	.10	.05	.025	.01	.005
7	1	58.9060	236.7684	948.2169	5928.3556	23714.5649
	2	9.3491	19.3532	39.3552	99.3564	199.3568
	3	5.2662	8.8867	14.6244	27.6717	44.4341
	4	3.9790	6.0942	9.0741	14.9758	21.6217
	5	3.3679	4.8759	6.8531	10.4555	14.2004
	6	3.0145	4.2067	5.6955	8.2600	10.7859
	7	2.7849	3.7870	4.9949	6.9928	8.8854
	8	2.6241	3.5005	4.5286	6.1776	7.6941
	9	2.5053	3.2927	4.1970	5.6129	6.8849
	10	2.4140	3.1355	3.9498	5.2001	6.3025
	11	2.3416	3.0123	3.7586	4.8861	5.8648
	12	2.2828	2.9134	3.6065	4.6395	5.5245
	13	2.2341	2.8321	3.4827	4.4410	5.2529
	14	2.1931	2.7642	3.3799	4.2779	5.0313
	15	2.1582	2.7066	3.2934	4.1415	4.8473
	16	2.1280	2.6572	3.2194	4.0259	4.6920
	17	2.1017	2.6143	3.1556	3.9267	4.5594
	18	2.0785	2.5767	3.0999	3.8406	4.4448
	19	2.0580	2.5435	3.0509	3.7653	4.3448
	20	2.0397	2.5140	3.0074	3.6987	4.2569
	25	1.9714	2.4047	2.8478	3.4568	3.9394
	30	1.9269	2.3343	2.7460	3.3045	3.7416
	40	1.8725	2.2490	2.6238	3.1238	3.5088
	60	1.8194	2.1665	2.5068	2.9530	3.2911
	100	1.7778	2.1025	2.4168	2.8233	3.1271
	1000	1.7228	2.0187	2.3002	2.6572	2.9190

Appendix V
F-Distributions
(Continued)

m = numerator degrees of freedom
n = denominator degrees of freedom

m	n	\.10	.05	.025	.01	.005
			AREA IN RIGHT-HAND TAIL			
8	1	59.4390	238.8827	956.6562	5981.0702	23925.4053
	2	9.3668	19.3710	39.3730	99.3742	199.3746
	3	5.2517	8.8452	14.5399	27.4892	44.1256
	4	3.9549	6.0410	8.9796	14.7989	21.3520
	5	3.3393	4.8183	6.7572	10.2893	13.9610
	6	2.9830	4.1468	5.5996	8.1017	10.5658
	7	2.7516	3.7257	4.8993	6.8400	8.6781
	8	2.5893	3.4381	4.4333	6.0289	7.4959
	9	2.4694	3.2296	4.1020	5.4671	6.6933
	10	2.3772	3.0717	3.8549	5.0567	6.1159
	11	2.3040	2.9480	3.6638	4.7445	5.6821
	12	2.2446	2.8486	3.5118	4.4994	5.3451
	13	2.1953	2.7669	3.3880	4.3021	5.0761
	14	2.1539	2.6987	3.2853	4.1399	4.8566
	15	2.1185	2.6408	3.1987	4.0045	4.6744
	16	2.0880	2.5911	3.1248	3.8896	4.5207
	17	2.0613	2.5480	3.0610	3.7910	4.3894
	18	2.0379	2.5102	3.0053	3.7054	4.2759
	19	2.0171	2.4768	2.9563	3.6305	4.1770
	20	1.9985	2.4471	2.9128	3.5644	4.0900
	25	1.9292	2.3371	2.7531	3.3239	3.7758
	30	1.8841	2.2662	2.6513	3.1726	3.5801
	40	1.8289	2.1802	2.5289	2.9930	3.3498
	60	1.7748	2.0970	2.4117	2.8233	3.1344
	100	1.7324	2.0323	2.3215	2.6943	2.9722
	1000	1.6764	1.9476	2.2045	2.5290	2.7663

m	n	\.10	.05	.025	.01	.005
9	1	59.8576	240.5433	963.2846	6022.4731	24091.0032
	2	9.3805	19.3848	39.3869	99.3881	199.3885
	3	5.2400	8.8123	14.4731	27.3452	43.8824
	4	3.9357	5.9988	8.9047	14.6591	21.1391
	5	3.3163	4.7725	6.6811	10.1578	13.7716
	6	2.9577	4.0990	5.5234	7.9761	10.3915
	7	2.7247	3.6767	4.8232	6.7188	8.5138
	8	2.5612	3.3881	4.3572	5.9106	7.3386
	9	2.4403	3.1789	4.0260	5.3511	6.5411
	10	2.3473	3.0204	3.7790	4.9424	5.9676
	11	2.2735	2.8962	3.5879	4.6315	5.5368
	12	2.2135	2.7964	3.4358	4.3875	5.2021
	13	2.1638	2.7144	3.3120	4.1911	4.9351

Appendix V
F-Distributions
(Continued)

m = numerator degrees of freedom
n = denominator degrees of freedom

		AREA IN RIGHT-HAND TAIL				
m	n	.10	.05	.025	.01	.005
9	14	2.1220	2.6458	3.2093	4.0297	4.7173
	15	2.0862	2.5876	3.1227	3.8948	4.5364
	16	2.0553	2.5377	3.0488	3.7804	4.3838
	17	2.0284	2.4943	2.9849	3.6822	4.2535
	18	2.0047	2.4563	2.9291	3.5971	4.1410
	19	1.9836	2.4227	2.8801	3.5225	4.0428
	20	1.9649	2.3928	2.8365	3.4567	3.9564
	25	1.8947	2.2821	2.6766	3.2172	3.6447
	30	1.8490	2.2107	2.5746	3.0665	3.4505
	40	1.7929	2.1240	2.4519	2.8876	3.2220
	60	1.7380	2.0401	2.3344	2.7185	3.0083
	100	1.6949	1.9748	2.2439	2.5898	2.8472
	1000	1.6378	1.8892	2.1264	2.4250	2.6429
m	n	.10	.05	.025	.01	.005
10	1	60.1950	241.8817	968.6274	6055.8466	24224.4859
	2	9.3916	19.3959	39.3980	99.3992	199.3996
	3	5.2304	8.7855	14.4189	27.2287	43.6858
	4	3.9199	5.9644	8.8439	14.5459	20.9667
	5	3.2974	4.7351	6.6192	10.0510	13.6182
	6	2.9369	4.0600	5.4613	7.8741	10.2500
	7	2.7025	3.6365	4.7611	6.6201	8.3803
	8	2.5380	3.3472	4.2951	5.8143	7.2106
	9	2.4163	3.1373	3.9639	5.2565	6.4172
	10	2.3226	2.9782	3.7168	4.8491	5.8467
	11	2.2482	2.8536	3.5257	4.5393	5.4183
	12	2.1878	2.7534	3.3736	4.2961	5.0855
	13	2.1376	2.6710	3.2497	4.1003	4.8199
	14	2.0954	2.6022	3.1469	3.9394	4.6034
	15	2.0593	2.5437	3.0602	3.8049	4.4235
	16	2.0281	2.4935	2.9862	3.6909	4.2719
	17	2.0009	2.4499	2.9222	3.5931	4.1424
	18	1.9770	2.4117	2.8664	3.5082	4.0305
	19	1.9557	2.3779	2.8172	3.4338	3.9329
	20	1.9367	2.3479	2.7737	3.3682	3.8470
	25	1.8658	2.2365	2.6135	3.1294	3.5370
	30	1.8195	2.1646	2.5112	2.9791	3.3440
	40	1.7627	2.0772	2.3882	2.8005	3.1167
	60	1.7070	1.9926	2.2702	2.6318	2.9042
	100	1.6632	1.9267	2.1793	2.5033	2.7440
	1000	1.6051	1.8402	2.0611	2.3386	2.5405

Appendix V
F-Distributions
(Continued)

m = numerator degrees of freedom
n = denominator degrees of freedom

				AREA IN RIGHT-HAND TAIL		
m	n	.10	.05	.025	.01	.005
12	1	60.7052	243.9060	976.7079	6106.3206	24426.3652
	2	9.4081	19.4125	39.4146	99.4159	199.4163
	3	5.2156	8.7446	14.3366	27.0518	43.3874
	4	3.8955	5.9117	8.7512	14.3736	20.7047
	5	3.2682	4.6777	6.5245	9.8883	13.3845
	6	2.9047	3.9999	5.3662	7.7183	10.0343
	7	2.6681	3.5747	4.6658	6.4691	8.1764
	8	2.5020	3.2839	4.1997	5.6667	7.0149
	9	2.3789	3.0729	3.8682	5.1114	6.2274
	10	2.2841	2.9130	3.6209	4.7059	5.6613
	11	2.2087	2.7876	3.4296	4.3974	5.2363
	12	2.1474	2.6866	3.2773	4.1553	4.9062
	13	2.0966	2.6037	3.1532	3.9603	4.6429
	14	2.0537	2.5342	3.0502	3.8001	4.4281
	15	2.0171	2.4753	2.9633	3.6662	4.2497
	16	1.9854	2.4247	2.8890	3.5527	4.0994
	17	1.9577	2.3807	2.8249	3.4552	3.9709
	18	1.9333	2.3421	2.7689	3.3706	3.8599
	19	1.9117	2.3080	2.7196	3.2965	3.7631
	20	1.8924	2.2776	2.6758	3.2311	3.6779
	25	1.8200	2.1649	2.5149	2.9931	3.3704
	30	1.7727	2.0921	2.4120	2.8431	3.1787
	40	1.7146	2.0035	2.2882	2.6648	2.9531
	60	1.6574	1.9174	2.1692	2.4961	2.7419
	100	1.6124	1.8503	2.0773	2.3676	2.5825
	1000	1.5524	1.7618	1.9577	2.2025	2.3800

m	n	.10	.05	.025	.01	.005
15	1	61.2203	245.9499	984.8668	6157.2845	24630.2042
	2	9.4247	19.4291	39.4313	99.4325	199.4329
	3	5.2003	8.7029	14.2527	26.8722	43.0847
	4	3.8704	5.8578	8.6565	14.1982	20.4383
	5	3.2380	4.6188	6.4277	9.7222	13.1463
	6	2.8712	3.9381	5.2687	7.5590	9.8140
	7	2.6322	3.5107	4.5678	6.3143	7.9678
	8	2.4642	3.2184	4.1012	5.5151	6.8143
	9	2.3396	3.0061	3.7694	4.9621	6.0325
	10	2.2435	2.8450	3.5217	4.5581	5.4707
	11	2.1671	2.7186	3.3299	4.2509	5.0489
	12	2.1049	2.6169	3.1772	4.0096	4.7213
	13	2.0532	2.5331	3.0527	3.8154	4.4600

Appendix V
F-Distributions
(Continued)

m = numerator degrees of freedom
n = denominator degrees of freedom

		AREA IN RIGHT-HAND TAIL				
m	n	.10	.05	.025	.01	.005
15	14	2.0095	2.4630	2.9493	3.6557	4.2468
	15	1.9722	2.4034	2.8621	3.5222	4.0698
	16	1.9399	2.3522	2.7875	3.4089	3.9205
	17	1.9117	2.3077	2.7230	3.3117	3.7929
	18	1.8868	2.2686	2.6667	3.2273	3.6827
	19	1.8647	2.2341	2.6171	3.1533	3.5866
	20	1.8449	2.2033	2.5731	3.0880	3.5020
	25	1.7708	2.0889	2.4110	2.8502	3.1963
	30	1.7223	2.0148	2.3072	2.7002	3.0057
	40	1.6624	1.9245	2.1819	2.5216	2.7811
	60	1.6034	1.8364	2.0613	2.3523	2.5705
	100	1.5566	1.7675	1.9679	2.2230	2.4113
	1000	1.4941	1.6764	1.8459	2.0565	2.2085
m	n	.10	.05	.025	.01	.005
20	1	61.7403	248.0131	993.1028	6208.7301	24835.9699
	2	9.4413	19.4458	39.4479	99.4492	199.4496
	3	5.1845	8.6602	14.1674	26.6898	42.7775
	4	3.8443	5.8025	8.5599	14.0196	20.1673
	5	3.2067	4.5581	6.3286	9.5526	12.9035
	6	2.8363	3.8742	5.1684	7.3958	9.5888
	7	2.5947	3.4445	4.4667	6.1554	7.7540
	8	2.4246	3.1503	3.9995	5.3591	6.6082
	9	2.2983	2.9365	3.6669	4.8080	5.8318
	10	2.2007	2.7740	3.4185	4.4054	5.2740
	11	2.1230	2.6464	3.2261	4.0990	4.8552
	12	2.0597	2.5436	3.0728	3.8584	4.5299
	13	2.0070	2.4589	2.9477	3.6646	4.2703
	14	1.9625	2.3879	2.8437	3.5052	4.0585
	15	1.9243	2.3275	2.7559	3.3719	3.8826
	16	1.8913	2.2756	2.6808	3.2587	3.7342
	17	1.8624	2.2304	2.6158	3.1615	3.6073
	18	1.8368	2.1906	2.5590	3.0771	3.4977
	19	1.8142	2.1555	2.5089	3.0031	3.4020
	20	1.7938	2.1242	2.4645	2.9377	3.3178
	25	1.7175	2.0075	2.3005	2.6993	3.0133
	30	1.6673	1.9317	2.1952	2.5487	2.8230
	40	1.6052	1.8389	2.0677	2.3689	2.5984
	60	1.5435	1.7480	1.9445	2.1978	2.3872
	100	1.4943	1.6764	1.8486	2.0666	2.2270
	1000	1.4280	1.5811	1.7223	1.8967	2.0219

Appendix V
F-Distributions
(Continued)

m = numerator degrees of freedom
n = denominator degrees of freedom

		AREA IN RIGHT-HAND TAIL				
m	n	.10	.05	.025	.01	.005
25	1	62.0545	249.2601	998.0808	6239.8250	24960.3395
	2	9.4513	19.4558	39.4579	99.4592	199.4596
	3	5.1747	8.6341	14.1155	26.5790	42.5910
	4	3.8283	5.7687	8.5010	13.9109	20.0024
	5	3.1873	4.5209	6.2679	9.4491	12.7554
	6	2.8147	3.8348	5.1069	7.2960	9.4511
	7	2.5714	3.4036	4.4045	6.0580	7.6230
	8	2.3999	3.1081	3.9367	5.2631	6.4817
	9	2.2725	2.8932	3.6035	4.7130	5.7084
	10	2.1739	2.7298	3.3546	4.3111	5.1528
	11	2.0953	2.6014	3.1616	4.0051	4.7356
	12	2.0312	2.4977	3.0077	3.7647	4.4115
	13	1.9778	2.4123	2.8821	3.5710	4.1528
	14	1.9326	2.3407	2.7777	3.4116	3.9417
	15	1.8939	2.2797	2.6894	3.2782	3.7662
	16	1.8603	2.2272	2.6138	3.1650	3.6182
	17	1.8309	2.1815	2.5484	3.0676	3.4916
	18	1.8049	2.1413	2.4912	2.9831	3.3822
	19	1.7818	2.1057	2.4408	2.9089	3.2867
	20	1.7611	2.0739	2.3959	2.8434	3.2025
	25	1.6831	1.9554	2.2303	2.6041	2.8981
	30	1.6316	1.8782	2.1237	2.4526	2.7076
	40	1.5677	1.7835	1.9943	2.2714	2.4823
	60	1.5039	1.6902	1.8687	2.0984	2.2697
	100	1.4528	1.6163	1.7705	1.9652	2.1080
	1000	1.3831	1.5171	1.6402	1.7915	1.8996
m	n	.10	.05	.025	.01	.005
30	1	62.2650	250.0951	1001.4144	6260.6485	25043.6267
	2	9.4579	19.4624	39.4646	99.4658	199.4662
	3	5.1681	8.6166	14.0805	26.5045	42.4658
	4	3.8174	5.7459	8.4613	13.8377	19.8915
	5	3.1741	4.4957	6.2269	9.3793	12.6556
	6	2.8000	3.8082	5.0652	7.2285	9.3582
	7	2.5555	3.3758	4.3624	5.9920	7.5345
	8	2.3830	3.0794	3.8940	5.1981	6.3961
	9	2.2547	2.8637	3.5604	4.6486	5.6248
	10	2.1554	2.6996	3.3110	4.2469	5.0706
	11	2.0762	2.5705	3.1176	3.9411	4.6543
	12	2.0115	2.4663	2.9633	3.7008	4.3309
	13	1.9576	2.3803	2.8372	3.5070	4.0727

Appendix V
F-Distributions
(Continued)

m = numerator degrees of freedom
n = denominator degrees of freedom

		AREA IN RIGHT-HAND TAIL				
m	n	.10	.05	.025	.01	.005
30	14	1.9119	2.3082	2.7324	3.3476	3.8619
	15	1.8728	2.2468	2.6437	3.2141	3.6867
	16	1.8388	2.1938	2.5678	3.1007	3.5389
	17	1.8090	2.1477	2.5020	3.0032	3.4124
	18	1.7827	2.1071	2.4445	2.9185	3.3030
	19	1.7592	2.0712	2.3937	2.8442	3.2075
	20	1.7382	2.0391	2.3486	2.7785	3.1234
	25	1.6589	1.9192	2.1816	2.5383	2.8187
	30	1.6065	1.8409	2.0739	2.3860	2.6278
	40	1.5411	1.7444	1.9429	2.2034	2.4015
	60	1.4755	1.6491	1.8152	2.0285	2.1874
	100	1.4227	1.5733	1.7148	1.8933	2.0239
	1000	1.3501	1.4706	1.5808	1.7158	1.8121
m	n	.10	.05	.025	.01	.005
40	1	62.5291	251.1432	1005.5981	6286.7819	25148.1523
	2	9.4662	19.4707	39.4729	99.4742	199.4746
	3	5.1597	8.5944	14.0365	26.4108	42.3082
	4	3.8036	5.7170	8.4111	13.7454	19.7518
	5	3.1573	4.4638	6.1750	9.2912	12.5297
	6	2.7812	3.7743	5.0125	7.1432	9.2408
	7	2.5351	3.3404	4.3089	5.9084	7.4224
	8	2.3614	3.0428	3.8398	5.1156	6.2875
	9	2.2320	2.8259	3.5055	4.5666	5.5186
	10	2.1317	2.6609	3.2554	4.1653	4.9659
	11	2.0516	2.5309	3.0613	3.8596	4.5508
	12	1.9861	2.4259	2.9063	3.6192	4.2282
	13	1.9315	2.3392	2.7797	3.4253	3.9704
	14	1.8852	2.2664	2.6742	3.2656	3.7600
	15	1.8454	2.2043	2.5850	3.1319	3.5850
	16	1.8108	2.1507	2.5085	3.0182	3.4372
	17	1.7805	2.1040	2.4422	2.9205	3.3108
	18	1.7537	2.0629	2.3842	2.8354	3.2014
	19	1.7298	2.0264	2.3329	2.7608	3.1058
	20	1.7083	1.9938	2.2873	2.6947	3.0215
	25	1.6272	1.8718	2.1183	2.4530	2.7160
	30	1.5732	1.7918	2.0089	2.2992	2.5241
	40	1.5056	1.6928	1.8752	2.1142	2.2958
	60	1.4373	1.5943	1.7440	1.9360	2.0789
	100	1.3817	1.5151	1.6401	1.7972	1.9119
	1000	1.3040	1.4063	1.4993	1.6127	1.6932

Appendix V
F-Distributions
(Continued)

m = numerator degrees of freedom
n = denominator degrees of freedom

		AREA IN RIGHT-HAND TAIL				
m	n	.10	.05	.025	.01	.005
60	1	62.7943	252.1957	1009.8001	6313.0299	25253.1359
	2	9.4746	19.4791	39.4812	99.4825	199.4829
	3	5.1512	8.5720	13.9921	26.3164	42.1494
	4	3.7896	5.6877	8.3604	13.6522	19.6107
	5	3.1402	4.4314	6.1225	9.2020	12.4024
	6	2.7620	3.7398	4.9589	7.0567	9.1219
	7	2.5142	3.3043	4.2544	5.8236	7.3088
	8	2.3391	3.0053	3.7844	5.0316	6.1772
	9	2.2085	2.7872	3.4493	4.4831	5.4104
	10	2.1072	2.6211	3.1984	4.0819	4.8592
	11	2.0261	2.4901	3.0035	3.7761	4.4450
	12	1.9597	2.3842	2.8478	3.5355	4.1229
	13	1.9043	2.2966	2.7204	3.3413	3.8655
	14	1.8572	2.2229	2.6142	3.1813	3.6552
	15	1.8168	2.1601	2.5242	3.0471	3.4803
	16	1.7816	2.1058	2.4471	2.9330	3.3324
	17	1.7506	2.0584	2.3801	2.8348	3.2058
	18	1.7232	2.0166	2.3214	2.7493	3.0962
	19	1.6988	1.9795	2.2696	2.6742	3.0004
	20	1.6768	1.9464	2.2234	2.6077	2.9159
	25	1.5934	1.8217	2.0516	2.3637	2.6088
	30	1.5376	1.7396	1.9400	2.2079	2.4151
	40	1.4672	1.6373	1.8028	2.0194	2.1838
	60	1.3952	1.5343	1.6668	1.8363	1.9622
	100	1.3356	1.4504	1.5575	1.6918	1.7896
	1000	1.2500	1.3318	1.4058	1.4953	1.5585
m	n	.10	.05	.025	.01	.005
100	1	63.0073	253.0411	1013.1748	6334.1099	25337.4492
	2	9.4812	19.4857	39.4879	99.4892	199.4896
	3	5.1443	8.5539	13.9563	26.2402	42.0216
	4	3.7782	5.6641	8.3195	13.5770	19.4970
	5	3.1263	4.4051	6.0800	9.1299	12.2996
	6	2.7463	3.7117	4.9154	6.9867	9.0257
	7	2.4971	3.2749	4.2101	5.7547	7.2165
	8	2.3208	2.9747	3.7393	4.9633	6.0875
	9	2.1892	2.7556	3.4034	4.4150	5.3223
	10	2.0869	2.5884	3.1517	4.0137	4.7721
	11	2.0050	2.4566	2.9561	3.7077	4.3585
	12	1.9379	2.3498	2.7996	3.4668	4.0368
	13	1.8817	2.2614	2.6715	3.2723	3.7795

Appendix V
F-Distributions
(Continued)

m = numerator degrees of freedom
n = denominator degrees of freedom

		AREA IN RIGHT-HAND TAIL				
m	n	.10	.05	.025	.01	.005
100	14	1.8340	2.1870	2.5646	3.1118	3.5692
	15	1.7929	2.1234	2.4739	2.9772	3.3941
	16	1.7570	2.0685	2.3961	2.8627	3.2460
	17	1.7255	2.0204	2.3285	2.7639	3.1192
	18	1.6976	1.9780	2.2692	2.6779	3.0093
	19	1.6726	1.9403	2.2167	2.6023	2.9131
	20	1.6501	1.9066	2.1699	2.5353	2.8282
	25	1.5645	1.7794	1.9955	2.2888	2.5191
	30	1.5069	1.6950	1.8816	2.1307	2.3234
	40	1.4336	1.5892	1.7405	1.9383	2.0884
	60	1.3576	1.4814	1.5990	1.7493	1.8609
	100	1.2934	1.3917	1.4833	1.5977	1.6809
	1000	1.1969	1.2596	1.3158	1.3835	1.4310

m	n	.10	.05	.025	.01	.005
1000	1	63.2960	254.1868	1017.7488	6362.6816	25451.7270
	2	9.4902	19.4947	39.4969	99.4982	199.4986
	3	5.1348	8.5292	13.9075	26.1367	41.8477
	4	3.7625	5.6317	8.2636	13.4745	19.3420
	5	3.1072	4.3690	6.0218	9.0314	12.1592
	6	2.7246	3.6732	4.8558	6.8908	8.8941
	7	2.4735	3.2343	4.1492	5.6601	7.0902
	8	2.2954	2.9324	3.6772	4.8694	5.9644
	9	2.1623	2.7116	3.3400	4.3211	5.2012
	10	2.0586	2.5430	3.0871	3.9196	4.6521
	11	1.9755	2.4098	2.8902	3.6131	4.2390
	12	1.9071	2.3017	2.7325	3.3716	3.9174
	13	1.8498	2.2121	2.6032	3.1763	3.6601
	14	1.8011	2.1365	2.4951	3.0150	3.4494
	15	1.7589	2.0718	2.4034	2.8795	3.2739
	16	1.7222	2.0157	2.3245	2.7641	3.1253
	17	1.6898	1.9666	2.2558	2.6644	2.9978
	18	1.6609	1.9232	2.1954	2.5775	2.8871
	19	1.6351	1.8845	2.1419	2.5009	2.7902
	20	1.6118	1.8497	2.0941	2.4329	2.7046
	25	1.5225	1.7181	1.9149	2.1818	2.3913
	30	1.4617	1.6299	1.7967	2.0192	2.1914
	40	1.3830	1.5175	1.6481	1.8189	1.9483
	60	1.2988	1.3994	1.4950	1.6169	1.7073
	100	1.2235	1.2958	1.3630	1.4468	1.5076
	1000	1.0845	1.1097	1.1320	1.1586	1.1771

Appendix VI   Table of Random Digits

50960	40163	81961	00843	24550	36522	39452	79917	98586	17521
72630	45181	52058	76764	18220	83033	57036	86600	71728	72333
60589	62357	03386	99275	88123	26521	20050	02695	26609	34169
54589	97714	44600	45043	52743	03018	37700	82690	63322	01915
35443	54474	41248	38339	20225	64986	17386	92055	13455	02781
02760	79500	83272	98876	31455	82773	86285	08893	14852	23231
34909	34768	90680	30766	62491	31826	20327	81683	21028	17206
40078	15933	04669	70936	21735	38584	26359	38503	95990	23622
90600	49418	84750	57037	33133	91190	41517	42265	94794	85845
45868	96616	95195	86485	23985	45892	47622	11304	75337	54682
06969	56737	61956	43382	94856	49786	75302	71354	47964	50766
89343	79183	89631	51299	69278	35013	14400	76360	33694	14115
42214	46401	85574	35076	00934	70478	84657	62188	42880	47766
85200	83665	55576	37988	84408	53648	06898	04419	32694	88302
21909	68892	32437	55567	87705	22638	87875	24268	88587	25071
66555	36076	03095	26490	53387	66699	56518	11847	43201	80755
45764	52347	43037	59347	11985	42296	15462	24165	34020	36122
78490	31181	43566	56600	74896	61496	76828	09063	50937	13891
37710	73871	97251	11926	68385	78055	17524	84727	79283	33652
26690	80042	97961	66031	10880	01264	55734	40502	62151	66529
72845	34959	14587	56200	31529	84536	90864	78110	60105	43172
61946	33833	19507	06006	84126	38045	92093	49265	68888	43544
03578	32398	90583	40531	33545	74017	51434	22521	78657	16986
25833	81557	40853	85116	97095	55739	02178	91182	55061	40148
33262	49673	38264	52881	20295	38399	26633	38876	21231	17876
87840	72801	26879	27046	12486	14732	16496	46441	53447	26679
40517	16564	44137	76289	35586	05925	69187	54135	09813	19009
77240	23861	44528	03035	29244	16032	05958	75696	02818	70855
71823	83501	11356	57658	08014	31502	44422	79503	53656	74133
27271	69648	48332	97709	60814	15250	56055	09976	21694	34299
07183	68954	42201	22979	09485	59191	00142	71063	85767	65472
88217	39439	92319	58339	31969	84814	90869	34498	38823	44718
94483	32185	34424	53284	43915	49574	07406	55179	60883	05521
88773	25562	96265	70398	98538	26428	84011	85880	66962	37190
92175	22746	58292	37741	92953	32708	68923	33573	11934	27532

source: The RAND Corporation (1955), p. 186.

Index